Organometallic Ion Chemistry

Understanding Chemical Reactivity

Volume 15

The titles published in this series are listed at the end of this volume.

Organometallic Ion Chemistry

Edited by

BEN S. FREISER

H.C. Brown Laboratory of Chemistry,
Purdue University,
West Lafayette,
Indiana,
USA

KLUWER ACADEMIC PUBLISHERS
DORDRECHT / BOSTON / LONDON

Library of Congress Cataloging-in-Publication Data

Organometallic ion chemistry / edited by Ben S. Freiser.
p. cm. -- (Understanding chemical reactivity ; v. 15)
Includes index.
ISBN 0-7923-3478-7 (hardbound : acid-free)
1. Organometallic chemistry. 2. Metal ions. I. Freiser, Ben S.
II. Series.
QD411.O734 1995
547'.0504572--dc20 95-12379

ISBN 0-7923-3478-7

Published by Kluwer Academic Publishers,
P.O. Box 17, 3300 AA Dordrecht, The Netherlands.

Kluwer Academic Publishers incorporates
the publishing programmes of
D. Reidel, Martinus Nijhoff, Dr W. Junk and MTP Press.

Sold and distributed in the U.S.A. and Canada
by Kluwer Academic Publishers,
101 Philip Drive, Norwell, MA 02061, U.S.A.

In all other countries, sold and distributed
by Kluwer Academic Publishers Group,
P.O. Box 322, 3300 AH Dordrecht, The Netherlands.

Printed on acid-free paper

Printed in the Netherlands

To my family, for their love and support

Table of Contents

Preface

The tremendous growth of organometallic chemistry over the past two decades has been paralleled by a similar rapid growth in the area of gas-phase transition metal ion chemistry. Catalyzing this growth has been the spectacular advances in the types of mass spectrometry instrumentation and methodology which are being brought to bear on this problem.

The study of transition-metal containing ions in the gas phase offers the opportunity to probe the intrinsic chemical and physical properties of these species in the absence of complicating factors such as solvation and ion-pairing effects. The chemistry of these highly electronically and coordinatively unsaturated species is not only inherently interesting, but can provide important clues as to mechanisms occurring on surfaces and in condensed phases by yielding a better understanding of key steps and potential intermediates. Furthermore, obtaining quantitative data on metal ion-ligand bond energies and studying the periodic properties of metal ions as a function of their ground and electronic state structures are important in rendering the outcome of an organometallic reaction predictable.

In *Organometallic Ion Chemistry*, several of these themes are struck by the authors who highlight topics of current interest in the field. Of note is the particular synergism between experiment and theory which has greatly accelerated our understanding in this area. Also included at the end of the book is the most comprehensive table of metal ion-ligand bond energies to date, many of which were determined by the authors. By gathering all of these values in the literature into one table, it is hoped that this will be invaluable resource to a broad range of scientists.

It should also be noted that this book by necessity does not cover all of the exciting work in this area. Not covered, for example, are cluster ion chemistry, metal ions with biologically interesting molecules, and analytical applications of metal ions as selective chemical ionization reagents. These topics will undoubtedly be the subjects of future books.

I would like to thank all of the authors for their outstanding contributions and for their patience. In addition I would like to thank my graduate students (particularly Ken Auberry) and my secretary Knox Clinton for their help in generating the bond energy table.

Ben S. Freiser
H. C. Brown Laboratory of Chemistry
Purdue University
West Lafayette, IN 47907,
U.S.A.

1. Gas-phase thermochemistry of transition metal ligand systems: reassessment of values and periodic trends

P. B. ARMENTROUT and BERNICE L. KICKEL
Department of Chemistry, University of Utah, Salt Lake City, Utah 84112, U.S.A.

1. Introduction

One of the key features of applying ion beam techniques to organometallic systems, recognized in its first application [1], has been the ability to determine quantitative thermodynamic information for coordinatively unsaturated organometallic molecules. Gas-phase methods are particularly useful for studies of such species, which are highly reactive because they are not stable 18–e^- complexes. This feature is used to advantage in examining the periodic trends in transition metal thermochemistry where the identity of the metal is systematically varied while maintaining the same ligands. Alternatively, trends where the type of ligand is varied for a particular metal allow the various contributions to the bonding to be more completely assessed. Finally, recent studies have examined how the bond energies change as the number of ligands on the metal is systematically increased.

A key feature of this particular review is a comprehensive reevaluation of all previous organometallic thermochemistry derived from studies in our laboratory. This reevaluation is necessitated by recent experiments that demonstrate that fundamental assumptions made in many of our early studies are incorrect. Although the changes in the thermochemistry are generally within the cited experimental errors, we believe that the numbers cited here are the best available from our laboratory. [A note on units: we have chosen to report our bond energies in kJ/mol, in keeping with recommended IUPAC nomenclature for thermodynamic information. Also, because our original data is obtained in units of electron volts in all cases, the use of kJ/mol avoids rounding errors that accompany conversion to kcal/mol, a unit often favored by those who still use miles instead of km.]

Ben S. Freiser (ed.), Organometallic Ion Chemistry, 1–45.

2. Experimental section

2.1. Instrumentation

The guided ion beam tandem mass spectrometer and the experimental methods used to measure gas-phase bond dissociation energies have been discussed in detail before [2–7]. In an ion beam experiment, reactant ions are created in the source region, mass selected by a mass spectrometer (a magnetic sector in our apparatus), and accelerated to a desired kinetic energy. Reactions then take place with a neutral reagent in a collision cell. In order to measure absolute reaction probabilities, this cell should have a well-defined length and the pressure of the neutral should be low enough that multiple collisions between the ion and neutral reagents are unlikely. In some reaction systems, the effects of secondary collisions cannot be avoided but can be easily removed by performing studies at several different neutral pressures and extrapolating to zero neutral pressure, rigorously single collision conditions [2, 8, 9]. This pressure extrapolation has been found to be particularly important in the analysis of collision-induced dissociation reactions.

In our apparatus, the interaction region is surrounded by an rf octopole ion beam "guide" [10], which ensures efficient collection of all ions. Further, the use of an octopole provides a more homogeneous potential than a quadrupole guide and thus provides much better control and less perturbation of the kinetic energy distribution of the reactant ions. After the reaction takes place, reactant and product ions drift to the end of the octopole, are mass separated in a second mass spectrometer (a quadrupole mass filter in our apparatus), and their absolute intensities detected. In our apparatus, ion intensities are measured by using a Daly detector [11], a high voltage (28 kV) first dynode, secondary electron scintillation ion counter, that provides sufficiently high sensitivity that all ions are detected with near 100% efficiency.

Reduction of the raw data to apparatus-independent results requires conversion of the ion intensities for the reactant and all products as a function of the ion kinetic energy in the laboratory frame to absolute reaction cross sections, $\sigma(E)$, as a function of the relative kinetic energy or energy in the center-of-mass (CM) frame. The cross section represents the probability of reaction and is directly related to a rate constant by $k = \sigma v$, where v is the relative velocity of the reactants. The relative energy is used because it does not include the kinetic energy of the reaction system moving through the laboratory and therefore unavailable for driving chemical reactions. Ion intensities are converted to cross sections by using a Beer's law type of formula [2]. Laboratory energies are converted to relative energies by the formula, $E(\text{CM}) = [E(\text{lab}) - E_z(\text{lab})]\ m/(m + M)$, where E_z is the absolute zero of energy and m and M are the masses of the neutral and ionic reactants, respectively. This stationary target approximation is correct at all but the lowest energies, where truncation of the ion beam must be explicitly accounted for [2]. In our laboratories, E_z is measured by a retarding potential analysis that has been verified by time-of-flight measurements [2] and comparisons with theoretical cross sections [12].

2.2. Ion sources

A critical aspect of acquiring accurate thermochemical data from any experiment is the characterization of the energies of the reactants. As described above, the kinetic energies of the ionic reactants are measured directly and those of the neutral reagents are described by a Maxwell-Boltzmann distribution at the temperature of the collision cell (~305 K in our apparatus). Likewise, the internal energy of the neutral reagents is accurately described by this same temperature. The internal energies of the ions are the most difficult quantities to determine. In our work, we have used a number of sources to overcome this difficulty.

2.2.1. *Atomic metal ions*

For atomic metal ions, it is important to realize that most low-lying electronic states are metastable. We first pointed this out in 1985 [13] by noting that all transitions between states having electrons only in s and d orbitals (which includes all ground and low-lying excited states of the transition metals) are parity forbidden, and thus the radiative lifetimes of the excited states should be on the order of seconds long [14]. Thus, very few excited ions are expected to radiatively relax before reaction. The metastability of these states has now been verified directly [15–17].

In much of the work reviewed here, atomic metal ions were produced by surface ionization (SI). In the SI source, the metals are introduced to the gas phase as a volatile organometallic vapor or by vaporizing a metal salt in a resistively heated oven. The metal containing vapor is directed toward a resistively heated rhenium filament where it decomposes and the resulting metal atoms are ionized. For many years, we assumed that ions produced by SI equilibrate at the temperature of the filament, typically 2000 K, and that the state populations are governed by a Maxwell-Boltzmann distribution. There are good reasons to believe this assumption, as discussed elsewhere [18], and recent experiments by Van Koppen *et al.* [19] have confirmed it for Co^+. Also, state-specific experiments by Weisshaar and coworkers involving Fe^+ are consistent with this assumption [20, 21]. Thus, the SI source generates a well-defined distribution of low-lying electronic states.

Atomic metal ions with colder electronic state distributions have been generated in high pressure sources. Extensive work with Fe^+ was performed with a drift cell [22], but more recent studies have utilized a more versatile flow tube source [7]. In the former case, ions are generated by electron impact and focused into a drift cell filled with Ar at pressures up to 0.25 Torr. In the latter source, a dc-discharge in Ar is used to sputter atomic metal ions from a cathode made of the desired metal. The resulting ions are swept down a meter long flow tube in a flow of helium and argon at a total pressure of ~0.5 Torr. In both sources, the metal ions encounter about 10^5 collisions with the bath gas. For many ions, these collisions are sufficient to quench many excited state ions to the ground state. The flow tube allows additional gases that are more efficient quenchers to be introduced as well.

2.2.2. *Metal-ligand complexes*

Most of our early studies involved atomic metal ions as the ionic reactant. This was because the only internal degrees of freedom of such species are electronic. In order to study more complex polyatomic species, the rotational and vibrational energies of the ions must also be characterized and controlled. The flow tube source was developed in order to permit such studies. Metal-ligand complexes can be generated either by starting with a stable organometallic precursor and ionizing it (either with electron impact or charge transfer from He^+) or by condensation of the ligand on the bare metal ion, generated by the dc-discharge source mentioned above. The ~10^5 collisions with the bath gas are sufficient to allow three-body stabilization of the metal ligand complexes and to thermalize the ions both rotationally and vibrationally. We assume ions produced in the flow tube source have internal energies that are well described by a Maxwell-Boltzmann distribution of states corresponding to 298 K. Although the internal temperature of the ions is difficult to measure precisely, previous work from this laboratory, including studies of N_4^+ [23], $Fe(CO)_x^+$ ($x = 1 - 5$) [9], $Cr(CO)_x^+$ ($x = 1 - 6$) [24], SiF_x^+ ($x = 1 - 4$) [25], SF_x^+ ($x = 1 - 5$) [26], $H_3O^+(H_2O)_x$ ($x = 1 - 5$) [27], and $Cu^+(H_2O)_x$ ($x = 1 - 4$) [28] have shown that these assumptions are reasonably accurate.

2.3. Reactions

Ion beam experiments are able to measure thermochemistry by examining endothermic chemical reactions and determining the threshold for processes of interest. We have used three types of reactions to determine metal-ligand bond dissociation energies (BDEs), reactions (1)–(3).

$$M^+ + RL \rightarrow ML^+ + R \tag{1}$$

$$M^+ + RL \rightarrow ML + R^+ \tag{2}$$

$$ML^+ + Xe \rightarrow M^+ + L + Xe \tag{3}$$

The neutral reagent in reactions (1) and (2), such as those listed elsewhere [29], is chosen such that these processes are endothermic. The collision-induced dissociation (CID) reaction (3) is intrinsically endothermic. If there is no activation energy in excess of the reaction endothermicity, then the thresholds measured for these reactions, E_T, can be converted to BDEs of interest by the following equations.

$$D_T(M^+ - L) = D_T(R - L) - E_T(1) \tag{4}$$

$$D_T(M - L) = D_T(R - L) + IE(R) - IE(M) - E_T(2) \tag{5}$$

$$D_T(M^+ - L) = E_T(3) \tag{6}$$

The subscript T refers to the temperature of the measurement, discussed in detail below. IE is the ionization energy of the appropriate species. The assumption of no reverse activation barriers is often a reasonable one for ion-molecule reactions due to the strong long-range ion-induced dipole potential [30]. Thus, exothermic ion-molecule reactions are generally observed to proceed without an activation energy, and endothermic ion-molecule reactions generally proceed once the available energy exceeds the thermodynamic threshold [6]. We have explicitly tested this assumption in several reactions where the thermochemistry is well established [4, 5, 31–34] although the observation of the true thermodynamic threshold can require extremely good sensitivity [4]. Exceptions do occur, however, and can be due to spin or orbital angular momentum conservation restrictions [6, 35]. More recently, restrictions involving C-H bond activation steps by atomic metal ions have also been characterized more completely [36, 37].

If possible, it is desirable to measure the same BDE in more than a single reaction system in order to verify that the thresholds observed do correspond to the thermodynamic limit. This has been done in many of the systems described here, but is not possible in all cases. Verification of this thermochemistry can also be attained by comparison with values from other experiments [38–43] and ab initio theory [44–49].

2.4. Data analysis

In order to determine the thresholds for reactions (1)–(3), we model the experimental cross sections with a mathematical expression justified by theory [50, 51] and experiment [3–6, 31–34, 52–54]. In early work, we used equation (7),

$$\sigma(E) = \sigma_o (E - E_{298})^n / E^m \tag{7}$$

and more recently we have used the modified form given by equation (8).

$$\sigma(E) = \sigma_o \Sigma g_i (E + E_i - E_0)^n / E^m \tag{8}$$

Here, σ_o is a scaling factor, E is the relative kinetic energy, and n and m are adjustable parameters. The latter equation involves an explicit sum of the contributions of individual reactant states (vibrational, rotational, and/or electronic), denoted by i, with energies E_i and populations g_i. E_{298} and E_0 are the reaction thresholds E_T required in equations (4)–(6), and the differences in these quantities are described in the following section. In early work, several reasonable values for the parameter m were explicitly considered. It was found that the experimental results could generally be described by $m = 1$, a value that is predicted for translationally driven reactions [51]. Our more recent work exclusively utilizes $m = 1$ unless this form fails to accurately model the data, which is rare.

For both equations (7) and (8), the model is convoluted with the explicit distributions of the kinetic energy of the neutral and ion reactants, as described previously [2, 55, 56], before comparison with the experimental data. The σ_0, n, and E_T parameters are then optimized by using a non-linear least squares analysis to give the best reproduction of the data. Error limits for E_T are calculated from the range of threshold values obtained for different data sets with different values of n and the error in the absolute energy scale. In cases where the internal energy is appreciable and the vibrational frequencies are not well-established, the error also includes variations in the calculated internal energy distribution of equation (8). The accuracy of the thermochemistry obtained by this modeling procedure is dependent on a variety of experimental parameters that have recently undergone an extensive discussion [6, 57].

As the complexity of the systems studied increases, the possibility that an endothermic reaction of interest does not occur on the available experimental time scale also increases. In our apparatus, this time scale is established by the time it takes the ions to travel from the reaction cell to the analysis quadrupole mass filter, measured to be about 10^{-4} s. All processes occurring in less time are observed, but as the lifetime of the reaction intermediates approaches this time scale, the apparent threshold for reaction will shift to higher translational energies, a kinetic shift. The extent of this shift ultimately depends on the sensitivity of the apparatus as well as the experimental time scale available, and can be quantified by using statistical kinetic theories, such as Rice-Ramsperger-Kassel-Marcus (RRKM) theory [58]. The means by which we combine RRKM theory with equation (8) are detailed elsewhere [24, 59]. For reactions (1) and (2), the systems we have examined have been sufficiently small that there does not appear to be a need for this kinetic shift correction, but larger systems could easily require such considerations. We have observed several CID reactions (3) where the presence of kinetic shifts is obvious [60–62], and several others where kinetic shifts are likely to be present [24, 27, 28]. In general, we have found that such effects start to be important when the total number of heavy atoms in the complex or cluster exceeds five.

2.5. Temperature assumptions in the thermochemical analysis

In early studies of reactions (1) and (2), we made what we will refer to as the "298 K assumption", that the reactant neutral and the products formed at the threshold of an endothermic reaction are characterized by a temperature of 298 K in their internal degrees of freedom. The situation is shown in

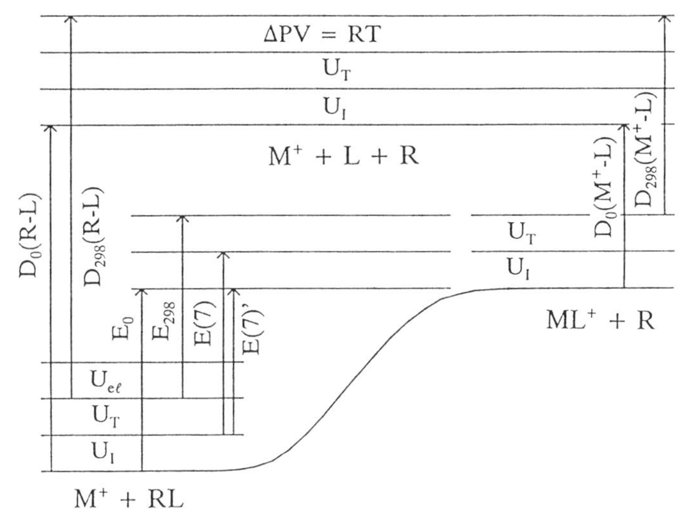

Fig. 1. Relationship between thermochemistry at 0 and 298 K for a simple exchange reaction. The internal, translational and electronic energy distributions of the various species at 298 K are indicated by U_I, U_T and $U_{e\ell}$, respectively. The four E values represent different thresholds for reaction.

Figure 1 for the example of reaction (1), where it should be remembered that the translational U_T and internal U_I energies of the reactants are really distributions. The 298 K assumption is that the threshold measured by using equation (7) is the quantity E(7) shown in Figure 1. E(7) is equivalent to E_{298} for bimolecular reactions because the translational energy, U_T, of the reactants and products is equal. Note that these thresholds are corrected for E_{el}, the energies of the electronic states of the atomic metal ion present in the beam. When combined with literature thermochemistry at 298 K, the BDEs derived by using equations (4) and (5) were reported as 298 K values.

The 298 K assumption is very practical because explicitly including the vibrational and rotational distributions of the reactants complicates the already complex data analysis procedure outlined above and in some cases the necessary molecular constants are not readily available. In addition, the 298 K assumption seems fairly reasonable. Careful consideration suggests that there are two possible rationales for this model. One, if rotational and most vibrational modes are not coupled well with the reaction coordinate, then products could be formed with energy retained in these modes at the same temperature as the reactants. Second, efficient formation of products may not begin until the products can be formed without transferring energy from these decoupled modes.

Although convenient and expedient, we now believe that the 298 K assumption is incorrect. In part this is because the two rationales noted above are not consistent with the statistical assumptions now generally recognized in modern kinetic theory. Further, experimental evidence is now firmly against this assumption. Early evidence came from several atom-diatom reactions where explicit consideration of the internal energies of the reactants was necessary in order to obtain the best agreement with precisely known literature thermochemistry [5, 33]. More compelling evidence comes from recent studies in our laboratories of several molecules having low vibrational frequencies, $Fe(CO)_x^+$ ($x = 1 - 5$) [9], $Cr(CO)_x^+$ ($x = 1 - 6$) [24], SF_x^+ ($x = 1 - 5$) [26], $H_3O^+(H_2O)_x$ ($x = 1 - 5$) [27], and $M^+(H_2O)_x$ ($x = 1 - 4$, M = Ti − Cu) [28]. For these systems, the internal energies of these species are nonnegligible even at 298 K, such that the effects of these internal energies on the thermochemistry is obvious. These results demonstrate that the internal energies of the reactant ions do couple with the reaction coordinate and that the threshold measured by the modeling procedure outlined is most consistent with formation of products at 0 K. Further, several of these studies have shown that significant systematic errors in the analysis are introduced if the threshold is simply corrected by the average internal energy, and very large errors are obtained if the internal energy is ignored.

Overall, the best procedure for modeling the data is to explicitly include the full distribution of electronic, vibrational and rotational states of both reactants in the analysis, as shown in equation (8). The use of this equation to analyze the threshold behavior of reaction cross sections makes the statistical assumption that the internal energy of the reactants is available to effect reaction. (It is tempting to worry about whether conservation of angular momentum limits the extent to which the rotational energy can couple into the reaction coordinate, but the bimolecular nature of the reactions studied allows such conservation to occur by compensating with the orbital angular momentum of the reactants' collision.) If this model truly measures the minimum energy necessary for reaction, then this threshold must correspond to the formation of products with no internal or kinetic energy. This is the value E_0, the threshold for reaction at 0 K, shown in Figure 1. Our most recent studies all utilize this form for modeling the cross sections for reactions (1)–(3).

For earlier work, this "0 K assumption" means that the thresholds measured by using equation (7) are actually the quantity $E(7)'$ in Figure 1. Comparison of this threshold with that previously assumed, $E(7)$, shows that the thermochemistry derived from the 298 K assumption differs from that derived from the 0 K assumption by U_I(products). The resultant 298 K BDEs, $D_{298}(M^+ - L)$, are too large by this amount. This error increases from a minimum value of $k_BT = 2.5$ kJ/mol for atom-diatom products to much larger values that depend on the complexity of the reaction system, e.g. 16 kJ/mol for $MCH_3^+ + t\text{-}C_4H_9$.

In this review, we reanalyze our previous results where the 298 K assumption was made. Ideally, this reanalysis would involve refitting the original data with equation (8), but this would require the analysis of tens of thousands of data sets, a task that is not practical. In most cases, however, the energy shifts involved are reasonably small (most are less than 0.1 eV) such that errors introduced by using the average internal energies rather than the explicit distribution of energies should be small. Therefore, we correct our previously reported thresholds to 0 K by adding the rotational and vibrational energy of the reactants. These E_0 values are then converted to 0 K BDEs by using equations (4)–(6). When BDEs are obtained from several reactions, the average value weighted by the uncertainties is reported [63]. In all cases, the values reported here correspond to 0 K bond energies with uncertainties that are one standard deviation of the mean.

It is also worth pointing out that assumptions regarding the temperature of thermochemistry obtained in other gas-phase experiments are often not considered explicitly or the effects are ignored. For example, thermochemistry obtained from bracketing reactions at thermal conditions should be considered as 298 K BDEs, while that obtained from photodissociation thresholds should probably be corrected for the internal energy and regarded as a 0 K threshold. In many cases, the accuracy of the methods has not been deemed to be sufficiently high that such considerations were worth dealing with. As the accuracy and precision of the various methods increases, this is no longer an excuse. In order to facilitate the conversion between 0 and 298 K thermochemistry for many of the transition metal species, for which there are no detailed molecular constants, we outline a uniform procedure appropriate for converting BDEs between these two temperatures in Appendix 1.

3. Periodic trends in covalent M-L bond energies

Periodic trends in the bond energies for transition metal ligands are most easily analyzed by examining the simplest of these systems, those containing only a single ligand. These periodic trends have been discussed several times previously [29, 43, 57, 64–68], but are repeated here in light of the revised numbers. We restrict ourselves to a discussion of first row transition metals because the data base for second and third row metal-ligand complexes is not as extensive and has not been evaluated thoroughly for systematic errors arising from electronic excitation of the metal. Those that have been measured have been discussed previously [29, 64–66].

3.1. Covalent metal ion-ligand single bonds

3.1.1. *Metal-hydride ions*

The simplest of the covalent metal ion-ligand systems are the diatomic metal hydrides. Table I reports our BDE values for MH^+ as determined exclusively from reactions of M^+ with H_2 and D_2 [13, 22, 69–74]. The values are unchanged from previous reports because the original BDEs were correctly interpreted with the 0 K model. For the first row transition metals, these

Table I. Bond energies of transition metal ions with ligands that can form one covalent bond (kJ/mol)[a]

M	*Ep*	M^+-H	M^+-CH_3	M^+-C_2H_5	M^+-NH_2	M^+-OH	M^+-C_2H_3
Ca	0	190(12)[b]					
Sc	15	235(9)[c]	233(10)[d]		347(5)[e]	499(9)[f]	
Ti	27	223(11)[g]	214(3)[h]	207(7)[h]	346(13)[e]	465(12)[f]	334(24)[h]
V	68	198(6)[i]	193(7)[j]	225(14)[k]	293(6)[l]	434(15)[f]	369(19)[j]
Cr	192	132(9)[m]	110(4)[n,o]	128(5)[o]	272(10)[p]	298(14)[q]	226(6)[o]
Mn	56	199(14)[r]	205(4)[s,t]		254(20)[q]	332(24)[q]	
Fe	47	204(6)[u]	229(5)[v,w]	233(9)[v]	309(10)[q]	366(12)[x]	238(11)[y]
Co	79	191(6)[z]	203(4)[aa,bb]	193(11)[cc]	247(7)[dd]	300(4)[bb]	203(8)[ee]
Ni	133	162(8)[z]	187(6)[aa]	222(25)[cc]	223(8)[dd]	235(19)[q]	
Cu	292	88(13)[z]	111(7)[cc]		192(13)[dd]		
Zn	0	228(13)[b]	280(7)[ff]				
Y		257(6)[c]	236(5)[gg]				
La		239(9)[c]	217(15)[gg]				
Lu		204(15)[c]	176(20)[gg]				

[a] Values are at 0 K with uncertainties in parentheses. [b] Ref. 69. [c] Ref. 70. [d] Ref. 52. [e] Ref. 103. [f] Ref. 108. [g] Ref. 71. [h] Ref. 89. [i] Ref. 13. [j] Ref. 3. [k] Ref. 102. [l] Ref. 104. [m] Ref. 72. [n] Ref. 86. [o] Ref. 92. [p] Ref. 106. [q] Ref. 107. [r] Ref. 73. [s] Ref. 87. [t] Ref. 93. [u] Ref. 22. [v] Ref. 90. [w] Ref. 94. [x] Ref. 109. [y] Ref. 120. [z] Ref. 74. [aa] Refs. 91 and 95. [bb] Ref. 96. [cc] Ref. 91. See text. [dd] Ref. 105. [ee] Ref. 121. [ff] Ref. 4. [gg] Ref. 88.

values agree nicely with those from theory, with average deviations of 2 ± 11 kJ/mol [46] (M = Sc − Cu) and 11 ± 12 kJ/mol (M = Ca − Zn) [44].

Our values are plotted across the periodic table in Figure 2. The most striking aspect of this plot is the weak bonds of CrH^+ and CuH^+, an observation that is rationalized by the stable $3d^5$ half-filled and $3d^{10}$ filled electron configurations, respectively, of these metals. The energetics associated with reorganizing such electron configurations in order to form a covalent bond is called the promotion energy *Ep*. A correlation between *Ep* and D(M^+-H) was first suggested by Armentrout, Halle, and Beauchamp [75] and later refined by Mandich *et al.* [76] and Elkind and Armentrout [66]. The definition of *Ep* that provides the best correlation with D(M^+-H) is the energy required to excite the ground state of the metal ion into an electronic state having a singly occupied 4s orbital that is spin decoupled from the 3d electrons. This spin-decoupling term accounts for the fact that snapping the M^+-H bond leaves the M^+ with a 4s electron that has an equal probability of being high-spin or low-spin coupled to the M^+(3d) electrons. This promotion energy, as calculated from atomic energies [77, 78] and compiled elsewhere [66, 79], is listed in Table I. Figure 3 shows that the correlation between D(M^+-H) for M = Sc − Zn and *Ep* is excellent and systematically accounts for the periodic trends observed in Figure 2.

Alternate definitions of *Ep* have also been considered, such as ignoring the spin decoupling energy [57] or bonding to the 3dσ orbital in a $3d^n$ or $4s^1 3d^{n-1}$ electron configuration on the metal ion [66]. Only in the latter case

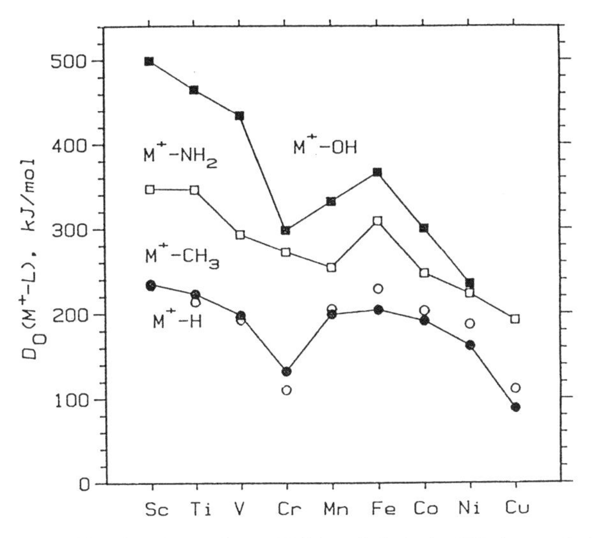

Fig. 2. Transition metal ion bond energies to H (closed circles), CH_3 (open circles), NH_2 (open squares), and OH (closed squares) across the first row of the periodic table.

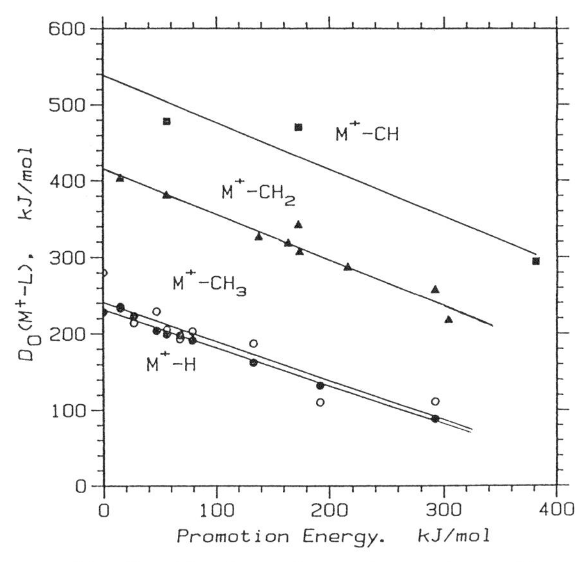

Fig. 3. Transition metal ion ligand bond energies vs. the atomic metal ion promotion energy to a $4s^1 3d^{n-1}$ spin-decoupled state (see text). Results for H (closed circles), CH_3 (open circles), CH_2 (closed triangles) and CH (closed squares) ligands are shown and taken from Tables 1, 2, and 3. The lines are linear regression fits to the data.

for $Ep(4s^13d^{n-1})$ was a reasonable correlation found. This suggests that the bonding orbital on the metal has both 4s and 3dσ character, in agreement with the results of ab initio calculations [44–46, 80, 81].

Another aspect of the correlation between $D(M^+\text{-}H)$ and Ep is the intercept at $Ep = 0$ eV. The value of 231 ± 7 kJ/mol corresponds to the metal-hydride ion BDE when uninfluenced by electronic effects. We have suggested [57, 64, 66] that this maximum BDE may be thought of as the "intrinsic" metal-hydrogen BDE, i.e. the likely BDE for *any* metal-hydride bond in the absence of electronic and steric effects. For ligated metals, metal clusters, and on metal surfaces, a directional and sterically unhindered metal orbital having a single electron that is electronically decoupled from other electrons must be prepared for efficient bonding to H. In such cases, the ligands and nearest neighbors surrounding the binding site can alter the electronic environment at this site in order to achieve this condition. Indeed, our intrinsic metal-hydride bond energy is comparable to several metal-hydride BDEs at 298 K in 18–e^- condensed phase complexes: for example, $HCo(CO)_4$ and $[HCo(CN)_5]_{aq}$ [82, 83], and $[H_2IrCl(CO)(PPh_3)_2]$ [84], and a value for metal-hydride BDEs of ~250 kJ/mol commonly cited as a reasonable first guess [85].

Finally, we note that the slope of this correlation does not equal unity. In other words, the sum of $D(M^+\text{-}H)$ and Ep does not yield a constant value. Rather the slope is close to -0.5, showing that $D(M^+\text{-}H) + Ep/2$ is nearly a constant equal to the intrinsic metal-hydrogen BDE. This slope is common to a number of different covalently bound metal-ligand species for first-row transition metals, as is evident in Figure 3, but a different value is found for such a correlation with second-row transition metal hydride ions [66]. The physical meaning of this deviation from unit slope is not completely clear, although it may be a further indication of the mixing of 4s and 3dσ orbitals to form the best covalent bond to first-row metals and a different degree of mixing for the 5s and 4dσ orbitals for second-row metals.

3.1.2. *Metal-methyl ion bond energies*

Complete thermodynamic information is available for the first-row transition metal methyl ions, Table I. The neutral reactants used include ethane in all systems [3, 86–88]; propane for several more [52, 89–91]; larger alkanes for Cr [92], Mn [93] and Zn [4]; the methyl halides for Fe [94], Co and Ni [95]; and ethene [3], acetone [93], and methanol [96] in one case each. The MCH_3^+ BDEs measured in several systems agree nicely, verifying the accuracy of the results. These results are in good agreement with theory [47], which predict values that average 8 ± 10 kJ/mol lower than the experimental values. Our value for $D(Co^+\text{-}CH_3)$ is also in reasonable agreement with the value of 223 ± 8 kJ/mol obtained by van Koppen *et al.* [97].

Table I and Figure 2 show that the MCH_3^+ BDEs are very similar to those for MH^+ for the early transition metals, M = Sc – Mn. For the later metals, M = Fe – Zn, the metal-methyl BDEs exceed the metal-hydride

BDEs by a small amount, 27 ± 15 kJ/mol. The similarity of $D(M^+\text{-}H)$ and $D(M^+\text{-}CH_3)$ is reasonable given that isolobal arguments [98] and ab initio calculations [47, 48, 99] find that both H and CH_3 can form only a single covalent bond with transition metals. The finding is surprising only because metal-carbon bonds are generally much weaker than metal-hydrogen bonds in condensed phase organometallic complexes. This difference in the relative BDEs in different phases suggests that steric effects are largely responsible for the weak M-C bond strengths for the condensed phase species [85].

The correlation between the metal-methyl ion BDEs and *Ep*, Figure 3, is comparable to that for $D(M^+\text{-}H)$, suggesting that the bonding character at the metal center is similar to that discussed above. The intrinsic MCH_3^+ BDE (linear regression analysis of data for Sc – Zn) is 241 ± 7 kJ/mol, slightly stronger than that for $D(M^+\text{-}H)$, Figure 3. One possible explanation [54] for this increase is the increased polarizability of the methyl group compared to a hydrogen atom, an effect that can contribute about 20–25 kJ/mol to the bond energy [48]. This agrees well with the difference between $D(M^+\text{-}CH_3)$ and $D(M^+\text{-}H)$ for Fe – Zn, but does not appear to be an important effect for the early metals, Sc – Cr. Effects due to agostic [100] M ·· H-C interactions have also been suggested [101], but these would be expected to be most important for the early metals where there are empty orbitals on the metal. Theory also does not find any evidence for agostic interactions in $M^+\text{-}CH_3$ [47].

3.1.3. *Metal-alkyl ion bond energies*

Limited data also exist for the BDEs of atomic metals ions with ethyl and propyl groups, Table I [89, 90, 92, 102]. For the Ti, V, Cr, and Fe species, the metal-ethyl ion BDEs are comparable to the metal-methyl ion BDEs. Although not originally reported, additional thermochemistry for $CoC_2H_5^+$ and $NiC_2H_5^+$ can be obtained from a reexamination of our data for the reactions of Co^+ and Ni^+ with propane [91]. This work suggests that $D(Co^+\text{-}C_2H_5)$ is weaker than $D(Co^+\text{-}CH_3)$ by 10 ± 10 kJ/mol. For the case of Ni^+, the data suggest that the Ni^+-ethyl bond is stronger than the Ni^+-methyl bond by 35 ± 24 kJ/mol, which may indicate a different structure, see below. The only reliable measurements that we have made for the propyl ligand concern Cr^+ [92]. Both the $Cr^+\text{-}1\text{–}C_3H_7$ BDE, 116 ± 6 kJ/mol, and the $Cr^+\text{-}2\text{–}C_3H_7$ BDE, 101 ± 5 kJ/mol, are comparable to $D(Cr^+\text{-}CH_3)$.

One complexity that should be considered is the structure of the $MC_2H_5^+$ and $MC_3H_7^+$ species. It is possible that the lowest energy structure is not the metal-alkyl ion but a hydrido-alkene complex, $H\text{-}M^+\text{-}C_nH_{2n}$. The thermochemistry does not differentiate between these two possibilities. Such a structural difference does not affect the BDE, which is just the energy needed for the $MC_nH_{2n+1}{}^+$ species, no matter what its structure, to dissociate to $M^+ + C_nH_{2n+1}$.

3.1.4. *Metal-amide and -hydroxide ion bond energies*
We have reported BDEs for M^+-NH_2 for M = Sc – V [103, 104] and Co – Cu [105]. We have also obtained preliminary values for M = Cr [106], Mn [107] and Fe [106], most of which have been listed previously [57]. These values are taken primarily from reactions with NH_3 or ND_3, although preliminary work also includes reactions with CH_3NH_2 [106]. Similarly, we have reported BDEs for M^+-OH for M = Sc – V [108], Fe [109] and Co [96], and obtained preliminary values for M = Cr and Mn, as listed previously [57]. These values are obtained from reactions with water, methanol, or reactions of the metal oxide ions with hydrogen.

The reinterpreted BDEs for M^+-NH_2 and M^+-OH are listed in Table I and shown in Figure 2. Our metal ion-amide BDEs are within experimental error of the two alternate literature measurements, $D_{298}(Fe^+\text{-}NH_2)$ = 280 ± 50 kJ/mol and $D_{298}(Co^+\text{-}NH_2)$ = 272 ± 33 kJ/mol [110]. Our metal ion-hydroxide BDEs agree well with values obtained by Michl and coworkers [40] for Ti, V, Cr, Mn, Fe, and Co, but our values differ from theirs for $ScOH^+$ (367 ± 13 kJ/mol) and $NiOH^+$ (177 ± 13 kJ/mol), both 298 K values. Our value for $D(Sc^+\text{-}OH)$ is in better agreement with a theoretical calculation [111], and it seems unlikely that $D(Ni^+\text{-}OH)$ is less than $D(Ni^+\text{-}CH_3)$ based on the discussion below. The Fe^+-OH BDE has been studied a number of times with mixed results. Our 0 K value of 366 ± 12 kJ/mol compares nicely to that of Michl and coworkers, 357 ± 13 [40], but exceeds those of Murad (0 K), 319 ± 19 kJ/mol [112], and Cassady and Freiser (CF), 307 ± 13 kJ/mol [113]. The latter authors also obtain $D(Co^+\text{-}OH)$ = 297 ± 13 kJ/mol, in good agreement with our results. It is possible that the results of Murad and CF are too low because they did not correct for internal excitation, Murad in his value for IE(FeOH) determined at 1800 K and CF in their photodissociation of $FeOH^+$. Because of these discrepancies, we have measured this BDE in a number of systems in our laboratory [109], and the value cited here is the end result.

The CH_3, NH_2, and OH ligands can form only a single covalent bond in organic molecules, and thus the CH_3-CH_3, CH_3-NH_2, and CH_3-OH BDEs are comparable, 367, 350 and 379 kJ/mol [114], respectively. In contrast, the metal ion-methyl BDEs are much smaller than the metal ion-amide BDEs which are much smaller than the metal ion-hydroxide BDEs, Table I. This is shown visually in Figure 2. The difference between the organic and metal-based systems is that the transition metals have nonbonding 3d orbitals that can accept electron density from the lone-pair electrons on the NH_2 and OH ligands. This augments the bonding by forming dative bonds. Because OH has two lone-pairs of electrons, it can form twice as many dative bonds as NH_2, which has only one lone-pair. Ab initio calculations on M^+-NH_2 (M = Sc – Mn) and Sc^+-OH confirm that the bonding in these molecules has both covalent and dative character [111, 115–117].

This bonding picture also explains why the differences between $D(M^+\text{-}CH_3)$, $D(M^+\text{-}NH_2)$ and $D(M^+\text{-}OH)$ are larger on the left side of the periodic

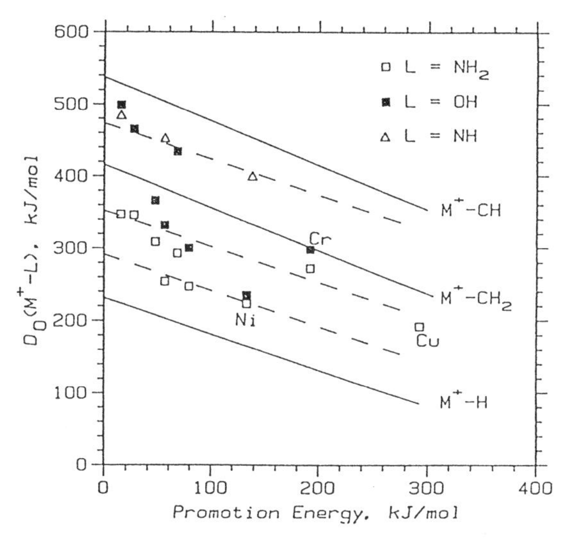

Fig. 4. Transition metal ion ligand bond energies vs. the atomic metal ion promotion energy to a $4s^1 3d^{n-1}$ spin-decoupled state (see text). Results for NH_2 (open squares) and OH (closed squares) ligands are taken from Table 1. Results for NH (open triangles) ligands are taken from Table 2. The full lines are the linear regression fits to the MH^+, MCH_2^+ and MCH^+ data shown in Figure 3. The dashed lines are the MH^+ correlation line plus 60, 120, and 240 kJ/mol, where 120 kJ/mol is the average energy for a full dative bond, see text.

table than on the right, Figure 2. The early metals have empty 3d orbitals, while all 3d orbitals for the late metals are at least half occupied. This also has implications for the correlation with *Ep* if this quantity is the same as for M^+-H and M^+-CH_3, which seem appropriate because this *Ep* successfully accounts for the periodic trends in the covalent bonding. Figure 4 shows this correlation. The early metals, Sc – V, have M^+-NH_2 BDEs that are an average of 115 ± 16 kJ/mol stronger than D(M^+-CH_3). The M^+-OH BDEs for these three metals are an average of 253 ± 13 kJ/mol stronger than D(M^+-CH_3). Thus, dative bonds appear to be worth an average of 126 ± 19 kJ/mol for the early transition metal species where there are at least two empty 3d orbitals for the $4s^1 3d^{n-1}$ electron configuration (used to form the covalent bond in this promotion scheme).

The later transition metals, Mn, Fe, Co, and Ni, have M^+-NH_2 BDEs that are an average of 52 ± 19 kJ/mol stronger than D(M^+-CH_3). For Mn, Fe and Co, the metal ion-hydroxide BDEs average 120 ± 21 kJ/mol higher than the metal ion-methyl BDEs. These enhancements are only half those observed for Sc – V, which is consistent with the fact that the metal orbitals accepting electron density for the late metals are half-filled. For these late transition metal species, dative bonds are worth an average of about 57 ± 17 kJ/mol, half that for the early transition metal species. For both early and late transition metals, this analysis quantifies the contribution of a full (two--

electron) dative bond to the bond energies in these metal ion-amides and hydroxides as 120 ± 27 kJ/mol.

The intermediate case of chromium deserves some additional comment. A preliminary value [106] for D(Cr^+-NH_2), Table I, is 162 ± 11 kJ/mol stronger than D(Cr^+-CH_3), consistent with a strong dative interaction. However, D(Cr^+-OH) is only 188 ± 17 kJ/mol stronger than D(Cr^+-CH_3), and thus falls in between the correlations for early and late transition metals, Figure 4. This suggests that the hydroxide forms only one and a half-dative bonds. This can be rationalized by noting that the promoted $Cr^+(4s^13d^4)$ can only have one empty 3dπ orbital. Thus, $CrNH_2^+$ can form a full dative bond, consistent with calculations [117]; however, the second dative bond in $CrOH^+$ involves a half-occupied 3dπ orbital.

At the right side of the transition metal series, the BDE of $NiOH^+$ seems particularly weak and that of $CuNH_2^+$ seems particularly strong. Because D(Ni^+-OH) is nearly equal to D(Ni^+-NH_2), Table I, the hydroxide ligand does not appear to enhance the bonding with a second dative interaction. Promoted Ni^+ has a $4s^13d^8$ electron configuration with two half-filled 3d orbitals. If one of these prefers to be the 3dσ orbital in order to form a strong covalent bond (consistent with 4s-3dσ hybridization), then Ni^+ can form a single half-dative bond but not a second. In contrast, D(Cu^+-NH_2) is stronger than expected based on the correlation with *Ep*, Figure 4. Because the promotion energy to form a covalent bond is so large for copper ion, the Cu^+-NH_2 bond may involve donation of the nitrogen lone-pair of electrons into the empty 4s orbital of $Cu^+(3d^{10})$ ground state. This hypothesis is consistent with the Cu^+-NH_2 BDE relative to that calculated for Cu^+-NH_3, 224 [118] or 206 [119] kJ/mol, a species where there is no covalent bonding interaction.

3.1.5. *Metal-vinyl ions*

Another ligand that can show both covalent and dative bonding interactions is vinyl, C_2H_3. We have measured metal ion-vinyl BDEs for several systems, Table I [3, 89, 92, 120, 121]. On average, the three early metal ions have M^+-C_2H_3 BDEs that are 137 ± 34 kJ/mol stronger than the MCH_3^+ BDEs, nearly the same dative bond enhancement found for the amide and hydroxide ligands. For the late transition metal ions, Fe and Co, the vinyl and methyl BDEs are nearly the same, indicating that there are no dative interactions for a vinyl ligand. This may be due to the distortion necessary to allow substantive overlap between the nonbonding orbitals on the metal and the pi orbital of C_2H_3.

3.2. Covalent metal ion-ligand multiple bonds

3.2.1. *Metal-methylidene ion bond energies*

Table II lists the BDEs that we have measured for M^+-CH_2. For the early transition metals, Sc – Cr, these values come from reactions with methane

Table II. Bond energies of transition metal ions with ligands that can form two covalent bonds (kJ/mol)[a]

M	*Ep*	M^+-CH_2	M^+-NH	M^+-O	M^+-C	M^+-$(CH_3)_2$	M^+-H_2
Ca				344(5)[b]			
Sc	15	402(23)[c]	483(10)[d]	689(6)[e,f]	322(6)[e]	464(5)[g]	467(5)[h]
Ti	56	380(9)[i]	451(12)[d]	664(7)[e]	391(23)[e]	472(25)[j]	
V	138	325(6)[k]	398(15)[l]	564(15)[e,m]	370(5)[e,n]	391(7)[o]	
Cr	304	217(4)[p]		359(12)[b]			
Mn	216	286(9)[q]		285(13)[b]			
Fe	173	341(4)[r]		335(6)[s]		409(12)[t]	
Co	164	317(5)[u,v]		314(5)[b,v]			
Ni	174	306(4)[v]		264(5)[b,v]			
Cu	292	256(5)[v]		156(15)[b]			
Zn				161(5)[b,w]			
Y		388(13)[c]					516(7)[c]
La		401(7)[c]					493(7)[c]
Lu		⩾230(6)[c]					394(18)[c]

[a] Values are at 0 K with uncertainties in parentheses. [b] Ref. 68. [c] Ref. 88. [d] Ref. 103. [e] Ref. 135. [f] Refs. 108 and 136. [g] Ref. 139. [h] Ref. 52. [i] Ref. 18. [j] Ref. 89. [k] Ref. 122. [l] Ref. 104. [m] Ref. 137. [n] Ref. 3. [o] Ref. 140. [p] Refs. 86 and 123. [q] Ref. 93. [r] Ref. 120. [s] Ref. 132. [t] Ref. 141. [u] Ref. 121. [v] Ref. 124. [w] Ref. 133.

[18, 88, 122, 123] because reactions to form MCH_2^+ with larger alkanes have barriers. Reactions with cyclopropane and ethylene oxide have been used to confirm the value obtained for Cr [86], and to determine the values for Mn – Cu [93, 120, 121, 124]. The values are in good agreement with theoretical calculations [125], with an average deviation of 9 ± 26 kJ/mol and an average absolute deviation of 22 ± 14 kJ/mol, and with experimental values for $FeCH_2^+$, 343 ± 21 kJ/mol, and $CoCH_2^+$, 305 ± 21 kJ/mol, determined by photodissociation [126]. The largest deviation occurs in the case of $ScCH_2^+$ where the experimental result is larger than the theoretical result by 46 ± 26 kJ/mol. Our experimental BDE in this case has the largest uncertainty because identification of the electronic state responsible for the reactivity observed is particularly difficult in the Sc^+ system. It is possible that our BDE could be lower by up to the excitation energy of the $Sc^+(^1D)$ state, 30 kJ/mol [77], such that the experimental and theoretical values would be much closer. Indeed, this is suggested by preliminary results from the reaction of SiH_4 with Sc^+ made in the flow tube and therefore much colder than Sc^+ generated by surface ionization [127].

For Sc – Cu, the metal-methylidene ion BDEs are an average of 1.73 ± 0.28 times stronger than the metal-methyl ion BDEs. This is comparable to the ratio of BDEs for CH_3-CH_3 and $CH_2 = CH_2$, 1.95 [114], suggesting that the metal-methylidenes have covalent double bonds [29, 128]. If true, then a correlation with a promotion energy is anticipated. To calculate the promotion energy for formation of a double bond to a metal ion, two electrons need to be spin-decoupled from the nonbonding electrons and placed in appropriate orbitals. For simplicity, the values of *Ep* listed in Table II are taken from the calculations of Carter and Goddard [79]. Correlations of the BDEs with *Ep* values for $4s^13d^{n-1}$ and $3d^n$ electron configurations have been considered previously, and only the former is reasonable [67]. This is shown in Figure 3. The intrinsic metal-carbon double bond energy is found to be 415 ± 13 kJ/mol, close to that for the 18–e^- $(CO)_5Mn^+ = CH_2$ species, 435 ± 13 kJ/mol [129]. Note that the strength of a covalent metal-carbon pi bond, 174 kJ/mol, is substantially greater than the strength of a dative pi bond, 120 kJ/mol.

3.2.2. *Metal-ethylidene and -propylidene ion bond energies*

Although the structural assignments are not unequivocal, we believe that we have observed metal-ethylidene ions, $M^+ = CHCH_3$, in five systems, Sc [52], V [102], Cr [92], Ni and Cu [91], as well as chromium-propylidene ions [92]. Useful thermochemical data could not be extracted in the scandium system, but the $V^+ = CHCH_3$ and $Cr^+ = CHCH_3$ BDEs are measured to be 280 ± 20 and 180 ± 8 kJ/mol, respectively. (This thermochemistry uses a 0 K heat of formation for $CHCH_3$ of 322 ± 8 kJ/mol, based on ab initio calculations of Pople *et al.* [130] and Trinquier [131] and roughly confirmed by experiment [120]. For the propylidenes, we assume that $D(H\text{-}CHCH_3) \approx D(H\text{-}CHC_2H_5) \approx D[H\text{-}C(CH_3)_2]$.) These values are slightly weaker than the values for the respective metal-methylidene ion BDEs, Table II, while values for the BDEs between Cr^+ and 1-propylidene, 113 ± 11, and 2-propylidene, 124 ± 9 kJ/mol, are much weaker. This difference could be due to barriers in the reactions that form the Cr^+-propylidene species [92]. Likewise, the BDEs for $Ni^+ = CHCH_3$, 224 ± 26 kJ/mol, and $Cu^+ = CHCH_3$, ≈78 kJ/mol, are considerably lower than those for $Ni^+ = CH_2$ and $Cu^+ = CH_2$ BDEs. These values and the chromium propylidene ion BDEs are best viewed as lower limits.

3.2.3. *Metal-imide, -oxide, and -carbide ion bond energies*

Complete thermodynamic information is available for the diatomic transition metal oxide ions, Table II. These are taken primarily from reactions with O_2 [68, 132], but reactions with c-C_2H_4O have been used to verify the FeO^+ [132], CoO^+ and NiO^+ BDEs [124], and reaction with NO_2 has verified the ZnO^+ BDE [133]. Bracketing reactions of Cr^+ with ethylene oxide and propylene oxide verify the CrO^+ BDE [134]. Reactions of Sc^+, Ti^+ and V^+ with O_2 are exothermic [68], therefore the metal-oxide ion BDEs in these systems have been measured by other reactions. For ScO^+, this involves reactions of Sc^+ with CO [135], D_2O [136], and the reaction of $ScO^+ + D_2 \rightarrow Sc^+ + D_2O$ [108]. The weighted average of these three values is reported in Table II and appears to be the best available. The value we obtain for $D(Ti^+\text{-}O)$ by reaction of Ti^+ with CO, 664 ± 7 kJ/mol, agrees very well with that determined from literature thermochemistry and equation (9), 668 ± 9 kJ/mol, as detailed elsewhere [135].

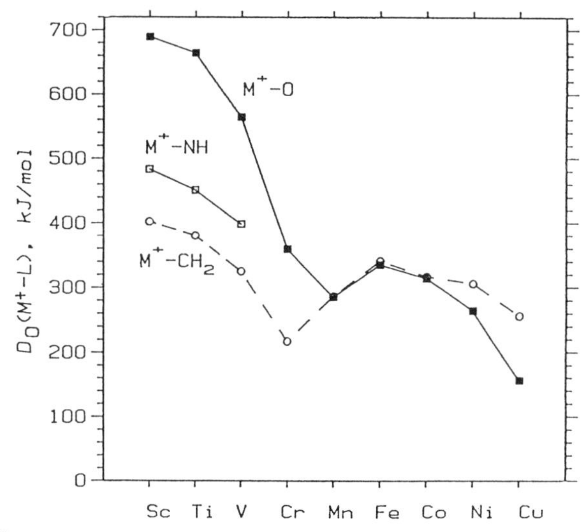

Fig. 5. Transition metal ion bond energies to CH_2 (open circles), NH (open squares), and O (closed squares) across the first row of the periodic table.

$$D(M^+\text{-}O) = D(M\text{-}O) + IE(M) - IE(MO) \tag{9}$$

We have determined $D(V^+\text{-}O)$ by reaction with CO [135] and by collision-induced dissociation [137]. The weighted average of these values, 564 ± 15 kJ/mol, is within experimental error of that calculated with equation (9) from literature thermochemistry, 578 ± 9 kJ/mol [135].

A few values for the imide ligand NH, determined exclusively from reactions with ammonia [103, 104], and for the carbon atom ligand, determined primarily from reactions with CO [3, 135], are also available, Table II. The V^+-NH BDE that we obtain agrees with that of Buckner *et al.*, 423 ± 29 kJ/mol [138]. The NH and O ligands are isoelectronic with CH_2, and their bonding with transition metal ions is compared in Figure 5. This comparison demonstrates that the imide and oxide BDEs are again enhanced compared to the methylidene BDEs for the early transition metals, presumably due to contributions of dative bonding. The M^+-NH BDEs for M = Sc, Ti, and V are stronger than the M^+-CH_2 BDEs by 75 ± 5 kJ/mol, indicating that the dative bond is somewhat weaker than in the systems where only a single covalent bond existed. This may be due to the relative orientation of the lone-pair of electrons in the doubly bonded MNH^+ systems vs. the MNH_2^+ and MOH^+ systems. When these BDEs are plotted vs *Ep*, Figure 4, the correlation is similar to that for M^+-OH (even though NH uses the *Ep* definition of Table II for two covalent bonds, while OH uses the *Ep* definition of Table II for one covalent bond). The near-equivalence of the NH and OH species, both effectively forming triple bonds, has been noted previously [57, 111]. Unfortunately, little information for late transition metal-imide

Table III. Bond energies of transition metal ions with ligands that can form triple covalent bnds (kJ/mol)[a]

M	*Ep*	M^+-CH	M^+-N
Ti	56	478(5)[b]	486(12)[c]
V	173	470(5)[d]	433(7)[e]
Cr	381	294(29)[f]	

[a] Values are at 0 K with uncertainties in parentheses. [b] Refs. 18 and 89. [c] Ref. 103. [d] Refs. 3 and 122. [e] Ref. 104. [f] Ref. 86.

ions is available. We have measured none of these BDEs and the value for $D_{298}(Fe^+$-NH) reported by Buckner *et al.* [138], 255 ± 21 kJ/mol, seems low compared with $D(Fe^+$-$CH_2)$ and $D(Fe^+$-O), Table II and Figure 5.

An analysis of the bonding in the transition metal oxide ions is complicated by significant contributions from ionic M^+-O^- character [68]. Figure 5 makes it clear that the enhancement of $D(M^+$-O) vs. $D(M^+$-$CH_2)$ is very significant for M = Sc, Ti, and V, about 274 ± 19 kJ/mol, well in excess of our estimate of a dative interaction. Because these three metal ions have empty 3d orbitals, their oxides may be viewed as having triple bonds, much like the CO molecule. Indeed, these early metal oxide ions have BDEs that exceed those for the covalent triple bonds in M^+-CH and M^+-N, Table III. This has been discussed in detail elsewhere [68].

The enhancement for CrO^+ vs. $CrCH_2^+$ is only 142 kJ/mol, consistent with the need to have at least one electron in an antibonding π or σ orbital. For M = Mn, Fe, and Co, the metal-methylidene and metal-oxide ion BDEs are nearly identical, indicating that no dative interaction is present. $D(Ni^+$-O) and $D(Cu^+$-O) are weaker than the respective MCH_2^+ BDEs, possibly due to more ionic character in the MO^+ species [68]. A comparison of the MC^+ and MO^+ BDEs, Table II, further illustrates the ability of the lone-pair of electrons on the oxygen atom to enhance the bonding. Both C and O have ^3P ground states with two-unpaired 2p electrons, but the MC^+ BDEs are much weaker than the MO^+ BDEs and more comparable in strength to the MCH_2^+ BDEs.

3.2.4. *Metal-methylidyne and -nitride ion bond energies*

We have measured only a few values for MCH^+ and MN^+ species [18, 86, 89, 103, 104, 122], largely because reactions of M^+ with acetylene [3] and nitrogen rise too slowly from threshold to provide accurate thermochemistry. Reactions involving other reagents (e.g. methane or ethane for CH, and ammonia for N) involve extensive rearrangement and therefore can lead to similar problems. Only for the early transition metals are the bonds strong enough to overcome these difficulties, Table III. We find that the M^+-CH and M^+-N BDEs are similar to one another, and are an average of 2.4 ± 0.2 times stronger than the M^+-CH_3 BDEs. This is comparable to the ratio between CH_3-CH_3 and CH ≡ CH bond energies of 2.6, consistent with the formation of a triple metal-ligand bond [29, 57, 128]. A plot of these BDEs vs *Ep*, as defined by Carter and Goddard [79], is shown in Figure 3, and shows a reasonable correlation. The data of Hettich and Freiser [126] for $D_{298}(Fe^+\text{-}CH) = 423 \pm 29$ kJ/mol and $D_{298}(Co^+\text{-}CH) = 418 \pm 29$ kJ/mol fits in this correlation quite nicely. Although the data are still incomplete, the information available indicates that the intrinsic bond energy for formation of a metal-ligand triple bond in the M^+-CH and M^+-N systems is 532 ± 16 kJ/mol.

3.3. Metal ion-ligand complexes with two covalently bound ligands

3.3.1. *Metal-dihydride and -dialkyl ions*

Atomic transition metal ions react with alkanes by oxidative addition and reductive elimination processes that involve the formation of species such as MH_2^+, HMR^+, and MR_2^+ where R is an alkyl. Unfortunately, the determination of the second covalent metal-ligand BDE is more difficult than the determination of the first BDE. Nevertheless, we have obtained several of these BDEs, all listed in Table II with the exception of $D[Sc^+\text{-}(H)(CH_3)] = 479 \pm 5$ [139] and $D[V^+\text{-}(H)(CH_3)] = 390 \pm 16$ kJ/mol [140]. For the values listed for M = Sc [52, 139], Y, La, Lu [88], Ti [89], V [140], and Fe [141], there are good reasons to believe that the information listed pertains to species containing two covalent metal ligand bonds. Other metal ions, such as Co^+ and Ni^+, have been observed to undergo exothermic decarbonylation reactions with acetone to form $MC_2H_6^+$ [142], a product that was assigned as the dimethyl metal ion. More recent studies of this $CoC_2H_6^+$ species indicate that it has the $Co^+ \cdot C_2H_6$ structure [143, 144], meaning that the thermochemistry for $Co(CH_3)_2^+$ previously derived from this observation is incorrect. At this point, no further studies have been performed to clarify the structure of $NiC_2H_6^+$ formed by decarbonylation of acetone.

Our value for ScH_2^+ can be compared to recent results of Kemper and Bowers [145]. They find that the sum of the scandium-hydride ion BDEs is 454.9 ± 0.8 kJ/mol, in reasonable agreement with our value of 468 ± 11 kJ/mol (the average of three independent determinations from ethane, propane, and cyclopentane) [52]. It is possible that the difference is due to electronic excitation in our results. Such an effect was explicitly looked for and not observed, but is difficult to ascertain in the scandium system. Because the reactions that produce ScH_2^+ are spin-forbidden from the ground state, these reactions are particulary susceptible to the presence of the ^1D state of Sc^+, lying 30 kJ/mol above the ground state [77]. Indeed, recent results for reaction of Sc^+ with SiH_4 find that ScH_2^+ is formed by the ^1D state but not by the ^3D ground state [127].

For Sc, Ti, V, and Fe, the sum of the two covalent BDEs in these species is 2.0 ± 0.2 times as large as the single covalent BDEs. We have previously found that the sum of the $M(CH_3)_2^+$ BDEs correlates reasonably well with the *Ep* defined in Table II for formation of two covalent bonds [29]. The intrinsic dimethyl BDE of this correlation is 480 ± 20 kJ/mol, about twice the intrinsic BDE for a metal ion-methyl BDE, 241 kJ/mol.

3.3.2. *Metallacyclobutane ions*

Other organometallic species with two covalent metal-ligand bonds are metallacyclobutanes, $\overline{M\text{-}CH_2CH_2CH_2^+}$. We have generated these species for M = Fe and Co by the decarbonylation of cyclobutanone [120, 121] and by collisional stabilization of the Fe^+-cyclopropane adduct [120]. CID spectra show that these species decompose much differently than the more stable metal ion-propene complexes, $M^+ \cdot C_3H_6$. Decomposition of the metallacycle to M^+ + cyclopropane requires 133 ± 4 and 125 ± 7 kJ/mol for M = Fe and Co, respectively. This thermochemistry is in good agreement with the 0 K values reported by van Koppen *et al.* [42, 146], 124 and 108 ± 21 kJ/mol, respectively. Taking the C-C bond energy in cyclopropane as 241 kJ/mol [147], we find that the sum of the two covalent bonds in these metallacycles is 374 ± 4 and 366 ± 7 kJ/mol, respectively. This can be compared with the sum for $Fe^+(CH_3)_2$ of 409 ± 12 kJ/mol. The difference can be attributed in part to strain energy in the cyclic compound [146].

3.4. Neutral metal-ligand bonds

3.4.1. *Metal-hydride bond energies*

If the intrinsic BDEs discussed above truly represent the strength of a typical metal-ligand bond devoid of electronic and steric effects, then the positive charge on the metals should not have a significant effect on the intrinsic BDE. This can be verified by examining the BDEs of neutral metal-ligand systems. In particular, we have made a concerted effort to obtain reliable BDEs for all of the diatomic metal hydrides of the first-row transition metals. Neutral transition metal hydride BDEs have been determined in the case of titanium [148], vanadium and chromium [149] by using mono-, di- and trimethylamine as hydride donors. For iron, five systems were studied: acetaldehyde, cyclopentane, butane, propane, and cyclopropane [150]. In the case of cobalt, nickel and copper, the hydride donors were ethane, propane, isobutane [91], and cyclopropane [124]. For manganese, the situation is somewhat more complex. The only result we have obtained comes from reaction with isobutane as a hydride donor [93]. This reaction was observed only for excited states of Mn^+. Depending on whether the observed reactivity is attributed to $Mn^+(^5S)$ or to both the ^5S and ^5D states, different values for D_0(Mn-H) of 163 ± 6 or 113 ± 18 kJ/mol, respectively, are obtained (reinterpreted as 0 K thresholds). Kant and Moon [151] failed to observe MnH, a result that they interpreted to mean that D(Mn-H) $\leqslant$ 134 kJ/mol. Based on this result, we chose the lower of our two values as the correct interpretation of our data. Based on a comparison with theoretical values

Table IV. Bond energies of neutral transition metal-ligand systems (kJ/mol)[a]

M	*Ep*	M-H[b]	M-CH_3	M-O
Sc	158		116(29)[c]	
Ti	108	189(6)[d]	174(29)[c]	
V	63	205(7)[e]	169(18)[f]	
Cr	45	186(7)[e]	140(7)[g]	433(6)[h]
Mn	241	163(6)[i]	>35(12)[i]	346(26)[i]
Fe	113	144(4)[j]	135(29)[c]	424(15)[k]
Co	65	180(5)[l]	178(8)[l]	409(20)[m]
Ni	22	236(8)[l]	208(8)[l]	
Cu	0	251(6)[l]	223(5)[l]	
Zn	387		70(10)[n]	155(4)[o]

[a] Values are at 0 K with uncertainties in parentheses. [b] These values are critically reviewed in reference 43. [c] Ref. 29. [d] Ref. 148. [e] Ref. 149. [f] Ref. 140. [g] Ref. 92. [h] Ref. 86. [i] Ref. 93. [j] Ref. 150. [k] Ref. 162. [l] Refs. 91 and 124. [m] Ref. 163. [n] Ref. 4. [o] Ref. 133.

[46, 152–154], however, the higher value appears to be the correct interpretation of our data and is reported here.

The final recommended values from the reanalysis of each of these systems are listed in Table IV. These values agree reasonably well with other experimental values as discussed elsewhere [43]. (The changes in the values instituted here are not large enough to alter the agreement.) They also agree nicely with values from the single systematic theoretical study of Chong *et al.* [152]. On average, our experimental values for D(M-H), which have an average error of 6 kJ/mol, are 2 ± 9 kJ/mol lower than the calculated values, and 9 ± 8 kJ/mol lower than values empirically corrected for basis set incompleteness in later references [46, 153].

It is tempting to directly compare ionic and neutral metal-hydrides that have the same number of valence electrons [155]. In some cases, such a comparison gives similar BDEs, e.g. TiH vs. VH^+ and CrH vs. MnH^+. In other cases, the differences in BDEs are large, e.g. VH vs. CrH^+ and NiH vs. CuH^+. Such a direct comparison fails to account for the differences in the electronic configuration of the neutral and ionic species, i.e. species that have the same number of valence electrons (e.g. V vs. Cr^{I} and Ni vs. Cu^{I}) do not necessarily have those electrons in the same orbitals (ground states of $4s^23d^3$ vs. $3d^5$ and $4s^23d^8$ vs. $3d^{10}$, respectively).

These differences in electronic configuration can be analyzed by again correlating with the promotion energy. Note that the strongest neutral metal-hydride BDEs, Table IV, are for Cu ($4s^13d^{10}$ ground state) and Ni ($4s^13d^9$ low-lying state), while the weakest are for Mn ($4s^23d^5$ ground state) and Fe ($4s^23d^6$ ground state). When *Ep* is defined in the same manner as for the ions, the plot of D(M-H) vs *Ep* shows the same general trends as Figure 3 [43, 64, 148], but the correlation is not nearly as good as that for the metal-hydride ions. Nevertheless, the intrinsic BDE for MH obtained from this correlation (excluding MnH) is 242 ± 10 kJ/mol. This is virtually the same as that for the metal-hydride ions, illustrating that charge has little effect on these covalent bond strengths.

More careful scrutiny of the correlation with *Ep* shows that the points for CrH, FeH, CoH, NiH and CuH fall nearly on a line with a slope of -1, i.e. D(M-H) + *Ep* is a constant of 248 ± 11 kJ/mol. These five metal hydrides have ground states that are reasonably well characterized by a $3d^{n+1}\sigma^2$ molecular configuration where the σ orbital is largely a 4s-1s bond [152]. In contrast, the hydrides of Mn and Zn have molecular configurations that are characterized as $4sp^1 3d^n \sigma^2$ where the 4sp nonbonding orbital is a 4s-4p hybrid and the σ orbital is largely 4sp-1s. The ground states of TiH (and VH to a lesser extent) are mixtures of these two types of configurations [152]. ScH has extensive 3d character present in its bonding orbital and the nonbonding orbital is a 4s-3d hybrid [156]. Therefore, the bond energies of ScH, MnH and ZnH, and to a lesser extent, TiH and VH, should not correlate with a promotion energy to a $4s^1 3d^{n+1}$ state of the metal atom. Deviations from this correlation increase as involvement of the $3d^{n+1}\sigma^2$ molecular configuration decreases, and as involvement of the 4p orbitals on the metal increases.

Another way of considering the periodic trends in the neutral metal hydrides is to compare them with the atomic metal anion, as discussed by Squires [157]. The correlation between D(M-H) and the electron affinities of the metals, EA(M), Table IV, is straightforwardly seen by converting this information to the proton affinities for M^-, $PA(M^-) \equiv \Delta H_{acid}(MH)$, by using equation (10),

$$PA(M^-) = D_0(M\text{-}H) + IE(H) - EA(M) \qquad (10)$$

where IE(H) = 1312 kJ/mol [114]. $PA(M^-)$ for Cr, Fe, Co, Ni, and Cu are found to be nearly identical and average to 1437 ± 7 kJ/mol. This is nearly the same as the 1423 ± 21 kJ/mol figure of Squires [157] and a 1444 ± 30 kJ/mol figure that includes first, second and third row metal hydrides [43], both corrected to 0 K values as outlined in Appendix 1. For these five metal hydrides, the $3d^{n+1}\sigma^2$ ground state molecular configurations are directly analogous with the $3d^{n+1}4s^2$ electron configurations of the metal anions. In contrast, the other metals have $PA(M^-)$ values that are higher in energy, by 56 kJ/mol for Ti, by 29 kJ/mol for V and by >38 kJ/mol for Mn. For these metals, the hydrides have electron configurations that differ from those of the atomic metal anions.

3.4.2. *Metal-methyl bond energies*

We have carefully measured neutral metal-methyl BDEs for Cr [92], Co, Ni, Cu [91], and Zn [4]. Only a lower limit has been obtained for $D(Mn\text{-}CH_3)$ [93], and only preliminary values have been obtained for Sc, Ti [29], V [140], and Fe [29]. These various values are listed in Table IV. Our values, excluding $ScCH_3$, deviate from theoretical results [47] by 1 ± 15 kJ/mol. For the five values that have been more rigorously obtained, $D(M\text{-}CH_3)$ is lower than D(M-H) by 22 ± 16 kJ/mol. Further, a plot of $D(M\text{-}CH_3)$ vs. *Ep* shows the same general trends as that for D(M-H) [43, 64, 148], but with a lower intrinsic BDE than for D(M-H). The observation that neutral metal-methyl BDEs are somewhat less than the MH BDEs (the opposite effect observed for the cations) is consistent with contributions to the bonding in the ions due to the polarizability of CH_3 vs. H. While the MCH_3 BDEs are somewhat less than the MH BDEs, the decrease is not nearly as great as is often observed in the condensed phase. Our results demonstrate that this discrepancy between gas-phase and condensed-phase results is not simply an effect of the charge on the metal.

3.4.3. *Metal-oxide bond energies*

In a few systems, we have also measured neutral metal oxide BDEs. Ethylene oxide has been used as an O^- donor in several systems: Cr [86], Mn [93], Co and Ni [124]. The BDE obtained for CrO, 433 ± 6 kJ/mol, agrees well with that recommended by Pedley and Marshall [158], 425 ± 29 kJ/mol, and that obtained by Balducci and coworkers, 441 ± 9 kJ/mol [159]. For MnO, we obtain 346 ± 25 kJ/mol, within experimental error of the value recommended by Smoes and Drowart, 369 ± 8 kJ/mol [160]. Values obtained for CoO and NiO BDEs are clearly too low, indicating that the thermodynamic threshold is not observed. For ZnO, we have used NO_2 as an O^- donor and obtained D(Zn-O) = 155 ± 4 kJ/mol [133], much lower than previous results but substantiated by recent high temperature studies [161].

We have also measured neutral metal oxide BDEs for FeO and CoO by examining reaction (11) [162, 163]. Combined with 0 K thermochemistry for $D(M^+\text{-}M)$, Table IX,

$$M_2^+ + O_2 \rightarrow MO^+ + MO \qquad (11)$$

and $D(M^+\text{-}O)$, Table II, the thresholds for reactions (11) corrected to 0 K lead to D(Fe-O) = 424 ± 15 kJ/mol and D(Co-O) = 409 ± 20 kJ/mol. The value for FeO has been critically reviewed by Smoes and Drowart [160] who recommend their own value of 401 ± 8 kJ/mol, within experimental error of our value. Our value for CoO agrees reasonably well with the value recommended by Pedley and Marshall, 380 ± 13 kJ/mol [158].

3.5. Miscellaneous metal-ligand systems

We have also determined a number of other metal ligand BDEs in our laboratories. Such thermochemistry has been determined in only a single reaction system and there are not enough values for other metals to determine the accuracy of the BDEs. They are listed here for completeness: $D(Sc^+\text{-}C_4H_4) = 422 \pm 6$ [139], $D(V^+\text{-}C_2) = 524 \pm 15$, $D(V^+\text{-}C_2H) = 491 \pm 8$ [3], $D(Co^+\text{-}C\ell) = 284 \pm 12$, and $D(Ni^+\text{-}C\ell) = 188 \pm 4$ kJ/mol [95]. Other miscellaneous thermochemistry includes that for a second oxo ligand: $D(OSc^+\text{-}O) = 166 \pm 18$, $D(OY^+\text{-}O) = 170 \pm 15$, $D(OLa^+\text{-}O) = 96 \pm 30$,

D(OSc-O) = 381 ± 32, D(OY-O) = 399 ± 21, and D(OLa-O) = 405 ± 32 kJ/mol.

4. Sequential metal-ligand bond energies

To better understand organometallic complexes encountered in solution-phase, it is important to characterize gas-phase complexes with a saturated (18–e^- or completed solvent shell) or near-saturated complement of ligands. In particular, it would be desirable to ascertain what types of ligand shells will adjust the electronic configuration of the metal center and thereby activate particular kinds of chemistry. The measurement of the sequential bond energies of metal ions with several ligands attached is now feasible because of technological developments in our laboratory [7] and others [164–167]. In all cases, the values reported here were determined recently enough that no reanalysis of the data are required.

Most of these studies involve ligands that bond primarily by dative (essentially, electrostatic) interactions, in part because of the relative ease with which they are prepared and characterized. In an electrostatic bonding mechanism, there are a number of phenomena that serve to reduce the BDEs as the number of ligands increases [168]. First, the charge becomes increasingly delocalized, resulting in decreased ion-dipole and ion-induced dipole interactions. Second, the ligands begin to crowd one another, resulting in increased ligand-ligand repulsion. Third, once the inner solvent shell is completely filled, additional ligands cannot bind directly to the metal ion. In addition, BDEs should decrease with increasing size of the metal ion because ligands cannot approach ions with larger radii as closely. Such monotonically decreasing BDEs have been observed for alkali [169] and coinage metal ions [38] bound to water molecules. In contrast to these expectations, transition metal ion-ligand BDEs are often found to vary nonmonotonically with increasing ligation. This observation suggests that BDEs can help identify when the metal has reorganized electronically in order to accommodate the ligand shell.

4.1. Metal ion-water systems

The binding of water molecules to transition metal ions is presently the most completely studied metal ion-ligand system in the gas phase. Two experimental [40, 41] and two theoretical [49, 170] studies have reported BDEs for $M^+(H_2O)_x$ (M = Ti − Cu, $x = 1$ and 2) and we have recently checked these results and extended them to cover $x = 3$ and 4 [28]. Our results for $M^+(H_2O)_x$ (M = Ti − Cu, $x = 1 - 4$), Table V, agree particularly well with the theoretical results, average deviations are less than 1 kJ/mol for $x = 1$ and 3 kJ/mol for $x = 2$ clusters, and help resolve the few discrepancies between the previous work [40, 41]. For instance, the experimental work

Table V. Sequential bond energies of $M^+(H_2O)_x$ (kJ/mol)[a]

M	M^+-H_2O	$(H_2O)M^+$-H_2O	$(H_2O)_2M^+$-H_2O	$(H_2O)_3M^+$-H_2O
Ti	154(6)	136(5)	67(7)	84(8)
V	147(5)	151(10)	68(5)	68(8)
Cr	129(9)	142(6)	50(5)	51(7)
Mn	119(6)	90(5)	108(6)	50(5)
Fe	128(5)	164(4)	76(4)	50(7)
Co	161(6)	162(7)	65(5)	58(6)
Ni	180(3)	168(8)	68(6)	52(6)
Cu	157(8)	170(7)	57(8)	54(4)

[a] Ref. 28. All values are at 0 K with uncertainties in parentheses.

found that $D(H_2ONi^+$-$OH_2)$ is slightly stronger than $D(Ni^+$-$OH_2)$, while theory predicted the opposite, in agreement with our results. Our BDEs for $Cu^+(H_2O)_x$ ($x = 3$ and 4) agree with values from equilibrium measurements of Castleman and coworkers [38] and the CID results of Michl and coworkers [164]. Good agreement is *not* obtained unless the vibrational energy of the ions is accounted for explicitly, as discussed in section 2.4.

Examination of the values in Table V show that several of the first row transition metal ions (V, Cr, Fe, Co, and Cu) have similar BDE patterns: the BDE for $x = 2$ is greater than that for $x = 1$; those for $x = 3$ and 4 are comparable to each other and much weaker than those for $x = 1$ and 2. For M = Ti and Ni, the $x = 1$ BDEs exceed the $x = 2$ BDEs and titanium exhibits an increase from $x = 3$ to 4. The pattern for the manganese BDEs is distinct from all others. The observation that the second water binding energy is greater than the first for most of the metal ions contrasts with expectations based on electrostatic interactions. As originally postulated by Marinelli and Squires [41] and confirmed by the calculations of Rosi and Bauschlicher [49], this increase is due to 4s-3d hybridization and 4s-4p polarization effects, electronic effects that are unavailable in alkali metals. In essence, the metal ion must be promoted to an electron configuration suitable for forming a strong dative metal-ligand bond (by removing metal electron density from the bond axis). This promotion energy can be "paid" by the first ligand such that no more electronic reorganization is necessary for binding the second ligand, and thus its bond energy is stronger. In the case of Mn^+, the $^7S(4s^13d^5)$ ground state of Mn^+ is sufficiently stable that hybridization and promotion cost too much energy, and thus the $Mn(H_2O)_2^+$ complex is forced into a bent configuration [49] to reduce repulsion between the ligands and the occupied 4s orbital. All other $M(H_2O)_2^+$ complexes have linear O-M-O bond angles [49].

In the Cu^+ system, theory has attributed the large decrease in metal ion-water BDEs from $x = 2$ to 3 to increased ligand-ligand repulsion and loss of 4s-3d hybridization [170]. The decreases observed for all other first row transition metal ions, except Mn^+, are likely to have similar origins. The observation that the $x = 3$ BDEs for most metal ions exceed that of Cu^+ may be because these other metals retain some 4s-3d hybridization because they have open 3d shells unlike $Cu^+(3d^{10})$. Not only is the BDE for $Mn^+(H_2O)_3$ greater than that for $Mn^+(H_2O)_2$, it is greater than any other $x = 3$ BDE, Table V. The only means by which we can rationalize this observation is to suggest that $Mn^+(H_2O)_3$ has changed spin to a quintet ground state. This removes electron density from the 4s orbital occupied in $Mn^+(H_2O)_x$ ($x = 1$ and 2), thereby reducing the metal-ligand repulsion and allowing 4s-3d hybridization. The energy necessary to promote to the quintet state is paid by the first two water ligands, as shown by the septet-quintet splitting energy of 18 kJ/mol in $Mn^+(H_2O)_2$ [49], compared to the splitting energy in the atomic ion of 113 kJ/mol [77].

Most $M^+(H_2O)_4$ BDEs are comparable or slightly lower than the $M^+(H_2O)_3$ BDEs, an effect attributable to increased ligand-ligand repulsion. The large decrease observed for Mn^+ is presumably due to the anomalously large $Mn^+(H_2O)_3$ BDE. The observation that the $Ti^+(H_2O)_4$ BDE is greater than the $Ti^+(H_2O)_3$ BDE and all other $x = 3$ BDEs suggests that a change in spin may have occurred for this cluster [28].

4.2. Metal ion-carbonyl systems

Table VI lists experimental sequential BDEs for $M(CO)_x^+$ where M = Cr [24], Mn [171], Fe [9], and Ni [172]. The accuracy of these values are confirmed by favorable comparison with the sum of the BDEs as calculated for the saturated species: $Cr(CO)_6^+$, $Mn(CO)_5^+$, $Fe(CO)_5^+$, and $Ni(CO)_4^+$. Work in progress in our laboratory also includes BDEs for $V(CO)_x^+$ and $Co(CO)_x^+$ [173, 174], and these are included for completeness. In all cases, the values were obtained in our laboratories by measuring the thresholds for the CID reactions (3). Values obtained from photoionization or electron impact appearance potentials appear to be flawed by kinetic shifts, as discussed in each of our reports. Values for the manganese system obtained by Dearden *et al.*

Table VI. Sequential bond energies of $M^+(CO)_x$ (kJ/mol)[a]

M	M^+-CO	$(OC)M^+$-CO	$(OC)_2M^+$-CO	$(OC)_3M^+$-CO	$(OC)_4M^+$-CO	$(OC)_5M^+$-CO
V[b]	113(3)	91(4)	69(4)	86(10)	91(3)	99(7)
Cr[c]	90(4)	95(3)	54(6)	51(8)	62(3)	130(8)
Mn[d]	25(10)	63(10)	74(10)	65(10)	121(10)	142(10)
Fe[e]	131(8)	151(14)	66(5)	103(7)	112(4)	
Co[f]	174(7)	152(9)	81(12)	75(6)	75(5)	
Ni[g]	175(11)	168(11)	92(6)	72(3)		

[a] Values are at 0 K with uncertainties in parentheses. [b] Ref. 173. [c] Ref. 24. [d] Ref. 171. [e] Ref. 9. [f] Ref. 174. [g] Ref. 172.

[175]. from analysis of kinetic energy release distributions are also unreliable for reasons discussed elsewhere [171].

4.2.1. *Nickel carbonyl ion bond energies*

In the nickel carbonyl system, the BDEs decrease monotonically as additional CO ligands are added to Ni^+. This is the trend that might be expected for purely electrostatic interactions, as noted above. One reason for this observation is that no changes in spin state are necessary as CO ligands are added to the atomic nickel ion. This is because Ni^+ has a $^2D(3d^9)$ ground state and $Ni(CO)_4^+$ is also expected to have a doublet ground state based on removing a single electron from the 18-e^- singlet $Ni(CO)_4$ complex.

4.2.2. *Iron carbonyl ion bond energies*

In contrast to the nickel system, extensive electronic changes are needed in the iron carbonyl ion system. Atomic Fe^+ has a $^6D(4s^13d^6)$ ground state and $Fe(CO)_5{}^+$ is expected to have a doublet ground state based on removing a single electron from the 18–e^- singlet $Fe(CO)_5$ species. Calculations indicate that the addition of a single carbonyl is sufficient to induce one change in spin, i.e. $FeCO^+$ has a $^4\Sigma^-$ ground state [176]. We find that dissociation to Fe^+ + CO occurs when 153 ± 8 kJ/mol is added at 0 K, in reasonable agreement with other work (see discussion in reference 9). Because the ground states of the $FeCO^+$ molecule and the Fe^+ + CO dissociation asymptote have different spins, there is a question whether the BDE we measure corresponds to the spin-forbidden adiabatic value (dissociation to $Fe^+(^6D)$) or the spin-allowed diabatic value. The situation is illustrated in Figure 6. (Note that *no* repulsive curves are indicated. This is the correct behavior for a system that dissociates heterolytically [177].) Our original assessment of this question suggested that our CID process corresponds to the spin-allowed dissociation pathway [9], and additional experiments appear to validate this conclusion [178]. Thus, the BDE reported in Table VI has been corrected for the splitting between the ^4F and ^6D states of Fe^+ [77], resulting in a Fe^+-CO bond energy of 131 ± 8 kJ/mol. This assignment also appears to be consistent with the observation that the diabatic BDE, 153 ± 8 kJ/mol, is nearly equivalent to $D[(OC)Fe^+\text{-}CO]$ = 151 ± 14 kJ/mol. This is reasonable because no spin change is required for the dissociation of $Fe(CO)_2^+$, which has a calculated ground state of $^4\Sigma_g^-$ [176], to ground state $FeCO^+$ + CO.

Examination of the BDEs for $Fe(CO)_3^+$, $Fe(CO)_4^+$ and $Fe(CO)_5^+$, Table VI, shows an interesting pattern. The latter two BDEs are stronger than the former, in contrast to expectations based on electrostatic binding. If we accept the proposition that $Fe(CO)_5{}^+$ has a doublet ground state, then the relatively strong bond observed for this species suggests that no spin change occurs upon its dissociation. Because $D[(OC)_3Fe^+\text{-}CO] \approx D[(OC)_4Fe^+\text{-}CO]$, this suggests that no spin change occurs upon dissociation of $Fe(CO)_4^+$ either, and thus that $Fe(CO)_3^+$, $Fe(CO)_4^+$ and $Fe(CO)_5^+$ all have doublet ground states. The weak BDE for $Fe(CO)_3^+$ can now be rationalized by referring to Figure

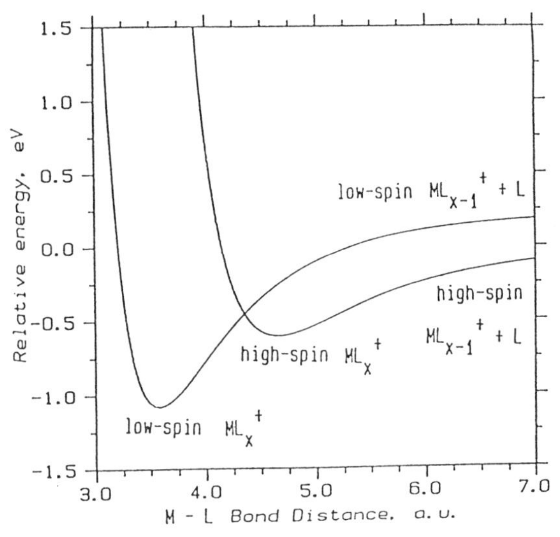

Fig. 6. Potential energy curves for dissociation of a metal-ligand complex with both a high-spin and low-spin electronic state. The curves are Morse potentials with molecular parameters appropriate for dissociation of $FeCO^+$ as calculated in reference 176.

6 again. If we imagine that the spin-allowed dissociation of $Fe(CO)_3^+$ to a doublet excited state of $Fe(CO)_2^+$ requires about 108 kJ/mol, then the measured BDE of 66 kJ/mol must correspond to the adiabatic BDE to form the quartet ground state of $Fe(CO)_2^+$. In essence, the first three CO ligands pay all the energy costs associated with reorganizing the electronic character at the Fe^+ center (the promotion energy), and the fourth and fifth CO ligands can bond with minimal interference. These ideas have yet to be verified by further experimental or theoretical work. [Note added in proof. For recent theoretical work, see A. Ricca and C.W. Bauschlicher, Jr. *J. Phys. Chem.* **98**, 12899 (1994).]

4.2.3. *Chromium carbonyl ion bond energies*

The chromium carbonyl ions also exhibit nonmonotonic changes in the BDEs with increasing ligation, Table VI. Indeed, $D[(OC)_5Cr^+\text{-}CO]$ is the strongest bond, a result incompatible with a system controlled by ligand-ligand repulsion and electrostatic bonding. The ground states of $Cr^+(^6S,3d^5)$, $CrCO^+(^6\Sigma^+)$, and $Cr(CO)_2^+(^6\Sigma_g^+)$ [176] all have the same spin, consistent with the similar BDEs for these two complexes. A spin change is easier to induce in the iron system than in the chromium system because the quartet state of Fe^+ is only 22 kJ/mol above the sextet ground state, while the lowest lying quartet state of Cr^+ is 233 kJ/mol above the sextet ground state [77].

We again anticipate that the saturated $Cr(CO)_6^+$ complex should have a doublet spin ground state based on removing a single electron from the 18–e^- singlet $Cr(CO)_6$ molecule. Therefore, the strong BDE for $Cr(CO)_6^+$ suggests that $Cr(CO)_5^+$ also has a doublet ground state. Identification of the spin states for $Cr(CO)_3^+$ and $Cr(CO)_4^+$ from the BDE patterns is more speculative and discussed elsewhere [24]. We conclude that either these species have sextet ground states (such that the BDE pattern follows that of the nickel carbonyl system) or that they have doublet and quartet spins, respectively. In either case, the dissociation of $Cr(CO)_5^+$ to form ground state $Cr(CO)_4^+$ is spin-forbidden, explaining its low BDE.

4.2.4. *Manganese carbonyl ion bond energies*

We have not yet analyzed the trends in the manganese carbonyl BDEs, but the variations do warrant some speculation. As carbonyl ligands are added to ground state $Mn^+(^7S,4s^13d^5)$, three spin changes must occur to eventually yield ground state $Mn(CO)_6^+$ ($^1A_{1g}$) [179, 180]. Theoretical results [176] indicate that $MnCO^+$ has a $^7\Sigma^+$ ground state and that $Mn(CO)_2^+$ has a $^5\Pi_g$ ground state, although a $^7A_1$ state is only 15 kJ/mol higher in energy. Because the lowest lying quintet state of Mn^+ is 113 kJ/mol above the ^7S state [77], an excitation energy that is intermediate compared with those for Fe^+ and Cr^+, it seems reasonable that two carbonyl ligands are required to induce a spin change for Mn^+ compared to one for iron and more than two for chromium. The strength of the $Mn(CO)_6^+$ and $Mn(CO)_5^+$ BDEs, Table VI, may suggest that $Mn(CO)_5^+$ and $Mn(CO)_4^+$ are both singlet ground states. The observation that $D[(OC)_2Mn^+\text{-}CO]$ is greater than $D[OCMn^+\text{-}CO]$, in contrast to all the other metal carbonyl systems, suggests that a spin change may occur here. This would mean that the tricarbonyl might have a triplet spin, such that dissociation of $Mn(CO)_4^+$ would be spin-forbidden, consistent with its relatively weak BDE.

4.3. METAL ION-HYDROCARBON SYSTEMS

4.3.1. *Metal-methane ion bond energies*

One means of systematically examining the influence of ligands on the electronic structure of the metal center is to change the strength of the metal-ligand interaction. Because gas-phase techniques are not limited to conventional ligands or 18-e^- complexes, nontraditional ligands can be included in such systematic studies. We have used methane, which has no lone-pair of electrons that is accessible for dative bonding, in this capacity. Table VII lists the binding of one to four molecules of methane to Fe^+ and Co^+ as

Table VII. Sequential bond energies of $M^+(CH_4)_x$ (kJ/mol)[a]

M^+	$M^+\text{-}CH_4$	$(CH_4)M^+\text{-}CH_4$	$(CH_4)_2M^+\text{-}CH_4$	$(CH_4)_3M^+\text{-}CH_4$
Fe^+[b]	57(3)	97(4)	99(6)	74(6)
Co^+[c]	90(6)	96(5)	40(5)	65(6)

[a] Values are at 0 K with uncertainties in parentheses. [b] Ref. 181. [c] Ref. 144.

determined by CID studies [57, 144, 181]. There are no indications that the metal ions have inserted into the CH bonds, and thus the methane ligands are believed to be intact. Our BDEs for the $Co^+(CH_4)$ and $Co^+(CH_4)_2$ complexes compare favorably with recent experimental results of Kemper *et al.* [167], 96 ± 3 and 104 ± 3 kJ/mol, and theoretical results of Perry *et al.* [182], 87 and 94 kJ/mol.

Comparison of these values to those for other sequential metal-ligand BDEs shows that the pattern of bond energies for cobalt is similar to that for $Co(H_2O)_x^+$, Table V. On average, the first three Co^+-CH_4 BDEs are 59 ± 4% of the Co^+-H_2O BDEs, suggesting that similar electronic considerations hold for both types of complexes. In contrast, the fourth methane is bound *more* strongly than the third methane, while the fourth water is bound less strongly than the third water. In the case of iron, the pattern of BDEs to methane vs those to water and CO are very different, to the point where $D[(CH_4)_2Fe^+\text{-}CH_4]$ exceeds $D[(H_2O)_2Fe^+\text{-}H_2O]$ and $D[(CO)_2Fe^+\text{-}CO]$, a rather unexpected result. These variations are likely to be due to electronic reorganization, as discussed above and elsewhere in the literature [181].

4.3.2. *Metal-ethane and -propane ion bond energies*

We have also used CID studies to measure the bond energies of Co^+ and Fe^+ ligated by the larger alkanes, ethane and propane [141, 144, 183, 185], Table VIII. Our values are in reasonable agreement with 0 K values for Co^+-C_2H_6 from Kemper *et al.* [167], 117 ± 7 kJ/mol, and theoretical values at 0 K from Perry *et al.* for the ethane and propane complexes with Co^+, 102 and 120 kJ/mol, respectively [182]. Again there are no indications that these BDEs refer to anything but simple adducts between the metal ion and an intact alkane molecule. For both the Fe^+ and Co^+ systems, the 0 K BDEs are found to increase systematically as the size of the alkane increases [141], consistent with an increase in the polarizability of the alkane, $\alpha = 2.6$, 4.4

Table VIII. Bond energies of M^+-hydrocarbon molecules (kJ/mol)[a]

M	M^+-C_2H_6	M^+-C_3H_8	M^+-C_2H_4	M^+-C_3H_6	M^+-C_2H_2
Sc			≥131[b]		240(20)[b]
Ti			≥119[c]		253(20)[c]
V			≥117[d]		205(20)[e]
Cr			≥125(19)[f]		184(20)[f]
Fe	64(6)[g]	75(4)[h,i]	145(6)[i]	145(7)[j]	
Co	100(5)[k]	129(6)[k]	179(7)[l]	180(7)[l]	>27(13)[m]
Ni			≥138(19)[m]		>7(18)[m]
Cu			≥95(11)[m]		>10(10)[m]
Y			≥109[n]		253(30)[n]
La			≥90[n]		262(30)[n]

[a] Values are at 0 K with uncertainties in parentheses. [b] Ref. 52. [c] Ref. 89. [d] Ref. 140. [e] Ref. 3. [f] Ref. 86. [g] Ref. 141. [h] Ref. 183. [i] Ref. 185. See text. [j] Ref. 120. See text. [k] Ref. 144. [l] Ref. 121. [m] Ref. 124. [n] Ref. 88.

and 6.2 Å^3 for CH_4, C_2H_6, and C_3H_8, respectively [184]. However, the BDEs are not directly proportional to the polarizability [181], suggesting that the metal-ligand bond length changes as the alkane changes.

4.3.3. *Metal-ethene and -propene ion bond energies*

Although metal-alkene ions are often observed as products in the reactions of atomic metal ions with alkanes, such reactions do not lead to precise thermochemistry. The most useful of these studies include formation of metal-ethene ions in the exothermic dehydrogenation of ethane and perdeutero ethane by Sc^+ and Ti^+ [52, 89], where it is assumed that $D(M^+\text{-}C_2H_4) \approx D(M^+\text{-}C_2D_4)$. Endothermic formation of metal-ethene ions is observed in reactions of ethane with V^+ [3, 140], ethylene oxide with Cr^+ [86], and cyclopropane with Co^+, Ni^+ and Cu^+ [124]. In the latter endothermic reactions, strong competition with more favorable channels may mean that the resulting BDEs are best viewed as lower limits. The results of these various studies are listed in Table VIII. Also many of the first row transition metal ions (Sc [139], Ti [89], V [140], Fe [90], Co and Ni [91]) are observed to dehydrogenate propane exothermically, indicating metal-propene ion BDEs greater than 117 kJ/mol.

More precise determinations of the metal-alkene BDEs have been accomplished by using the collision-induced dissociation (CID) reaction (3). These studies lead to BDEs for Fe^+ and Co^+ bound to ethene and propene [120, 121, 185]. The values for Co^+, Table VIII, compare favorably to those measured by kinetic energy release distribution (KERD) studies [42]: 176 and 184 ± 21 kJ/mol for ethene and propene, respectively. The BDEs we measure for Fe^+ bound to ethene and propene are 167 ± 6 and 167 ± 7 kJ/mol, respectively, somewhat higher than the KERD measurements of 146 and 156 ± 21 kJ/mol, although not outside of experimental error. Jacobson and Freiser have measured $D(Fe^+\text{-}C_2H_4) = 142 \pm 8$ kJ/mol [186]. The discrepancies in the iron-ethene ion BDEs can be reconciled if our BDEs correspond to diabatic dissociation to $Fe^+(^4F)$, similar to the behavior of $FeCO^+$ discussed above. To obtain the adiabatic BDE we reduce our measured values by 22 kJ/mol to yield 145 kJ/mol for both ethene and propene bound to Fe^+, as listed in Table VIII. Given this interpretation, we find that both the Co^+ and Fe^+-alkene BDEs are 10% stronger than the Co^+ and Fe^+-carbonyl BDEs, Table VI, and their relative values agree reasonably well with the calculated values of Sodupe *et al.* [187]. In general, it appears that the calculated values are about 33 kJ/mol lower than the experimental values, equivalent to the estimated correction to the theoretical numbers of 29–38 kJ/mol.

Comparison of the metal-alkene ion BDEs with the metal-alkane BDEs, Table VIII, shows that the former are substantially higher even though the polarizabilities of the alkenes are slightly lower than those of the comparable alkanes [184]. This highlights the enhancement in the bonding based on the ability of the alkenes to donate a pair of π electrons to the metal center.

4.3.4. *Metal-ethyne and -C_3H_4 ion bond energies*

We have reported several M^+-C_2H_2 BDEs in the past. In several cases, M = Sc [52], Y, La [88], and Ti [89], we observed production of this species in reaction with ethane. The energy dependence was unusual (distinct exothermic and endothermic cross section features were observed) and unusual isotope effects were observed upon deuteration. In light of recent work by Freiser and coworkers [188], it now appears that these unusual effects are anomalous: the exothermic reactivity observed in these systems is due to reaction with ethene, a contaminant in the ethane. The endothermic formation of M^+-C_2H_2 in these systems is now believed to be due to reaction with ethane. Estimates of the thresholds in these systems leads to the BDEs listed for these metals in Table VIII.

Other metal-ethyne ion BDEs come from reaction of V^+ with ethene to eliminate 2 H atoms [3], and from reaction of Cr^+ with cyclopropane to eliminate $CH_2 + H_2$ (although the identity of these products is equivocal) [86]. Neither of these values are particularly reliable. Limits for Co^+ and Ni^+-ethyne BDEs come from reactions with cyclopropane to eliminate methane, where it should be noted that the values cited originally are grossly in error due to arithmetic errors [124]. These reevaluated values are also listed in Table VIII. For the early metals, Sc – V, the M^+-C_2H_2 BDEs listed here are much lower than our previous values, but still well above recent theoretical calculations, by an average of 82 ± 18 kJ/mol. Part of this is the estimated error in the theoretical values, ~29 kJ/mol, but part could easily be due to inaccuracies in the experimental determinations.

In a few systems, we have also measured the bond energies between atomic metal ions and C_3H_4, likely to be propyne but possibly allene. The values obtained, $D(Sc^+\text{-}C_3H_4) = 256 \pm 20$ [139], $D(Ti^+\text{-}C_3H_4) = 251 \pm 14$ [89], $D(Cr^+\text{-}C_3H_4) = 140 \pm 7$ [92], and $D(Co^+\text{-}C_3H_4) \geqslant 78 \pm 9$ kJ/mol [121], are similar to those for these metal ions bound to ethyne.

4.4. METAL ION-RARE GAS SYSTEMS

In a number of CID reaction systems, we have also measured the bond energy between transition metal ions and rare gas atoms. These values include $D(V^+\text{-Ar}) = 20 \pm 20$ kJ/mol, $D(V^+\text{-Kr}) = 39 \pm 20$ kJ/mol, $D(V^+\text{-Xe}) = 81 \pm 17$ kJ/mol [137], $D(Cr^+\text{-Xe}) = 68 \pm 10$ kJ/mol [24], and $D(Fe^+\text{-Xe}) = 38 \pm 6$ kJ/mol [9, 181]. These compare reasonably well with experimental and theoretical values for other metal ion-rare gas systems [166, 189, 190].

5. Transition metal clusters

In addition to the systems discussed above, we have used CID studies to measure the BDEs for clusters of transition metal atoms. Our results now include values for Ti_x^+ ($x = 2 - 22$) [191], V_x^+ ($x = 2 - 20$) [192], Cr_x^+ ($x = 2 - 21$) [193, 194], Mn_2^+ [195, 196], Fe_x^+ ($x = 2 - 19$) [59, 197, 198], Co_x^+ ($x = 2 - 18$) [163, 199], Ni_x^+ ($x = 2 - 18$) [200, 201], Nb_x^+ ($x = 2–11$) [8], and preliminary results for Cu_2^+ [202], and Ta_x^+ ($x = 2–4$) [203]. In all cases but Mn_2^+, the ions were generated in a laser vaporization-supersonic expansion source detailed elsewhere [204].

In our earlier metal cluster CID studies (Ti, Fe, Ni, and Nb), equation (7) was used to analyze the thresholds, while in our more recent work (V, Cr, and Co), equation (8) has been used. The differences in these approaches have been assessed in reference 194. There we find that for $x < 6$, the thresholds increase by approximately the internal energy in the cluster, while for $x \geqslant 6$, the thresholds determined with equation (8), which are equivalent to 0 K BDEs, are about 20 kJ/mol larger than those determined with equation (7). This shift is nearly constant due to a near cancellation between two competing effects as the cluster size increases: increasing internal energies lower the thresholds, while increased kinetic shifts raise the thresholds. Our results for Ti, Fe, Ni and Nb have been adjusted by the internal energy of the cluster for $x < 6$ and by 20 kJ/mol for $x \geqslant 6$, such that all values listed in Table IX are believed to correspond to 0 K. This table summarizes all of our results except $D(Cu^+\text{-}Cu) = 177 \pm 10$ kJ/mol and $D(Mn^+\text{-}Mn) = 109 \pm 20$ kJ/mol [196].

The accuracy of these results can be verified by comparison to the few alternate determinations available, as discussed in more detail elsewhere [60]. In addition, we have recently measured BDEs for Ti_2^+, V_2^+, Co_2^+ and Co_3^+ by photodissociation, a method that provides much more precise values (uncertainties of 0.2 kJ/mol or less) [205, 206]. In all cases, the values are within experimental error of the CID results and are reported in Table IX. Our value for $D(Cu^+\text{-}Cu)$ agrees well with the best literature value, 178 ± 8 kJ/mol [207]. Our value for $D(Mn^+\text{-}Mn)$ is just below a lower limit of 134 kJ/mol determined by photodissociation [208], but this is attributed to internal excitation left by the electron impact source used to generate Mn_2^+ (a problem not encountered for any other cluster and discussed in detail elsewhere) [163].

For some of the transition metal clusters for which we have determined cluster ion BDEs, ionization energies (IEs) have also been measured by others: vanadium [209], iron [210–212], cobalt [210, 212], nickel [212, 213], and niobium [209, 214]. In these cases, equation (12) can be used to calculate neutral transition metal cluster BDEs, also listed in Table X.

$$D(M_{x-1}\text{-}M) = D(M_{x-1}{}^{+}\text{-}M) + IE(M_x) - IE(M_{x-1}) \tag{12}$$

For the neutral nickel clusters, these results can be compared to the theoretical results of Stave and DePristo for Ni_x ($x = 5 - 23$) [215]. Not only do the experiment and theory agree regarding the patterns in the BDEs as the cluster size varies, but the deviation in the absolute values (theoretical minus experimental values) is only 20 ± 29 kJ/mol. Note that without the correction

Table IX. Bond energies of M_{x-1}^{+}-M (kJ/mol)[a]

x	Ti[b,c]	V[d]	Cr[e]	Fe[c,f]	Co[g]	Ni[c,h]	Nb[c,i]	Ta[j]
2	234.9[k]	303.0[k]	125	268	266.8[l]	204	569	666
3	230	219	194	169	201.4[k]	235	490	644
4	338	341	100	216	205	199	590	744
5	340	313	215	260	274	223	557	
6	354	398	171	315	319	275	568	
7	400	372	246	320	283	296	650	
8	278	385	217	252	303	265	592	
9	346	354	249	281	283	283	578	
10	335	382	232	284	285	283	604	
11	344	382	248	308	314	293	600	
12	406	398	258	334	329	335		
13	470	449	289	408	351	352		
14	321	388	294	281	303	299		
15	400	452	267	377	380	334		
16	360	376	282	319	355	338		
17	344	404	273	323	346	354		
18	318	381	208	311	371	341		
19	450	408	233	373				
20	409	443	289					
21	395		253					
22	410							

[a] Values are at 0 K. Absolute uncertainties gradually increase from ~10 kJ/mol for small cluster ions to 50 kJ/mol for larger cluster ions. Relative uncertainties are less than 15 kJ/mol for all clusters. [b] Ref. 191. [c] Values have been adjusted to 0 K as discussed in the text. [d] Ref. 192. [e] Ref. 194. [f] Ref. 198. [g] Ref. 199. [h] Ref. 201. [i] Ref. 8. [j] Preliminary values from ref. 203. [k] Ref. 206. [l] Ref. 205.

to 0 K discussed above, the deviation would be double this. Both the cation and neutral transition metal cluster BDEs have been reviewed recently [60], and therefore will not be discussed further here.

Acknowledgements

The authors thank their colleagues N. Aristov, Y.- M. Chen, D. E. Clemmer, N. F. Dalleska, J. L. Elkind, E. R. Fisher, R. Georgiadis, C. L. Haynes, K. Honma, F. Khan, S. K. Loh, R. H. Schultz, and L. S. Sunderlin for their contributions to these studies. The National Science Foundation and Department of Energy, Office of Basic Energy Sciences have provided continuing support for this work.

Table X. Bond energies of M_{x-1}-M (kJ/mol)[a]

x	V[b]	Fe[c]	Co[c]	Ni[d]	Nb[e]
2	266[f]	114[g]	≤127[h]	200[i]	509
3	137	184[g]	≥140[j]	87	448
4	354	211[g]	233[k]	158	571
5	297	217[g]	274[k]	272	539
6	389	306	319[k]	334[k]	561
7	360	301	256	227	647
8	397	225	283	271	610
9	339	282	279	274	546, 519
10	379	276	294	275	631
11	365	311	301	283	528
12	395	340	329	334	
13	447	417	355	353	
14	391	290	301	298	
15	<426	363	371	333	
16	>404	326	369	340	
17	405	310	332	355	
18	372	302	371	339	
19	<388	352			

[a] Values are at 0 K, and except where noted, are calculated by using equation (12) with cluster ion bond energies from Table 9 and ionization energies (IEs) from the references given. Absolute uncertainties gradually increase from 10 kJ/mol for small cluster ions to 50 kJ/mol for larger clusters. Relative uncertainties are less than 15 kJ/mol for all clusters. [b] IEs from ref. 209. [c] IEs from ref. 210, except as noted. [d] IEs from ref. 213, except as noted. [e] IEs from ref. 214. For Nb_9, two isomers have been identified. [f] E. M. Spain and M. D. Morse, *Int. J. Mass Spectrom. Ion Processes* **102**, 183 (1990). [g] IEs from ref. 211. [h] $IE(Co_2) \leq 6.42$ eV, M. D. Morse and Z.-W. Fu, personal communication. [i] M. D. Morse, G. P. Hansen, P. R. R. Langridge-Smith, L.-S. Zheng, M. E. Geusic, D. L. Michalopoulos, and R. E. Smalley, *J. Chem. Phys.* **80**, 5400 (1984). [j] The sum of D(Co-Co) and D(Co_2-Co) equals D(Co^+-Co) + D(Co_2^+-Co) + IE(Co_3) − IE(Co) = 281 ± 14 kJ/mol. [k] IEs from ref. 212.

Appendix: Conversion of metal-ligand bond energies from 0 K to 298 K

The heats of formation of a molecule at some temperature T and 0 K are related by equation (A1) [216]. The bond dissociation energy (more properly,

$$\Delta_f H^\circ_T = \Delta_f H^\circ_0 + [H^\circ_T - H^\circ_0]_{molecule} - \Sigma[H^\circ_T - H^\circ_0]_{elements} \qquad (A1)$$

the bond dissociation enthalpy at nonzero temperatures) for AB at some temperature T is given by equation (A2).

$$D_T(AB) = \Delta_f H^\circ_T(A) + \Delta_f H^\circ_T(B) - \Delta_f H^\circ_T(AB) \qquad (A2)$$

Substituting equation (A1) into this expression yields equation (A3).

$$D_T(AB) = D_0(AB) + [H^\circ_T - H^\circ_0]_A + [H^\circ_T - H^\circ_0]_B - [H^\circ_T - H^\circ_0]_{AB} \qquad (A3)$$

Thus conversion of 0 K bond energies to bond enthalpies at temperature T requires the enthalpy changes for the molecule and its dissociation products. The enthalpy change for a species (in a high rotational temperature limit) is given by equation (A4),

$$[H_T^\circ - H_0^\circ] \approx (5/2)RT + (n_{rot}/2)RT + [H_T^\circ - H_0^\circ]_{vib} + [H_T^\circ - H_0^\circ]_{e\ell} \tag{A4}$$

where $n_{rot} = 0$ for an atom, 2 for a linear molecule and 3 for a nonlinear molecule. The (5/2)RT term in equation (A4) has contributions of 3RT/2 from translation and RT from $\Delta PV = \Delta nRT$ for one mole of an ideal gas. $[H_T^\circ - H_0^\circ]_{e\ell}$ is the contribution to the enthalpy change from electronic states. $[H_T^\circ - H_0^\circ]_{vib}$ is the contribution from vibrational states and equals RT Σ $u/(e^u - 1)$ where $u = h\nu_i/k_BT$ and the summation is over the vibrational frequencies of the molecule, ν_i. Substituting expression (A4) into equation (A3) yields equation (A5),

$$\Delta D_T = D_T(AB) - D_0(AB) = (5 + \Delta n_{rot})RT/2 + \Delta H_{vib} + \Delta H_{e\ell} \tag{A5}$$

where Δn_{rot} are the number of rotational degrees of freedom created upon dissociation, and ΔH_{vib} and $\Delta H_{e\ell}$ are the vibrational and electronic enthalpies of the A + B dissociation products minus those of the AB molecule. In the limit that the vibrational and electronic enthalpy terms are negligible, the difference between bond energies at T and 0 K is easy to calculate. For species discussed in this article, these differences are listed in Table XI for T = 298.15 K.

For most covalently bound species, the neglect of vibrational contributions is a reasonable approximation, but metal-ligand complexes and metal clusters have many low frequency vibrations with nonnegligible populations even at 298 K. In such cases, accurate values for ΔD_{298} require that the vibrational frequencies be estimated. This can be done by making use of recent ab initio calculations for metal-ligand complexes, and in other cases, the frequencies can be estimated from analogous species where calculations have been done or from stable organometallic species where vibrational frequencies have been measured (e.g. the metal carbonyls). For metal clusters, we have calculated the distribution of metal cluster vibrational frequencies by using a method outlined by Jarrold and Bower [217]. An illustration of the importance of these vibrational contributions is given by the metal-water ion complexes. Neglecting vibrational and electronic contributions, $\Delta D_{298}(M^+\text{-}H_2O)$ equals 6.2 kJ/mol and for larger clusters ($x = 2 - 4$), $\Delta D_{298}[(H_2O)_{x-1}M^+\text{-}H_2O] = 9.9$ kJ/mol. Estimates of the vibrational frequencies based on ab initio calculations for $Na^+(H_2O)_x$ [168] and $Cu^+(H_2O)_x$ [170] alter these values to 3.3, 0.3, 0.3 and −0.7 kJ/mol for $x = 1 - 4$, respectively [28]. Estimates of ΔH_{vib} for most of the species discussed in this article are also listed in Table XI for T = 298.15 K.

The electronic contributions are harder to estimate and calculate accu-

Table XI. Conversion between 298 and 0 K bond dissociation energies

system[a]	Δn_{rot}	ΔD_{298}[b] (kJ/mol)	metal-ligand systems for (x ≥ 1)[c]
L → A + A	−2	3.7	M^+-H (0), M-H (0), M^+-O (−0.7[d]), M-O (−0.5[d]), M^+-C (0.2)[e], M^+-N (0.2), M^+-Rg (1.7), M^+-Cℓ (0.2)[f], M^+-M (0.8)
N → A + L	−1	5.0	HM^+-H (0), M^+-C_2H_2 (2.9), M^+-C_2 (0), M_2-M (2.8), M_2^+-M (2.8)
L → A + L	0	6.2	M^+-OH (1.2), M^+-NH (1.2), M^+-CH (1.2), M^+-CO (4.4), M^+-C_2H (1.0)
N → A + N	0	6.2	M^+-CH_3 (1.0), M-CH_3 (1.0), M^+-alkyl (1.0), M^+-C_2H_3 (1.0), M^+-CH_2 (2.3), M^+-alkylidene (2.3), M^+-NH_2 (2.3), M^+-H_2O (2.9), M^+-alkane (2.9), M^+-alkene (2.9), M^+-propyne (2.9), $(CH_3)M^+$-H (1.1), M^+-$(CH_2)_3$ (1.9), M_{x+2}^+-M (4.1), M_{x+2}-M (4.1)
N → L + L	1	7.4	$(OC)_2M^+$-CO (3.8)
L → L + L	2	8.7	$(OC)M^+$-CO (8.2)
N → N + L	2	8.7	$(OC)_{x+2}M^+$-CO (2.6, 6.8, 4.4), HM^+-CH_3 (2.0)
N → N + N	3	9.9	$(CH_3)M^+$-CH_3 (6.3), $(CH_4)_xM^+$-CH_4 (10.3, 8.0, 10.4), $(H_2O)_xM^+$-H_2O (9.6, 9.6, 10.6)

[a] A = atom, L = linear molecule, N = nonlinear molecule. [b] Value calculated by using equation (A5) assuming no contributions from vibrational and electronic states, i.e. $\Delta D_T = (5 + \Delta n_{rot})RT/2$. [c] Geometries of some complexes may differ from those indicated here. Numbers in parentheses are average values of $-\Delta H_{vib}$ in equation (A5) for the indicated species as estimated from calculated vibrational frequencies or comparisons with analogous molecules. Multiple values are for incrementally increasing values of *x*. [d] These values are $-(\Delta H_{vib} + \Delta H_{e\ell})$ where the enthalpy functions for neutral metal oxides are taken from ref. 158. Enthalpy functions for ionic metal oxides are assumed to be equivalent to the neutral analogues. [e] $[H_{298}\text{-}H_0](C,g) = 5RT/2 + 0.34$ kJ/mol. [f] $[H_{298}\text{-}H_0](C\ell,g) = 5RT/2 + 0.1$ kJ/mol.

rately. These electronic contributions may be substantial because of the many low-lying electronic states available to transition metals. Such information is available for the diatomic neutral transition metal oxides [158]. However, until more detailed information is available, a uniform procedure is to assume that the electronic degrees of freedom of the metal containing molecule and its dissociation products are comparable and therefore that $\Delta H_{e\ell} \approx 0$.

References

1. P. B. Armentrout and J. L. Beauchamp, *J. Am. Chem. Soc.* **102**, 1736 (1980).
2. K. M. Ervin and P. B. Armentrout, *J. Chem. Phys.* **83**, 166 (1985).
3. N. Aristov and P. B. Armentrout, *J. Am. Chem. Soc.* **108**, 1806 (1986).
4. R. Georgiadis and P. B. Armentrout, *J. Am. Chem. Soc.* **108**, 2119 (1986).
5. K. M. Ervin and P. B. Armentrout, *J. Chem. Phys.* **84**, 6738 (1986).
6. P. B. Armentrout, in *Advances in Gas Phase Ion Chemistry*, Vol. 1, edited by N. G. Adams and L. M. Babcock (JAI, Greenwich, 1992), pp. 83–119.
7. R. H. Schultz and P. B. Armentrout, *Int. J. Mass Spectrom. Ion Processes* **107**, 29 (1991).

8. D. A. Hales, L. Lian, and P. B. Armentrout, *Int. J. Mass Spectrom. Ion Processes* **102**, 269 (1990).
9. R. H. Schultz, K. Crellin, and P. B. Armentrout, *J. Am. Chem. Soc.* **113**, 8590 (1991).
10. E. Teloy and D. Gerlich, *Chem. Phys.* **4**, 417 (1974).
11. N. R. Daly, *Rev. Sci. Instrum.* **31**, 264 (1959).
12. J. D. Burley, K. M. Ervin, and P. B. Armentrout, *Int. J. Mass Spectrom. Ion Processes* **80**, 153 (1987).
13. J. L. Elkind and P. B. Armentrout, *J. Phys. Chem.* **89**, 5626 (1985).
14. R. H. Garstang, *Mon. Not. R. Astron. Soc.* **124**, 321 (1962); (personal communication).
15. F. Strobel and D. P. Ridge, *J. Phys. Chem.* **93**, 3635 (1989).
16. P. R. Kemper and M. T. Bowers, *J. Am. Chem. Soc.* **112**, 3231 (1990); *J. Phys. Chem.* **95**, 5134 (1991).
17. J. V. B. Oriedo and D. H. Russell, *J. Phys. Chem.* **96**, 5314 (1992).
18. L. S. Sunderlin and P. B. Armentrout, *J. Phys. Chem.* **92**, 1209 (1988).
19. P. A. M. van Koppen, P. R. Kemper, and M. T. Bowers, *J. Am. Chem. Soc.* **114**, 10941 (1992).
20. S. D. Hanton, R. J. Noll, and J. C. Weisshaar, *J. Phys. Chem.* **94**, 5655 (1990).
21. S. D. Hanton, R. J. Noll, and J. C. Weisshaar, *J. Chem. Phys.* **96**, 5176 (1992).
22. J. L. Elkind and P. B. Armentrout, *J. Am. Chem. Soc.* **108**, 2765 (1986); *J. Phys. Chem.* **90**, 5736 (1986).
23. R. H. Schultz and P. B. Armentrout, *J. Chem. Phys.* **96**, 1046 (1992).
24. F. A. Khan, D. C. Clemmer, R. H. Schultz, and P. B. Armentrout, *J. Phys. Chem.* **97**, 7978 (1993).
25. E. R. Fisher, B. L. Kickel, and P. B. Armentrout, *J. Phys. Chem.* **97**, 10204 (1993).
26. E. R. Fisher, B. L. Kickel, and P. B. Armentrout, *J. Chem. Phys.* **97**, 4859 (1992).
27. N. F. Dalleska, K. Honma, and P. B. Armentrout, *J. Am. Chem. Soc.* **115**, 12125 (1993).
28. N. F. Dalleska, K. Honma, L. S. Sunderlin, and P. B. Armentrout, *J. Am. Chem. Soc.* **116**, 3519 (1994).
29. P. B. Armentrout, *ACS Symp. Ser.* **428**, 18 (1990).
30. V. L. Talrose, P. S. Vinogradov, and I. K. Larin, in *Gas Phase Ion Chemistry*, Vol. 1, edited by M. T. Bowers (Academic, New York, 1979), p. 305.
31. K. M. Ervin and P. B. Armentrout, *J. Chem. Phys.* **86**, 2659 (1987).
32. M. E. Weber, J. L. Elkind, and P. B. Armentrout, *J. Chem. Phys.* **84**, 1521 (1986).
33. J. L. Elkind and P. B. Armentrout, *J. Phys. Chem.* **88**, 5454 (1984).
34. B. H. Boo and P. B. Armentrout, *J. Am. Chem. Soc.* **109**, 3549 (1987).
35. P. B. Armentrout, L. F. Halle, and J. L. Beauchamp, *J. Chem. Phys.* **76**, 2449 (1982).
36. P. A. M. van Koppen, J. Brodbelt-Lustig, M. T. Bowers, D. V. Dearden, J. L. Beauchamp, E. R. Fisher, and P. B. Armentrout, *J. Am. Chem. Soc.* **112**, 5663 (1990); P. A. M. van Koppen, J. Brodbelt-Lustig, M. T. Bowers, D. V. Dearden, J. L. Beauchamp, E. R. Fisher, and P. B. Armentrout, *J. Am. Chem. Soc.* **113**, 2359 (1991).
37. C. L. Haynes, Y.-M. Chen, and P. B. Armentrout *J. Phys. Chem.* **91**, 9110 (1995).
38. P. M. Holland and A. W. Castleman, Jr., *J. Am. Chem. Soc.* **102**, 6174 (1980); *J. Chem. Phys.* **76**, 4195 (1982); K. I. Peterson, P. M. Holland, R. G. Keesee, N. Lee, T. D. Mark, and A. W. Castleman, Jr., *Surf. Sci.* **106**, 136 (1981).
39. S. W. Buckner and B. S. Freiser, *Polyhedron* **7**, 1583 (1988).
40. T. F. Magnera, D. E. David, and J. Michl, *J. Am. Chem. Soc.* **111**, 4100 (1989).
41. P. J. Marinelli and R. R. Squires, *J. Am. Chem. Soc.* **111**, 4101 (1989).
42. P. A. M. van Koppen, M. T. Bowers, J. L. Beauchamp, and D. V. Dearden, *ACS Symp. Ser.* **428**, 34 (1990).
43. P. B. Armentrout and L. S. Sunderlin, in *Transition Metal Hydrides*, edited by A. Dedieu (VCH, New York, 1992), pp. 1–64.
44. J. B. Schilling, W. A. Goddard, and J. L. Beauchamp, *J. Am. Chem. Soc.* **108**, 582 (1986); **109**, 5565 (1986); *J. Phys. Chem.* **91**, 5616 (1987).
45. A. K. Rappe and T. H. Upton, *J. Chem. Phys.* **85**, 4400 (1986).

46. L. G. M. Pettersson, C. W. Bauschlicher, S. R. Langhoff, and H. Partridge, *J. Chem. Phys.* **87**, 481 (1987).
47. C. W. Bauschlicher, S. R. Langhoff, H. Partridge, and L. A. Barnes, *J. Chem. Phys.* **91**, 2399 (1989).
48. J. B. Schilling, W. A. Goddard, and J. L. Beauchamp, *J. Am. Chem. Soc.* **109**, 5573 (1987).
49. M. Rosi and C. W. Bauschlicher, *J. Chem. Phys.* **90**, 7264 (1989); **92**, 1876 (1990).
50. See discussion in reference 3.
51. W. J. Chesnavich and M. T. Bowers, *J. Phys. Chem.* **83**, 900 (1979).
52. L. Sunderlin, N. Aristov, and P. B. Armentrout, *J. Am. Chem. Soc.* **109**, 78 (1987).
53. P. B. Armentrout and J. L. Beauchamp, *J. Chem. Phys.* **74**, 2819 (1981).
54. P. B. Armentrout and J. L. Beauchamp, *J. Am. Chem. Soc.* **103**, 784 (1981).
55. P. J. Chantry, *J. Chem. Phys.* **55**, 2746 (1971).
56. C. Lifshitz, R. L. C. Wu, T. O. Tiernan, and D. T. Terwilliger, *J. Chem. Phys.* **68**, 247 (1978).
57. P. B. Armentrout and D. E. Clemmer, in *Energetics of Organometallic Species*, edited by J. A. M. Simoes (Kluwer, Dordrecht, 1992), pp. 321–356.
58. P. J. Robinson and K. A. Holbrook, *Unimolecular Reactions*, (Wiley, London, 1972).
59. S. K. Loh, D. A. Hales, Li Lian, and P. B. Armentrout, *J. Chem. Phys.* **90**, 5466 (1989).
60. P. B. Armentrout, D. A. Hales, and L. Lian, *Advances in Metal and Semiconductor Clusters*, Vol. II, edited by M. A. Duncan (JAI, Greenwich, 1994), pp. 1–39.
61. K. Honma, L. S. Sunderlin, and P. B. Armentrout, *Int. J. Mass Spectrom. Ion Processes* **117**, 237 (1992).
62. K. Honma, L. S. Sunderlin, and P. B. Armentrout, *J. Chem. Phys.* **99**, 1623 (1993).
63. J. R. Taylor, *An Introduction to Error Analysis* (University Science, Mill Valley, CA, 1982).
64. P. B. Armentrout and R. Georgiadis, *Polyhedron* **7**, 1573 (1988).
65. P. B. Armentrout, in *Selective Hydrocarbon Activation: Principles and Progress*, edited by J. A. Davies, P. L. Watson, J. F. Liebman, and A. Greenberg (VCH, New York, 1990), p. 467.
66. J. L. Elkind and P. B. Armentrout, *Inorg. Chem.* **25**, 1078 (1986).
67. P. B. Armentrout, L. S. Sunderlin, and E. R. Fisher, *Inorg. Chem.* **28**, 4436 (1989).
68. E. R. Fisher, J. L. Elkind, D. E. Clemmer, R. Georgiadis, S. K. Loh, N. Aristov, L. S. Sunderlin, and P. B. Armentrout, *J. Chem. Phys.* **93**, 2676 (1990).
69. R. Georgiadis and P. B. Armentrout, *J. Phys. Chem.* **92**, 7060 (1988).
70. J. L. Elkind, L. S. Sunderlin, and P. B. Armentrout, *J. Phys. Chem.* **93**, 3151 (1989).
71. J. L. Elkind and P. B. Armentrout, *Int. J. Mass Spectrom. Ion Processes* **83**, 259 (1988).
72. J. L. Elkind and P. B. Armentrout, *J. Chem. Phys.* **86**, 1868 (1987).
73. J. L. Elkind and P. B. Armentrout, *J. Chem. Phys.* **84**, 4862 (1986).
74. J. L. Elkind and P. B. Armentrout, *J. Phys. Chem.* **90**, 6576 (1986).
75. P. B. Armentrout, L. F. Halle, and J. L. Beauchamp, *J. Am. Chem. Soc.* **103**, 6501 (1981).
76. M. L. Mandich, L. F. Halle, and J. L. Beauchamp, *J. Am. Chem. Soc.* **106**, 4403 (1984).
77. C. Corliss and J. Sugar, *J. Chem. Phys. Ref. Data* **11**, 1 (1982).
78. C. E. Moore, *Natl. Stand. Ref. Data Ser., Natl. Bur. Stand.* **35**, Vols. I–III (1971).
79. E. A. Carter and W. A. Goddard, *J. Phys. Chem.* **92**, 5679 (1988). Table III of this reference lists values for E(lost), equivalent to E_p in this paper.
80. A. E. Alvarado-Swaisgood, J. Allison, and J. F. Harrison, *J. Phys. Chem.* **89**, 2517 (1985); A. E. Alvarado-Swaisgood and J. F. Harrison, *J. Phys. Chem.* **89**, 5198 (1985); *Ibid.* **92**, 2757 (1988).
81. M. A. Vincent, Y. Yoshioka, and H. F. Schaefer, *J. Phys. Chem.* **86**, 3905 (1982).
82. F. Ungvary, *J. Organomet. Chem.* **36**, 363 (1972).
83. B. de Vries, *J. Catal.* **1**, 489 (1962).
84. L. Vaska, *Acc. Chem. Res.* **1**, 335 (1968).

85. J. Halpern, *Acc. Chem. Res.* **15**, 238 (1982); *Inorgan. Chim. Acta* **100**, 41 (1985).
86. R. Georgiadis and P. B. Armentrout, *Int. J. Mass Spectrom. Ion Processes* **89**, 227 (1989).
87. R. Georgiadis and P. B. Armentrout, *Int. J. Mass Spectrom. Ion Processes* **91**, 123 (1989).
88. L. S. Sunderlin and P. B. Armentrout, *J. Am. Chem. Soc.* **111**, 3845 (1989).
89. L. S. Sunderlin and P. B. Armentrout, *Int. J. Mass Spectrom. Ion Processes* **94**, 149 (1989).
90. R. H. Schultz, J. L. Elkind, and P. B. Armentrout, *J. Am. Chem. Soc.* **110**, 411 (1988).
91. R. Georgiadis, E. R. Fisher, and P. B. Armentrout, *J. Am. Chem. Soc.* **111**, 4251 (1989).
92. E. R. Fisher and P. B. Armentrout, *J. Am. Chem. Soc.* **114**, 2039 (1992).
93. L. S. Sunderlin and P. B. Armentrout, *J. Phys. Chem.* **94**, 3589 (1990).
94. E. R. Fisher, R. H. Schultz, and P. B. Armentrout, *J. Phys. Chem.* **93**, 7382 (1989).
95. E. R. Fisher, L. S. Sunderlin, and P. B. Armentrout, *J. Phys. Chem.* **93**, 7375 (1989).
96. Y.-M. Chen, D. E. Clemmer, and P. B. Armentrout, *J. Am. Chem. Soc.* **116**, 7815 (1994).
97. P. A. M. van Koppen, P. R. Kemper, and M. T. Bowers, *J. Am. Chem. Soc.* **115**, 5616 (1993).
98. J.-Y. Saillard and R. Hoffmann, *J. Am. Chem. Soc.* **106**, 2006 (1984).
99. T. Ziegler, V. Tschinke, and A. Becke, *J. Am. Chem. Soc.* **109**, 1351 (1987).
100. M. Brookhart and M. L. H. Green, *J. Organomet. Chem.* **250**, 395 (1983).
101. M. J. Calhorda and J. A. Simoes, *Organometallics* **6**, 1188 (1987).
102. N. Aristov, Thesis, University of California, Berkeley (1986).
103. D. E. Clemmer, L. S. Sunderlin, and P. B. Armentrout, *J. Phys. Chem.* **94**, 3008 (1990).
104. D. E. Clemmer, L. S. Sunderlin, and P. B. Armentrout, *J. Phys. Chem.* **94**, 208 (1990).
105. D. E. Clemmer and P. B. Armentrout, *J. Phys. Chem.* **95**, 3084 (1991).
106. Y.-M. Chen and P. B. Armentrout (work in progress).
107. D. E. Clemmer and P. B. Armentrout (work in progress).
108. D. E. Clemmer, N. Aristov, and P. B. Armentrout, *J. Phys. Chem.* **97**, 544 (1993).
109. Y.-M. Chen, D. E. Clemmer, and P. B. Armentrout (work in progress).
110. S. W. Buckner and B. S. Freiser, *J. Am. Chem. Soc.* **109**, 4715 (1987).
111. J. L. Tilson and J. F. Harrison, *J. Phys. Chem.* **95**, 5097 (1991).
112. E. Murad, *J. Chem. Phys.* **73**, 1381 (1980).
113. C. J. Cassady and B. S. Freiser, *J. Am. Chem. Soc.* **106**, 6176 (1984).
114. S. G. Lias, J. E. Bartmess, J. F. Liebman, J. L. Holmes, R. D. Levin, and W. G. Mallard, *J. Phys. Chem. Ref. Data* **17**, Supp. 1 (GIANT tables) (1988).
115. A. Mavridis, K. Kunze, J. F. Harrison, and J. Allison, *ACS Symp. Ser.* **428**, 263 (1990).
116. A. Mavridis, F. L. Herrera, and J. F. Harrison, *J. Phys. Chem.* **95**, 6854 (1991).
117. S. Kapellos, A. Mavridis, and J. F. Harrison, *J. Phys. Chem.* **95**, 6860 (1991).
118. C. W. Bauschlicher, S. R. Langhoff, and H. Partridge, *J. Phys. Chem.* **94**, 2068 (1991).
119. H. J. Hoffman, P. Hobza, R. Cammi, J. Tomasi, and R. Zahradnik, *J. Mol. Struct.* (*THEOCHEM*) **201**, 339 (1989).
120. R. H. Schultz and P. B. Armentrout, *Organometallics* **11**, 828 (1992).
121. C. L. Haynes and P. B. Armentrout, *Organometallics* **13**, 3480 (1994).
122. N. Aristov and P. B. Armentrout, *J. Phys. Chem.* **91**, 6178 (1987).
123. R. Georgiadis and P. B. Armentrout, *J. Phys. Chem.* **92**, 7067 (1988).
124. E. R. Fisher and P. B. Armentrout, *J. Phys. Chem.* **94**, 1674 (1990).
125. C. W. Bauschlicher, H. Partridge, J. A. Sheehy, S. R. Langhoff, and M. Rosi, *J. Phys. Chem.* **96**, 6969 (1992).
126. R. L. Hettich and B. S. Freiser, *J. Am. Chem. Soc.* **108**, 2537 (1986). The value for $D(CoCH_2^+)$ corresponds to the 390 nm threshold observed by these authors. This value was discounted on the basis of the presumed exothermic reaction of Co^+ with ethylene oxide to form $CoCH_2^+$, a reaction that is actually slightly endothermic [124].
127. B. L. Kickel and P. B. Armentrout *J. Am. Chem. Soc.* **117**, 4057 (1995).
128. N. Aristov and P. B. Armentrout, *J. Am. Chem. Soc.* **106**, 4065 (1984).
129. A. E. Stevens, Ph.D. Thesis, Caltech, Pasadena, CA (1981).
130. J. A. Pople, K. Raghavachari, M. J. Frisch, J. S. Binkly, and P. v. R. Schleyer, *J. Am. Chem. Soc.* **105**, 6389 (1983).

131. G. Trinquier, *J. Am. Chem. Soc.* **112**, 2130 (1990).
132. S. K. Loh, E. R. Fisher, L. Lian, R. H. Schultz, and P. B. Armentrout, *J. Phys. Chem.* **93**, 3159 (1989).
133. D. E. Clemmer, N. F. Dalleska, and P. B. Armentrout, *J. Chem. Phys.* **95**, 7263 (1991).
134. H. Kang and J. L. Beauchamp, *J. Am. Chem. Soc.* **108**, 5663 (1986).
135. D. E. Clemmer, J. L. Elkind, N. Aristov, and P. B. Armentrout, *J. Chem. Phys.* **95**, 3387 (1991).
136. Y.-M. Chen, D. E. Clemmer, and P. B. Armentrout, *J. Phys. Chem.* **98**, 11490 (1994).
137. N. Aristov and P. B. Armentrout, *J. Phys. Chem.* **90**, 5135 (1986).
138. S. W. Buckner, J. R. Gord, and B. S. Freiser, *J. Am. Chem. Soc.* **110**, 6606 (1988).
139. L. S. Sunderlin and P. B. Armentrout, *Organometallics* **9**, 1248 (1990).
140. M. R. Sievers and P. B. Armentrout (work in progress).
141. R. H. Schultz and P. B. Armentrout, *J. Phys. Chem.* **96**, 1662 (1992).
142. L. F. Halle, W. E. Crowe, P. B. Armentrout, and J. L. Beauchamp, *Organometallics* **3**, 1694 (1984).
143. P. A. M. van Koppen (personal communication).
144. C. L. Haynes, E. R. Fisher, and P. B. Armentrout (work in progress); C. L. Haynes, P. B. Armentrout, J. K. Perry, and W. A. Goddard, III, *J. Phys. Chem.* **99**, 6340 (1995).
145. P. R. Kemper and M. T. Bowers (personal communication); P. A. M. van Koppen (elsewhere in this volume).
146. P. A. M. van Koppen, D. B. Jacobson, A. Illies, M. T. Bowers, M. Hanratty, and J. L. Beauchamp, J. Am. Chem. Soc. **111**, 1991 (1989).
147. This estimate assumes that D(H-$CH_2CH_2CH_2$) = D(H-1–C_3H_7).
148. Y.-M. Chen, D. E. Clemmer, and P. B. Armentrout, *J. Chem. Phys.* **95**, 1228 (1991).
149. Y.-M. Chen, D. E. Clemmer and P. B. Armentrout, *J. Chem. Phys.* **98**, 4929 (1993).
150. R. H. Schultz and P. B. Armentrout, *J. Chem. Phys.* **94**, 2262 (1991).
151. A. Kant and K. A. Moon, *High. Temp. Sci.* **14**, 23 (1981); **11**, 52 (1979).
152. D. P. Chong, S. R. Langhoff, C. W. Bauschlicher, Jr., S. P. Walch, and H. Partridge, *J. Chem. Phys.* **85**, 2850 (1986).
153. C. W. Bauschlicher, Jr. and S. R. Langhoff, in *Transition Metal Hydrides*, edited by A. Dedieu (VCH, New York, 1992) pp. 103–126.
154. S. P. Walch and C. W. Bauschlicher, *J. Chem. Phys.* **78**, 4597 (1983).
155. H. A. Skinner and J. A. Connor, in *Molecular Structure and Energetics: Physical Measurements*, Vol. 2, edited by J. F. Liebman and A. Greenberg (VCH, New York, 1987) p. 233.
156. C. W. Bauschlicher, Jr. and S. P. Walch, *J. Chem. Phys.* **76**, 4560 (1982).
157. R. R. Squires, *J. Am. Chem. Soc.* **107**, 4385 (1985).
158. J. B. Pedley and E. M. Marshall, *J. Phys. Chem. Ref. Data* **12**, 967 (1983).
159. G. Balducci, G. Gigli, and M. Guido, *J. Chem. Soc. Faraday Trans.* 2 **77**, 1107 (1981).
160. S. Smoes and J. Drowart, *High Temp. Sci.* **17**, 31 (1984).
161. L. R. Watson, T. L. Thiem, R. A. Dressler, R. H. Salter, and E. Murad, *J. Phys. Chem.* **97**, 5577 (1993).
162. S. K. Loh, L. Lian, and P. B. Armentrout, *J. Chem. Phys.* **91**, 6148 (1989).
163. D. A. Hales and P. B. Armentrout, *J. Cluster Sci.* **1**, 127 (1990).
164. T. F. Magnera, D. E. David, D. Stulik, R. G. Orth, H. T. Jonkman, and J. Michl, *J. Am. Chem. Soc.* **111**, 5036 (1989).
165. L. S. Sunderlin and R. R. Squires, in *Energetics of Organometallic Species*, edited by J. A. M. Simoes (Kluwer, Dordrecht, 1992) pp. 269–286.
166. P. R. Kemper, M.-T. Hsu, and M. T. Bowers, *J. Phys. Chem.* **95**, 10600 (1991).
167. P. R. Kemper, J. Bushnell, P. van Koppen, and M. T. Bowers, *J. Phys. Chem.* **97**, 1810 (1993).
168. C. W. Bauschlicher, S. R. Langhoff, H. Partridge, J. E. Rice, and A. Komornicki, *J. Chem. Phys.* **95**, 5142 (1991).

Can. J. Chem. **47**, 2619 (1969).
170. C. W. Bauschlicher, Jr., S. R. Langhoff, and H. Partridge, *J. Chem. Phys.* **94**, 2068 (1991).
171. F. A. Khan and P. B. Armentrout (work in progress).
172. F. A. Khan, D. A. Steele, and P. B. Armentrout, *J. Phys. Chem.* **99**, 7819 (1995).
173. M. Sievers and P. B. Armentrout, *J. Phys. Chem.* **99**, 8135 (1995).
174. S. Goebel, C. L. Haynes, F. A. Khan, and P. B. Armentrout, *J. Am. Chem. Soc.* **117**, 6994 (1995).
175. D. V. Dearden, K. Hayashibara, J. L. Beauchamp, N. J. Kirchner, P. A. M. van Koppen, and M. T. Bowers, *J. Am. Chem. Soc.* **111**, 2401 (1989).
176. L. A. Barnes, M. Rosi, and C. W. Bauschlicher, *J. Chem. Phys.* **93**, 609 (1990).
177. P. B. Armentrout and J. Simons, *J. Am. Chem. Soc.* **114**, 8627 (1992).
178. N. F. Dalleska and P. B. Armentrout (work in progress).
179. N. A. Beach and H. B. Gray, *J. Am. Chem. Soc.* **90**, 5713 (1968).
180. J. K. Burdett, *J. Chem. Soc. Faraday Trans.* 2 **70**, 1599 (1974).
181. R. H. Schultz and P. B. Armentrout, *J. Phys. Chem.* **97**, 596 (1993).
182. J. K. Perry, G. Ohanessian, and W. A. Goddard, *J. Phys. Chem.* **97**, 5238 (1993). The values given here are those without the $+8 \pm 8$ kJ/mol estimated "correction" suggested by the authors.
183. R. H. Schultz and P. B. Armentrout, *J. Am. Chem. Soc.* **113**, 729 (1991).
184. E. W. Rothe and R. B. Bernstein, *J. Chem. Phys.* **31**, 1619 (1959).
185. R. H. Schultz and P. B. Armentrout (manuscript in preparation).
186. D. B. Jacobson and B. S. Freiser, *J. Am. Chem. Soc.* **105**, 7492 (1983).
187. M. Sodupe, C. W. Bauschlicher, S. R. Langhoff, and H. Partridge, *J. Phys. Chem.* **96**, 2118, 5670 (1992).
188. Y. A. Ranasinghe, T. J. MacMahon, and B. S. Freiser, *J. Am. Chem. Soc.* **114**, 9112 (1992).
189. D. Lessen and P. J. Brucat, *J. Chem. Phys.* **91**, 4522 (1989).
190. C. W. Bauschlicher and S. R. Langhoff, *Int. Rev. Phys. Chem.* **9**, 149 (1990).
191. L. Lian, C.-X. Su, and P. B. Armentrout, *J. Chem. Phys.* **97**, 4084 (1992).
192. C.-X. Su, D. A. Hales, and P. B. Armentrout, *J. Chem. Phys.* **99**, 6613 (1993).
193. C.-X. Su, D. A. Hales, and P. B. Armentrout, *Chem. Phys. Lett.* **201**, 199 (1993).
194. C.-X. Su and P. B. Armentrout, *J. Chem. Phys.* **99**, 6506 (1993).
195. K. Ervin, S. K. Loh, N. Aristov, and P. B. Armentrout, *J. Phys. Chem.* **87**, 3593 (1983).
196. P. B. Armentrout, in *Laser Applications in Chemistry and Biophysics*, edited by M. El-Sayed, Proc. SPIE **620**, 38 (1986).
197. S. K. Loh, L. Lian, D. A. Hales, and P. B. Armentrout, *J. Phys. Chem.* **92**, 4009 (1988).
198. L. Lian, C.-X. Su, and P. B. Armentrout, *J. Chem. Phys.* **97**, 4072 (1992).
199. D. A. Hales, C.-X. Su, L. Lian, and P. B. Armentrout, *J. Chem. Phys.* **100**, 1049 (1994).
200. L. Lian, C-X Su, and P. B. Armentrout, *Chem. Phys. Lett.* **180**, 168 (1991).
201. L. Lian, C.-X. Su, and P. B. Armentrout, *J. Chem. Phys.* **96**, 7542 (1992).
202. L. Lian, R. H. Schultz, and P. B. Armentrout (work in progress).
203. D. A. Hales, Thesis, University of California, Berkeley, 1990.
204. S. K. Loh, D. A. Hales, and P. B. Armentrout, *Chem. Phys. Lett.* **129**, 527 (1986).
205. L. M. Russon, S. A. Heidecke, M. K. Birke, J. Conceicao, P. B. Armentrout, and M. D. Morse, *Chem. Phys. Lett.* **204**, 235 (1993).
206. L. M. Russon, S. A. Heidecke, M. K. Birke, J. Conceicao, M. D. Morse, and P. B. Armentrout, *J. Chem. Phys.* **100**, 4747 (1994).
207. A. D. Sappey, J. E. Harrington, and J. C. Weisshaar, *J. Chem. Phys.* **91**, 3854 (1989).
208. M. F. Jarrold, A. J. Illies, and M. T. Bowers, *J. Am. Chem. Soc.* **107**, 7339 (1985).
209. D. M. Cox, R. L. Whetten, M. R. Zakin, D. J. Trevor, K. C. Reichmann, and A. Kaldor, in *AIP Conference Proceedings* 146, *Optical Science and Engineering Ser.* 6, *Advances in Laser Science*, Vol. I, edited by W. C. Stwalley and M. Lapp (Eds.), AIP, New York, 1986, p. 527.
210. S. Yang and M. B. Knickelbein, *J. Chem. Phys.* **93**, 1533 (1990).

211. E. A. Rohlfing, D. M. Cox, A. Kaldor, and K. H. Johnson, *J. Chem. Phys.* **81**, 3846 (1984).
212. E. K. Parks, T. D. Klots, and S. J. Riley, *J. Chem. Phys.* **92**, 3813 (1990).
213. M. B. Knickelbein, S. Yang, and S. J. Riley, *J. Chem. Phys.* **93**, 94 (1990).
214. M. B. Knickelbein and S. Yang, *J. Chem. Phys.* **93**, 5760 (1990).
215. M. S. Stave and A. E. DePristo, *J. Chem. Phys.* **97**, 3386 (1992).
216. M. W. Chase, C. A. Davies, J. R. Downey, D. J. Frurip, R. A. McDonald, and A. N. Syverud, *J. Phys. Chem. Ref. Data* **14**, Suppl. No. 1. (1985).
217. M. F. Jarrold and J. E. Bower, *J. Chem. Phys.* **87**, 5728 (1987).

2. *Ab initio* calculations applied to problems in metal ion chemistry

CHARLES W. BAUSCHLICHER, JR., STEPHEN R. LANGHOFF and HARRY PARTRIDGE
NASA Ames Research Center, Moffett Field, CA 94035, U.S.A.

1. Introduction

The study of the interaction of metal ions (M) with a large variety of ligands (L) is an area of active experimental research [1]. Metal ion systems are often more amenable to experimental study than the corresponding neutral systems, due to the ease of detection of metal ions and the fact that endothermic reactions can be driven to threshold by accelerating the ions. A wide variety of experimental methods such as the guided-ion beam method [2], collision induced dissociation [3], and photodissociation techniques [4] are capable of giving accurate metal ion ligand binding energies. The electrostatic interactions result in association reactions with no barrier. While there may be a barrier to go from the initial complex to the final products, for example when the metal ion has inserted into a bond, it is common that these barriers are below the $M^+ + L$ asymptote. Thus the reactivity is thermodynamically instead of kinetically controlled, so that considerable information is obtained directly from the metal ion ligand binding energies.

We have recently reviewed [5] the available theoretical metal ion-ligand binding energies ranging from the weakly bound rare gas atoms to transition metal-C double bonds. The theoretical values are generally in very good agreement with experiment. The calculation of metal ion-ligand binding energies for an entire transition metal row has been particularly useful for identifying and explaining trends in the experimental data. Experimental results which are inconsistent with the theoretical trends have commonly been found to be in error. Calculations have shown that the transition metal ions have a variety of ways to reduce metal ligand repulsion that are not available to simple metal ions such as the alkali metals. The complexity of the bonding often leads to trends in the experimental data that are difficult to explain and analogics based on the well-known alkali systems are often inappropriate.

The major thrust of this chapter is to give the reader some insight into metal ion ligand bonding by considering several examples where theory has been able to explain unexpected trends in the experimental data or to explain

Ben S. Freiser (ed.), Organometallic Ion Chemistry, 47–87.

the discrepancy in experimental values determined with different techniques. Most of the examples involve transition metal ions, but we also discuss simple metal ions in some cases to contrast the bonding. The recent success of theoretical studies of metal ion ligand chemistry is due in large part to the rapid improvement in both methodology and computer architectures. Furthermore, the availability of full configuration-interaction calculations to benchmark electron correlation methods [6] has given new insight into the accuracy of different correlation methods. This, combined with a dramatic improvement in the quality of one-particle Gaussian basis sets [7,8], is leading to unprecedented accuracy in the calculated spectroscopic constants and the ability to assign realistic error bars on many computed binding energies [9].

This chapter is organized as follows. In section 2 we give an overview of the electronic structure methods discussed in this work. In section 3 we discuss the competing effects that determine the optimal bonding mechanism and illustrate through several examples the subtleties that make the optimal bonding state difficult to determine *a priori*. In section 4 we discuss the $MgFe^+$, Ni_2^+, and $LaFe^{2+}$ diatomic ions. We illustrate how *ab initio* calculations have resolved discrepancies in the experimental binding energies. In section 5 we illustrate the utility of theory in computing optimal structures, vibrational frequencies, and entropies of association. Calculations were able to explain the trend in the experimental $AlL^+ - L$ entropies of association for a variety of ligands. In section 6 we present several examples where calculations have helped resolve unexpected trends in the experimental data. In section 7 we present several examples where calculations have aided in the explanation of spectroscopic studies of metal ion-ligand systems. Finally, our conclusions are presented in section 8.

2. Methods

In this section we briefly discuss the methods referred to in this work to define the notation and indicate their expected accuracy. The original literature or recent review articles [10,11] should be consulted for more detailed discussions of the methods. We seek an approximate solution to the electronic Schrödinger equation

$$\hat{H}\Psi = E\Psi,$$

where Ψ is the wave function, E is the energy, and $\hat{H}$ is the electronic Hamiltonian, by employing a double basis set expansion. The accuracy of the calculation depends on the completeness of both basis set expansions. In general, the computational requirements (the basis set expansions) for a given accuracy are greater for metal containing systems than for first and second row systems. The expected accuracy of a given calculation is estimated based on calibration calculations, where both basis sets are systematically expanded, for a system or systems representative of the bonding in molecules of interest. Additional information about the accuracy of *ab initio* calculations has been obtained by the comparison of the results for numerous systems with accurate experimental data.

The one-particle basis sets employed are of at least double-zeta plus polarization quality and are usually of triple-zeta plus polarization quality or better. For the first transition row metal atoms, we generally employ an [8*s* 6*p* 4*d* 2*f*] contracted basis set derived from Wachters' (14*s* 9*p* 5*d*) primitive set [12], supplemented with *p* functions to describe the 4*p* orbital, a diffuse *d* function to describe the $3d^{n+1}4s^1$ occupation [13], and polarization functions. For the second transition row metal atoms, we employ the relativistic effective-core potentials (RECP) developed by Hay and Wadt [14] with the supplemented valence basis set and polarization functions of Ref. [15]. The RECP's, which include the outer core orbitals in the valence shell, account for the dominant relativistic effects, namely the Darwin and mass velocity terms [16], which are important for the second and third row transition metal atoms. When higher accuracy is required and for describing weakly bound systems, we employ large generally contracted basis sets. For the metal atoms we employ atomic natural orbital (ANO) contractions [7] of large primitive sets [17] with the outer functions uncontracted to give the added flexibility necessary to describe properties such as the polarizabilities [18]. For the first and second row atoms, we employ either the ANO basis sets or the correlation consistent sets developed by Dunning and co-workers [8].

Most of the metal-ion ligand ground state systems are studied employing correlation methods based on a single reference self-consistent-field (SCF) zeroth-order wave function. For many systems involving the simple metal ions, the SCF provides a good description of the system and accurate geometries can be determined at this level. However, to obtain accurate binding energies it is usually necessary to include electron correlation. For first transition row metal containing systems an SCF description is generally inadequate and electron correlation must be included even in the geometry optimization. Recent work [19] on the transition metal ion-ligand systems indicates that a second-order Møller-Plesset perturbation theory (MP2) treatment [20] gives reliable geometries, making this approach particularly valuable for larger systems and for estimating zero-point effects. The optimization of geometries at the SCF and MP2 levels is particularly straightforward due to the availability of analytical first and second derivatives for these methods [21]. For accurate energies we have usually employed the modified coupled-pair functional (MCPF) method [22]. The method is essentially size-extensive and approximately accounts for the effect of higher than double excitations. It has been shown to be reliable for electrostatic and single bonds, but it can significantly underestimate the binding energy for multiply bonded systems [23]. For these multiply bonded systems it is necessary to use still higher levels of correlation treatment, such as the single and double excitation coupled-cluster method [24] with a perturbational estimate of the triple excitations (CCSD(T)) [25]. For example, the dissociation energies of the doubly bonded MCH_2^+ systems at the MCPF level are in error by up to 12 kcal/mol [23], but the CCSD(T) results agree with our best descriptions to within about 1 kcal/mol [26]. It is important to emphasize that the triples correction is necessary, as the CCSD results are generally slightly inferior to the MCPF results. Considering that efficient implementations of this method for open-shell systems have recently become available [27], it is likely that this will become the single reference based method of choice.

For systems that are very poorly described by an SCF description (i.e. where non-dynamical correlation effects are large due to, for example, near degeneracy effects), the complete-active-space SCF (CASSCF) method [28] is used to obtain the zeroth-order description of the system. For spectroscopic studies we generally employ the state-averaged (SA) CASSCF approach [29] to obtain a balanced and orthogonal set of zeroth-order wave functions. More extensive correlation is added using the multireference single and double configuration-interaction (MRCI) method [30]. The effect of higher excitations are included using a generalized Davidson correction ($+Q$) [31] or using the averaged coupled-pair functional (ACPF) method [32]. To reduce the size of the resulting CI expansions we often employ the internally-contracted [33] MRCI (ICMRCI) or the size extensive ICACPF method. These approaches give energies and properties very close to the uncontracted results provided the CASSCF active space is adequate. We should emphasize that the CASSCF/MRCI methods are very general and accurate. However, the CI expansion can easily become prohibitively long, even if internal contraction is used. This is, in part, the motivation for the development of efficient single reference methods.

3. An overview of the bonding

Electronic structure calculations have not only yielded accurate binding energies for many metal ion systems, they have also elucidated the important interactions that affect the bonding. On the basis of detailed analyses of the trends for $M^+ - L$ systems for an entire transition row for several representative ligands, it is clear that many effects in addition to the obvious electrostatic terms must be considered in identifying the optimal bonding mechanism. Other factors such as the loss of $d - d$ or $d - s$ exchange energy, ligand-ligand and metal-ligand repulsion, the cost of promotion to the bonding state, dative interactions (including both metal to ligand and ligand to metal donation), etc. are also important in determining the optimal structures and bonding state. In this section we elaborate on these different effects and illustrate through several examples the subtleties that make transition-metal systems so diverse.

The variation in the metal-ligand binding energies ($M^+ - L$) for a given ligand generally does not reveal any systematic trends across an entire transition row. For example, the bonding in most of the metal methyl [34] (or

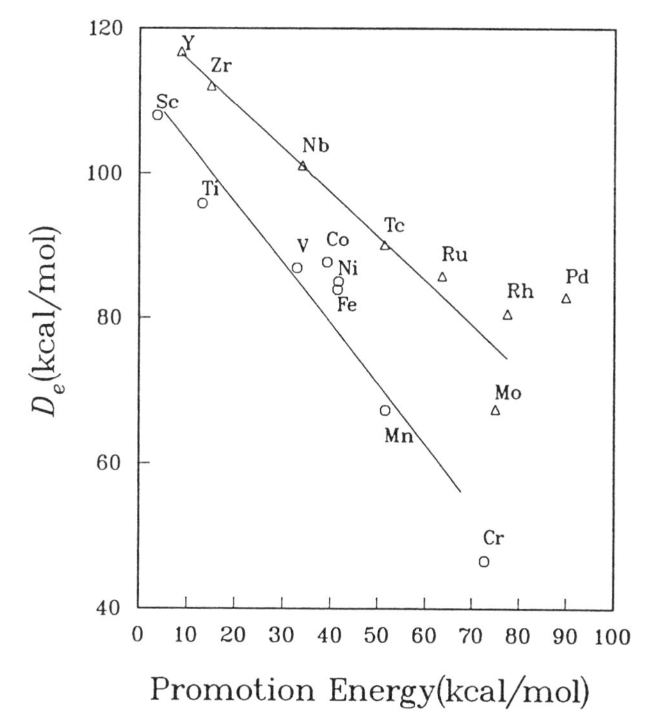

Fig. 1. The $M^+ - CH_3$ binding energies plotted versus the promotion energy corrected for loss of exchange energy. The lines are least squares fits to the first and second transition row atoms.

hydride [35]) ions involves the metal *s* orbital and C *sp* hybrid orbital (or hydrogen *s* orbital). While some ions such as Sc^+ and Fe^+ have a $3d^n4s^1$ ground state, others such as V^+ and Co^+ must promote from the $3d^{n+1}$ ion ground state occupation to the $3d^n4s^1$ excited state before bonding occurs. Thus the large variation in the $3d^n4s^1 - 3d^{n+1}$ separations [36] is a major factor that masks the trends in the binding energies. Another important factor is the extent to which $d - d$ and/or $d - s$ exchange energy is lost when a bond is formed [37]. This factor, which depends both on the number of open-shell orbitals and the number of bonds, is more important for the first transition row than either the second or third transition row, because the correlation and exchange energies in the $3d$ shell are significantly larger than those in the $4d$ and $5d$ shells. The differences in the metal ions can be approximately accounted for by adding the promotion energy required to reach the correct asymptote for bonding to the exchange energy lost when the bond is formed [38]. If the metal-methyl binding energy is plotted against the promotion energy corrected for the loss of exchange energy [34] (see Figure 1), the binding energies for both the first and second transition rows have an approximate straight line dependence, thus exposing the underlying similarity of the bonding.

Another factor affecting the bonding is the relative size of the valence *s* and *d* orbitals. The propensity for *sd* hybridization increases as the radial extent of these orbitals becomes more similar. Because the *d* orbitals contract with respect to the *s* orbitals with increasing Z for each transition row, *sd* hybridization is more favorable on the left side of the row. Furthermore, it is more favorable in the second transition row, because of the smaller difference in radial extent of the *s* and *d* orbitals. This coupled with the smaller exchange energies, results in stronger bonds and more *d* involvement in the bonding of the second transition row metal ions.

Maximizing the dative interactions (both ligand to metal and metal to ligand donation) is another important factor that determines the optimal structure and electronic configuration for transition metal ion systems. Ligand to metal donation is particularly important for those metal ions on the left side of the row where there are empty *d* orbitals to accept charge. This has been used to explain the much larger M^+-OH and M^+-NH_2 than M^+-H and M^+-CH_3 binding energies [39] for the transition metal atoms on the left side of the transition row. This ligand to metal donation is also found for the neutrals [40]. While metal to ligand donation is well known for neutral systems such as $Ni(CO)_4$ [41], it can still be important for metal ions despite the positive charge on the metal.

Minimizing metal-ligand repulsion is also an important consideration in identifying the optimal bonding mechanism. This usually involves occupying the *d* orbitals in a manner that minimizes the repulsion, but may also result in $sd\sigma$ or *sp* hybridization or promotion of a d^ns^1 ground state atom to the d^{n+1} occupation to reduce the repulsion. In Figure 2 we illustrate both $sd\sigma$ and *sdxz* hybridization. One effect of $sd\sigma$ hybridization is to minimize metal-ligand repulsion by reducing the electron density along the z axis. The occupation of the other hybrid orbital, $s + d\sigma$, usually occurs in conjunction with the formation of a strong covalent bond. The *sdxz* hybridization is also usually found in cases of covalent bonding.

Another and somewhat more subtle factor that affects the strength of metal-ligand bonds is the atomic composition of the metal ion in the complex. For example, in CoL^+ some of the $3d^8$ occupations are derived from the ^3F ground state, while others are a mixture of ^3F and ^3P [42]. This promotion energy can affect the strength of the bonding.

The competition between the various factors affecting the bonding can result in very different bonding mechanisms for the same metal ion as the ligand is varied. These points are well illustrated by considering several Fe^+-L systems. The ground state of Fe^+ is $^6D(3d^64s^1)$, but the $^4F(3d^7)$ excited state is only 5.8 kcal/mol higher in energy [36]. The bonding in $FeAr^+$ is due to a charge-induced dipole interaction of Ar with the Fe^+ ^6D state [43]. In addition to Ar polarization, the Fe^+ 4*s* orbital polarizes away from the Ar atom, by mixing in some 4*p* character to minimize the repulsion, thereby

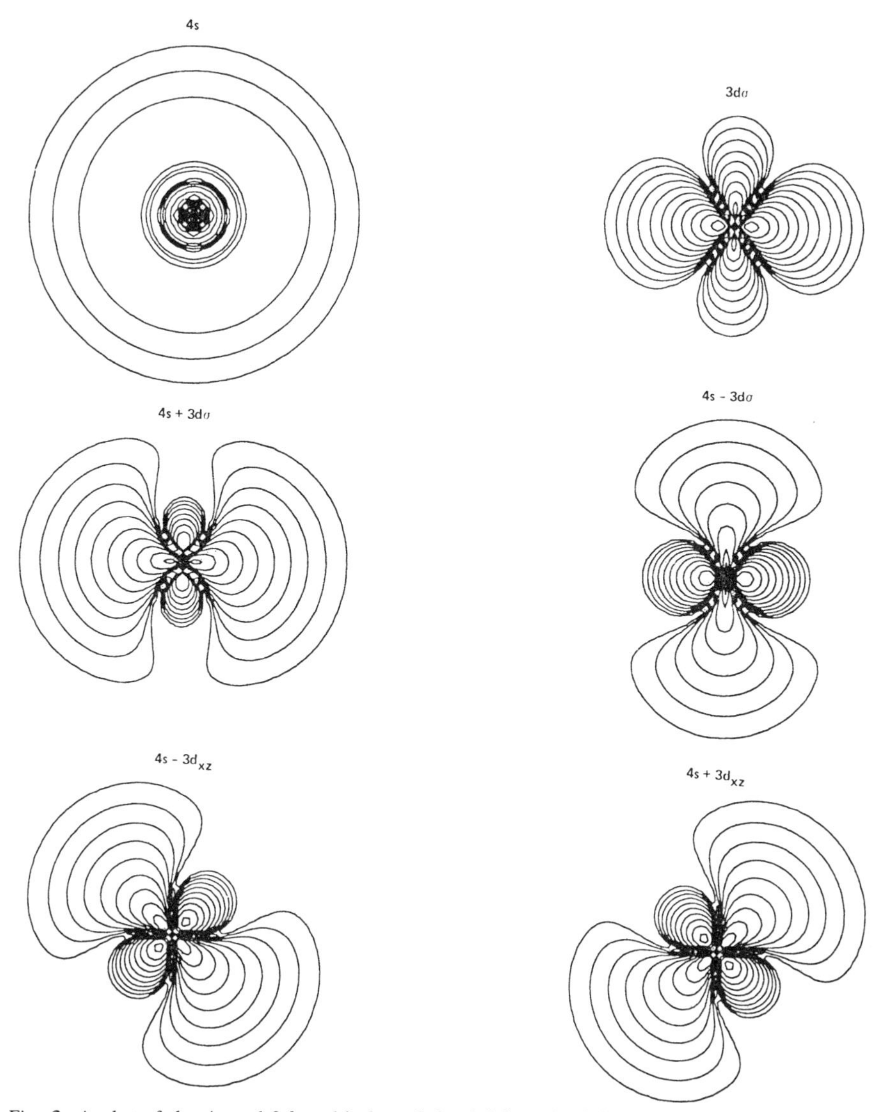

Fig. 2. A plot of the 4*s* and 3*dσ* orbitals and the $4s3d_\sigma$ and $4s3d_{xz}$ hybrid orbitals of Cr atom.

enhancing the bonding. However, the bonding is rather weak as manifested by the very small binding energy of only 4.0 kcal/mol. Promotion to the ^4F state of Fe^+ is found to be unfavorable, because the increase in the electrostatic binding due to the shorter bond length is not sufficient to compensate for the promotion energy.

The bonding in the ground state of FeH_2O^+ also arises from the ^6D ground state of Fe^+. As for Ar, Fe 4*s* polarization enhances the bonding.

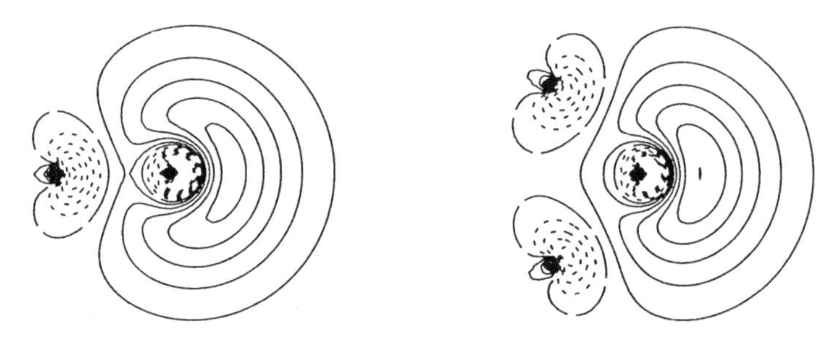

Fig. 3. The SCF open-shell orbital for MgH_2O^+ and $Mg(H_2O)_2^+$.

The much stronger charge-dipole interaction leads to a binding energy [44] of 33.7 kcal/mol, which is much larger than that found for $FeAr^+$.

The electrostatic components to the bonding in $FeC_6H_6^+$ are principally charge-induced dipole and charge-quadrupole. However, dative interactions, particularly metal 3*d* to π^* donation, significantly increase the binding energy [45]. Promotion to the ^4F state of Fe^+ increases the electrostatic binding by reducing the metal-ligand repulsion and hence decreasing the metal-ligand distance. Moreover, it also increases the metal to ligand dative interaction by increasing the number of 3*d* electrons available for donation and increases the *d* orbital overlap with the benzene π^* orbitals. Thus, unlike $FeAr^+$ or FeH_2O^+, the bonding in the ground state of $FeC_6H_6^+$ is derived from the ^4F state of Fe^+. The computed $Fe^+ - C_6H_6$ binding energy is 51.1 kcal/mol, which is significantly larger than that for $FeAr^+$ or FeH_2O^+.

The bonding in the ground state of $FeCO^+$ is also derived from the ^4F state of Fe^+. In this case promotion to the $3d^7$ asymptote enhances the 3*d* to $2\pi^*$ donation. It should be noted, however, that the $Fe^+ - CO$ binding energy [46] of 21.9 kcal/mol is smaller than that found for FeH_2O^+, where Fe^+ did not promote to the $3d^7$ occupation.

For both $FeAr_2^+$ [47] and $Fe(H_2O)_2^+$ [44] the bonding is derived from the ^4F state. This arises because the polarization of the 4*s* orbital with the first ligand leads to an area of high electron density on the side of the Fe^+ away from the ligand. Therefore, two ligands bind to the same side of $Fe^+(^6D)$, giving rise to a bent excited sextet state for FeL_2^+. To reduce the ligand-ligand and metal-ligand repulsion, it becomes favorable to promote Fe^+ to the ^4F state; the cost of which is now shared by two ligands, thereby leading to a quartet ground state with the ligands on opposite sides of the Fe^+ ion.

The ML_2^+ systems are found to be nonlinear only when *s* to *d* promotion is not favorable. For example, this is the case for $Mn(H_2O)_2^+$ [44], where the promotion energy to the $3d^6$ occupation is larger than the enhancement in the bonding, and for $Mg(H_2O)_2^+$ [48], where there is no low-lying *d* orbital. The plot of the MgH_2O^+ and $Mg(H_2O)_2^+$ open-shell orbitals, shown in Figure 3, illustrate the polarization of the 3*s* orbital away from the ligands.

The FeL^+ results show the diversity of the bonding that can occur for one metal atom with different ligands. In addition, they show that the bonding mechanism cannot be deduced *a priori* on the basis of the binding energy alone. Instead, one must consider the specific metal-ligand interactions in order to deduce the optimal bonding mechanism.

4. Diatomic metal ions

4.1. $MgFe^+$

The upper bound for the dissociation energy of $MgFe^+$ determined by Roth *et al.* [49] from photodissociation threshold experiments ($\leq 44 \pm 3$ kcal/mol) was considerably larger than the 34 ± 5 kcal/mol value they determined from ion-molecule reactions. Photodissociation experiments yield only an upper bound to bond dissociation energies, because the height of the absorbing dissociative state above the asymptote is, in general, not known. However, this method is often reasonably accurate for the binding energies of transition metal ion systems [4], due to the high density of electronic states. The determination of bond energies by ion-molecule reactions, such as ligand displacement, relies on bracketing the unknown bond energy by known $M^+ - L'$ bond energies. This approach is subject to error if there are reaction barriers or if one of the "known" bond energies is incorrect. Thus it was not clear whether the $MgFe^+$ bond energy determined from photodissociation experiments was too large because the dissociative state lies above the dissociation threshold or whether the value determined from ion-molecule reactions was in error.

It should be noted that $Fe^+ + Mg$ is the principal product in the photodissociation experiment, even though this corresponds to the second asymptote. In contrast, collision induced dissociation (CID) experiments, in which bond rupture is achieved by the collision of the ion with an inert gas target, yields mostly $Mg^+ + Fe$ at low energies. This suggests that the upper state in the photodissociation experiment does not correlate to the lowest asymptote.

To resolve these questions, we performed SA-CASSCF/MRCI calculations [49] for $MgFe^+$, which allows for a rigorous treatment of states of the same symmetry and facilitates the determination of the electronic transition moments. The effect of higher excitations was accounted for using the multireference analog of the Davidson correction. This level of theory is expected to yield a dissociation energy accurate to about 5 kcal/mol.

The lowest asymptote for $MgFe^+$ is Fe $^5D(3d^64s^2)$ + Mg^+ $^2S(3s^1)$, but the Fe^+ $^6D(3d^64s^1)$ + Mg $^1S(3s^2)$ asymptote is only 5.2 kcal/mol higher in energy. The Fe^+ $^4D(3d^7)$ + Mg $^1S(3s^2)$ asymptote also lies only 10.9 kcal/mol above the lowest asymptote. The calculations show that the ground state correlates to the Fe(^5D) + Mg^+(^2S) asymptote, but that the bonding arises from a mixing of the two lowest asymptotes, giving a $\sigma^2\sigma^{*1}$ occupation. The six essentially non-bonding $3d$ electrons lead to three nearly degenerate states,

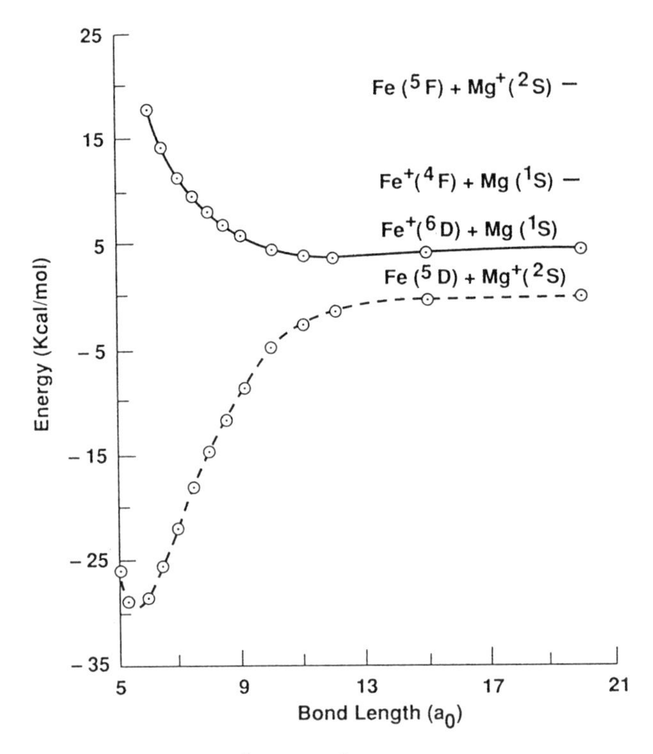

Fig. 4. $X^6\Delta$ and $(2)^6\Delta$ states of $MgFe^+$.

namely $^6\Delta$, $^6\Pi$, and $^6\Sigma^+$. For example, the $^6\Pi$ state is only 0.12 kcal/mol above the $^6\Delta$ ground state. Thus for clarity, only the $X^6\Delta$ state is plotted in Figure 4. While the bond order of these three bound states is formally only 0.5, polarization of the σ^* orbital by mixing in p character gives it significant nonbonding character – see Figure 5 where the σ and σ^* valence s orbitals are plotted. This enhancement in the binding by polarization of the σ^* orbital leads to a sizable binding energy of 29.4 kcal/mol for the $X^6\Delta$ state. Considering that this value has an estimated uncertainty of $\pm$ 5 kcal/mol, it agrees with the experimental value of 34 $\pm$ 5 kcal/mol deduced from an analysis of the ion-molecule reactions.

The other combination of the two lowest asymptotes leads to a $\sigma^1\sigma^{*2}$ occupation, which produces three nearly degenerate repulsive states of the same symmetries as the bound states. Again, only the lowest of these, the $(2)^6\Delta$ state, is shown for clarity in Figure 4. The computed vertical excitation energy is 50.8 kcal/mol, and the transition moment is more than 1 a.u. Thus it is clear that the photodissociation experiment corresponds to an excitation from the $X^6\Delta$ state to the $(2)^6\Delta$ state and subsequent dissociation to Fe^+ + Mg. The computed vertical excitation energy is in excellent agreement with the experimental photon energy of 49 kcal/mol and also explains why

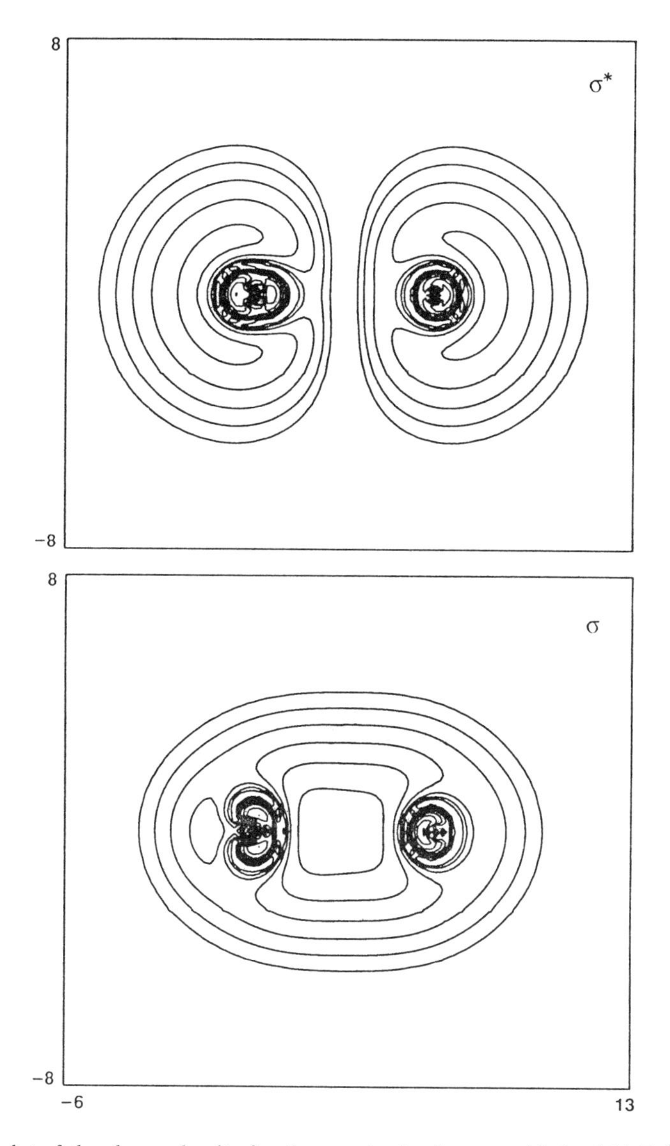

Fig. 5. Log plot of the charge density for the σ and σ^* valence s orbitals of $MgFe^+$. Fe is at the origin.

the predominant (>95%) product observed is Fe^+(+Mg) in spite of this being the second asymptote.

The calculations show that both experimental determinations are correct. The photodissociation threshold is much larger than the true dissociation energy, because the lowest dissociative excited state lies significantly above the asymptote in the Franck-Condon region. The calculations also show that the bonding involves only the Fe 4*s* electron and that the 3*d* orbitals are nonbonding. The orbital plots clearly show that the σ^* orbital contains significant nonbonding character. Thus the calculations provide considerable insight into the bonding, that is not apparent from the experimental data alone.

4.4. Ni_2^+

The Ni_2^+ metal ion is another system where a controversy has arisen over its dissociation energy. For this ion, the CID experiments of Lian, Su, and Armentrout [50] yield a binding energy of 2.08 ± 0.07 eV, while the photodissociation experiments of Lessen and Brucat [51] have been interpreted in terms of a binding energy between 3.0 and 3.5 eV. The photodissociation experiments are actually performed for Ni_2Ar^+, where at 3.0 eV Ni_2^+ + Ar is observed, while at 3.5 eV Ni^+ + Ni + Ar is observed. The appearance potential measurement of Kant [52] yields a dissociation energy of 3.30 ± 0.2 eV, which is consistent with the value of Lessen and Brucat.

Electronic structure calculations [53,54] have shown that the bonding in Ni_2 arises from two $3d^94s^1$ Ni atoms, which form a 4*s*-4*s* σ bond, similar to that found in Cu_2, with very little 3*d*-3*d* involvement in the bonding. This is consistent with the fact that the experimental dissociation energies for Ni_2 and Cu_2 are very similar, 2.068 ± 0.01 eV [55] and 2.078 eV [56], respectively. The ground state of Cu_2^+ is $^2\Sigma_g^+$ where one of the bonding 4*s* σ electrons has been removed. Nevertheless, the dissociation energy [57] of Cu_2^+ is 92% of that found for Cu_2, because the 4*s*4*p* near degeneracy allows a strong polarization of the single bonding electron between the two Cu atoms [58]. The small Ni_2^+ binding energy of Lian, Su, and Armentrout could imply that there is very limited 3*d*-3*d* contribution to the bonding in Ni_2^+, as in Cu_2^+. A large Ni_2^+ binding energy clearly implies a dramatic increase in the 3*d*-3*d* bonding relative to Ni_2. Thus a determination of the Ni_2^+ binding energy is important not only to resolve the disagreement over the binding energy, but to clarify the importance of 3*d*-3*d* bonding in the ion.

The bonding for selected low-lying states of Ni_2 was studied at the SA-CASSCF/ICACPF level [59]. The spectroscopic constants for these states are summarized in Table I. Because all of the low-lying states contain two open-shell 3*d* orbitals, they are classified by the location of the holes. For example, a $3d\delta$ hole on each Ni atom would be classified as $\delta\delta$. Because of the weak 3*d*-3*d* interaction, all states arising from a given set of holes tend to have similar energies – see Table I and the figures in Refs. [53] and [54]. All of the calculations have found that the most stable states arise from the $\delta\delta$ configuration. However, the $\pi\delta$ hole states are lower lying in our calculations than in the earlier work of Noell *et al.* [54] or Upton and Goddard [53]. The $\delta\delta$ occupation gives rise to six states, $^3\Gamma_u$, $^3\Sigma_g^-$, $^3\Sigma_u^+$, $^1\Gamma_g$, $^1\Sigma_u^-$, and $^1\Sigma_g^+$. The experiments [55,60] were initially interpreted in terms of a ground state with $\Omega = 4$, which implies that the ground state is either $^3\Gamma_u$ or $^1\Gamma_g$.

However, after realizing that ligand field theory predicted [61] a $^3\Sigma_g^-$ ground state, Morse and co-workers [62] reinvestigated Ni_2 and at lower temperatures found the spectrum to be completely different from that initially recorded. Thus, the identity of the ground state has still not been definitively determined experimentally. While the separation between the $\delta\delta$ states is too small for a reliable prediction of the ground state based on the calculations to date, the small separation between the states is consistent with very little *3d-3d* involvement in the bonding.

The computed r_e values are larger and the D_e values are significantly smaller than the experimental values. However, the errors are comparable for Cu_2 at the same level of theory [63]. Very high levels of theory are required to accurately compute the bond energy of transition metal diatomics. This is beyond the scope of our calculations, which are designed to provide an equivalent description of Ni_2 and Ni_2^+.

For Ni_2^+, we performed SA-CASSCF/ICACPF calculations [59] for both *4s* and $3d\sigma$ ionization. For *4s* ionization, the lowest ion states are derived from the $\delta\delta$ states of the neutral. While our lowest state for the neutral is a singlet, the extra *4s-3d* exchange leads to the quartet states being lower than the doublet states for the ion. The computed binding energies are very similar to that found for the neutral. That is, ionizing the *4s* electron leads to little change in the dissociation energy, as expected based on the Cu_2 and Cu_2^+ results.

Upton and Goddard [53] predicted that removing a $3d\sigma_u$ electron leads to a state that is 0.8 eV more stable than that which is derived from *4s* ionization. The extra stability of this $3d\sigma$ hole state was due to *3d-3d* bonding which can occur without significantly changing the *4s-4s* bonding. This *3d-3d* bonding also resulted in Ni_2^+ being more than an eV more strongly bound than Ni_2. However, their calculations were only qualitative in nature, producing, for example, an Ni_2 binding energy that was about 0.8 eV larger than experiment.

The lowest states arising from $3d\sigma$ ionization are derived from the $\sigma\delta\delta$ configuration, see Table I. These states have shorter bond lengths and larger vibrational frequencies than the *4s* hole states. In fact, the vibrational frequencies are larger than those for the neutral, which probably indicates some *3d-3d* bonding. The binding energies of these states are 2.05–2.09 eV if directly computed to the ground state asymptote. However, this procedure is inaccurate, because there is a larger correlation error for the transition metal states with more *3d* electrons [64]. This error can be largely avoided by computing the dissociation energy with respect to the Ni $^3F(3d^84s^2)$ + Ni^+ $^2D(3d^9)$ asymptote and correcting this value to the ground state asymptote using the experimental atomic separations [36]. This leads to the binding energies reported in Table I. Our computed Ni_2^+ binding energy is only 0.17 eV larger than that of Ni_2. Thus we conclude that while the ground state of Ni_2^+ is derived from $3d\sigma$ ionization, there is little *3d-3d* enhancement to the bonding. Our best estimate for the Ni_2^+ binding energy (2.26 eV) is derived

Table I. Spectroscopic constants for the low-lying electronic states of Ni_2 and Ni_2^+ computed using the ICACPF method

	r_e (Å)	D_e (eV)	ω_e (cm^{-1})
Ni_2			
$^1\Delta_u$ ($\sigma\delta$)	2.334	1.504	248
$^3\Phi_g$ ($\pi\delta$)	2.330	1.529	250
$^1\Phi_g$ ($\pi\delta$)	2.329	1.549	251
$^3\Delta_g$ ($\sigma\delta$)	2.292	1.553	238
$^1\Phi_u$ ($\pi\delta$)	2.315	1.559	244
$^3\Pi_u$ ($\pi\delta$)	2.290	1.599	242
$^3\Gamma_u$ ($\delta\delta$)	2.297	1.642	256
$^1\Sigma_u^-$ ($\delta\delta$)	2.296	1.642	256
$^3\Sigma_g^-$ ($\delta\delta$)	2.289	1.654	253
$^3\Sigma_u^+$ ($\delta\delta$)	2.297	1.667	255
$^1\Gamma_g$ ($\delta\delta$)	2.291	1.691	253
Ground State (Expt.)	2.200[a,b]	2.092[a]	≈330[c]
Ni_2^+			
$^2\Sigma_u^-$ ($4s\delta\delta$)	2.479	1.608	182
$^2\Gamma_g$ ($4s\delta\delta$)	2.476	1.610	181
$^4\Gamma_u$ ($4s\delta\delta$)	2.473	1.673	184
$^4\Sigma_g^-$ ($4s\delta\delta$)	2.469	1.675	183
$^4\Gamma_g$ ($\sigma\delta\delta$)	2.244	1.82[d]	293
$^4\Sigma_u^-$ ($\sigma\delta\delta$)	2.239	1.84[d]	285
$^4\Sigma_g^+$ ($\sigma\delta\delta$)	2.244	1.86[d]	297

[a] Ref. [55]. [b] The experimental r_0 value. [c] Ref. [60] – the assignment of ω_e is uncertain. [d] Computed with respect to the Ni $^3F(3d^84s^2)$ + Ni$^+$ $^2D(3d^9)$ asymptote and corrected to the ground state Ni $^3D(3d^94s^1)$ + Ni$^+$ $^2D(3d^9)$ asymptote using the experimental ^3F-^3D separation.

by adding our computed difference in the Ni_2^+ and Ni_2 binding energies to the experimental result for Ni_2. The uncertainty in our estimated Ni_2^+ binding energy is probably ±0.1 eV and is almost certainly less than ±0.2 eV. Thus our calculations rule out the higher dissociation energies and favor the lower value of Lian, Su, and Armentrout.

Considering that the open-shell 3*d* orbitals give rise to a high density of states, it is unclear why the photodissociation experiments overestimate the D_e value of Ni_2^+. It is possible that the state (or states) above 3 eV are the first that predissociate rapidly compared with the time scale of the experiment and, therefore, dissociation to an excited state of Ni or Ni^+ is observed. As noted by Lian, Su, and Armentrout [50], dissociation to the ^4F state of Ni^+ could explain the difference. Lessen and Brucat [65] studied resonant two-photon dissociation of Ni_2^+. In this experiment they observe what they described as a relatively simple and unperturbed spectrum up to 2.16 eV. Above this energy it becomes more congested. In addition, they found a small production of Ni^+ from Ni_2^+ at about 2.16 eV, which they attributed to two-photon ionization of Ni_2^+ on the basis of a non-linear laser fluence dependence. It is possible, however, that these observations are due to the weak coupling to dissociative states at about 2.16 eV. This would bring the photodissociation experiments into good agreement with theory and the CID experiments.

4.3. $LaFe^{2+}$

The existence of metastable dipositive diatomic ions has been known for a long time [66]. *Ab initio* calculations [67] have shown that even singly bonded dipositive ions, such as Be_2^{2+}, can have very long lifetimes. Recently, Huang and Freiser [68] found that $LaFe^{2+}$ had a very long lifetime in their ion cyclotron resonance apparatus. This dipositive ion is potentially unique, because the small second ionization potential of La could allow significant contributions to the bonding from both the $La^+Fe^+$and $La^{2+}Fe$ configurations. This latter configuration has the advantage that the repulsive charge-charge interaction is replaced by an attractive charge-induced dipole interaction.

The bonding is not *a priori* obvious considering the number of atomic states from which the ground state could be derived. For example, the ground state of La^+ is $^3F(5d^2)$, but the ^1D and ^3D states derived from the $5d^16s^1$ occupation are both less than 0.2 eV higher in energy. The ground state of La^{2+} is $^2D(5d^1)$. For Fe^+ the $^4F(3d^7)$ state is only 0.25 eV above the $^6D(3d^64s^1)$ ground state, while for Fe the $^5F(3d^74s^1)$ state is 0.87 eV above the $^5D(3d^64s^2)$ ground state. Thus several bonding mechanisms derived from either La^+Fe^+or $La^{2+}Fe$ can be envisioned. We therefore undertook a series of calculations to determine the character of the bonding in this ion [69].

CASSCF/MRCI calculations [69] were performed for $LaFe^{2+}$ as well as the valence isoelectronic $LaRu^{2+}$ and YRu^{2+} ions. As noted in the methods section, the Y, Ru, and La atoms were described using an RECP with the semicore orbitals in the valence space. Calibration calculations for $LaFe^{2+}$ showed that if the La $5s$ and $5p$ orbitals were not included in the valence space, the calculations "collapsed" at short r values. This probably indicates that the repulsion between La $5s$ and $5p$ orbitals and the Fe orbitals is important in determining the bond length. This underscores the importance of using RECP's that include the semicore orbitals in the valence space.

The $LaFe^{2+}$ SA-CASSCF calculations indicate that the lowest doublet and quartet states are:

$$^2\Delta \quad [\mathrm{La}(5d\sigma) + \mathrm{Fe}(3d\sigma)]^2\,[\mathrm{La}\,(5d\pi_x) + \mathrm{Fe}(3d\pi_x)]^2\,\mathrm{Fe}(3d\pi_y^2\,3d\delta^3) \tag{1}$$

and

$$^4\Sigma^- \quad [\mathrm{La}\,(5d\sigma) + \mathrm{Fe}(3d\sigma)]^2\,[\mathrm{La}\,(5d\pi_x) + \mathrm{Fe}(3d\pi_x)]^2 \times \mathrm{Fe}(3d\pi_y^2\,3d\delta^2\,4s^1). \tag{2}$$

Table II. Valence population analysis for the natural orbitals of the MRCI wave function for the computed point closest to the minimum

	$LaFe^{2+}$				$LaRu^{2+}$		YRu^{2+}	
	$^4\Sigma^-$		$^2\Delta$		$^2\Delta$		$^2\Delta$	
	La	Fe	La	Fe	La	Ru	Y	Ru
Net	+1.56	+0.44	+1.40	+0.60	+1.54	+0.46	+1.16	+0.84
s	0.22	1.15	0.16	0.42	−0.04	0.07	0.23	−0.08
p	0.05	0.16	0.07	0.22	−0.01	0.11	0.06	0.00
d	1.14	6.24	1.35	6.75	1.44	7.31	1.48	7.19

In both states a σ and π_x bond are formed; these double bonds are highly polarized toward the Fe, but back donation from the doubly occupied Fe $3d\pi_y$ orbital into the empty La $5d\pi_y$ orbital gives some triple bond character. While both states have some similarities in the bonding, they are derived from different asymptotes. The $^2\Delta$ state is derived from Fe^+ $3d^7$, while the $^4\Sigma^-$ state is derived from $3d^64s^1$ – this is reflected in the populations in Table II. One factor limiting the Fe 4*s* population, and hence the $La^{2+}Fe$ character, is that the optimal *s-s* bond length is significantly longer than the optimal $d-d$ bond length. However, in spite of this, the populations show that there is significant $La^{2+}Fe$ character in the wave function.

Details of the potential curves are given in Table III for the $X^4\Sigma^-$ and excited $^2\Delta$ states of $LaFe^{2+}$. The dissociation energy is given as the difference between the bottom of the molecular well and the top of the barrier. For both states there is a sizable barrier to dissociation, which increases with the level of theoretical treatment. The height of the molecular well above the dissociation limit also decreases with the level of theory. While $LaFe^{2+}$ is clearly metastable, the large barrier makes it extremely stable towards unimolecular dissociation. For example, using the WKB [70] approximation, the lowest ro-vibrational level is computed [69] to have a lifetime with respect to unimolecular decay of 10^{600} sec. The binding energies for the $X^4\Sigma^-$ and $^2\Delta$ states are comparable. The $^2\Delta$ state is found to lie 1.58, 1.28, and 1.25 eV above the ground state at the CASSCF, MRCI, and MRCI + Q levels. Calculations [64] on Fe and Fe^+ show that the states with fewer 3*d* electrons are easier to describe theoretically. Therefore, we expect the T_e value of the $^2\Delta$ state to decrease slightly with further improvements in the level of theory.

To identify other potentially long-lived metastable dipositive ions, we carried out similar calculations for the $^2\Delta$ ground states of $LaRu^{2+}$and YRu^{2+} – see Table III. The energy difference between the $M(d^7s^1)$ and $M^+(d^7)$ states is very similar for Ru and Fe, 7.37 vs 7.25 eV. However, since the $4d^8$ occupation of Ru is relatively low-lying, the *d* population is larger for $LaRu^{2+}$, even though the net metal charges are similar for Ru and Fe. Considering that Ru^+generally forms stronger bonds than Fe^+, it is not surprising that the dissociation energy of $LaRu^{2+}$ is significantly larger than

Table III. Spectroscopic constants for the $^4\Sigma^-$ and $^2\Delta$ states of $LaFe^{2+}$ and the $^2\Delta$ states of $LaRu^{2+}$ and YRu^{2+}

	r_e (Å) inner well	D_e (eV) inner well	r (Å) barrier	Inner well relative to asymptotes (eV)
$LaFe^{2+}(^4\Sigma^-)$				
CASSCF	3.38	0.61	5.49	1.70
MRCI	2.96	1.29	5.92	0.95
MRCI + Q	2.88	1.53	6.15	0.65
$LaFe^{2+}(^2\Delta)$				
CASSCF	2.85	0.54	4.04	2.37[a]
MRCI	2.76	1.42	4.42	1.32[a]
MRCI + Q	2.76	1.60	4.56	1.06[a]
$LaRu^{2+}(^2\Delta)$				
CASSCF	2.40	1.64	3.85	0.87
MRCI	2.35	2.03	5.25	0.14
MRCI + Q	2.36	2.18	5.33	−0.02
$YRu^{2+}(^2\Delta)$				
CASSCF	2.39	1.05	3.81	1.78
MRCI	2.28	1.47	4.26	1.16
MRCI + Q	2.32	1.66	4.45	0.95

[a] Dissociates to Fe^+ $^4D(3d^64s^1)$ instead of $^4D(3d^7)$ because the calculations are biased toward states with fewer $3d$ electrons.

$LaFe^{2+}$. In fact, at the MRCI + Q level, the well lies slightly below the dissociation limit. The population analysis in Table II shows that there is more polarization of the charge for $LaRu^{2+}$, because the charge can be accepted into the more compact d orbital instead of the s orbital. The potential curves for the $^2\Delta$ state of $LaRu^{2+}$ shown in Figure 6 illustrate the very large barrier to dissociation that is typical of this class of molecules.

Details of the $^2\Delta$ ground state of $YRu^{2+}$are also given in Table III. Because the second IP [36] of Y is larger than that of La, 12.24 vs 11.43 eV, we expect less Y^{2+}character in the wave function. Also, since the ground state of Y^+ is $^1S(5s^2)$ with the $^3D(4d^15s^1)$ and $^3F(4d^2)$ states being 0.15 and 1.05 eV higher in energy, respectively, we expect a larger Y s population. This is borne out by the populations in Table II. Furthermore, the binding energy of YRu^{2+} is less than $LaRu^{2+}$, because Y^+ must promote to the excited state to bond.

These calculations for three dipositive ions illustrate the utility of theory for understanding the bonding based on the net charges and populations. Even though the binding energies may be underestimated at this level of theory, the calculations definitely establish the ground states and establish trends in the spectroscopic constants.

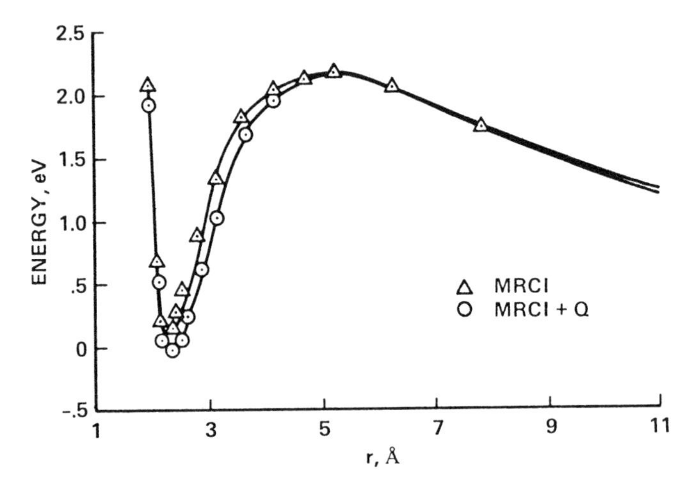

Fig. 6. MRCI and MRCI + Q potentials for the $^2\Delta$ state of $LaRu^{2+}$. The zero of energy is $La^+(^3F) + Ru^+(^4F)$.

5. The entropy of association for acetone with Al^+

In general very little structural information can be extracted from most current experimental techniques used to study ions. Some indirect information about structure comes from the entropy of association. This quantity, as well as the binding energy, can be deduced from a van't Hoff plot, where the equilibrium constant is determined as a function of temperature using a high pressure mass spectrometer. Such experiments have been performed for many systems, especially those containing alkali ions [71,72]. It has been found that the entropy of association for the second ligand tends to be -25 ± 2 cal mol^{-1} K^{-1}. This value has been interpreted as showing that the second ligand is on the opposite side of the alkali ion and that there is little ligand-ligand interaction. This has been confirmed by *ab initio* calculations [73]; see for example the optimal structures for $Na(H_2O)_n^+$.

Since Al^+ has a $^1S(3s^2)$ ground state, it was initially thought that the entropies of association for AlL_n^+ would be similar to those of the alkali ions. However, Bouchard, Hepburn, and McMahon [74] found that the entropy of association for the second acetone to Al^+ was -45 cal mol^{-1} K^{-1}. This value is indicative of an $Al(acetone)_2^+$ geometry that involves significant ligand-ligand interaction. Thus this value would appear to be consistent with structure (I) (see Figure 7), instead of structure (II) that would be expected based on a comparison with the alkali ions.

Freiser and co-workers [75] were able to deduce some structural information by performing CID experiments for the Al^+ – acetone system using both acetone and acetone-d_6. Regardless of whether they first formed Al(acetone)$^+$ and then added acetone-d_6, or first formed Al(acetone-d_6)$^+$ and then

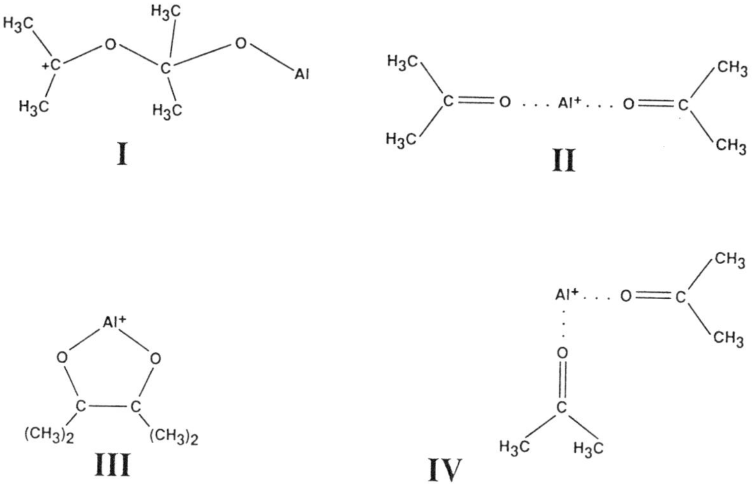

Fig. 7. The proposed stuctures of $Al(acetone)_2^+$.

added acetone, there was an equal probability of dissociating an acetone or acetone-d_6. This clearly rules out structure (I), because the two ligands must be equivalent. They suggested structure (III) as an alternative, since it has equivalent ligands and would have an entropy of association that is much larger in magnitude than that of structure (I). However, Freiser and co-workers were reluctant to accept this structure and suggested that electronic structure calculations might be able to resolve this issue.

Due to the fact that $Al(acetone)_2^+$ is a large molecular system to treat with *ab initio* calculations, we first carried out calculations [75] for the smaller, but analogous $Al(formaldehyde)_2^+$ system. The geometries were optimized at the SCF level using a basis set of triple-zeta quality augmented with two sets of polarization functions (TZ2P). The optimal structure was found to be analogous to (IV), which was 12 kcal/mol more stable than a structure like (II), which would have been expected based on analogy with the alkali systems. The calculations also showed that the (I)- and (III)-like structures for $Al(formaldehyde)_2^+$ are less stable by about 80 kcal/mol, even when electron correlation is added at the MCPF level. The bent structure is more stable than the linear one, because of the polarization of the Al 3*s* orbital, as shown for $Mg(H_2O)_2^+$ in Figure 3. The alkali ions are not sufficiently polarizable for bending to be energetically favorable.

On the basis of the formaldehyde calculations, only structure (IV) was considered for the acetone system. Because acetone is a larger ligand, the calculations were performed using a double-zeta (DZ) basis set. The results in Table IV show that the DZ basis gives D_e values that are uniformly larger than for the TZ2P basis, mostly due to basis set superposition error. However, improving the basis set reduces the dissociation energies of both

Table IV. Dissociation energies (in kcal/mol) for AlL^+ and AlL_2^+ for L = H_2CO and acetone

	D_e		D_0
	DZ	TZ2P	DZ
Al^+ – H_2CO	34.4	27.2	33.2
AlH_2CO^+ – H_2CO	22.8	18.6	21.2
Al^+ – acetone	48.3	41.7	47.4
$Al(acetone)^+$ – acetone	27.8		26.8

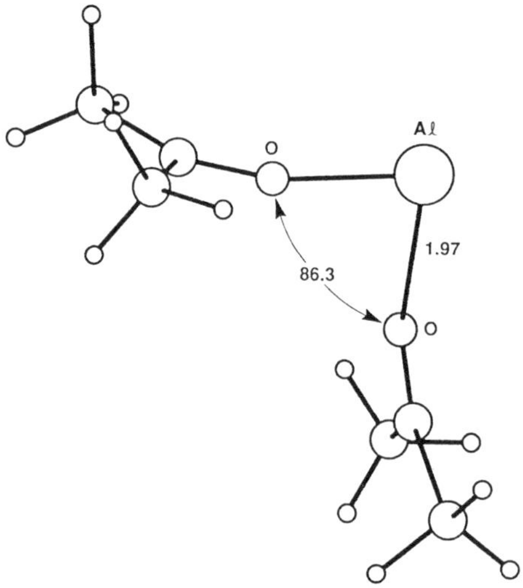

Fig. 8. The optimal SCF structure for $Al(acetone)_2^+$. This is equivalent to structure IV in Figure 7. The bond length is in Å and the angle is in degrees. The Al and O atoms are labelled, while the C and H atoms are not.

Al^+ – formaldehyde and Al^+ – acetone by approximately the same amount. For both systems, the second ligand binding energy is significantly smaller than the first due to ligand-ligand repulsion – see Figure 8. The reduction in the second ligand binding energy is also much larger than found for systems such as $Na(H_2O)_n^+$ where the ligands are on the opposite sides of the metal ion. On the basis of the results in Table IV, the second acetone binding energy was estimated [75] to be 23 kcal/mol, which is in reasonable agreement with experiment [75] (28.3 kcal/mol), and consistent with the level of theory used. We should note that correlated calculations for $Al(acetone)_2^+$ using the TZ2P basis set are now possible using a direct MP2 approach [76]. This illustrates how rapidly computational chemistry methods are advancing.

The harmonic frequencies can be computed straightforwardly at the SCF level using analytical second derivatives. The entropy of association is then computed using these harmonic frequencies and a rigid rotor approximation.

Table V. Computed entropies (in cal mol^{-1} K^{-1}) for selected ions

	100 K	200 K	300 K
$NaH_2O^+ - H_2O$(TZ2P)	−21.3	−20.9	−20.0
$AlH_2CO^+ - H_2CO$(TZ2P)	−29.8	−29.3	−28.2
$AlH_2CO^+ - H_2CO$(DZ)	−31.6	−30.3	−29.0
Al(acetone)$^+$ − acetone(DZ)	−31.9	−30.0	−28.6

Theoretical values for selected species are summarized in Table V. At 667 K, the computed value for $Na(H_2O)_2^+$ is about 80% of the experimental value [71], but our approximations are expected to be more accurate at lower temperatures. The entropies for Al^+ – formaldehyde and Al^+ – acetone are significantly larger in magnitude than that of Na^+ – water, which is consistent with the greater degree of hindered vibrations and rotations caused by the larger ligand-ligand interaction for the bent AlL_2^+ structures. The values are not very sensitive to basis set quality. Nevertheless, the calculated values for Al(acetone)$_2^+$ are still significantly less in magnitude than the energy of association measured by Bouchard *et al.* [74]. The difference between theory and experiment was resolved when the entropy of association was remeasured [75] as −30.7 cal mol^{-1} K^{-1}, which is in excellent agreement with the computed value.

Further confirmation for non-linear AlL_2^+ structures is provided by the measured [75] second entropy of association for Al^+ with acetonitrile and diethylether. The magnitude of the entropy is expected to increase with the size of the ligands, because of increasing ligand-ligand repulsion. The experimental values of −20 for acetonitrile, −31 for acetone and −45 cal mol^{-1} K^{-1} for diethylether, corroborates the theoretical prediction of non-linear AlL_2^+ structures. This again illustrates the utility of using *ab initio* calculations to explain trends in the experimental measurements. In turn, accurate experimental values help calibrate theoretical approaches to determine these quantities.

6. Explanation of unexpected experimental results

Theory has contributed significantly to an understanding of the bonding of metal ion systems and has successfully explained several unexpected experimental results. Often it is *not* necessary to carry out high-level calculations to explain trends in, for example, the $M^+ - L$ binding energies across a transition row. In this section we present several examples that illustrate the utility of theory in aiding the interpretation of experimental results.

Table VI. Equilibrium structures for $ScC_2H_2^+$, $ScC_2H_2^{2+}$, and the relevant free ligands. The bond lengths are in Å, the bond angles are in degrees, and the binding energies (BE) are in kcal/mol

	r(C-C)	r(C-H)	r(M-BMP[a])	H-Bend[b]	BE
C_2H_2	1.21	1.06			
Expt.[c]	1.203	1.061			
C_2H_4	1.33	1.08			
Expt.[c]	1.330	1.076			
$ScC_2H_2^+$	1.33	1.08	1.93	47.5	44.3
$ScC_2H_2^{2+}$	1.22	1.08	2.42	9.8	47.4

[a] BMP is the bond mid-point of the C-C bond. [b] H-Bend is the angle for H bending away from the metal. [c] Ref. [81].

6.1. Covalent vs electrostatic bonding

For a long time it was assumed that charge exchange would be the only product of reactions involving dipositive ions. However, experiments by Tonkyn and Weisshar [77] demonstrated that product formation could occur for Ti^{2+}. Reactions can occur for dipositive ions on the far left side of the row, because the second IP of the metal is less than the first IP of many ligands. Recently, the doubly charged metal ions have become an area of active research. For example, Freiser and co-workers [78,79] determined the binding energies of $La^+$and La^{2+} with several unsaturated hydrocarbons. They found that La^+ was more strongly bound than La^{2+} to both C_2H_2 and C_3H_6. For example the C_3H_6 binding energies were 63 ± 6 and 41 ± 6 kcal/mol for La^+ and La^{2+}, respectively. The bonding was clearly not electrostatic in origin, as the La^{2+} binding energies would be much larger than those for La^+. This larger binding energy for the monopositive ion implies bond insertion has occured for La^+, which results in a significant covalent contribution to the bonding. La^{2+} does not insert because it can form only one covalent bond as a result of having only one valence electron.

To compare the bonding in the singly and doubly charged ions, we carried out electronic structure calculations [80] to determine the equilibrium structures and binding energies for the mono- and dipositive ions of Sc, Y, and La interacting with C_2H_2, C_2H_4, and C_3H_6. As expected, the monopositive ions insert into the π bond and the dipositive ions bind electrostatically. This difference in the bonding mechanism is reflected by the geometries [80,81] given in Table VI. The C-C bond distance in $ScC_2H_2^+$ is very close to the double bond distance in C_2H_4 while the C-C bond distance in $ScC_2H_2^{2+}$ is nearly the same as in free C_2H_2. The change in the C-C-H angle is also consistent with the expected C hybridization. For the electrostatically bonded systems, the H atoms are only slightly bent away from the metal, because the C is still essentially *sp* hybridized. Conversely the C-C-H bond angle of 132.5° for $ScC_2H_2^+$ indicates a C hybridization close to sp^2.

It is interesting to note that while the computed C_2H_2 binding energy for La^+ is larger than that for La^{2+}, 52.9 vs 38.6 kcal/mol, the opposite trend is observed for Sc, see Table VI. The short C-C bond length and the large size of the metal atom results in a very small C-M-C angle for the $MC_2H_2^+$ systems. At this small angle the metal must promote to the d^2 occupation to form strong M-C bonds. Although this is the ground state occupation for La^+, it is an excited state for Sc^+. Thus the covalent bond is weaker in $ScC_2H_2^+$ than $LaC_2H_2^+$, due to the promotion energy required to reach the Sc^+d^2 occupation. In addition to this change for the monopositive ions, the binding energy for $Sc^{2+}$is larger than that for La^{2+}due the smaller size of Sc^{2+}. Therefore, while the bonding in both the mono- and dipositive Sc and La systems is qualitatively similar, there are significant differences in the relative binding energies for the mono- and dipositive systems.

The binding energies for the first and second transition row $MC_2H_2^+$ systems [82] and the first transition row $MC_2H_4^+$ systems [83] are compared with experiment [82–86] in Table VII. The type of bonding is denoted as "E" for electrostatic and as "C" for covalent. For C_2H_2, covalent bonding is favored for Sc^+, Ti^+, and V^+ in the first transition row and for Y^+, Zr^+, Nb^+, and probably Mo^+ in the second transition row. For C_2H_4, only Sc^+ and Ti^+ favor covalent bonding. As we show next, theory offers a straightforward explanation of these results, which is not apparent from the experimental dissociation energies alone.

Due to the more comparable radial extent of the *s* and *d* orbitals, *sd* hybridization is more favorable on the left side of the row. This results in a decrease in the covalent bond strength with increasing *Z*. Also decreasing the strength of the covalent bonding with increasing *Z* for the first half of the row is the loss of exchange energy when bonds are formed. Conversely, the strength of the electrostatic bonding increases from left to right paralleling the decrease in radial extent of the ion with increasing *Z*. These opposite trends result in electrostatic bonding becoming favored over covalent bonding at Cr^+. The transition from covalent to electrostatic bonding occurs one atom later in the second tranition row, because the second transition row atoms form stronger covalent bonds as a result of having smaller exchange energies and more comparable radial extents for the *s* and *d* orbitals.

The M-C bonds are considerably weaker for $MC_2H_4^+$ than $MC_2H_2^+$ when bond insertion occurs, because the π bond in C_2H_4 is stronger (by 28 kcal/mol at the MCPF level) than that in C_2H_2. Thus it is not surprising that $VC_2H_2^+$ favors covalent bonding while $VC_2H_4^+$ favors electrostatic bonding. The experimental results for $MC_2H_4^+$ are believed to be much more accurate than those for $MC_2H_2^+$. Thus the generally good agreement between the experimental and theoretical $MC_2H_4^+$ binding energies in Table VII give us considerable confidence that theory can be used to identify incorrect

Table VII. Comparison of the experimental and theoretical dissociation energies (in kcal/mol) for $MC_2H_2^+$ and $MC_2H_4^+$. The type of bonding, either covalent, denoted "C", or electrostatic, denoted "E", is also given

		$MC_2H_2^+$		$MC_2H_4^+$	
Metal	Type	Theory	Expt	Theory	Expt
Sc	C	51	52 ± 3[a], 78.0[b]	32	⩾35.1 ± 1.2[b], 40 ± 5[c]
	E	23		21	
Ti	C	47	71.0[b]	31	⩾32[b]
	E	26		23	
V	C	35	50.8[b]	18	~50[b], 16–28[b]
	E	29		29	
Cr	E	25	48.5[b]	26	33 ± 5[b]
Mn	E	19		20	
Fe	E	28		30	42 ± 9[b], 34 ± 2[d]
Co	E	37	33.9[b]	40	>40 ± 5[b], 37 ± 2[c], 46 ± 8[c]
Ni	E	39	30.0[b]	41	>35 ± 5[b], 48 ± 8[c], 37 ± 2[d]
Cu	E	36	30.5[b]	40	26 ± 3[b], >28 ± 1[d]
Y	C	56	52 ± 3[a], 70.8[b]		
	E	27			
Zr	C	68			
	E	34			
Nb	C	59			
	E	32			
Mo	E	24	74.6[e]		
	C	23			
Tc	E	25			
Ru	E	32			
Rh	E	33			
Pd	E	35			
Ag	E	28			

[a] Ref. [84]. [b] Ref. [85]. [c] Ref. [86]. [d] Ref. [87]. [e] Ref. [88].

experimental measurements. For example, theory demonstrated that the original [85] $ScC_2H_2^+$ binding energy was incorrect. This has now been confirmed by recent experiments [84]. The error in the original experiment has been attributed to ethylene contamination in the ethane source.

6.2. Relative binding energies of singly and doubly charged ions

Hill *et al.* [89] measured the binding energies of Y^{2+} to a series of saturated and unsaturated hydrocarbons. They had expected that Y^{2+}, with only one valence electron, would bind electrostatically to hydrocarbons, so that the binding energy should increase with the ligand polarizability (and hence increase with the ligand size), but that all unsaturated ligands, regardless of the number of carbon atoms, should be more strongly bound than the saturated ligands. Thus the expected ordering for electrostatic bonding is

$$CH_4 < C_2H_6 < C_3H_8 < C_4H_{10} < C_2H_2 < C_2H_4 < C_3H_6 < C_4H_8.$$

Instead, for Y^{2+}, they observed the following order:

$$CH_4,\ C_2H_6 < C_2H_2,\ C_2H_4 < C_3H_8 < C_3H_6 < C_4H_{10} < C_4H_8.$$

These binding energies are ordered principally by the number of carbon atoms and for a given number of carbon atoms, the unsaturated systems are more strongly bound than the saturated ones.

To help explain the trends observed in the experimental binding energies, calculations [89] were carried out for Sc^{2+} and Y^{2+}, as well as Cu^+ to contrast the singly and doubly charged ions. The computed binding energies are given in Figure 9 along with the structures for ScL^{2+}. The ScL^{2+} and YL^{2+} potentials for $L = C_2H_2$ and C_2H_4 are so flat with respect to bending that we cannot definitively determine if the metal ion is above the C-C bond midpoint at the level of theory employed. Also, the computed binding energies reported in Ref. [89] were systematically too small. For example, the $CuCH_4^+$ binding energy was reported as 14.8 kcal/mol with an $\eta 3$ geometry, whereas recent highly accurate calculations [19] show that the most stable configuration is $\eta 2$ with a binding energy of about 21 kcal/mol. Nevertheless, the theoretical binding energies agree with the experimental trends for Cu^+ and Y^{2+}, and are sufficiently accurate to give considerable insight into why the binding energies follow the observed trends.

The calculations show that the bonding in all of the systems is principally electrostatic, resulting in larger binding energies for the doubly charged ions than the singly charged ions. The difference in binding energies is larger for the saturated systems than for the unsaturated ones. For example, the $Y^{2+} - C_3H_8$ binding energy is 20.2 kcal/mol larger than Cu^+-C_3H_8, whereas the difference is only 9.3 kcal/mol for the C_3H_6 system. This is a consequence of metal to π^* donation that occurs for the singly charged Cu^+, but not for the doubly charged ions. The different trend observed for the Cu^+ ion is due to the fact that the dative bonding interaction is larger than the increase in the charge-induced dipole interaction with increasing ligand size. Since the dative interaction is not important for the doubly charged ions, the trend is dictated principally by the magnitude of the charge-induced dipole interaction. The large metal to π^* donation causes Cu^+ to strongly favor the C-C bond midpoint for the unsaturated systems, whereas the potential energy surface for moving the metal parallel to the C-C bond is very flat for the dipositive ions.

The importance of metal to π^* donation for the singly charged ions is well illustrated by comparing calculations on $Sc^+ - C_2H_2$ and $Sc^+ - C_2H_6$. To prevent Sc^+ insertion into the C-C π bond, calculations were carried out for the triplet state of Sc^+. The calculated binding energies of 17.8 and 9.2 kcal/mol for $Sc^+ - C_2H_2$ and $Sc^+ - C_2H_6$, respectively, differ by an amount that is similar to Cu^+ rather than Sc^{2+}. This is further illustrated by the ratio of the C_2H_2/C_2H_6 binding energies for Cu^+ and Sc^+ (2.1 and 1.9, respec-

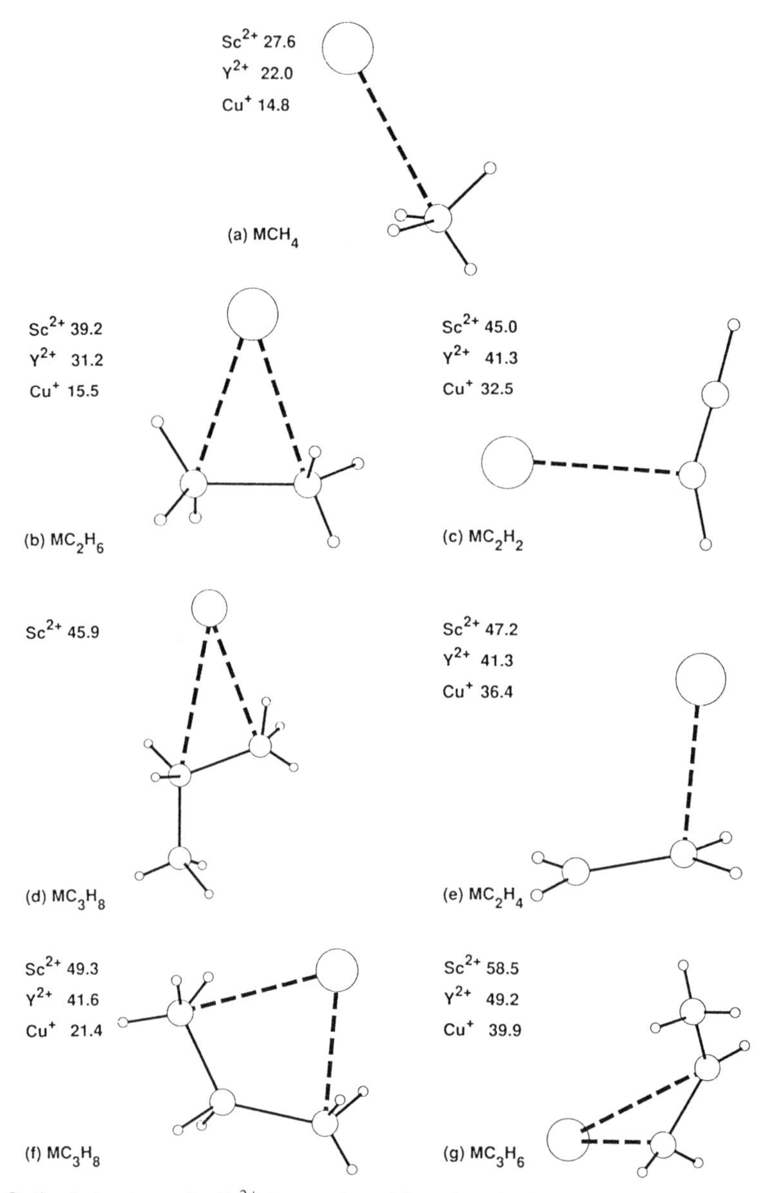

Fig. 9. Optimal structures for Sc^{2+} interaction with various hydrocarbons. The structures are drawn to scale for Sc^{2+}. The MCPF binding energy (in kcal/mol) for Sc^{2+}, Y^{2+}, and $Cu^{+}$are given. The Y^{2+} and Cu^{+} structures are similar to those for Sc^{2+} except for MC_2H_2 and MC_2H_4, where the metal sits at the bond midpoint.

tively) which are very different from those of Sc^{2+} and Y^{2+} (1.2 and 1.3, respectively).

For $Sc^{2+} - C_3H_8$ we compared the binding energies where Sc^{2+} interacts with only one C_2H_5 unit, see Figure 9(d), and where it interacts with two groups, see Figure 9(f). By comparison with C_2H_6, we see that while replacing an H atom with a CH_3 group increases the binding energy, this structure is more than 3 kcal/mol less stable than structure 9(f) where the Sc^{2+} ion is able to interact with the entire C_3H_8 ligand.

It should be noted that the trends for the Sc^{2+} and Y^{2+} binding energies with hydrocarbons are not typical of all doubly charged ions. For example, Nb^{2+}, with three valence electrons, forms a double bond with CH_2 and inserts into the π bonds of C_2H_2. Calculations [90] were performed to decide between two disparate experimental values for the $NbCH_2^{2+}$ binding energy. The larger of the experimental values [91] was deduced from the $NbCH_2^+$ binding energy and the ionization potential (IP) of $NbCH_2^+$. The IP was deduced using a bracketing approach for the reaction $NbCH_2^{2+} + A \leftarrow NbCH_2^+ + A^+$. While such bracketing approaches work well for monopositive ions, they are difficult to apply to dipositive ions, because the $1/r$ repulsive interaction of the two singly charged ions must be accounted for. This correction requires knowledge of the r value where charge exchange occurs, which is not currently available from experiment. The calculations were able to show that the IP deduced from bracketing reactions was too small, thereby resolving the discrepancy between the two measured values [91] of the $NbCH_2^{2+}$ binding energy.

6.3. The unexpected order of the $M^+ - He$ and $M^+ - Ne$ binding energies.

Bowers and co-workers [92] recently developed the "electronic state chromatography" technique which allows them to separate various states of the metal ions (at least with different *d* occupations), because of their different mobilities in a high temperature drift cell. Accurate binding energies can be determined by equilibrium measurements. This was confirmed by the excellent agreement for the $CrAr^+$ dissociation energy found by this technique with that determined by accurate spectroscopic experiments [93]. Average binding energies and potential energy curves can be deduced by fitting the mobility data as a function of temperature. Using these approaches [92] they have been able to determine the binding energies for both the ground and excited states of a number of weakly bound metal ion-rare gas (Rg) systems.

The MRg^+ binding energy is largely determined by the charge-induced dipole interaction, which is proportional to α/r^4, where α is the ligand polarizability and r is the internuclear separation. The polarizabilities [94] and radial expectation values [95] of the Rg atoms increase with Z; the polarizabilities (radial expectation values), in units of $a_0^3(a_0)$ are:

Table VIII. Spectroscopic constants for metal rare gas ions[a]

Ion	State[b]	He		Ne		Ar	
		r_e	D_e	r_e	D_e	r_e	D_e
Na^+	$^1\Sigma^+(2s^22p^6)$	4.563	0.033	4.688	0.063	5.408	0.143
V^+	$^5\Sigma^+(3d^4)$	3.903	0.132	4.748	0.087	5.019	0.313
Cr^+	$^6\Sigma^+(3d^5)$	4.455	0.044	5.089	0.052	5.083	0.238
Co^+	$^3\Delta\ (3d^8)$	3.536	0.173	4.438	0.114	4.597	0.429
	$^5\Phi\ (3d^74s^1)$	6.073	0.016	5.795	0.032	5.286	0.188
Ni^+	$^2\Sigma^+(3d^9)$	3.464	0.193	4.248	0.146	4.487	0.492
	$^2\Delta\ (3d^9)$	3.725	0.096	4.668	0.076	4.508	0.400
	$^4\Delta\ (3d^84s^1)$	6.042	0.019	5.587	0.038	5.103	0.219
Cu^+	$^1\Sigma^+(3d^{10})$	3.718	0.110	4.489	0.096	4.487	0.405

[a] r_e in a_o and D_e with respect to the ground state of the metal ion in eV. [b] The metal occupation for the state is given in parentheses.

1.387(0.927), 2.64(0.963) and 11.13(1.656) for He, Ne, and Ar, respectively. Because of the similar size of He and Ne, one expects the M^+ – Ne binding energies to be uniformly larger than for M^+ – He, because of the significantly larger polarizability of Ne. While the binding energies determined by Bowers and co-workers followed the trend Ar > Ne > He for Cr^+, for Co^+ and Ni^+ they unexpectedly found that the He binding energies were larger than the Ne binding energies. Because of the relatively simple ion – Rg interaction and the weak Rg-Rg interactions, it was assumed that the metal ion-Rg systems would provide ideal models for clustering reactions. It was therefore disconcerting that their results were inconsistent with existing models. Hence, high quality *ab initio* calculations [43] were performed to assess the reliability of the experimental determinations and to explain the experimental trends.

The computed r_e and D_e values for selected metal ion Rg systems are reported in Table VIII. These results were obtained using very large basis sets and an MCPF level treatment of electron correlation. The binding energies for $NaRg^+$ show the expected trend. While the r_e values of $NaHe^+$ and $NaNe^+$ are comparable, the binding energy of $NaNe^+$ is almost twice as large. The $NaAr^+$ ion has a significantly longer bond length and larger binding energy than $NaNe^+$. The computed binding energies for $CrRg^+$ also follow the expected order and are in reasonable agreement with experiment; the computed values are about 80% of the experimental results, as expected for weakly bound systems at this level of theory. However, the $CrRg^+$ systems are different from $NaRg^+$ in that the $CrHe^+$ bond length is significantly shorter than $CrNe^+$, which leads to comparable binding energies for $CrHe^+$ and $CrNe^+$.

The remaining ions included in Table VIII have a $3d^{n+1}$ ground state occupation. For all of these ions, the M^+ – He binding energy is larger than M^+ – Ne, consistent with the experimental results of Bowers and co-workers [92] for Co^+ and Ni^+. The excited states of Co^+ and Ni^+ with the $4s$ orbital

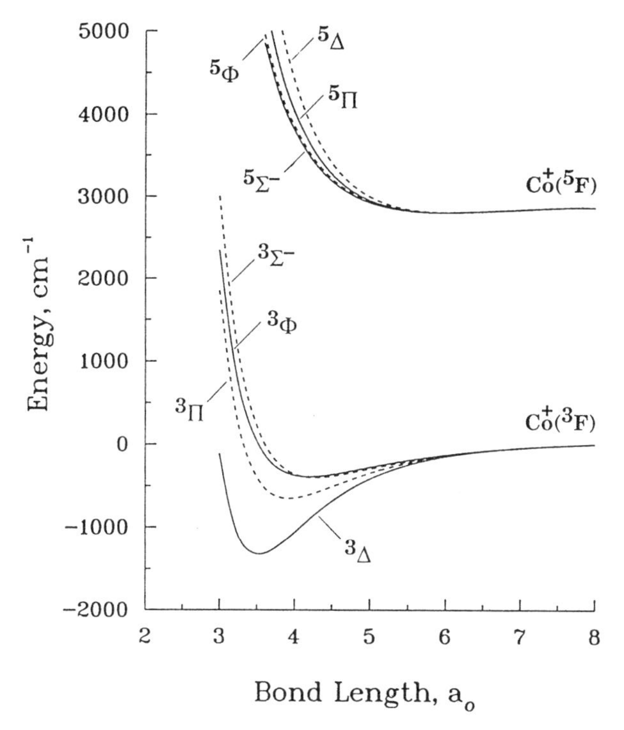

Fig. 10. The MRCI+Q potential energy curves for $CoHe^+$. Alternate potential energy curves are dashed to help distinguish the curves. The $^3F(3d^8)$ and $^5F(3d^74s^1)$ asymptotes for Co^+ are given.

occupied show relative He and Ne binding energies similar to those for $NaRg^+$, but the relative transition metal-Rg bond lengths are very different from those of $NaRg^+$.

The theoretical and experimental observations can be understood in terms of the mechanisms available to the metal ions for reducing metal ligand repulsion. For the metal ions with a $3d^{n+1}$ occupation, $sd\sigma$ hybridization and the orientation of the $3d$ holes are both important factors affecting the metal-ligand repulsion. These effects are illustrated by the potential energy curves [96] for He and Ar interacting with $Co^+(^3F)$ and $Co^+(^5F)$ shown in Figures 10 and 11. There is a significant variation in the D_e and r_e values for the four states arising from the Rg-$Co^+(^3F)$ interaction due to the different orientations of the $3d$ holes. The metal ligand repulsion is minimized by singly occupying the $3d\sigma$ orbital since the $3d\sigma$-Rg overlap is larger than the overlap of the Rg atom with the metal $3d\pi$ or $3d\delta$ orbitals. Thus the $^3\Delta$ state, with a $3d\sigma^13d\pi^43d\delta^3$ occupation, is the most strongly bound. The $^3\Pi$ state, which is described by the configurations $3d\sigma^13d\pi^33d\delta^4$ and $3d\sigma^23d\pi^33d\delta^3$, is the next most strongly bound, because it has some $3d\sigma$

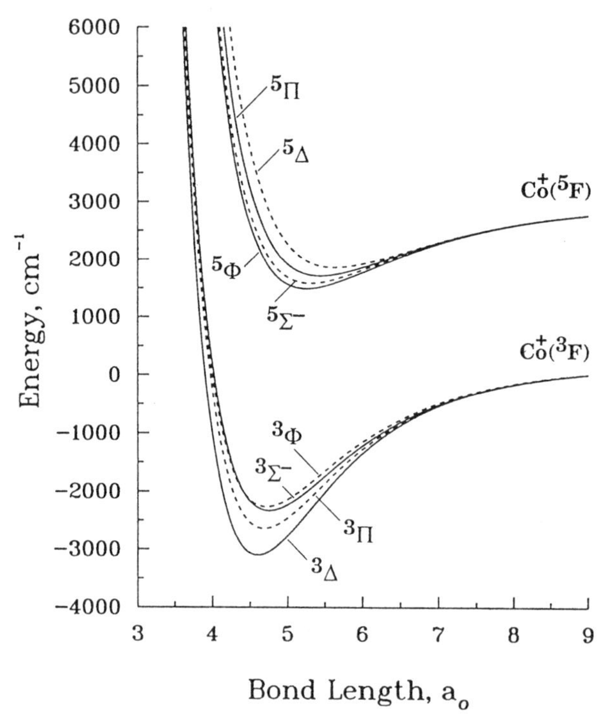

Fig. 11. The MRCI + Q potential energy curves for $CoAr^+$. Alternate potential energy curves are dashed to help distinguish the curves. The $^3F(3d^8)$ and $^5F(3d^74s^1)$ asymptotes for Co^+ are given.

hole character. The $^3\Sigma^-$ and $^3\Phi$ states, which have the $3d\sigma$ orbital doubly occupied, are the least strongly bound and are nearly degenerate.

For Cr^+ and Cu^+ the $3d$ shell is singly and doubly occupied, respectively. Thus there is no flexibility in the occupation of the $3d$ orbitals without mixing in much higher lying states. Therefore, the only mechanism for reducing metal-ligand repulsion is $sd\sigma$ hybridization. Since this mechanism must mix in some excited state character and does not remove an entire electron from the $3d\sigma$ orbital, it is less effective at reducing the repulsion than localizing a hole in the $3d\sigma$ orbital.

If the $4s$ orbital is occupied, sp or $sd\sigma$ hybridization can reduce the metal-ligand repulsion by polarizing the $4s$ electron density away from the ligand. The magnitude of this effect increases with the strength of the interaction. This leads to the inversion in the MRg^+ bond lengths (He > Ne > Ar), with respect to those observed for $NaRg^+$ or the states derived from the $3d^{n+1}$ occupation, for the $^5\Phi$ state of $CoRg^+$ and the $^4\Delta$ state of $NiRg^+$, see Table VIII. The states arising from a metal ion with the $4s$ orbital occupied show less variation in the D_e values with $3d$ occupation, because the much larger $4s$ orbital dominates the interaction. For example, the bonding between

$Co^+(^5F)$ and He is too weak to significantly differentiate the states of $CoHe^+$(see Figure 10). The stronger interaction with Ar results in some splitting of the quintet states of $CoAr^+$ (see Figure 11), but this splitting is still much smaller than for the lowest asymptote where the metal has a $3d^{n+1}$ occupation.

The observed and calculated binding energies can be explained in terms of the mechanisms that reduce metal-Rg repulsion. For the lowest states derived from the $3d^{n+1}$ occupation, the bonding is enhanced by either $sd\sigma$ hybridization or by minimizing the occupation of the $3d\sigma$ orbital. Because the $3d$ orbital is relatively compact, the slightly smaller size of the He atom results in a larger reduction in metal-ligand repulsion than for the Ne atom. Thus the MHe^+ bond length is always significantly smaller than for the corresponding MNe^+ systems. Assuming the interaction is entirely charge-induced dipole, when the MNe^+ bond length is more than 1.175 times longer than the MHe^+ bond length, the $1/r^4$ term will overcome the difference in the He and Ne polarizabilities and the $M^+ - He$ binding energy will exceed the $M^+ - Ne$ binding energy. This occurs for all of the systems where the $3d\sigma$ occupation is less than that of the $3d\pi$ orbital, such as for VRg^+, where the $3d\sigma$ orbital is empty, and for the states of $CoRg^+$ and $NiRg^+$ where the $3d\sigma$ orbital is singly occupied. The inversion in the binding energies also occurs for $NiRg^+(^2\Delta)$ and $CuRg^+$, where only $sd\sigma$ hybridization is available to reduce the metal ligand repulsion. However, as $sd\sigma$ hybridization is less efficient at reducing the repulsion, the ratio is closer to 1.175 and the MHe^+ systems are only slightly more strongly bound. For CrL^+, $sd\sigma$ hybridization reduces the He repulsion and hence shortens the bond length relative to Ne. However, the bond length shortening is smaller than for other systems, because the $3d^44s^1$ state is rather high lying and the larger size of the Cr^+ ion reduces the effectiveness of the hybridization. The $CrHe^+$ bond length is only 1.142 times shorter than the $CrNe^+$ bond length, resulting in a slightly larger binding energy for $CrNe^+$.

The *ab initio* results fully corroborate the experimental observations that the $M^+ - He$ binding energies can be larger than the $M^+ - Ne$ binding energies. The calculations show that the slightly smaller He atom benefits more from the reduction in the metal-ligand repulsion associated with $sd\sigma$ hybridization or with the orientation of the $3d$ holes. The calculations demonstrate that these effects are not sufficiently large for $Cr^+ - Rg$ to result in an inversion of the He and Ne binding energies.

The computed binding energies are compared with the experimental values [92,93] in Table IX. In general, *ab initio* calculations (corrected for BSSE) will give binding energies smaller than the experimental values because improvements in the calculations will lower the molecular complex more than the asymptotes. Surprisingly, however, we found some of the computed binding energies for the $M^+ - Rg$ systems to be larger than the experimental values. The difference arises [92] because the spin-orbit splitting is comparable to the $M^+ - Rg$ binding energies, particularly for $Ni^+$and

Table IX. Comparison of the computed and experimental D_0 values (eV) for selected transition metal noble-gas ions[a]

	He	
	Expt	MCPF
Cr^+	0.042 ± 0.004	0.035
Co^+	0.131 ± 0.002	0.152(0.123)
Ni^+	0.129 ± 0.002	0.170(0.095)
Ni^{+} [c]	0.016 ± 0.003	0.015
	Ne	
	Expt	MCPF
Cr^+	0.060 ± 0.004	0.047
Co^+	0.095 ± 0.004	0.106(0.077)
Ni^+	0.103 ± 0.004	0.136(0.061)
Ni^{+} [c]	0.032 ± 0.009	0.035
	Ar	
	Expt	MCPF
V^+	0.369[b]	0.303(0.287)
Cr^+	0.284 ± 0.018, 0.29 ± 0.04[b]	0.230
Co^+	0.508[b]	0.418(0.402)
Ni^+	0.55[b]	0.480(0.405)

[a] The theoretical ω_e values are used for the zero-point correction. The experiments are from Ref. [92] unless otherwise noted. The values in parentheses include a correction for the spin-orbit interaction. [b] Reference [93]. [c] The $^4\Delta$ excited state.

Co^+. Thus, it is necessary to include a correction for the spin-orbit interaction for these very weakly interacting systems. After this correction has been made, the theoretical values are consistently about 80% of experiment, as expected for this level of theory.

6.4. The successive binding energies of $M(H_2O)_n^+$

Marinelli and Squires [97] and Michl and co-workers [98] found that the second water ligand was more strongly bound than the first for some transition metals. This is in distinct contrast to the alkali metals, where each successive water is less strongly bound, as the interaction is dominated by the increasing ligand-ligand repulsion. To explain why the trends were different for transition metals, calculations [44] were carried out for the ions in the first transition row.

The bonding in the alkali systems, such as $Na(H_2O)_n^+$, is entirely electrostatic in origin. This interaction is well described even at the SCF level [73]; the first four computed binding energies are in excellent agreement with experiment [71], see Table X. The addition of correlation at the MCPF level has little effect on the binding energies. The trends [96–99] in the successive

Table X. Successive $M(H_2O)_n^+$ binding energies (in kcal/mol) for M = Na and Cu

	D_0		Expt
	SCF[a]	MCPF[a]	
$Na^+ - H_2O$	23.3	23.4	23.4[b]
$Na(H_2O)^+ - H_2O$	20.6	20.7	19.2[b]
$Na(H_2O)_2^+ - H_2O$	16.7		15.2[b]
$Na(H_2O)_3^+ - H_2O$	13.9		13.2[b]
	SCF[c]	MCPF[c]	
$Cu^+ - H_2O$	32.4	38.8	35 ± 3[d] 37.6 ± 2.6[e]
$Cu(H_2O)^+ - H_2O$	30.0	39.4	39 ± 3[d] 39.0 ± 0.6[e]
$Cu(H_2O)_2^+ - H_2O$	17.2	15.4	17 ± 2[d] 14.7 ± 0.4[e] 15.8 ± 0.2[f]
$Cu(H_2O)_3^+ - H_2O$	12.9	13.3	15 ± 2[d] 15.2 ± 1.1[e] 16.1 ± 0.2[f]

[a] Ref. [73]. [b] Ref. [71]. [c] Ref. [101]. [d] Ref. [98]. [e] Ref. [100]. [f] Ref. [99].

binding energies are quite different for $Cu(H_2O)_n^+$. There is a significant correlation effect on the first two binding energies and the second ligand is more strongly bound than the first, in agreement with experimental observations. The calculations show that unlike the alkali ions, the Cu^+ ion has the flexibility of reducing the repulsion through $sd\sigma$ hybridization. As this effect is not well described at the SCF level, there is a large increase in the binding energy of the first two waters with the addition of electron correlation. Since $sd\sigma$ hybridization reduces the repulsion on both sides of Cu^+ at the same time, two ligands can share the cost of the hybridization leading to a second ligand binding energy that is larger than the first. When more ligands are added, it is impossible to arrange them in a fashion where all of the ligands can benefit from $sd\sigma$ hybridization. Thus with the addition of the third water, this effect is lost and the closed-shell Cu^+ acts much like Na^+.

The calculations [44] also showed that for systems like $Mn(H_2O)_n^+$ where the $3d^6$ excited state lies too high in energy to contribute significantly, the second ligand is less strongly bound than the first, in agreement with experiment [97]. The 7.3 kcal/mol reduction in the second binding energy for $Mn(H_2O)_2^+$ is much larger than that found for $Na(H_2O)_2^+$, because the polarization of the Mn 4*s* orbital leads to a bent structure in analogy with that of AlL_2^+ or MgL_2^+. As noted in section 3, the bonding in $Fe(H_2O)_n^+$ changes between the first and second water ligands. The first water bonds to Fe^+ in its $^6F(3d^64s^1)$ ground state, while two water ligands bond to the excited Fe^+ $3d^7$ occupation. The smaller repulsion and larger electrostatic attraction due to the smaller size of the $3d^7$ occupation leads to the second ligand being more strongly bound than the first in spite of having to pay the promotion energy. The larger second ligand binding energy is consistent with experiment. The $Ni(H_2O)_n^+$ calculations illustrate another interesting effect. For NiH_2O^+, the 3*d* hole is in the $3d\sigma$ orbital, but for $Ni(H_2O)_2^+$ $sd\sigma$ hybridization is sufficiently efficient at reducing the repulsion in the σ space, that the $3d$ hole moves to the $3d\pi$ orbital to reduce the repulsion between the Ni and the H_2O out-of-plane lone pair electrons.

The calculations are again able to explain the trends observed experimentally. They illustrate that several competing effects act to reduce metal-ligand repulsion. Since these effects are unique to the transition metals, it is generally inappropriate to make comparisons based on the well-known alkali systems.

7. Metal ion spectroscopy

Absorption or emission spectroscopic studies of systems containing metal ions are generally complicated by the very low concentrations that can be generated. However, bound-bound transitions have been studied by detecting fragment ions produced either by photodissociation of the excited state by another photon [102] or by crossing from the excited state onto the ground state potential above the dissociation limit [103,104]. For closed-shell ligands the excited states of $M^+ - L$ usually correspond to different occupations of the metal orbitals. Because these metal orbitals may have very different overlaps with the ligand, the changes in the bonding between the various states can offer insight into how different metal orbitals contribute to the bonding. In this section we show how theory has been used to help interpret experimental spectra.

The $^2D(4d^1)$ and $^2P(5p^1)$ states of Sr^+ lie 14,700 and 24,316 cm^{-1} above the $^2S(5s^1)$ ground state [36], respectively. The bonding in the SrL^+ground state is expected to be primarily electrostatic for closed-shell ligands. The bonding will be enhanced by polarization of the $5s$ orbital away from the ligand. However, unlike Al^+ or Mg^+, where this involves sp polarization, the low-lying ^2D state of Sr^+ results in the metal-ligand repulsion being preferentially reduced by $sd\sigma$ hybridization. As a consequence, the repulsion on both sides of the Sr^+ ion is reduced at the same time. This causes SrL_2^+ to be linear [48] rather than bent like the AlL_2^+ or MgL_2^+ systems.

Farrar and co-workers [103,104] measured the spectra for $Sr(H_2O)_n^+$ and $Sr(NH_3)_n^+$. In order to assign their spectra, they considered molecular states derived from the ^2S, ^2P, and ^2D states of Sr^+. The shifts in these atomic ion states due to the ligands was estimated by considering how the ligands overlapped with the Sr^+ orbitals. One interesting feature of their study was the dramatic change in the $Sr(NH_3)_3^+$ spectra relative to those for $SrNH_3^+$ or $Sr(NH_3)_2^+$. They attributed this to a change in character from $Sr^+(NH_3)_3$ in the ground state to $Sr^{2+}(NH_3)_3^-$ in the excited states.

The bonding in the states of SrH_2O^+ derived from the ^2S, ^2P, and ^2D states of Sr^+ is analyzed [105] in Table XI. The overlaps of the Sr^+ and H_2O separated fragment orbitals are computed for the ground state Sr-O distance; note that the Sr^+ orbitals are taken from a SA-CASSCF calculation that includes the three states of interest in the averaging procedure. As expected,

Table XI. Analysis of the SrH_2O^+ excitation energies (in cm^{-1})

Overlap of Sr^+ orbitals with H_2O				
Symmetry				
a_1	$5s = 0.335$	$5p\sigma = 0.475$	$4d\sigma = 0.438$	$4d\delta = 0.002$
b_2	$5p\pi = 0.104$	$4d\pi = 0.123$		
b_1	$5p\pi = 0.122$	$4d\pi = 0.157$		

		frozen orbitals[a]		
	Free Sr^{+b}	SCF[c]	VCI	SA-CASSCF
$(1)^2A_1(5s)$	0	0	0	0
$(2)^2A_1(4d\delta)$	13316	10475	12544	13124
$(1)^2A_2(4d\delta)$	13316	10481	12557	13156
$(1)^2B_2(4d\pi)$	13316	11971	13717	14434
$(1)^2B_1(4d\pi)$	13316	13501	14580	15282
$(3)^2A_1(4d\sigma)$	13316	22072	15308	15841
$(2)^2B_2(5p\pi)$	21301	16680	19084	18561
$(2)^2B_1(5p\pi)$	21301	17694	20767	20299
$(4)^2A_1(5p\sigma)$	21301	25050	38047	28424

[a] The orbitals are from Sr^+ and H_2O and are unchanged except for orthogonalization. [b] The experimental [36] ^2D-^2S and ^2P-^2S separations are 14,700 and 24,316 cm^{-1}, respectively. [c] Evaluated using the single configuration energies.

the overlaps are significantly larger for the "σ" orbitals than for the "π" orbitals. The a_1 "δ" orbital overlap is very small. Thus the overlaps are consistent with the Sr^+ being in a pseudolinear environment. It is important to note that the $4d\sigma$ overlap is comparable to the other σ orbitals. The b_2 overlaps are slightly smaller than for the out-of-plane b_1 symmetry, because the O-H bonds are bent away from Sr^+.

Using these fragment orbitals, except for orthogonalization, the energy of each single configuration is evaluated – the energy separation at this level is summarized in Table XI under the column "SCF". The separations are significantly different from free Sr^+. Note that the ^2D and ^2P states of free Sr^+ are computed to be 1400 and 3000 cm^{-1} too low relative to the ^2S state, due to the neglect of relativistic and core-valence correlation effects. The shifts in the molecular states relative to the atomic states from which they are derived are consistent with the Sr^+ – water overlaps. For example, the states derived by promoting the $5s$ orbital to the $5p\sigma$ and $4d\sigma$ orbitals are higher in the molecule than the atom, because the $5s$ orbital has a smaller overlap with the water ligand than do the $5p\sigma$ and $4d\sigma$ orbitals. Excluding the $(1)^2B_1$ state, the remaining states are shifted down relative to the atomic state from which they are derived, because the π and δ orbitals have smaller overlaps with the ligands than does the $5s$ orbital. The $(1)^2B_1$ state is slightly shifted up in energy relative to the ^2D state of Sr^+, despite the fact that the $4d\pi(b_1)$ orbital has a smaller overlap with the water than does the $5s$ orbital. It therefore appears that there is more energy associated with the $4d$-water

Table XII. Vertical excitation energies[a] for $SrNH_3^+$

State	Separation			TM^2	Population		
	CASSCF	Est[b]	Expt[c]	CASSCF	$5s$	$5p$	$4d$
$SrNH_3^+$							
$(1)^2A_1$	0	0	0		0.93	0.04	0.09
$(2)^2A_1$	16268	17700	17300	2.82	0.03	0.43	0.59
$(3)^2A_1$	30370	<28900		2.11	0.12	0.68	0.41
$(1)^2E$	13681	14700		0.00	0.02	0.01	1.03
$(2)^2E$	15380	16400		2.34	0.02	0.11	0.94
$(3)^2E$	19060	21000	20700	12.88	0.02	0.92	0.14

[a] Separations in cm^{-1}, and the square of the transition moment (TM^2) in a.u.^{2}. [b] Our estimated values; these include a correction based on the error in the atomic separations and the molecular populations. [c] The molecular results are deduced from the spectra of Shen and Farrar [103,104].

overlap than the $5s$-water overlap. Similarily, the $5p\pi(b_1)$ and $4d\pi(b_2)$ orbitals have essentially the same overlap with water, but the $5p\pi(b_1)$ derived state is shifted to lower energy. Thus both the kind of orbital and its overlap are important factors in determining the repulsion.

In the next step, the occupations of the same symmetry are allowed to mix in a valence CI (VCI) calculation. The mixing occurs in both the σ and the π spaces, but there is a larger energetic effect associated with the mixing of the $5s$, $5p\sigma$, and $4d\sigma$ orbitals. This results in a stabilization of the ground state with respect to all states except the $(3)^2A_1$ state, where the strong interaction in the σ space stabilizes the $(3)^2A_1(4d\sigma)$ state by even more than the ground state. Fully relaxing the orbitals in the SA-CASSCF calculation further stabilizes the ground state with respect to the states derived from the $4d$ orbital, but by less than the $5p$ derived states. Thus the stabilization follows the orbital polarizability.

SA-CASSCF calculations were performed for $Sr(H_2O)_n^+$ and $Sr(NH_3)_n^+$, for $n = 1$–3; the results [48] of these calculations are consistent with the overlap and hybridization concepts illustrated in the analysis of SrH_2O^+. The vertical excitation energies for $SrNH_3^+$ given in Table XII agree well with experiment, if we assign the $^2E(5p\pi)$ state to the transition at 20,700 cm^{-1} and the $(2)^2A_1(5p\sigma + 4d\sigma)$ state to the transition at 17,300 cm^{-1}. Shen and Farrar had assigned the ^2E state to the lower transition and the $(3)^2A_1$ state to the transition at higher energy. To identify the molecular states, they considered the atomic states and the metal-ligand overlaps, which is similar in spirit to our analysis for SrH_2O^+. While Shen and Farrar correctly identified the character of the bonding in the various states they underestimated the efficiency of sp and $sd\sigma$ hybridization at reducing the Sr-water repulsion and as a consequence placed the $5s$-derived ground state about 1 eV too high in energy. With three ligands surrounding the Sr^+ ion, there is strong mixing of the atomic states in all of the molecular states. This combined with the lower symmetry results in many strong transitions extending to lower frequency than observed for either $SrNH_3^+$ or $Sr(NH_3)_2^+$. Thus the calculations suggest that the change in the character of the observed spectra does not necessarily mean that the bonding has changed character.

The MgL^+ spectrum is expected to be simpler than SrL^+, because while both have a strong $p \leftarrow s$ transition, MgL^+ has no low-lying ^2D state. However, differences arise between MgL^+ and SrL^+ even for the $p \leftarrow s$ transition. For SrH_2O^+ the excited states derived from the $5p\pi \leftarrow 5s$ transition of Sr^+ are shifted down by about 3000 cm^{-1} compared with free Sr^+. The binding energy in the excited state, relative to its asymptote, is larger than in the ground state, because of the smaller metal-ligand repulsion for the $5p\pi$ orbital. Based on a preliminary report of the theoretical results in Ref. [48], Duncan and co-workers [106] started their search for the $3p \leftarrow 3s$ derived transition in MgH_2O^+ at about 3500 cm^{-1} below the atomic transition. As the spectra were extremely complicated, they requested the use of electronic structure calculations in the analysis. Thus SA-CASSCF calculations [48] were performed to determine the vertical excitation energies for MgH_2O^+, at the SCF optimized geometry. These calculations showed that the lowest excited state of MgH_2O^+ was almost 20 kcal/mol more strongly bound than the ground state. Thus instead of a 3500 cm^{-1} shift, as found for SrH_2O^+, the shift was more than 6000 cm^{-1}. This theoretical prediction quickly lead to the experimental assignment of the bandhead for this transition. A more complete assignment of the spectra was aided by performing SA-CASSCF calculations [107] to determine the vibrational frequencies of the ground and excited states. Thus again the synergism of experiment and theory resulted in a definitive assignment of the MgH_2O^+ spectra, as well as an explanation for the differences between the MgH_2O^+ and SrH_2O^+ systems.

Another system where theory has helped confirm the assignment of the experimental spectrum is MgH_2^+. Recently Stwalley and co-workers [108] observed spectra of MgH_2^+, which they tentatively assigned as transitions from the (000) level of the ground state to high vibrational levels of the lowest excited state. This suggested a strongly non-vertical transition, which is surprising using MH_2O^+ or MNH_3^+ as a model. Calculations [109] for MgH_2^+ showed that the lowest excited state arose from promoting an electron from the $3s$ orbital to the in-plane $3p\pi(b_2)$ orbital. However, unlike MH_2O^+ or MNH_3^+, where the bonding is electrostatic, Mg^+ inserts into the H_2 bond. This results in a very different geometry: r(H-H) changes from $1.414a_0$ in the ground state to $4.441a_0$ in the $^2B_2$ state, and the Mg-H_2 bond midpoint distance changes from 5.145 to $2.567a_0$. Given the large geometry change, transitions to the higher vibrational levels of the excited state can be fully rationalized.

In this section we have shown that calculations can be useful in assigning or confirming the assignment of experimental spectra. In addition, calculations contribute to an understanding of how the bonding varies with the different occupations of the metal. For example, for SrH_2O^+ the calculations helped elucidate the role of $sd\sigma$ and sp hybridization in stabilizing the ground state.

The combination of experimental and theoretical studies often give insight unobtainable by either approach alone.

Conclusions

In this chapter we have illustrated how modern electronic structure calculations can contribute not only to an elucidation of the bonding and an explanation of trends in the experimental data, but to resolving the disagreement between disparate experimental measurements. Sufficient progress has been made in recent years in benchmarking theoretical methods that realistic error bars can now be placed on the calculated spectroscopic constants. Improvements in both electronic structure methods and codes, combined with the increasing availability of high-speed computing, has made it possible to obtain accuracy that is comparable to experiment for many systems. Many of the examples presented in this chapter illustrate the benefits of having electronic structure calculations to aid in the interpretation of experimental measurements. We expect that theory will continue to be an increasingly valuable tool in understanding the reactivity of metal ions with ligands. Furthermore, application to neutral systems has considerable potential for contributing to an understanding of homogeneous catalysis.

References

1. D. H. Russell (Ed.), "Gas Phase Inorganic Chemistry" (Plenum Press, New York, 1989); S. W. Buckner and B. S. Freiser, *Polyhedron* **7**, 1583 (1988); P. B. Armentrout and R. Georgiadis, *Polyhedron* **7**, 1573 (1988).
2. K. M. Ervin and P. B. Armentrout, *J. Chem. Phys.* **83**, 166 (1985).
3. L. S. Sunderlin, D. Wang, and R. R. Squires, *J. Am. Chem. Soc.* **114**, 2788 (1992); F. A. Khan, D. E. Clemmer, R. H. Schultz, and P. B. Armentrout, *J. Phys. Chem.* **97**, 7978 (1993).
4. R. L. Hettich, T. C. Jackson, E. M. Stanko, and B. S. Freiser, *J. Am. Chem. Soc* **108**, 5086 (1986).
5. C. W. Bauschlicher, S. R. Langhoff, and H. Partridge, in *Modern Electronic Structure Theory*, edited by D. R. Yarkony (World Scientific Publishing Company Co, London, 1995); "The calculation of accurate metal-ligand bond dissociation energies" C. W. Bauschlicher, H. Partridge, and S. R. Langhoff, "Advances in metal and semiconductor clusters" Ed. M. A. Duncan, (JAI Press, Inc. Greenwich 1994); C. W. Bauschlicher and S. R. Langhoff, *Int. Rev. Phys. Chem.* **9**, 149 (1990).
6. C. W. Bauschlicher, S. R. Langhoff, and P. R. Taylor, *Adv. Chem. Phys.* **77**, 103 (1990).
7. J. Almlöf and P. R. Taylor, *J. Chem. Phys.* **86**, 4070 (1987); P.-O. Widmark, J. B. Persson, and B. O. Roos, *Theor. Chim. Acta* **79**, 419 (1991); P.-O. Widmark, P. A. Malmqvist, and B. O. Roos, *Theor. Chim. Acta* **77**, 291 (1990).
8. T. H. Dunning, *J. Chem. Phys.* **90**, 1007 (1989); R. A. Kendall, T. H. Dunning, and R. J. Harrison, *J. Chem. Phys.* **96**, 6796 (1992).
9. C. W. Bauschlicher and H. Partridge, *Chem. Phys. Lett.* **208**, 241 (1993).
10. K. P. Lawley (Ed.), *Adv. Chem. Phys.* Vols. 67 and 69 (Wiley, New York, 1987).

11. D. R. Yarkony (Ed.), *Modern Electronic Structure Theory*, (World Scientific Publishing Company Co, London 1995).
12. A. J. H. Wachters, *J. Chem. Phys.* **52**, 1033 (1970).
13. P. J. Hay, *J. Chem. Phys.* **66**, 4377 (1977).
14. P. J. Hay and W. R. Wadt, *J. Chem. Phys.* **82**, 299 (1985).
15. S. R. Langhoff, L. G. M. Pettersson, C. W. Bauschlicher, and H. Partridge, *J. Chem. Phys.* **86**, 268 (1987).
16. R. D. Cowan and D. C. Griffin, *J. Opt. Soc. Am.* **66**, 1010 (1976).
17. H. Partridge, *J. Chem. Phys.* **90**, 1043 (1989); H. Partridge and K. Faegri, *Theor. Chim. Acta* **82**, 207 (1992).
18. C. W. Bauschlicher, *Chem. Phys. Lett.* **142**, 71 (1987).
19. P. Maitre and C. W. Bauschlicher, *J. Phys. Chem.* **97**, 11912 (1993).
20. J. A. Pople, J. S. Binkley, and R. Seeger, *Int. J. Quantum Chem. Symp.* **10**, 1 (1976).
21. N. C. Handy, R. D. Amos, J. F. Gaw, J. E. Rice, and E. D. Simandiras, *Chem. Phys. Lett.* **120**, 151 (1985).
22. D. P. Chong and S. R. Langhoff, *J. Chem. Phys.* **84**, 5606 (1986). Also see R. Ahlrichs, P. Scharf, and C. Ehrhardt, *J. Chem. Phys.* **82**, 890 (1985).
23. C. W. Bauschlicher, H. Partridge, J. A. Sheehy, S. R. Langhoff, and M. Rosi, *J. Phys. Chem.* **96**, 6969 (1992).
24. R. J. Bartlett, *Annu. Rev. Phys. Chem.* **32**, 359 (1981).
25. K. Raghavachari, G. W. Trucks, J. A. Pople, and M. Head-Gordon, *Chem. Phys. Lett.* **157**, 479 (1989).
26. C. W. Bauschlicher, H. Partridge, and G. E. Scuseria, *J. Chem. Phys.* **97**, 7471 (1992).
27. J. D. Watts, J. Gauss, and R. J. Bartlett, *J. Chem. Phys.* **98**, 8718 (1993).
28. B. O. Roos, *Adv. Chem. Phys.* **69**, 399 (1987).
29. K. K. Docken and J. Hinze, *J. Chem. Phys.* **57**, 4928 (1972).
30. V. R. Saunders and J. H. van Lenthe, *Mol. Phys.* **48**, 923 (1983); I. Shavitt, in *Methods of Electronic Structure Theory*, edited by H. F. Schaefer (Plenum Press, New York, 1977, p. 189.
31. S. R. Langhoff and E. R. Davidson, *Int. J. Quantum Chem.* **8**, 61 (1974); M. R. A. Blomberg and P. E. M. Siegbahn, *J. Chem. Phys.* **78**, 5682 (1983).
32. R. J. Gdanitz and R. Ahlrichs, *Chem. Phys. Lett.* **143**, 413 (1988).
33. H.-J. Werner and P. J. Knowles, *J. Chem. Phys.* **89**, 5803 (1988); P. J. Knowles and H.-J. Werner, *Chem. Phys. Lett.* **145**, 514 (1988).
34. C. W. Bauschlicher, S. R. Langhoff, H. Partridge, and L. A. Barnes, *J. Chem. Phys.* **91**, 2399 (1989).
35. L. G. M. Pettersson, C. W. Bauschlicher, S. R. Langhoff, and H. Partridge, *J. Chem. Phys.* **87**, 481 (1987).
36. C. E. Moore, *Atomic Energy Levels*, *Natl. Bur. Stand.* (*US*) *circ.* **467** (1949).
37. E. A. Carter and W. A. Goddard, *J. Phys. Chem.* **92**, 5679 (1988).
38. J. L. Elkind and P. B. Armentrout, *Inorg. Chem.* **25**, 1078 (1986); P. B. Armentrout, L. F. Halle, and J. L. Beauchamp, *J. Am. Chem. Soc.* **103**, 6501 (1981); M. L. Mandich, L. F. Halle, and J. L. Beauchamp, *J. Am. Chem. Soc.* **106**, 4403 (1984).
39. See Figure 2 in the chapter by P. B. Armentrout and B. L. Kickel.
40. P. E. M. Siegbahn, *Theor. Chim. Acta* **86**, 219 (1993).
41. C. W. Bauschlicher and P. S. Bagus, *J. Chem. Phys.* **81**, 5889 (1984).
42. S. P. Walch and C. W. Bauschlicher, *J. Chem. Phys.* **78**, 4597 (1983).
43. C. W. Bauschlicher, H. Partridge, and S. R. Langhoff, *J. Chem. Phys.* **91**, 4733 (1989); H. Partridge, C. W. Bauschlicher, and S. R. Langhoff, *J. Phys. Chem.* **96**, 5350 (1992).
44. M. Rosi and C. W. Bauschlicher, *J. Chem. Phys.* **90**, 7264 (1989); and M. Rosi and C. W. Bauschlicher, *J. Chem. Phys.* **92**, 1876 (1990).
45. C. W. Bauschlicher, H. Partridge, and S. R. Langhoff, *J. Phys. Chem.* **96**, 3273 (1992); C. W. Bauschlicher and H. Partridge, *J. Phys. Chem.* **95**, 9694 (1991).
46. L. A. Barnes, M. Rosi, and C. W. Bauschlicher, *J. Chem. Phys.* **93**, 609 (1990).

47. C. W. Bauschlicher, H. Partridge, and S. R. Langhoff, *Chem. Phys. Lett.* **165**, 272 (1990).
48. C. W. Bauschlicher, M. Sodupe, and H. Partridge, *J. Chem. Phys.* **96**, 4453 (1992).
49. L. M. Roth, B. S. Freiser, C. W. Bauschlicher, H. Partridge, and S. R. Langhoff, *J. Am. Chem. Soc.* **113**, 3274 (1991).
50. L. Lian, C.-X. Su, and P. B. Armentrout, *Chem. Phys. Lett.* **180**, 168 (1991).
51. D. Lessen and P. J. Brucat, *Chem. Phys. Lett.* **149**, 473 (1989).
52. A. Kant, *J. Chem. Phys.* **41**, 1872 (1964).
53. T. H. Upton and W. A. Goddard, *J. Am. Chem. Soc.* **100**, 5659 (1978).
54. J. O. Noell, M. D. Newton, P. J. Hay, R. L. Martin, and F. W. Bobrowicz, *J. Chem. Phys.* **73**, 2360 (1980).
55. M. D. Morse, G. P. Hansen, P. R. R. Langridge-Smith, L. S. Zheng, M. E. Geusic, D. L. Michalopoulos, and R. E. Smalley, *J. Chem. Phys.* **80**, 2866 (1984).
56. E. A. Rohlfing and J. J. Valentini, *J. Chem. Phys.* **84**, 6560 (1986).
57. D. E. Powers, S. G. Hansen, M. E. Geusic, D. L. Michalopoulos, and R. E. Smalley, *J. Chem. Phys.* **78**, 2866 (1983) for the molecular ionization potential (IP); C. E. Moore, reference [36], for the atomic IP.
58. H. Partridge and C. W. Bauschlicher, *Theor. Chim. Acta* **83**, 201 (1992).
59. C. W. Bauschlicher, H. Partridge, and S. R. Langhoff, *Chem. Phys. Lett.* **195**, 360 (1992).
60. M. D. Morse, *Chem. Rev.* **86**, 1049 (1986).
61. E. M. Spain and M. D. Morse, *J. Chem. Phys.* **97**, 4641 (1992).
62. J. C. Pinegar, J. D. Langenberg, C. A. Arrington, E. M. Spain and M. D. Morse, *J. Chem. Phys.* **102**, 666 (1995).
63. C. W. Bauschlicher and S. R. Langhoff (unpublished).
64. C. W. Bauschlicher, S. P. Walch, and H. Partridge, *J. Chem. Phys.* **76**, 1033 (1982); C. W. Bauschlicher, P. Siegbahn, and L. G. M. Pettersson, Theor. Chim. Acta **74**, 479 (1988); C. W. Bauschlicher, *J. Chem. Phys.* **86**, 5591 (1987).
65. D. Lessen and P. J. Brucat, *Chem. Phys. Lett.* **160**, 609 (1989).
66. T. T. Tsong, *J. Chem. Phys.* **85**, 639 (1986).
67. C. W. Bauschlicher and M. Rosi, *Chem. Phys. Lett.* **159**, 485 (1989); C. W. Bauschlicher and M. Rosi, *Chem. Phys. Lett.* **165**, 501 (1990).
68. Y. Huang and B. S. Freiser, *J. Am. Chem. Soc.* **110**, 4435 (1988).
69. C. W. Bauschlicher and S. R. Langhoff, *Chem. Phys. Lett.* **161**, 383 (1989).
70. L. I. Schiff, *Quantum Mechanics*, McGraw-Hill, New York, 1968.
71. I. Dzidic and P. Kebarle, *J. Phys. Chem.* **74**, 1446 (1970).
72. W. R. Davidson and P. Kebarle, *J. Am. Chem. Soc.* **98**, 6125 (1976); W. R. Davidson and P. Kebarle, *J. Am. Chem. Soc.* **98**, 6133 (1976); B. C. Guo, B. J. Conklin, and A. W. Castleman, *J. Am. Chem. Soc.* **53**, 560 (1978); A. W. Castleman, *Chem. Phys. Lett.* **53**, 560 (1978).
73. C. W. Bauschlicher, S. R. Langhoff, H. Partridge, J. E. Rice, and A. Komornicki, *J. Chem. Phys.* **95**, 5142 (1991).
74. F. Bouchard, J. W. Hepburn, and T. B. McMahon, *J. Am. Chem. Soc.* **111**, 8934 (1989).
75. C. W. Bauschlicher, F. Bouchard, J. W. Hepburn, T. B. McMahon, P. Surjasamita, L. Roth, J. R. Gord, and B. S. Freiser, *Int. J. Mass Spectrom. Ion Proc.* **109**, 15 (1991).
76. J. Almlöf, K. Faegri, and K. Korsell, *J. Comput. Chem.* **3**, 385 (1982); S. Saebo and J. Almlöf, *Chem. Phys. Lett.* **154**, 521 (1987).
77. R. Tonkyn and J. C. Weisshar, *J. Am. Chem. Soc.* **108**, 7128 (1986).
78. T. MacMahon and B. S. Freiser (personal communication).
79. Y. Haung, M. B. Wise, D. B. Jacobson, and B. S. Freiser, *Organometallics* **6**, 346 (1987).
80. C. W. Bauschlicher and S. R. Langhoff, *J. Phys. Chem.* **95**, 2278 (1991); M. Rosi and C. W. Bauschlicher, *Chem. Phys. Lett.* **166**, 189 (1990).
81. J. A. Pople, in *Applications of Electronic Structure Theory*, edited by H. F. Schaefer (Plenum Press, New York, 1977), pp. 1–27.
82. M. Sodupe and C. W. Bauschlicher, *J. Phys. Chem.* **95**, 8640 (1991).

83. M. Sodupe, C. W. Bauschlicher, S. R. Langhoff, and H. Partridge, *J. Phys. Chem.* **96**, 2118 (1992).
84. Y. A. Ranasinghe and B. S. Freiser, *Chem. Phys. Lett.* **200**, 135 (1992).
85. (a) L. S. Sunderlin and P. B. Armentrout, *J. Am. Chem. Soc.* **111**, 3845 (1989); (b) L. S. Sunderlin, N. Aristov, and P. B. Armentrout, *J. Am. Chem. Soc.* **109**, 78 (1987); (c) L. S. Sunderlin and P. B. Armentrout, *Int. J. Mass. Spec. Ion. Proc.* **94**, 149 (1989); (d) N. Aristov and P. B. Armentrout, *J. Am. Chem. Soc.* **108**, 1806 (1986); (e) R. Georgiadis and P. B. Armentrout, *Int. J. Mass. Spec. Ion. Proc.* **89**, 227 (1989); (f) R. H. Schultz, J. L. Elkind, and P. B. Armentrout, *J. Am. Chem. Soc.* **110**, 411 (1988); (g) E. R. Fisher and P. B. Armentrout, *J. Phys. Chem.* **94**, 1674 (1990).
86. (a) M. A. Tolbert and J. L. Beauchamp, J. Am. Chem. Soc. **106**, 8117 (1984); (b) P. B. Armentrout and J. L. Beauchamp, *J. Am. Chem. Soc.* **103**, 6628 (1981); (c) M. A. Hanratty, J. L. Beauchamp, A. J. Illies, P. van Koppen, and M. T. Bowers, *J. Am. Chem. Soc.* **110**, 1 (1988).
87. (a) D. B. Jacobson and B. S. Freiser, *J. Am. Chem. Soc.* **105**, 7492 (1983); (b) R. C. Burnier, G. D. Byard, and B. S. Freiser, *Anal. Chem.* **52**, 1641 (1980).
88. J. B. Schilling and J. L. Beauchamp, *Organometallics* **7**, 194 (1988).
89. Y. D. Hill, B. S. Freiser, and C. W. Bauschlicher, *J. Am. Chem. Soc.* **113**, 1507 (1991).
90. C. W. Bauschlicher, S. R. Langhoff, and H. Partridge, *J. Phys. Chem.* **95**, 6191 (1991).
91. S. W. Buckner and B. S. Freiser, *J. Am. Chem. Soc.* **109**, 1247 (1987); J. R. Gord, B. S. Freiser, and S. W. Buckner, *J. Chem. Phys.* **91**, 7530 (1989).
92. P. R. Kemper, M.-T. Hsu, and M. T. Bowers, *J. Phys. Chem.* **95**, 10600 (1991); G. von Helden, P. R. Kemper, M.-T. Hsu, and M. T. Bowers, *J. Chem. Phys.* **96**, 6591 (1992).
93. (a) D. E. Lessen, R. L. Asher, and P. J. Brucat, *Chem. Phys. Lett.* **177**, 380 (1991); (b) D. Lessen and P. J. Brucat, *J. Chem. Phys.* **90**, 6296 (1989); (c) D. Lessen and P. J. Brucat, *J. Chem. Phys.* **91**, 4522 (1989); (d) D. Lessen and P. J. Brucat, *Chem. Phys. Lett.* **152**, 473 (1988).
94. T. M. Miller and B. Bederson, *Adv. Atom. Mol. Phys.* **13**, 1 (1978).
95. P. Desclaux, *Atomic Data and Nuclear Data Tables* **12**, 312 (1973).
96. H. Partridge and C. W. Bauschlicher, *J. Phys. Chem.* **98**, 2301 (1994).
97. P. J. Marinelli and R. R. Squires, *J. Am. Chem. Soc.* **111**, 4101 (1989).
98. T. F. Magnera, D. E. David, and J. Michl, *J. Am. Chem. Soc.* **111**, 4100 (1989); T. F. Magnera, D. E. David, D. Stulik, R. G. Orth, H. T. Jonkman, and J. Michl, *J. Am. Chem. Soc.* **111**, 5036 (1989).
99. P. M. Holland and A. W. Castleman, *J. Chem. Phys.* **76**, 4195 (1982).
100. K. Honma, N. F. Dalleska, and P. B. Armentrout (personal communication). This work is reviewed in P. B. Armentrout and D. E. Clemmer, *Energetics of Organometallic Species*, edited by J. A. M. Simoes (Kluwer, Dordrecht, 1992), p. 321.
101. C. W. Bauschlicher, S. R. Langhoff, and H. Partridge, *J. Chem. Phys.* **94**, 2068 (1991).
102. K. F. Willey, C. S. Yeh, D. L. Robbins, and M. A. Duncan, *Chem. Phys. Lett.* **192**, 179 (1992).
103. M. H. Shen and J. M. Farrar, *J. Chem. Phys.* **94**, 3322 (1991).
104. M. H. Shen and J. M. Farrar, *J. Chem. Phys.* **93**, 4386 (1989).
105. M. Sodupe and C. W. Bauschlicher, *Chem. Phys. Lett.* **212**, 624 (1993).
106. C. S. Yeh, K. F. Willey, D. L. Robbins, J. S. Pilgrim, and M. A. Duncan, *Chem. Phys. Lett.* **196**, 233 (1992).
107. M. Sodupe and C. W. Bauschlicher, *Chem. Phys. Lett.* **195**, 494 (1992).
108. L. N. Ding, M. A. Young, P. D. Kleiber, and W. C. Stwalley, *J. Phys. Chem.* **97**, 2181 (1993).
109. C. W. Bauschlicher, *Chem. Phys. Lett.* **201**, 11 (1993).

3. The electronic structure of the neutral, mono and di positive transition metal nitrides, ScN, TiN, VN and CrN

JAMES F. HARRISON and KATHRYN L. KUNZE*

Department of Chemistry and Center for Fundamental Materials Research, Michigan State University, East Lansing, MI 48824-1322, U.S.A.

1. Introduction

Understanding the nature of the transition metal (TM)-main-group-element bond is important in many areas of science. Information on the bonding in neutral TM-main group systems is obtained primarily from matrix isolation experiments [1], gas phase electronic spectroscopy [2] and theoretical calculations [3], while information on the TM ion – main group element bond has been obtained from studies of gas-phase bimolecular reactions [4] of TM ions with organic and inorganic molecules as well as theoretical [5, 6] calculations. The resulting bond energies permit one to systematize and interpret the observed chemistry of these systems and have immediate impact in organometallic chemistry, surface science, catalysis and atmospheric chemistry. Theoretical calculations on the TM-main group bond in coordinately unsaturated molecules have lagged behind the experiments because the extent of electron correlation needed for even qualitatively correct results is significantly raised relative to molecules containing only main group-main group bonds. For example, absent electron correlation, the energy separations between the terms of the low lying s^2d^N, sd^{N+1} and d^{N+2} configurations of the neutral transition metal atoms are seriously flawed. In a molecular calculation this atomic misalignment affects the relative mixture of the TM s, p and d orbitals in the bond and therefore the bond distance and energy. Studies over the last decade by several groups [3, 5, 6] have shown that theoretical studies can provide usefully accurate descriptions of these systems provided one takes into account the structural correlation energy in the transition-metal-main-group bonds as well as the various spin couplings associated with those bonds and the valence spectator electrons.

Here we describe the electronic structure of the neutral, monopositive and dipositive nitrides of the early transition metals, ScN, TiN, VN and CrN. Our emphasis will be on developing insights into their electronic structure

* Deceased, March 28, 1991.

Ben S. Freiser (ed.), Organometallic Ion Chemistry, 89–121.

by analyzing the electron distribution in the ground and low lying states as predicted by our calculations. The theoretical methods and computational strategies have been described [3c] in detail elsewhere. Briefly, the calculations use the Wachters [7] basis set for the TM, augmented by two additional p functions and an extra d function as recommended by Hay [8]. The N basis functions are those of Almlof and Taylor [9]. The MCSCF calculations include the structural correlation of the bonds in a generalized valence bond (GVB) fashion plus all possible spin couplings among the valence electrons (bonding and spectator) consistent with the desired spin state. The CI calculations include valence single and double excitations from the entire MCSCF reference space. These CI calculations are often referred to as MCSCF + 1 + 2. When we discuss the charge on an atom it has been obtained via a Mulliken population analysis on the MCSCF wavefunction. All calculations were done using the COLUMBUS [10] electronic structure codes.

2. The neutral nitrides

The experimental energy separation [11] between the lowest states of Sc, Ti, V and Cr corresponding to the configurations $4s^23d^N$, $4s3d^{N+1}$ and $3d^{N+2}$ are shown in Figure 1. Note that the $4s^23d^N$-$4s3d^{N+1}$ separation decreases as one goes from Sc to Cr and, indeed, for Cr the ground state is a ^7S state arising from the $4s3d^5$ configuration. In Figure 2 we compare the experimental energy separations [11] between the low lying states of Sc and N. Note that the first three electronic states of N correspond to the same $2p^3$ electronic configuration while in a comparable energy range Sc visits the four configurations, $4s^23d$, $4s3d^2$, $4s4p3d$ and $3d^3$. Clearly, when Sc and N interact we need to consider several configurations of Sc but only the $2p^3$ of N.

2.1. States of ScN that dissociate to the ground state of Sc(^2D) and N(^4S)

When ground state Sc(^2D) interacts with N(^4S) we expect Σ^-, Π and Δ states of triplet and quintet multiplicity [3c]. The qualitative character of these states is easily predicted. Since both Sc and N have three valence electrons (the N 2s pair is essentially doubly occupied in the diatomics) a triplet state will require two singlet coupled electron pairs and one triplet coupled pair. The singlet pairs could be bonds or lone pairs and thus we anticipate triplet states with single or double bonds. The quintet states require four high spin electrons leaving one singlet pair which could be either a bond or a lone pair. The quintets will, at most, be singly bonded species.

The $^3\Sigma^-$ state corresponds to maintaining the Sc 4s pair doubly occupied and forming a σ bond between the Sc d_σ and N p_σ orbitals. The triplet

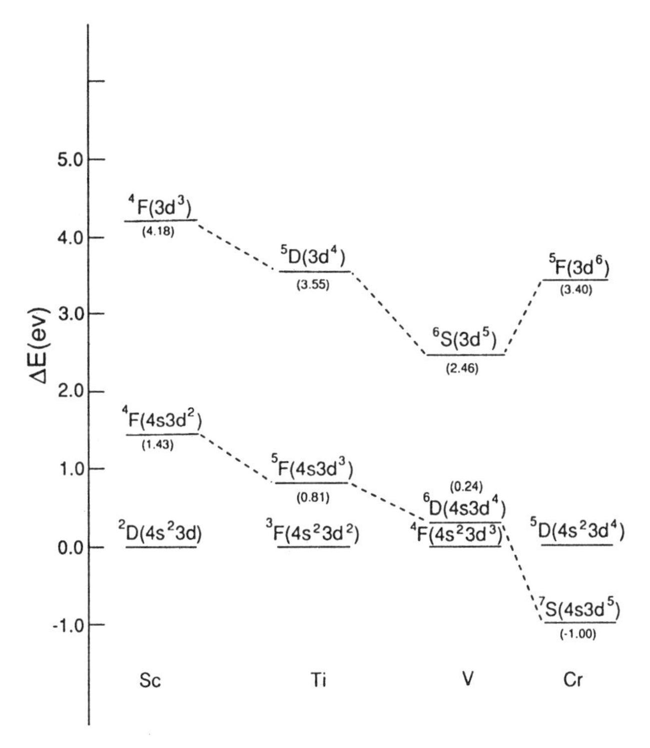

Fig. 1. Experimental energy separations between the lowest states of Sc, Ti, V and Cr corresponding to the configurations $4s^2 3d^N$, $4s3d^{N+1}$ and $3d^{N+1}$.

coupled electrons are both on N and we have a single bond between Sc and N

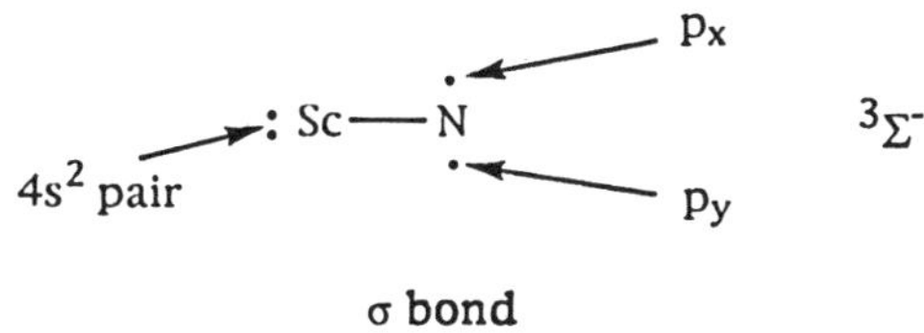

The asymptotic form of the $^3\Pi$ state corresponds to the Sc $4s^2$ pair being intact while the $3d_\pi$ couples to one of the N p electrons. At this point, both unpaired electrons are on N. As the Sc and N come closer, the $4s^2$ pair can uncouple and bond to the N p_σ leaving one unpaired electron on each center and essentially a ScN double bond.

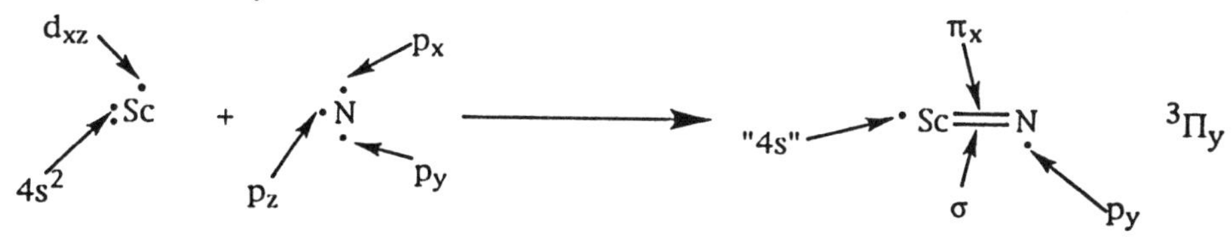

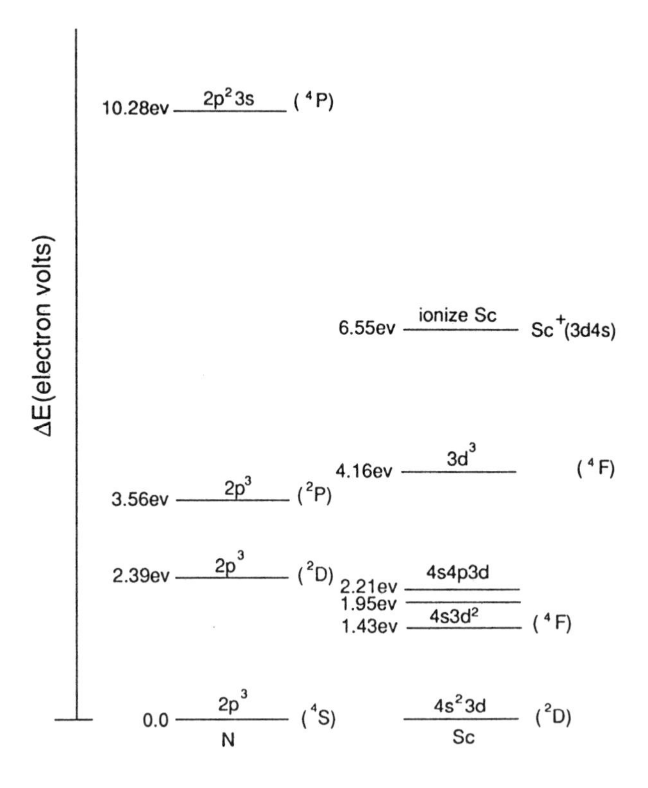

Fig. 2. Experimental energy separations between the low lying states of Sc and N.

The asymptotic form of the $^3\Delta$ state is very similar to the $^3\Pi$ with the Sc d_δ orbital singlet coupled to one of the N p electrons. The difference is of course that the Sc d_δ and N p_π orbitals are orthogonal by symmetry and will not form a bond. The σ system can still respond as in the $^3\Pi$ state and we anticipate a single (σ) bond between Sc and N

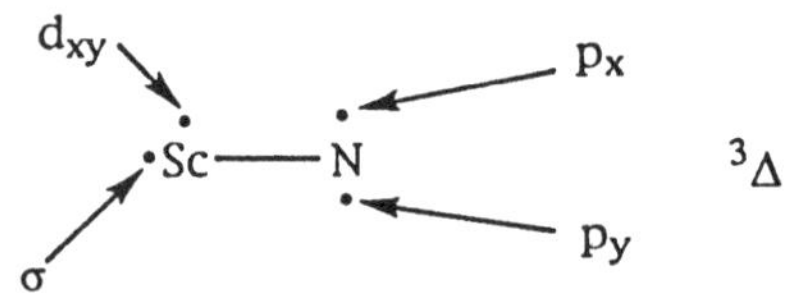

Because of the weak Sc, N spin coupling we expect the $^5\Delta$ state to have a similar energy.

The $^5\Sigma^-$ state corresponds asymptotically to maintaining the a σ^2 pair (the $4s^2$), keeping the Sc d_σ orbital occupied and high spin coupled the to the three unpaired N electrons. As the Sc and N atoms approach one another one can uncouple the Sc $4s^2$ pair (hybridize), form a σ bond with N and leave two high spin electrons on both N and Sc.

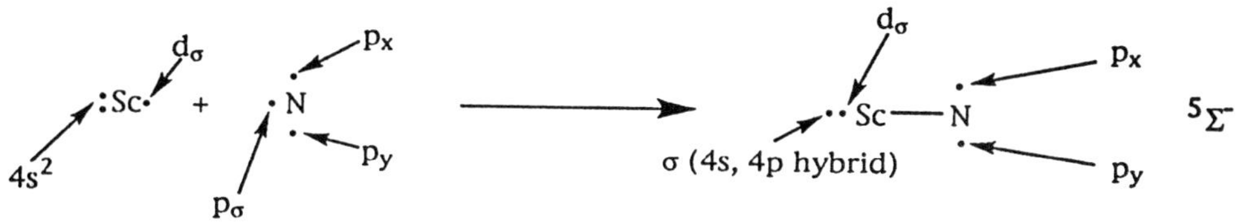

The $^5\Pi$ state begins as

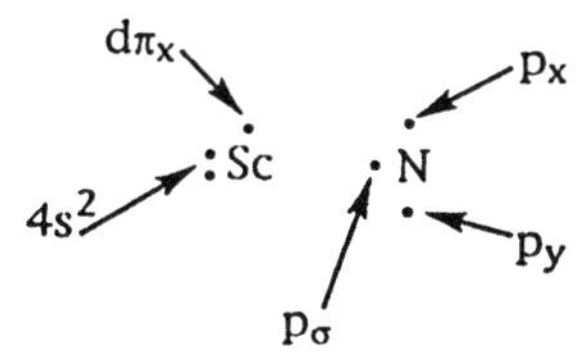

and evolves into

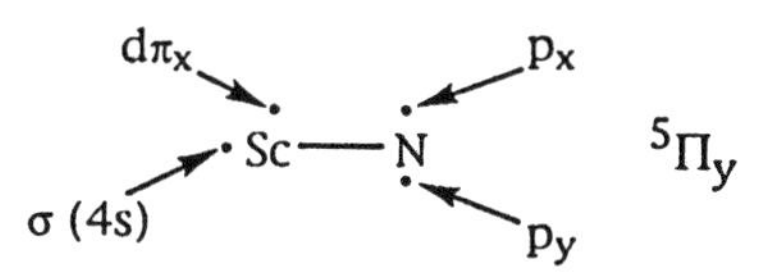

Among the triplet states we expect the $^3\Pi$ to be lowest because it is doubly bonded and the $^3\Sigma^-$ to be next in energy because the σ bond requires no uncoupling of the $4s^2$ pair. The $^3\Delta$ will be the highest triplet because it requires that the $4s^2$ pair uncouple (hybridize) in order to form the lone σ bond.

The quintets are all singly bonded and in each case the single bond is between a σ orbital on Sc and a p_σ orbital on N. The energy difference in the quintets is primarily due to the exchange coupling between the high spin electrons on Sc and this results in the order $^5\Delta < {^5\Pi} < {^5\Sigma^-}$. Since the $^3\Delta$ and $^5\Delta$ are expected to have comparable energies we have the order $^3\Pi < {^3\Sigma^-} < {^3\Delta} \sim {^5\Delta} < {^5\Pi} < {^5\Sigma^-}$.

2.2. States of ScN that correlate with the $^4F(4s3d^2)$ state of Sc

The ^4F state of Sc can combine with the ^4S state of N to produce Σ^+, Π, Δ and Φ states with spin multiplicities 7, 5, 3 or 1. The septets correspond to 6 unpaired electrons and these states will not be bound and we have discussed the $^3\Pi$, $^5\Pi$, $^3\Delta$ and $^5\Delta$ previously. The $^1\Sigma^+$ state corresponds to three singlet coupled electron pairs, i.e. a triple bond.

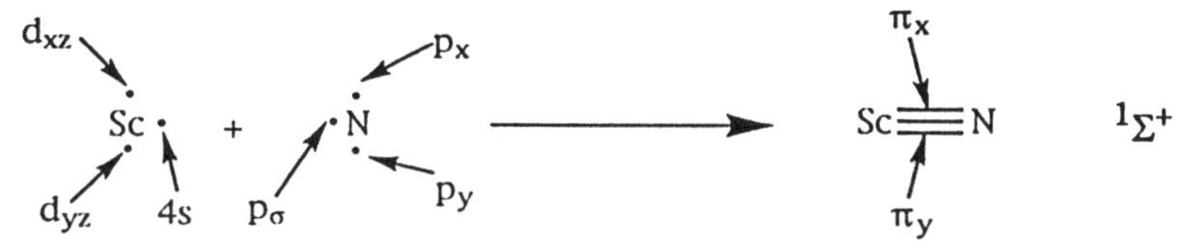

The triplet state of Σ^+ symmetry is obtained by breaking the σ bond in the $^1\Sigma^+$ state and maintaining the two π bonds

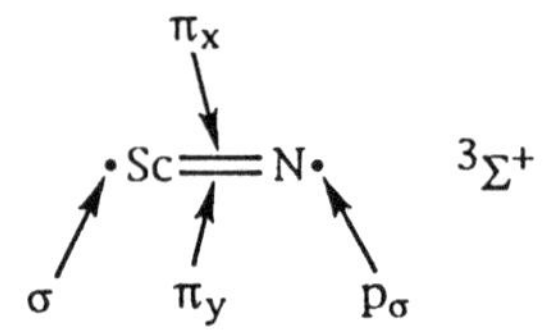

The quintet state of Σ^+ symmetry is obtained by breaking the two π bonds in the $^1\Sigma^+$ state and maintaining the σ bond

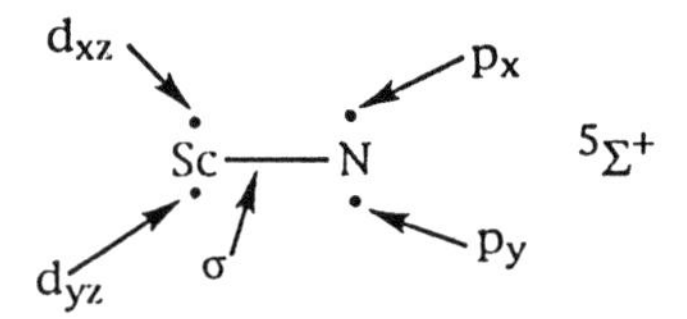

2.3. Computational results for ScN

The calculated [3c] CI potential energy curves for several electronic states are collected in Figure 3. In Figure 4 we compare the bond energies of several representative states as a function of the level of calculation. The bond lengths and vibrational frequency of these states are shown in Table I. Note that the ground state of ScN is indeed the triply bonded $^1\Sigma^+$ and that it separates to the excited ^4F state of Sc. This state is bound by 106 kcal/mol relative to the ^4F state and by 64 kcal/mol relative to the ^2D state of Sc. At 10 au the two atoms are essentially non interacting and the Sc configuration is 4s $3d_{xz}$ $3d_{yz}$. Figure 5 shows the electron population in various orbitals of σ symmetry as a function of internuclear separation. The 4s on Sc remains singly occupied until R ~ 5 au when its population begins to drop. Simultaneously the Sc d_σ and p_σ orbitals begin to be populated and at equilibrium we have the in situ σ valence occupancy

Sc		N	
4s	$4p_\sigma$	$3d_\sigma$	$2p_\sigma$
0.17	0.20	0.50	1.15

A similar analysis on the π system reveals

Sc		N
$4p_x$	$3d_{zx}$	$2p_x$
0.10	0.61	1.27

with the π_y occupancies being identical with the π_x. At equilibrium Sc has

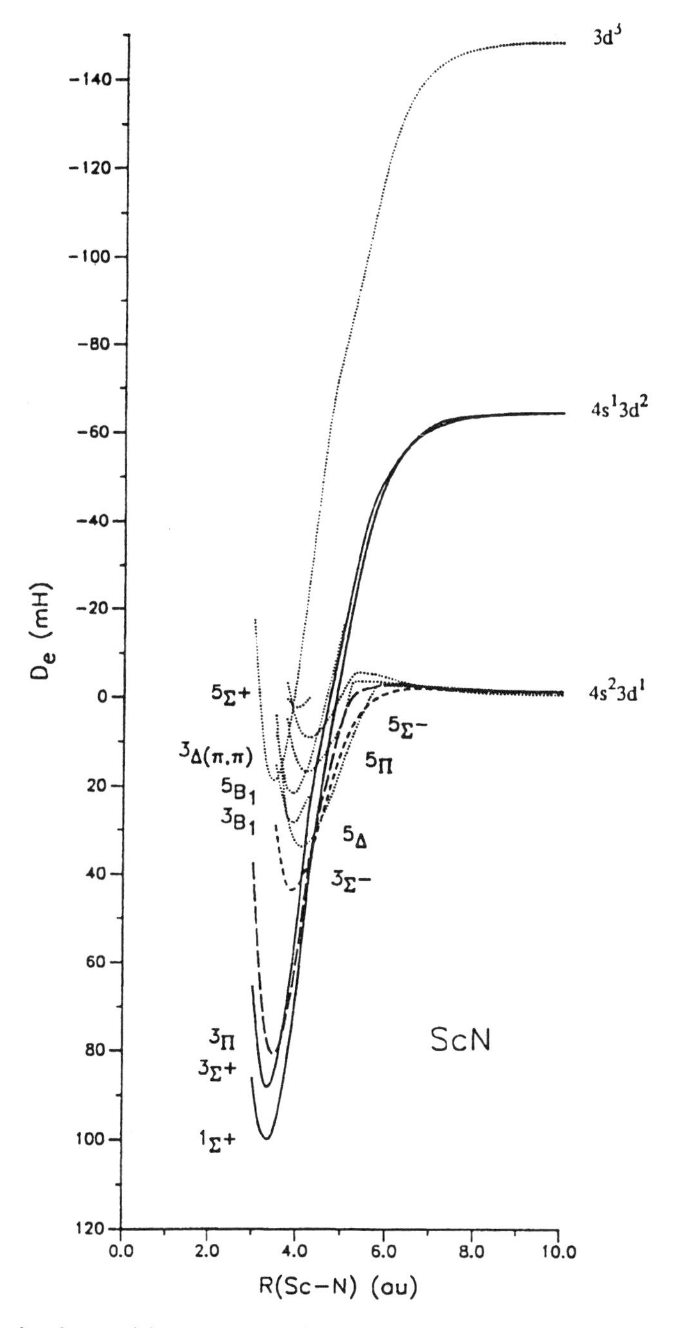

Fig. 3. Calculated potential energy curves for ScN. Energy units are milli hartrees (mh) and one mh is equivalent to 0.0272 ev or 0.626 kcal/mol.

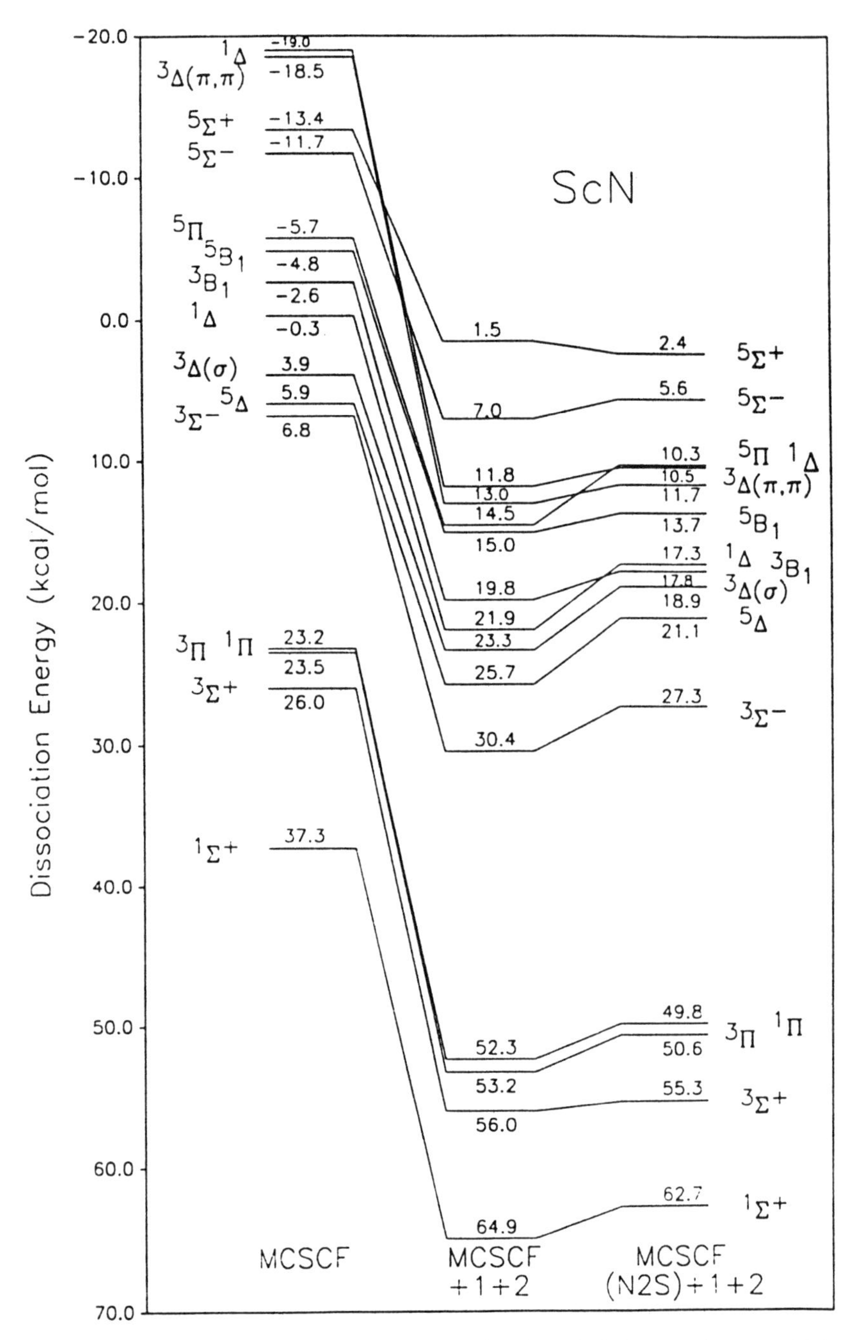

Fig. 4. Energies of several representative single, double and triple bonded states of ScN.

Table I. ScN bond energies, bond lengths and vibrational frequencies at the MCSCF + 1 + 2 level [3c]

State	bond energy (ev)	bond length (Å)	Vibrational Frequencies (cm^{-1})
		σ, π, π triple bond	
$^1\Sigma^+$	2.72	1.768	726
		π, π double bonds	
$^3\Sigma^+$	2.40	1.769	861
$^3\Delta$	0.51	1.830	790
$^1\Delta$	0.46	1.843	785
		σ, π double bonds	
$^3\Pi$	2.19	1.839	802
$^1\Pi$	2.16	1.845	789
		π single bond	
$^5\Pi$	0.59	2.059	576
$^5\Sigma^+$	0.10	2.102	530
		σ single bond	
$^3\Sigma^-$	1.18	2.058	577
$^5\Sigma^-$	0.24	2.250	400
$^5\Pi$	0.45	2.231	438
$^5\Delta$	0.92	2.173	489
$^3\Delta$	0.82	2.178	491
$^1\Delta$	0.75	2.179	496

a charge of +0.59 and the σ bonding orbital associated with Sc is predominately d_σ.

The next two electronic states are the $^3\Sigma^+$ and $^3\Pi$ and, although both are doubly bonded, the $^3\Sigma^+$ has two π bonds while the $^3\Pi$ has a more conventional σ and π bond. Although the $^3\Sigma^+$ dissociates to excited state Sc and the $^3\Pi$ to ground state Sc, the $^3\Sigma^+$ is more strongly bound. Indeed, the $^3\Sigma^+$ state is only 7.4 kcal/mol above the $^1\Sigma^+$. Although breaking the σ bond in $^1\Sigma^+$

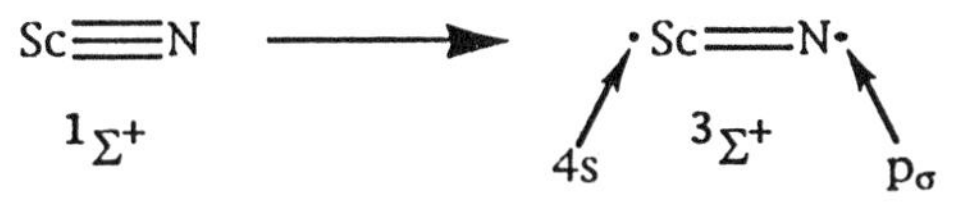

requires only 7.4 kcal/mol we see from Figure 4 we see that Sc-N σ bonds have dissociation energies of ~25 kcal/mol. The reason for this difference becomes apparent when we compare the valence orbital populations in the two states.

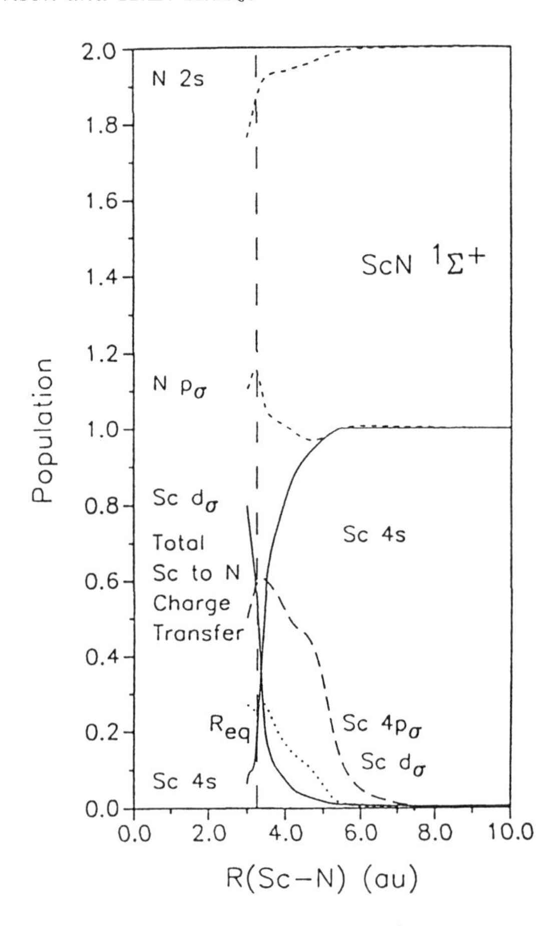

Fig. 5. Electron population in various orbitals of ScN ($^1\Sigma^+$) as a function of internuclear separation.

	Sc							N		
	4s	$4p_\sigma$	$3d_\sigma$	$4p_x$	$3d_{xz}$	$4p_y$	$3d_{yz}$	$2p_\sigma$	$2p_x$	$2p_y$
$^3\Sigma^+$	0.77	0.21	0.02	0.07	0.54	0.07	0.54	0.89	1.37	1.37
$^1\Sigma^+$	0.17	0.20	0.50	0.10	0.61	0.10	0.61	1.15	1.27	1.27

The Sc σ orbital in $^1\Sigma^+$ is dominated by the $3d_\sigma$ while in $^3\Sigma^+$ it is dominated by the 4s. Uncoupling the σ bonding pair allows the Sc σ orbital to relax, taking the in situ Sc configuration from predominately $3d^3$ character in the $^1\Sigma^+$ state to predominately $4s3d^2$ in the $^3\Sigma^+$.

2.4. Consequences of the small $^1\Sigma^+$-$^3\Sigma^+$ energy difference

The very weak σ bond in Sc ≡ N means that a reaction with R· will be exothermic and will occur with little or no barrier. Indeed, when R· is H the reaction

$$\text{Sc}\equiv\text{N}\ (^1\Sigma^+) + \text{H}\cdot\ (^2\text{S}) \longrightarrow \cdot\text{Sc}\equiv\text{N—H}\quad ^2\Sigma^+$$

(with σ and π arrows labelled on the Sc≡N bond of the product)

is exothermic by 98 kcal/mol [12], making the N-H bond approximately 22 kcal/mol stronger than our computed 76 kcal/mol bond in imidogen, N-H ($^3\Sigma^-$). This enhanced stability of the N-H bond in ScNH obtains because the 2s and 2p orbitals on N hybridize in forming the ScNH molecule. A consequence of this is that the companion sp hybrid on N which is not bonding to H can form a dative bond with the empty σ orbitals (4s, 4p and $3d_\sigma$) on Sc.

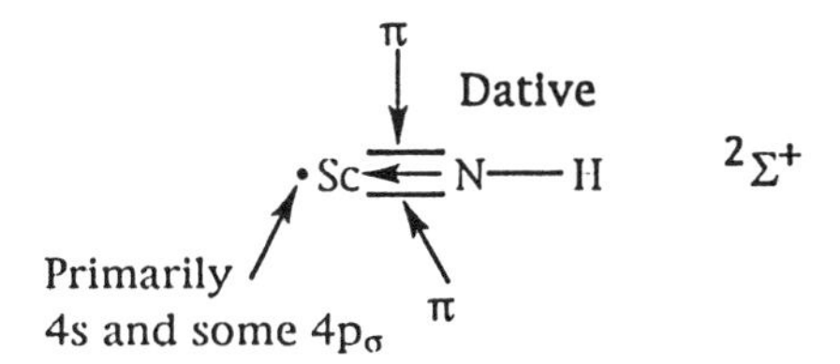

A measure of the strength of this dative bond is the dissociation energy of Sc ≡ NH relative to Sc ≡ N. We calculate that

$$\cdot\text{Sc}\equiv\text{NH}\ (^2\Sigma^+) \longrightarrow \text{Sc}\cdot\ (^2\text{D}) + \text{NH}\ (^3\Sigma^-)$$

requires 85 kcal/mol, 22 kcal/mol larger than the calculated dissociation energy of Sc ≡ N ($^1\Sigma^+$). These energetics are summarized in Figure 6.

The small σ bond energy of Sc ≡ N also permits the reaction

$$\text{Sc}\equiv\text{N—H}\ (^2\Sigma^+) + \text{H}_2\ (^1\Sigma_g{}^+) \longrightarrow \text{H—Sc}\equiv\text{N—H}$$

to be exothermic by 54 kcal/mol.

2.5. The remaining nitrides, TiN, VN and CrN

Knowing the ground state of ScN the ground states of the neutral nitrides TiN, VN and CrN are easily predicted [13]. The ground state of TiN has $^2\Sigma^+$ symmetry, dissociates to ground state Ti and N and has the unpaired electron in an sp hybrid on Ti

$$\cdot\text{Ti}\equiv\text{N}\ (^2\Sigma^+)$$

(with a σ arrow pointing to the unpaired electron on Ti)

The low lying states maintaining a triple bond between Ti and N are of $^2\Delta$ and $^2\Pi$ symmetry. The unpaired electron in the $^2\Delta$ state is in a Ti d_δ orbital

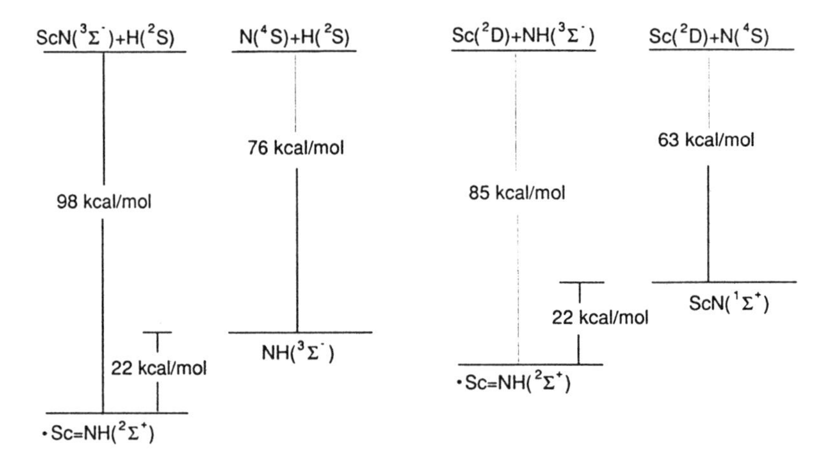

Fig. 6. Effect of N to Sc dative bond on the ScN bond energy.

Table II. In situ atomic occupation of various TiN states [13]

State	Orbital(s)	Ti 4s	Ti 4p	Ti 3d	N2s	Nsp	Total
$^2\Sigma^+$	σ bond	0.00	0.02	0.78	0.00	1.20	2.00
	π_x or π_y bond		0.05	0.75		1.20	2.00
	unpaired σ	0.79	0.17	0.04	0.0	0.0	1.00
	total	0.79	0.29	2.32	0.0	3.60	7.00
$^2\Delta$	σ bond	0.03	0.07	0.72	0.01	1.17	2.00
	π_x or π_y bond		0.08	0.66		1.26	2.00
	unpaired δ			1.00			1.00
	total	0.03	0.23	3.04	0.01	3.69	7.00
$^2\Pi$	σ bond	0.00	0.02	0.84	0.01	1.13	2.00
	π_x or π_y bond		0.04	0.77		1.19	2.00
	unpaired π		0.80	0.18		0.02	1.00
	total	0.00	0.90	2.56	0.01	3.53	7.00
$^4\Delta$	unpaired 8σ	0.78	0.19	0.03	0.0	0.0	1.00
	unpaired 9σ	0.00	0.00	0.07	0.02	0.91	1.00
	π_x or π_y bond		0.05	0.61		1.34	2.00
	unpaired δ			1.00			1.00
	total	0.78	0.29	2.32	0.02	3.59	7.00

while the unpaired electron in $^2\Pi$ is in a Ti $4p_\pi$ orbital because the Ti d_π orbitals are involved in the triple bond. We collect in Table II the calculated electron distribution characteristic of these three states. The σ and π bonds in all three states are essentially between the d's on Ti and p's on N. Most interestingly, the σ bond has no 4s or $4p_\sigma$ character. Although each of these states dissociate to Ti in the $4s^2 3d^2$ configuration the in situ Ti d population is always greater than 2 reflecting the $4s3d^3$ and $4p3d^3$ contributions.

The lowest state of TiN having a double bond is the $^4\Delta$. We can imagine this being formed from the $^2\Delta$ by triplet coupling the electrons in the σ bond.

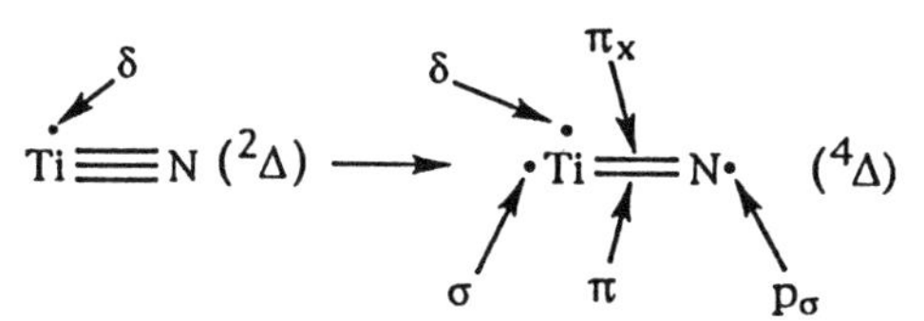

As with Sc ≡ N the character of the Ti σ electron changes from $3d_\sigma$ when it is singlet coupled to the N p_σ to 4s when it is triplet coupled (see Table II). In TiN breaking this σ bond in situ requires 18 kcal/mol whereas in ScN it required 7.2 kcal/mol.

The ground state of VN has a triple bond, is of $^3\Delta$ symmetry and dissociates to the ground states of V and N

δ

σ → •V≡N $^3\Delta$

The first excited state is of $^1\Sigma^+$ symmetry

4s → :V≡N $^1\Sigma^+$

with the two singlet coupled electrons in the "4s" orbital on V.

The ground state of CrN has a triple bond, is of $^4\Sigma^-$ symmetry and dissociates to the ground states of Cr and N.

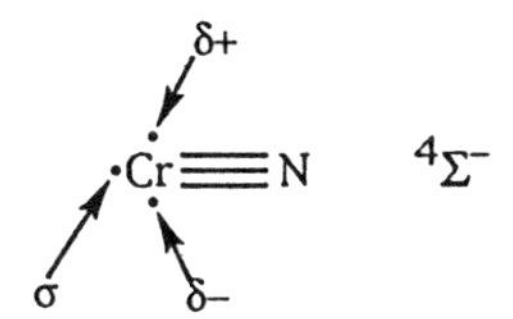

Characteristics of these four neutral nitrides [13] in their ground states at both the MCSCF and MCSCF + 1 + 2 level are collected in Table III. Note that the bond lengths in Ti, V, and Cr nitride are very similar while that of Sc ≡ N is substantially longer. This is consistent with the in situ atomic character of the bond as shown in Table IV. Sc is the only nitride with significant 4s, 4p character in the σ bonds and it also has the smallest d_π character in the π bonds. Interestingly, although the bond lengths in the Ti, V and Cr nitrides are very similar, their bond energies are substantially different. It has been suggested by us [3c, 5g] and others [14] that when a high-spin metal forms a bond, the calculated bond energy is the stabilization energy that remains after payment of the differential intraatomic exchange energy associated with uncoupling pairs of like-spin electrons on the metal. Ab initio SCF calculations of the exchange integrals for the $4s3d^N$ states of Ti, V and Cr combined with the calculated $4s^23d^N - 4s3d^{N+1}$ separations permit us to estimate this exchange energy loss. When one corrects the calculated bond energies in this way one obtains the intrinsic bond energies

Table III. Characteristics of neutral nitride ground states [13]

Molecule	State	R_e (Å)		D_e(ev)		ω_e(cm^{-1})		
		MCSCF	CI[a]	MCSCF	CI[a]	MCSCF	CI[a]	Q(Metal)
•Cr≡N (with two additional dots on Cr)	$^4\Sigma^-$	1.597	1.619	1.97	2.75	909	854	+0.51
•V≡N (with one additional dot on V)	$^3\Delta$	1.588	1.608	3.11	3.74	1028	974	+0.47
•Ti≡N	$^2\Sigma^+$	1.602	1.613	3.64	4.18	1066	1024	+0.50
Sc≡N	$^1\Sigma^+$	1.748	1.762	3.30[b]	4.56[b]	708	774	+0.59

[a] The configuration interaction (CI) includes all single and double excitations from the valence orbitals in the MCSCF reference space.
[b] Relative to symmetry allowed asymptote Sc ($4s3d^2$). When measured relative to the ground state of Sc ($4s^23d$) these energies are 1.62 and 2.82 respectively.

Table IV. Equilibrium populations of valence orbitals in neutral nitride ground states [13]

Molecule	Orbital(s)	Metal			Nitrogen		Total
		4s	4p	3d	2s	2p	
ScN ($^1\Sigma^+$)	σ bond	0.17	0.18	0.50	0.02	1.13	2.00
	π bond		0.12	0.61		1.27	4.00
TiN ($^2\Sigma^+$)	σ bond	0.00	0.02	0.78	0.00	1.20	2.00
	π bond		0.05	0.75		1.20	4.00
	unpaired σ	0.79	0.17	0.04			1.00
VN ($^3\Delta$)	σ bond	0.00	0.02	0.80	0.01	1.17	2.00
	π bond		0.04	0.77		1.19	4.00
	unpaired σ	0.79	0.16	0.05			1.00
	unpaired δ			1.00			1.00
CrN ($^4\Sigma^-$)	σ bond	0.01	0.02	0.77	0.01	1.19	2.00
	π bond		0.02	0.80		1.18	4.00
	unpaired σ	0.78	0.15	0.07			1.00
	unpaired $\delta_\pm$			1.00			2.00

6.6 ev (CrN), 6.6 ev(VN) and 6.3 ev (TiN). While this simple model brings some order to the bond strengths and bond lengths it must be emphasized that the directly calculated and not the augmented bond energies should be compared with experiment.

3. The monopositive nitrides, ScN^+, TiN^+, VN^+ and CrN^+

The experimental [11] energy separation between the lowest states of Sc^+, Ti^+, V^+ and Cr^+ corresponding to the configurations $4s3d^N$ and $3d^{N+1}$ are

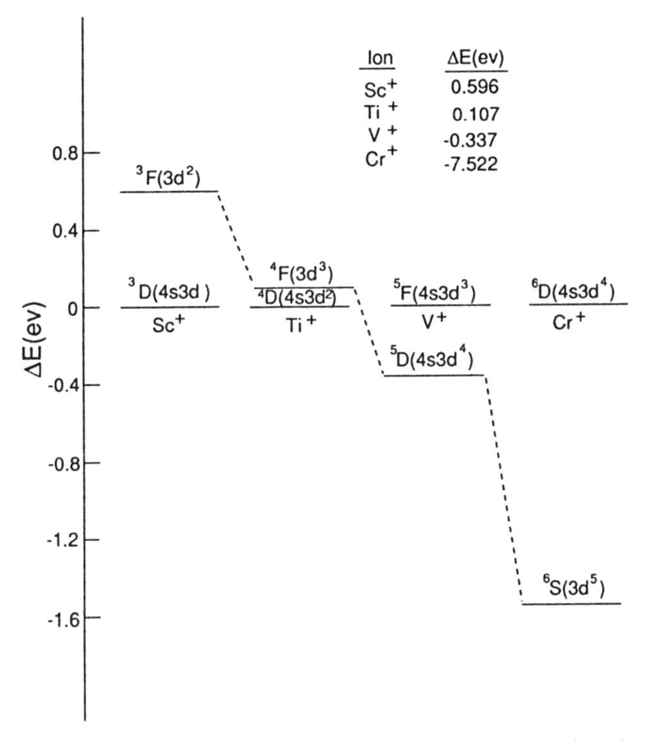

Fig. 7. Experimental energy separation between the lowest states of Sc^+, Ti^+, V^+ and Cr^+ corresponding to the configurations $4s3d^N$ and $3d^{N+1}$.

shown in Figure 7. Note that there are no low lying states in which the 4s orbital is doubly occupied and, relative to the neutrals (Figure 1) the $4s3d^N - 3d^{N+1}$ separation is considerably reduced. In fact, the ground configuration of V^+ and Cr^{N+1} is $3d^{N+1}$.

3.1. The low lying states of ScN^{N+1}

When $Sc^+(^3D)$ interacts with $N(^4S)$ we anticipate states of Σ^-, Π and Δ symmetry with multiplicities 6, 4 and 2 permitting zero, one or two formal chemical bonds [5i]. Only the $^2\Pi$ corresponds to a doubly bonded species

π_y p_x

$+Sc{=}\dot{N}$ $\quad ^2\Pi_x$

σ

This molecular state obtains from the $^1\Sigma^+$ state of ScN by breaking the π_x bond and removing the Sc $d\pi_x$ electron. The low lying ^3F $(3d^2)$ state of Sc^+

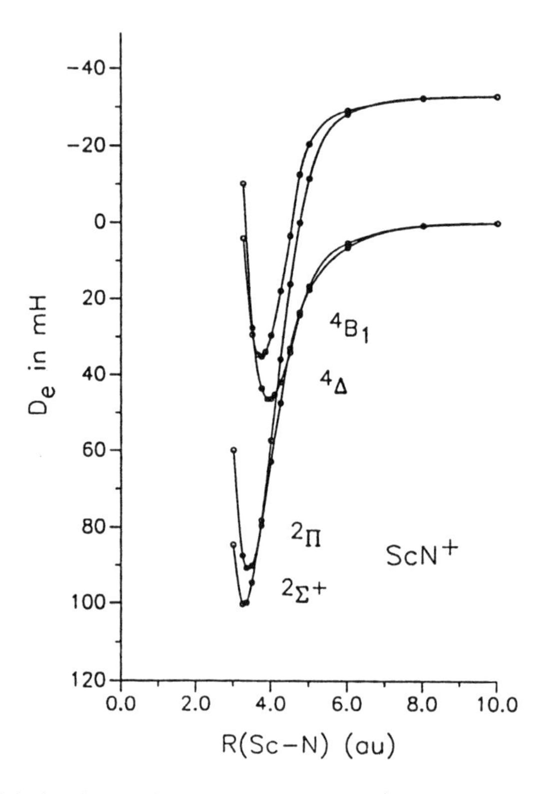

Fig. 8. Calculated potential energy curves of ScN^+ in the two lowest states.

will form Σ^+, Π, Δ and Φ states when it interacts with N. The Σ^+ symmetry also presents the opportunity for the (π, π) doubly bonded species

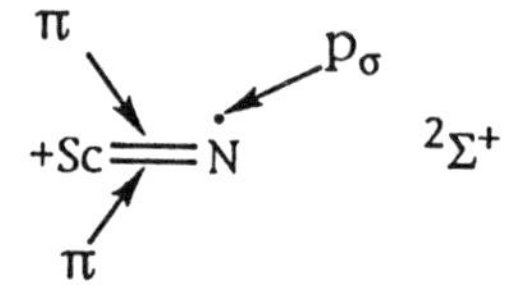

We show in Figure 8 the potential curves [5i] for these two states. Note that the $^2\Sigma^+$ is indeed lower than the $^2\Pi$ even though the $^2\Sigma^+$ dissociates to the first excited state of Sc^+. This is very similar to the situation in neutral ScN in which the ground $^1\Sigma^+$ state dissociates to the excited ^4F state of Sc.

We collect in Table V some characteristics of these doubly bonded states along with representative examples of singly bonded states. Note the monotonic decrease in the equilibrium bond length and vibrational frequency with the monotonic increase in the dissociation energy (referred to the symmetry allowed asymptotes). A more detailed view of the nature of the bonding is given in Figures 9 and 10. From Figure 9 we see that at large separation in the $^2\Sigma^+$ state there is one electron in both of the Sc d_π and all of the N p

Table V. Some characteristics of the low lying states of ScN^+ [5i]

State	Lewis Structure	D_e*(ev)	R_e(Å)	ω_e(cm^{-1})	Q(Sc)
$^2\Sigma^+$ (d^2)	π, π, p_σ; +Sc═N·	3.61/2.73	1.738	871	+1.51
$^2\Pi$ (sd)	π, σ, p_x; +Sc═N	2.40	1.804	811	+1.51
$^4\Phi$ (sd)	δ, p_x, p_y; +Sc—N	1.79/0.92	1.979	674	+1.47
$^4\Delta$ (d^2)	δ, π, p_x, p_y; +Sc—N	1.16	2.101	534	+1.46

* Dissociation energies for $^2\Sigma^+$ and $^4\Phi$ are given relative to both the symmetry allowed asymptote (^{3}F) and the thermodynamic limit (^{3}D). The $^2\Pi$ and $^4\Delta$ are relative to the ^{3}D.

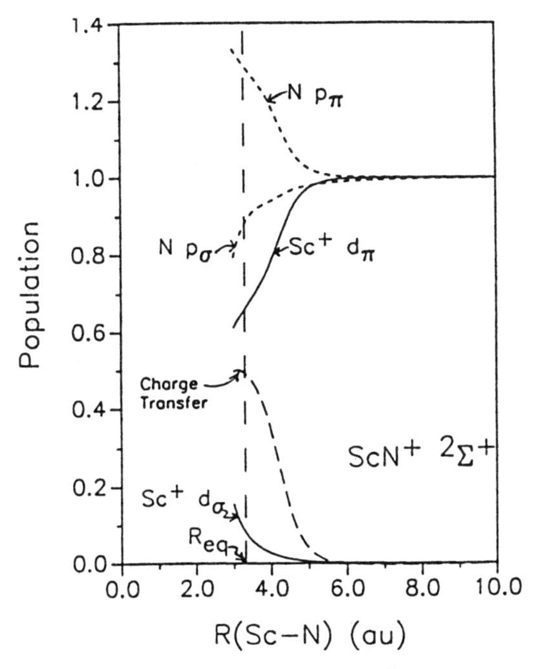

Fig. 9. Electron population in various orbitals of ScN^+ ($^2\Sigma^+$) as a function of internuclear separation.

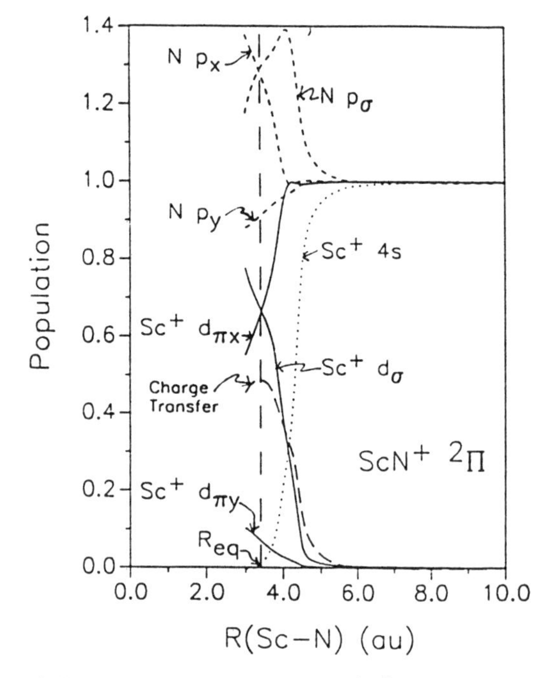

Fig. 10. Electron population in various orbitals of ScN^+ ($^2\Pi$) as a function of internuclear separation.

orbitals. As we approach the equilibrium separation there is a gradual transfer of electrons from the Sc d_π to the N p_π orbitals and a slight back donation from the singly occupied N p_σ to the empty d_σ. Contrast this with the rather abrupt charge transfer shown for the $^2\Pi$ state in Figure 10. At large separation we have three singly occupied N p orbitals and one electron in a Sc^+ 4s and 3dπ. In the σ system we see 4s to 3d_σ promotion as well as charge transfer to the N p_σ orbital. After considerable charge has been transferred in the σ system the π bond undergoes an abrupt charge transfer from Sc to N. Finally, the singly occupied N p_π orbital delocalizes some charge into the empty d_π on Sc. The net result is that, in situ, Sc^+, in both the $^2\Sigma^+$ and $^2\Pi$ states has no 4s character and 1.49 d electrons. In the $^2\Sigma^+$ we have the in situ configuration $d_{\pi x}^{0.70}\ d_{\pi y}^{0.70}\ d_\sigma^{0.09}$ while in the $^2\Pi$ state we have $d_\sigma^{0.70}\ d_{\pi x}^{0.70}\ d_{\pi y}^{0.09}$.

3.2. The remaining mono positive nitrides TiN^+, VN^+ and CrN^+

The ground state of TiN^+ has a triple bond, is of $^1\Sigma^+$ symmetry and separates to ground state products.

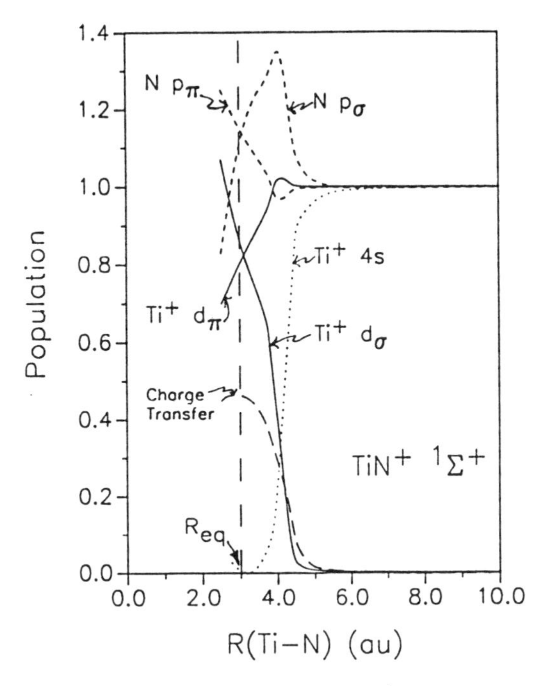

Fig. 11. Electron population in various orbitals of TiN^+ ($^1\Sigma^+$) as a function of internuclear separation.

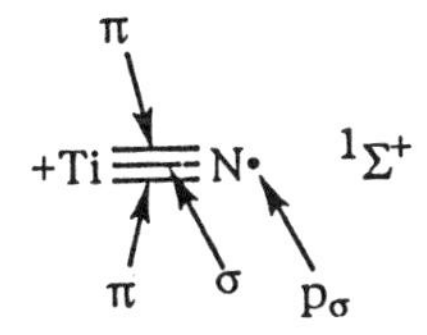

The electron population of selected valence orbitals as a function of the TiN bond length is shown in Figure 11. The orbital populations vary qualitatively as in the isoelectronic neutral ScN (Figure 5). At large separation Ti^+ is in the $4s3d^2$ configuration and as the nuclei approach one another charge is transferred from the $Ti^+\sigma$ system to the N p_σ orbital while simultaneously the 4s electron is being promoted to the $3d_\sigma$. At smaller internuclear separations the π bonds begin to form and the d_π orbitals on Ti begin to transfer electrons to the p_π orbitals on N. At equilibrium there are no 4s electrons on Ti and the charge transfer has peaked. The valence orbital populations at equilibrium are

		Ti				N		
4s	$3d_\sigma$	$4p_\sigma$	$3d\pi_x$	$3d\pi_y$	2s	$2p\sigma$	$2p_x$	$2p_y$
0.00	0.90	0.07	0.08	0.08	1.90	1.13	1.20	1.20

resulting in Ti carrying a formal charge of +1.43 with N hosting a charge of −0.43. The remaining bound states of TiN^+ have either double or single bonds. The π, π double bond is obtained by triplet coupling the two electrons that were singlet coupled in the σ bond of $^1\Sigma^+$. The lowest doubly bonded π, π state is the $^3\Delta$

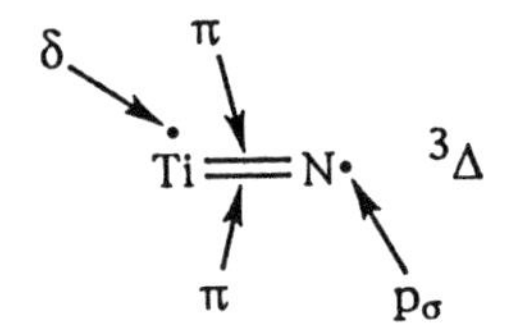

which is 43 kcal/mol above the $^1\Sigma^+$, dissociates to the d^3 configuration of Ti^+ and is bound relative to the ground ^4F ($4s3d^2$) by 54 kcal/mol.

Interestingly, the $^3\Sigma^+$ state

π

•Ti═N•+ $^3\Sigma^+$

π

lies 80 kcal/mol above the $^1\Sigma^+$ which is considerably larger than the 7 kcal/mol $^1\Sigma^+$-$^3\Sigma^+$ gap in the isoelectronic ScN ($^3\Sigma^+$). Additionally, the unpaired σ electron on Ti in the $^3\Sigma^+$ state of TiN^+ has a large $3d\sigma$ contribution.

			Ti				N		
	4s	$4p_\sigma$	$3d\sigma$	$3d\pi_x$	$3d\pi_y$	2s	$2p\sigma$	$2p_x$	$2p_y$
$^1\Sigma^+$	0.0	0.0	0.86	0.80	0.80	2.00	1.13	1.18	1.18
$^3\Sigma^+$	0.50	0.0	0.60	0.80	0.80	1.97	0.95	1.18	1.18

This behavior is consistent with the small $sd^2 - d^3$ separation [11] in Ti^+ (0.1 ev) which precludes a large in-situ relaxation effect in going from the singlet to the triplet. This effect is enormous in ScN because the $sd^2 - d^3$ separation [11] is 2.75 ev.

The lowest $\sigma\pi$ bonded state is the $^3\Pi$

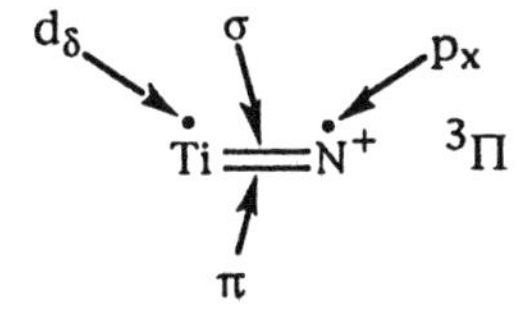

where the two unpaired electrons are in a d_δ on Ti and $2p_x$ or $2p_y$ on N. This state is 42 kcal/mol above the $^1\Sigma^+$ and is bound by 56 kcal/mol. It is energetically very similar to the $^3\Delta$ (π, π) state.

VN^+ has a $^2\Delta$ ground state [5i] in contrast to the $^2\Sigma^+$ ground state of the isoelectronic TiN.

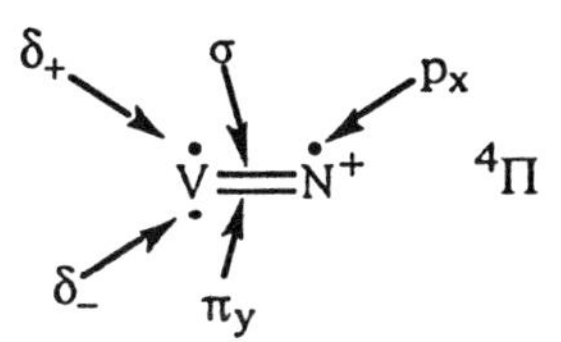

This state is bound by 87 kcal/mol and dissociates to ground state V^+. This preference for an in-situ d^4 configuration is a consequence of the ground state of V^+ being $^5D(3d^4)$ whereas Ti is $^3F(4s^23d^2)$. Since three of the d electrons on V are tied up in the triple bond the remaining triply bonded doublets ($^2\Pi$ and $^2\Sigma^+$) are much higher in energy because the unpaired electron is in a $4p_\pi$ or 4s, $4p_\sigma$ hybrid. There are a variety of valence excitations that result in a double bond. The lowest is the σ, π bound $^4\Pi$ state

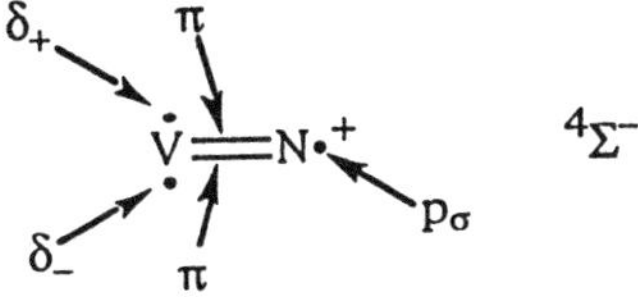

which is 48 kcal/mol above the $^2\Delta$ and is bound by 39 kcal/mol. The companion state in which the unpaired electron on N is in a p_σ orbital has $^4\Sigma^-$ symmetry,

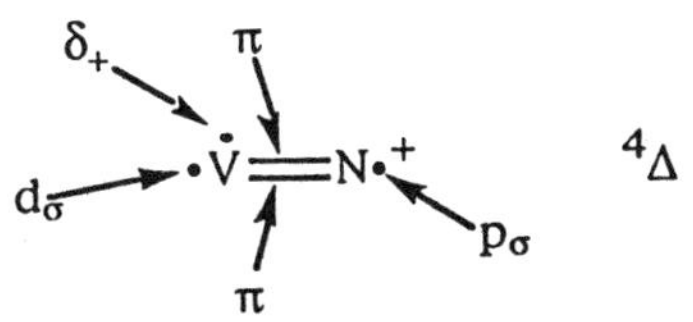

is 53 kcal/mol above the $^2\Delta$ and is bound by 34 kcal/mol. Finally, the $^4\Delta$ is obtained from the ground $^2\Delta$ by triplet coupling the two electrons in the σ bond

δ+ π
dσ •V═N+ pσ $^4\Delta$
π

The ground state of CrN^+ is of $^3\Sigma^-$ symmetry, is bound by 50 kcal/mol and separates to ground state products.

Table VI. Characteristics of mono positive nitrides in their ground states [5i]

Molecule	State	R_e(Å)		D_e(ev)		$\omega_e(cm^{-1})$		
		MCSCF	CI[a]	MCSCF	CI[a]	MCSCF	CI[a]	Q(Metal)
$\dot{Cr}\equiv N^+$	$^3\Sigma^-$	1.574	1.596	1.49	2.15	924	864	+1.33
$\dot{V}\equiv N^+$	$^2\Delta$	1.553	1.574	2.90	3.78	1043	1005	+1.38
$Ti\equiv N^+$	$^1\Sigma^+$	1.575	1.585	3.47	4.24	1075	1039	+1.43
$Sc{=}N\cdot^+$	$^2\Sigma^+$	1.752	1.740	2.63[b]	3.53[b]	872	871	+1.49

[a] The configuration interaction calculation (CI) includes all single and double excitations from the valence orbitals in the MCSCF reference space. [b] Relative to symmetry allowed asymptote Sc^+ ($3d^2$). When measured relative to the ground state of Sc^+ (4s3d) these energies are 1.47 and 2.74 respectively.

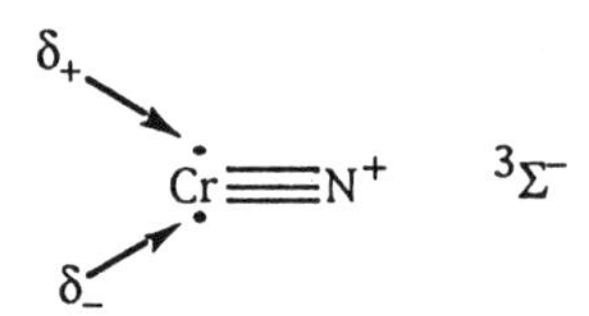

The lowest double bonded species is obtained by high spin coupling the σ bonding electrons.

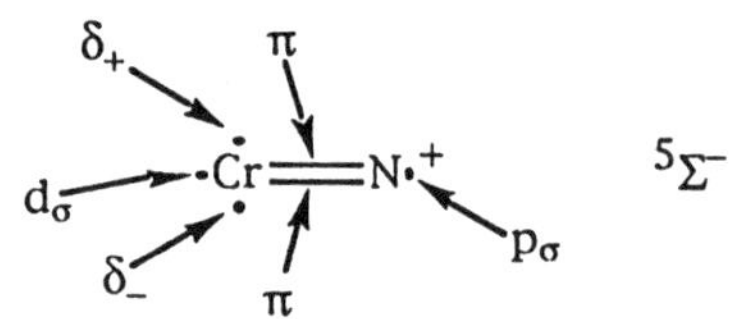

We collect in Table VI the bond lengths, bond energies and vibrational frequencies of the ground states of the monopositive nitrides [5*i*]. Note that all of the molecules are strongly bound and in each molecule there has been substantial charge transfer from the metal to nitrogen, giving significant $M^{++}N^-$ character. As with the neutrals the bond lengths for the triple bonded Ti, V and Cr nitrides are very similar while the bond energies decrease rapidly in going from Ti to Cr. Using the exchange energy loss concept we may derive intrinsic bond energies of 5.5 ev (CrN^+), 5.7 ev (VN^+) and 5.3 ev (TiN^+). These similar intrinsic bond strengths are consistent with the constancy of the calculated bond lengths and electron distribution shown in Table VII.

4. The dipositive ions ScN^{++}, TiN^{++}, VN^{++} and CrN^{++}

The character of the electronic structure of diatomic dipositive ions is dominated by the relative energies of the asymptotic fragments. For a diatomic AB^{++} the fragments are either $A^+ + B^+$ or $A^{++} + B^\circ$. If the asymptote is

Table VII. Electron population in the ground states of the mono positive nitrides [5i]

Molecule	Orbital(s)	Metal 3d	N 2p	Total
$ScN^+(^2\Sigma^+)$	unpaired σ	0.12	0.88	1.00
	π bond	0.67	1.32	4.00
$TiN^+(^1\Sigma^+)$	σ bond	0.86	1.14	2
	π bond	0.83	1.17	4
$VN^+(^2\Delta)$	σ bond	0.86	1.11	2
	π bond	0.87	1.12	4
	unpaired δ	1.0	0	1
$CrN^+(^3\Sigma^-)$	σ bond	0.86	1.11	2
	π bond	0.90	1.10	4
	unpaired $\delta_\pm$	2	0	2

$A^+ + B^+$ the resulting molecule must overcome a substantial Coulomb repulsion before the atoms can get close enough to bond. Often the resulting bond energy is not enough to recover the Coulomb cost and the electronic state is quasi-stationary. If however, the asymptote is $A^{++} + B^\circ$, the initial interaction between A and B is attractive and the resulting molecule is thermodynamically stable. Because the second ionization energy of Sc, Ti and V is larger than the first ionization energy of N (see Figure 12) ScN^{++}, TiN^{++} and VN^{++} belong to the latter class and will be thermodynamically stable [5h]. While the asymptotic lowest energy products of CrN^{++} are Cr^+ and N^+ the $Cr^{++} + N$ asymptote is only 1.95 ev higher which means that the asymmetrically charged system becomes the ground state at a Cr-N separation of ~7 Å.

4.1. The dipositive ion ScN^{++}

When Sc^{++} (3d) approaches N (^4S) we expect triplets and quintets of Σ^- and Δ symmetry according to whether the d electron on Sc^{++} is in a d_σ, d_π or d_δ orbital [5h]. The quintet states have no opportunity for a covalent bond and are expected to be bound electrostatically. In contrast, the triplet states permit both electrostatic and covalent coupling. The electrostatic component of the triplet arises when the intact N (^4S) couples to the doublet on Sc^{++} to produce a triplet. The covalent component of the triplet arises when the Sc^{++} d electron couples into a singlet with a N 2p orbital, leaving the triplet multiplicity to be carried by the two remaining N 2p electrons.

Schematically

$$|\ \text{Electrostatic Triplet}\ \rangle = |\ \underbrace{\text{Sc}\boxdot}_{S=1/2}\ \overbrace{\boxed{\cdot\dot{\underset{\cdot}{N}}}}^{S=3/2}\ \rangle$$

and

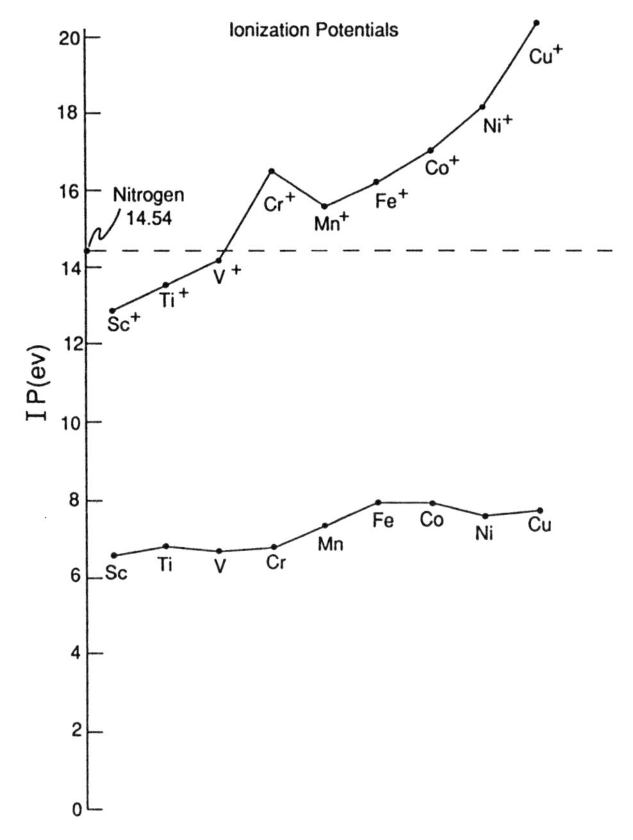

Fig. 12. First and second ionization energies of the first transition series compared to the first ionization energy of N.

S = 1

| Covalent Triplet > = | Sc[• •] [N] >

S = 0

In Figure 13 we show the potential energy curves for the $^3\Sigma^-$, $^3\Pi$ and $^3\Delta$ states. At large internuclear separations the electrostatic interaction is dominant and the states order as $^3\Delta < {}^3\Pi < {}^3\Sigma^-$. However as the bond length decreases the energy of the $^3\Delta$ state increases rapidly while the $^3\Sigma^-$ and $^3\Pi$ continue to decrease. This is because the δ_+ orbital in the $^3\Delta$ state cannot form an effective bond with any of the 2p orbitals on N, while the d_σ orbital in the $^3\Sigma^-$ state can form σ bond with the $2p_\sigma$ and the $3d_\pi$ orbital in the $^3\Pi$ state can form a π bond with the $2p_\pi$ on N. Interestingly the sigma bonded and pi bonded states have comparable D_e's and R_e's, suggesting that the single σ or π bonds are equally effective in stabilizing the molecule. The quintet states are purely electrostatic and as we can see from Figure 14

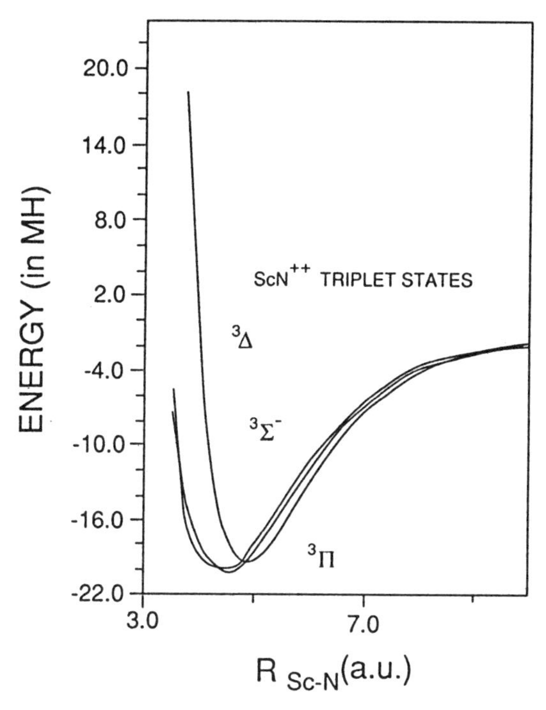

Fig. 13. Potential energy curves calculated at the MCSCF + 1 + 2 level for the three states ($^3\Sigma^-$, $^3\Pi$, and $^3\Delta$), which dissociate to the ground state products Sc^{+2} (^2D) + N (^4S).

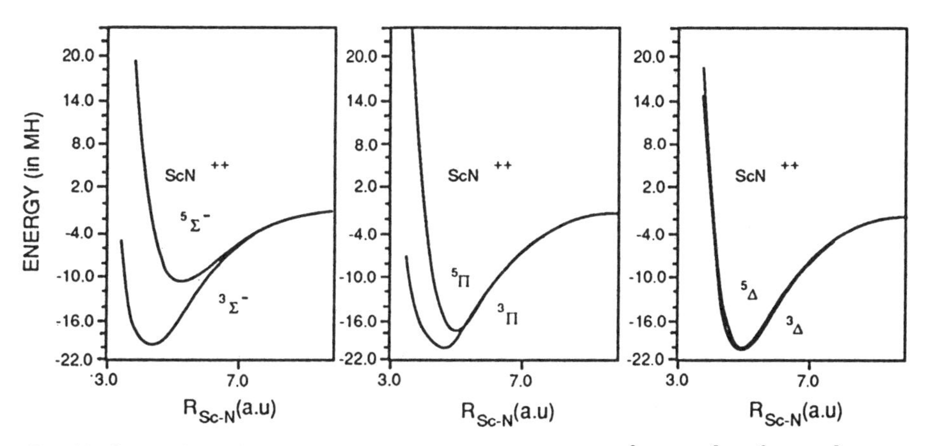

Fig. 14. Comparison of the potential energy curves of the states $^3\Sigma^-$ and $^5\Sigma^-$; $^3\Pi$ and $^5\Pi$; and $^3\Delta$ and $^5\Delta$.

Table VIII. Bond energies, bond lengths, and vibrational frequencies of ScN^{++} [5h]

State	R_e(Å)	D_e(ev)	$\omega(cm^{-1})$
$^3\Sigma^-$	2.323	0.53	178
$^3\Pi$	2.391	0.54	202
$^3\Delta$	2.570	0.52	234
$^5\Sigma^-$	2.825	0.30	168
$^5\Pi$	2.605	0.47	234
$^5\Delta$	2.552	0.53	239

parallel the triplets at large separations. Since the quintet spin couplings preclude the formation of a covalent bond the minimum energies are determined by the Pauli repulsion of the Sc^{++}d electron and the valence electrons on N. This results in the state order $^5\Delta < {}^5\Pi < {}^5\Sigma^-$ at all internuclear distances. The increased stability of the $^3\Sigma^-$ versus the $^5\Sigma^-$ and the $^3\Pi$ versus $^5\Pi$ are apparent from the plots. Additionally, although neither the $^5\Delta$ or $^3\Delta$ can form a covalent chemical bond they are both contenders for the ground state of ScN^{++}. The orientation of the Sc^{++} $3d_\delta$ electron in these states results in a differentially smaller intermolecular electron-electron repulsion while allowing the full electrostatic stabilization to obtain. Note that although the $^3\Sigma^-$ and $^3\Pi$ states appear to be forming a covalent bond there is little charge transfer. For example, the electron population in the valence orbitals of the $^3\Sigma^-$ state is

Sc				N		
4*s*	3*d*σ	3dπx	3dπy	2pσ	2pπx	2pπy
0.	0.91	0.02	0.02	1.05	0.98	0.98

These results suggest that at equilibrium, ScN^{++} is essentially a dication of Sc electrostatically bound to neutral N. We collect the bond energies, bond lengths and vibrational frequencies for these states in Table VIII.

4.2. The remaining dipositive ions, TiN^{++}, VN^{++} and CrN^{++}

As with Sc^{++}, the 4s orbital on Ti^{++}, V^{++} and Cr^{++} will play no role in the low lying electronic states of these dipositive ions [15]. The two valence d electrons on Ti^{++} permit double and single bonds to N. The doubly bonded states are either $^2\Sigma^+$ or $^2\Pi$ according to whether the bond is π, π or π, σ respectively

Table IX. Bond energies, bond lengths and vibrational frequencies of TiN^{++} [15]

State	Bond	R_e(Å)	D_e(ev)	ω_e(cm^{-1})
$^2\Sigma^+$	π, π double	1.74	0.82	552
$^2\Pi$	σ, π double	2.35	0.65	238
$^4\Delta$	σ	2.48	0.53	220
$^4\Pi$	σ	2.41	0.63	254

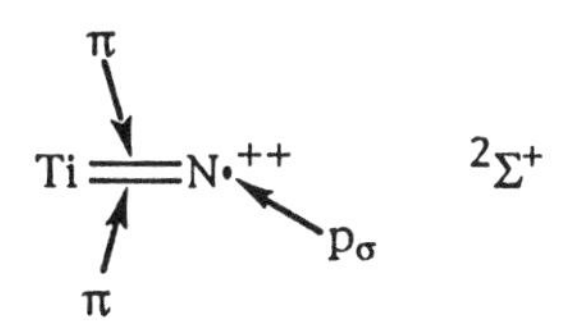

σ pπ

Ti N ++ $^2\Pi$

π

The $^2\Delta$ state obtains by forming a single σ bond and coupling the spins of the Ti $3d_{\delta\pm}$ and the N $2p_x$ and $2p_y$ electrons into a doublet. Obviously, given the weakness of the Ti $3d_\delta$ and N $2p_\pi$ interactions, the $^4\Delta$ state will be of comparable energy.

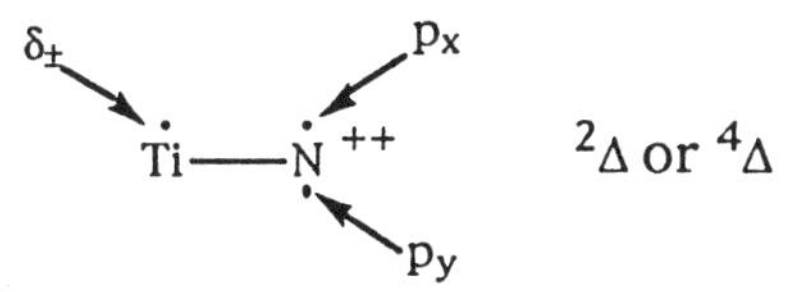

Additionally, we may form a $^4\Pi$ state from the $^2\Pi$ by uncoupling the electrons in the π bond

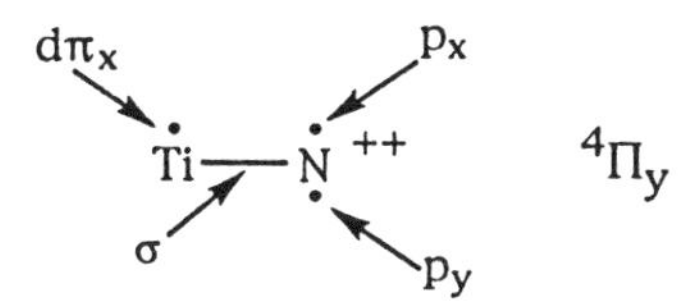

With the exception of the $^2\Sigma^+$ state, all of these states have very similar bond lengths and bond energies (see Table IX). The potential energy curves for all of these states are also similar being dominated by induced dipole effects at large separations and Pauli repulsion at small separations. The $^2\Sigma^+$ state has a significantly smaller bond length (1.72 Å) because $d\pi$, $d\pi$ configuration on Ti cause the Pauli repulsion effects to become important at smaller internuclear separation and then they are somewhat offset by the differential stabilization due to the π, π bonds.

Table X. Ground state characteristics of MN^{++} [15]

Molecule	State	R_e(Å)	D_e(ev)	ω_e(cm^{-1})	Metal d
CrN^{++}	$^2\Delta$	1.632	1.05[a]	613	3.98
VN^{++}	$^1\Sigma^+$	1.580	1.07	906	2.98
TiN^{++}	$^2\Sigma^+$	1.743	0.82	552	1.98
ScN^{++}	$^3\Sigma^{-}$[b]	2.323	0.53	178	0.95

[a] CrN^+ is not bound relative to the Cr^+, N^+ asymptote. This energy is relative to the Cr^+ + N asymptote.
[b] Several other states of ScN^+ have very similar bond energies.

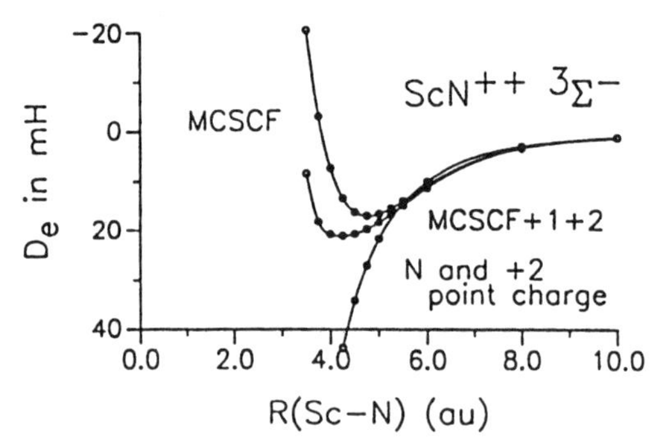

Fig. 15. Comparison of the ab-initio potential energy curves for ScN^{++} in the $^3\Sigma^-$ state and the interaction energy of an N atom with a +2 point charge.

The triply bonded state of VN^{++} has $^1\Sigma^+$ symmetry, is bound by 1.06 ev and has a bond length of 1.58 Å. The lowest π, π doubly bonded state is obtained by uncoupling the electron in the σ bond and placing the V electron in a d_δ orbital. This $^1\Delta$ state has a bond length of 2.12 Å and a bond energy of 0.82 ev.

$$V{\equiv}N^{++}\ (^1\Sigma^+) \longrightarrow \dot{V}{=}N^{++}\ (^1\Delta) \quad (\delta,\ \pi,\ \pi,\ p_\sigma)$$

CrN^{++} also has a triple bond with the extra d electron on Cr in a d_δ orbital. The resulting $^2\Delta$ state has a bond length of 1.63 Å and a bond energy of 1.05 ev, very similar to VN^{++}. The remaining low lying states of CrN^{++} have larger bond lengths (~2.2 Å) and somewhat smaller bond energies (~0.8 ev) – again rather similar to the excited state of VN^{++}. We collect in Table X the characteristics of the ground states of the dipositive ions.

The bonding in the dipositive ions is distinct from that in the neutral and monopositive ions in that it is dominated by electrostatics. Figure 15 compares the ab-initio potential energy curves for ScN^{++} in the $^3\Sigma^-$ state with the interaction energy of a N atom and a +2 point charge. Clearly the

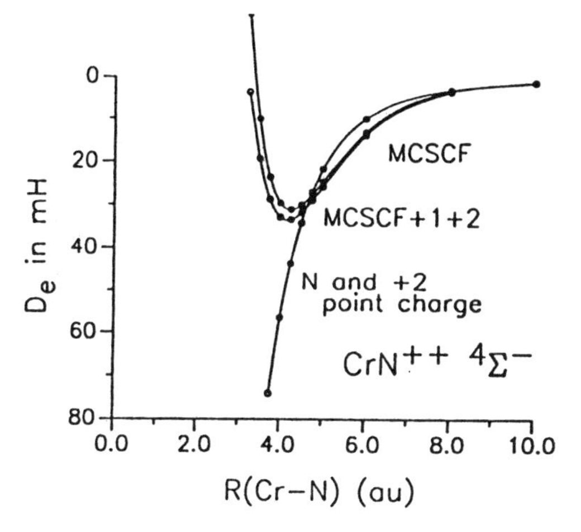

Fig. 16. Comparison of the ab-initio potential energy curves for CrN^{++} in the $^4\Sigma^-$ state and the interaction energy of an N atom with a +2 point charge.

charge induced dipole interaction stabilizes the molecule until the atoms are close enough for Pauli repulsion to become important. Figure 16 makes the same comparison for double bonded CrN^{++} in the $^4\Sigma^-$ state. Here, however, we see the incremental stabilization due to the incipient triple bond.

5. Effect of charge on the bonding in the early transition metal nitrides

The theoretical results for the neutral, mono and dipositive nitrides permit a variety of comparisons of the effect of charge on the nature of the metal-nitrogen bond.

5.1. The CrN, CrN^+ and CrN^{++} series

The ground state of CrN is of $^4\Sigma^-$ symmetry and contains three formally non bonding electrons

$$\delta_+ \rightarrow,\ \sigma \rightarrow \cdot\dot{\mathrm{Cr}}\equiv\mathrm{N}\ (^4\Sigma^-),\ \delta_- \rightarrow$$

The lowest lying positive ion obtains by removing the non bonding σ electron to form the $^3\Sigma^-$ state,

$$\dot{\underset{\bullet}{\mathrm{Cr}}}\equiv\mathrm{N}^+\ (^3\Sigma^-)$$

The lowest lying dipositive ion obtained by removing one of the non bonding δ electrons to form a $^2\Delta$ state,

$$\dot{C}r \equiv N^{++} \; (^2\Delta)$$

Although we have kept the triple bond intact during these ionizations, the calculated D_e drops from 2.75 ev to 2.15 ev to 1.05 ev. This is consistent with the bonding changing from a conventional covalent-ionic mixture in the neutral to predominately electrostatic in the dipositive ion. While the monopositive ion enjoys a charge induced dipole interaction at long distances, the bonding at equilibrium has a large covalent component. The calculated bond lengths in these three molecules are 1.62 Å, 1.60 Å and 1.63 Å. This remarkable constancy, in the light of the highly variable D_e's, suggests that the size of the Cr 3d orbitals is a dominant factor in determining the Cr-N bond lengths.

5.2. The CrN^{++}, VN^{+}, TiN $^2\Delta$ series

These three molecules have a formal metal-nitrogen triple bond and an unpaired electron in a d_δ metal orbital. The bond energies, bond lengths and charges are

Molecule	R_e(Å)	D_e(ev)	Q (Metal)
$\dot{C}r \equiv N^{++}$ $(^2\Delta)$	1.632	1.05	+2.02
$\dot{V} \equiv N^{+}$ $(^2\Delta)$	1.574	3.78	+1.38
$\dot{T}i \equiv N$ $(^2\Delta)$	1.657	3.23	+0.46

TiN's D_e is less than VN^+'s because the $^2\Delta$ state of TiN is an excited state while it is the ground state of VN^+. This reflects the increased stability of the d electrons as we go from left to right across the transition series and the increased stability of the d's in the positive ions. VN^+ prefers the extra electron in a $3d_\delta$ rather than a 4s, $4p_\sigma$ hybrid while TiN favors the reverse.

5.3. The VN^{++}, TiN^{+}, ScN $^1\Sigma^+$ series

These three molecules have formal metal-nitrogen triple bonds and no unpaired electrons. The bond energies, bond lengths and charges are

Molecule	R_e(Å)	D_e(ev)	Q (Metal)
$V \equiv N^{++}$ $(^1\Sigma^+)$	1.580	1.07	+1.93
$Ti \equiv N^{+}$ $(^1\Sigma^+)$	1.585	4.24	+1.43
$Sc \equiv N$ $(^1\Sigma^+)$	1.762	4.56	+0.59

Each of these molecules has a $^1\Sigma^+$ ground state. The striking feature of these data is the large ScN bond length relative to TiN^+. This probably reflects the effect of the d orbitals contracting as we go from left to right in the transition series and as we go from a neutral to a positive TM.

6. Comparison with experiment and other theoretical results

There seem to be no published experimental results on the mono and dipositive nitrides of the early transition metals. However, several gaseous neutral diatomic nitrides including ScN, TiN, VN and CrN have been characterized experimentally as well as theoretically and we collect, in Table XI, the available data for the ground states of these neutral nitrides. The experimental estimates for D_e are all from thermochemical measurements and with the exception of ScN are larger than our calculated values. These quantities are notoriously difficult to measure and to calculate, ab-initio, and the poor agreement reflects this. The larger value for ScN is probably because we have referred our calculated value to the symmetry constrained asymptotic products Sc(^{4}F) and N(^{4}S). While our calculated bond lengths are always larger than experiment they do track the variation in R_e as we go from ScN to VN, suggesting that CrN will have a slightly larger bond length than VN. The variations among the theoretical estimates of D_e, R_e and ω_e along with the properties of various excited states are discussed elsewhere [13].

Table XI. Experimental and theoretical constants for ground state neutral nitrides

System	R_e(Å)	D_e(ev)	ω_e(cm^{-1})	Reference
ScN(X$^1\Sigma^+$)	1.687		795	20-exp
		4.56		2p-exp
	1.768	4.86	726	13-theory
TiN(X$^2\Sigma^+$)	1.582		1049	3c, 2f, exp
		4.90		16-exp
	1.613	4.19	1024	13-theory
	1.630	3.78	1010	3n-theory
	1.568	8.05	1139	3j-theory, 3q-theory
VN(X$^3\Delta$)	1.566			2g-exp
	1.574		1033	2h-exp
		4.9		16-exp
	1.608	3.74	974	13-theory
CrN(X$^4\Sigma^-$)		3.9		16-exp
	1.619	2.75	854	13-theory
	1.62	3.18		3p-theory

References

1. Z. L. Xiao, R. H. Hauge, and J. L. Margrave, *J. Phys. Chem.* **96**, 636 (1992); R. J. van Zee and W. Weltner, Jr., *J. Am. Chem. Soc.* **111**, 4519 (1989); Y. M. Hamrick and W. Weltner, Jr., *J. Chem. Phys.* **94**, 3371 (1991).
2. Publications of the experimental groups on neutral metal nitride dimers include: (a) J. C. Howard and J. G. Conway, *J. Chem. Phys.*, **43**, 3055 (1965); (b) T. M. Dunn and K. M. Rao, *Nature (London)* **222**, 266 (1969); (c) T. M. Dunn, L. K. Hanson, and K. A. Rubinson, *Can. J. Phys.*, **48**, 1657 (1970); (d) J. K. Bates, N. L. Ranieri, and T. M. Dunn, *Can. J. Phys.*, **54**, 915, (1976); (e) T. C. DeVore and T. N. Gallaher, *J. Chem. Phys.* **70**, 3497 (1979); (f) A. E. Douglas and P. M. Veillette, *J. Chem. Phys.* **72**, 5378 (1980); (g) T. M. Dunn and S. L. Peter, *J. Chem. Phys.*, **90**, 5333 (1989). (h) B. Simard, C. Masoni, and P. A. Hackett, *J. Mol. Spectrosc.* **136**, 44 (1989); (i) J. K. Bates and D. M. Gruen, *J. Mol. Spectrosc.* **78**, 284 (1979); (j) L. B. Knight and J. Steadman, *J. Chem. Phys.* **76**, 3378 (1982); (k) R. C. Carlson, J. K. Bates, and T. M. Dunn, *J. Mol. Spectrosc.* **110**, 215 (1985); (l) J.-L. Femenias, C. Athenour, K. M. Rao, and T. M. Dunn, *J. Mol. Spectrosc.* **130**, 269 (1988); (m) J. K. Bates and D. M. Gruen, *J. Chem. Phys.* **70**, 4428 (1979); (n) D. W. Green and G. T. Reedy, *J. Mol. Spectrosc.* **74**, 423 (1979); (o) R. S. Ram and P. F. Bernath, *J. Chem. Phys.* **96**, 6344 (1992); (p) K. A. Gingerich, *J. Chem. Phys.* **49**, 19 (1968).
3. See, for example (a) J. F. Harrison, *J. Phys. Chem.* **87**, 1312 (1983); (b) J. F. Harrison, *J. Phys. Chem.* **87**, 1323 (1983); (c) K. L. Kunze and J. F. Harrison, *J. Am. Chem. Soc.* **112**, 3812 (1990); (d) S. R. Langhoff and C. W. Bauschlicher, Jr., *Annu. Rev. Chem.* **39**, 181 (1988); (e) S. R. Langhoff, C. W. Bauschlicher, Jr., and H. Partridge, *J. Chem. Phys.* **89**, 396 (1988); (f) S. R. Langhoff, C. W. Bauschlicher, Jr., L. G. M. Pettersson, and P. E. M. Siegbahn, *Chem. Phys.* **132**, 49 (1989); (g) P. G. Jasien and W. J. Stevens, *Chem. Phys. Lett.* **147**, 72 (1988); (h) T. R. Cundari and M. S. Gordon, *J. Am. Chem. Soc.* **113**, 5231 (1991); (i) K. Balasubramanian and P. Y. Feng, *J. Chem. Phys.* **96**, 2881 (1992); (j) S. M. Mattar, *J. Phys. Chem.* **97**, 3171 (1993); (k) C. W. Bauschlicher and S. P. Walch, *J. Chem. Phys.* **76**, 4560 (1982); (l) S. P. Walch and C. W. Bauschlicher, Jr., *Chem. Phys. Lett.* **85**, 66 (1982); (m) S. P. Walch and C. W. Bauschlicher, Jr., *J. Chem. Phys.* **78**, 4597 (1983); (n) C. W. Bauschlicher, Jr., *Chemical Physics Letters* **100**, 515 (1983); (o) P. E. M. Siegbahn and M. R. A. Blomberg, *Chem. Phys.* **87**, 189 (1984); (p) M. R. A. Blomberg and P. E. M. Siegbahn, *Theoretica Chimica Acta* **81**, 365 (1992); (q) K. D. Carlson, C. R. Claydon, and C. Moser, *J. Chem. Phys.* **46**, 4963 (1967).
4. Recent publications of the experimental groups currently active in this area include: (a) J. L. Elkind and P. B. Armentrout, *J. Chem. Phys.* **86**, 1868 (1987); (b) L. Sunderlin, N. Aristov, and P. B. Armentrout, *J. Am. Chem. Soc.* **109**, 78 (1987); (c) N. Aristov and P. B. Armentrout, *J. Phys. Chem.* **91**, 6178 (1987); (d) W. D. Reents, F. Strobel, R. B. Freas, J. Wronka, and D. P. Ridge, *J. Phys. Chem.* **89**, 5666 (1985); (e) R. L. Hettich and B. S. Freiser, *J. Am. Chem. Soc.* **109**, 3543 (1987); (f) B. D. Radecki and J. Allison, *Organometallics* **5**, 411 (1986); (g) S. W. McElvany and J. Allison, *Organometallics* **5**, 1219 (1986); (h) M. A. Hanratty, J. L. Beauchamp, A. J. Illies, P. van Koppen, and M. T. Bowers, *J. Am. Chem. Soc.* **110**, 1 (1988); (i) J. L. Schilling and J. L. Beauchamp, *J. Am. Chem. Soc.* **110**, 15 (1988); (j) J. L. Schilling and J. L. Beauchamp, *Organometallics* **7**, 194 (1988); (k) M. A. Tolbert, M. L. Mandich, L. F. Halle, and J. L. Beauchamp, *J. Am. Chem. Soc.* **108**, 5675 (1986); (l) H. Kang and J. L. Beauchamp, *J. Am. Chem. Soc.* **108**, 5663 (1986); (m) H. Kang and J. L. Beauchamp, *J. Am. Chem. Soc.* **108**, 7502 (1986); (n) C. B. Lebrilla, C. Schulze, and H. Schwarz, *J. Am. Chem. Soc.* **109**, 98 (1987); (o) C. B. Lebrilla, T. Drewello, and H. Schwarz, *Int. J. Mass Spectrom. Ion Proc.* **79**, 287 (1987); (p) C. Schulze and H. Schwarz, *J. Am. Chem. Soc.* **110**, 67 (1988); (q) R. Stepnowski and J. Allison, *Organometallics* **7**, 2097 (1988); (r) D. E. Clemmer, N. Aristov, and P. B. Armentrout, *J. Phys. Chem.* **97**, 544 (1993); (s) P. B. Armentrout, *Annu. Rev. Phys. Chem.* **41**, 313 (1990);

(t) P. R. Kemper, J. Bushnell, G. von Helden, and M. T. Bowers, *J. Phys. Chem.* 97, 52 (1993); (u) Y. Huang, D. Hill, M. Sodupe, C. W. Bauschlicher, and B. S. Freeser, *J. Am. Chem. Soc.* **114**, 9106 (1992); (v) S. D. Hanton, R. J. Noll, and J. C. Weisshaar, *J. Chem. Phys.* **96**, 5176 (1992); (w) A. D. Sappey, G. Eiden, J. E. Harrington, and J. C. Weisshaar, *J. Chem. Phys.* **90**, 1415 (1989).

5. Publications in this series include: (a) A. E. Alvarado-Swaisgood, J. Allison, and J. F. Harrison, *J. Phys. Chem.* **89**, 2517 (1985); (b) A. E. Alvarado-Swaisgood and J. F. Harrison, *J. Phys. Chem.* **89**, 5198 (1985); (c) J. F. Harrison, *J. Phys. Chem.* **90**, 3313 (1986); (d) A. Mavridis, A. E. Alvarado-Swaisgood, and J. F. Harrison, *J. Phys. Chem.* **90**, 2548 (1986); (e) A. E. Alvarado-Swaisgood and J. F. Harrison, *J. Phys. Chem.* **92**, 2757 (1988); (f) A. E. Alvarado-Swaisgood and J. F. Harrison, *J. Phys. Chem.* **92**, 5896 (1988); (g) A. E. Alvarado-Swaisgood and J. F. Harrison, *THEOCHEM.* **46**, 155 (1988); (h) K. L. Kunze and J. F. Harrison, *J. Phys. Chem.* **95**, 6418 (1991); (i) K. L. Kunze and J. F. Harrison, *J. Phys. Chem.* **93**, 2983 (1989).
6. Recent publications of the theoretical groups besides our own currently active in this area include: (a) J. B. Schilling, J. L.Beauchamp, and W. A. Goddard, III, *J. Am. Chem. Soc.* **109**, 4470 (1987); (b) J. B. Schilling, W. A. Goddard, III, and J. L. Beauchamp, *J. Am. Chem. Soc.* **109**, 5573 (1987); (c) J. B. Schilling, J. L. Beauchamp, and W. A. Goddard, III, *J. Am. Chem. Soc.* **109**, 5565 (1987); (d) E. A. Carter and W. A. Goddard, III, *J. Am. Chem. Soc.* **108**, 2180, 4746 (1986); (e) L. G. M. Pettersson, C. W. Bauschlicher, Jr., S. R. Langhoff, and H. Partridge, *J. Chem. Phys.* **87**, 481 (1987); (f) M. R. A. Blomberg, P. E. M. Siegbahn, and J.-E. Backvall, *J. Am. Chem. Soc.* **109**, 4450 (1987); (g) T. R. Cundari and M. S. Gordon, *J. Phys. Chem.* **96**, 631 (1992); (h) D. G. Musaev, N. Koga, and K. Morokuma, *J. Phys. Chem.* **97**, 4064 (1993); (i) R. A. Blomberg, P. E. M. Siegbahn, and M. Svensson, *Inorg. Chem.* **32**, 4218 (1993).
7. A. J. H. Wachters, *J. Chem. Phys.* **52**, 1033 (1970).
8. P. J. Hay, *J. Chem. Phys.* **66**, 4377 (1977).
9. J. Almlöf and P. R. Taylor, *J. Phys. Chem.* **86**, 4070 (1987).
10. R. Shepard, I. Shavitt, R. M. Pitzer, D. C. Comeau, M. Pepper, H. Lischka, P. G. Szalay, R. Ahlrichs, F. B. Brown, and J. G. Zhao, *Internat. J. Quantum Chem.* **S22**, 149 (1988).
11. C. E. Moore, C.E. *Atomic Energy Levels*, Vols. I and II (Nat. Stand. Ref. Data Ser., Nat. Bur. Stand., Cir. 35, Washington, D.C., 1971).
12. J. F. Harrison, (unpublished work).
13. J. F. Harrison, "The Electronic Structure of ScN, TiN, VN and CrN" (to be published).
14. E. A. Carter and W. A. Goddard, III, *J. Phys. Chem.* **88**, 1485 (1984); *J. Phys. Chem.* **92**, 5679 (1988).
15. J. F. Harrison, "The Electronic Structure of ScN^{++}, TiN^{++}, VN^{++} and CrN^{++}" (to be published).
16. K. P. Huber and G. Herzberg, *Molecular Spectra and Molecular Structure IV, Constants of Diatomic Molecules* (Van Nostrand, New York, 1979).

4. Fundamental studies of functionalized hydrocarbons with transition-metal ions

KARSTEN ELLER
BASF-AG, ZAK/Z – Q200, D-67056 Ludwigshafen, Germany

1. Introduction

The widespread interest in gas-phase organometallic chemistry is evidenced by the increasing number of review articles published on the topic [1–10]. The present article is concerned with some of the work done at the Technical University of Berlin in the group of Professor Schwarz [5], and for the very interesting results from other laboratories the reader is referred to the other chapters of this book and the individual accounts published before [6–10]. The reason for that anybody would be interested in the reactions of individual metal ions lies in the fact that only in a high-vacuum apparatus in the absence of disturbing influences presented by solvent, neighbouring species, etc., the unique properties of these most simple reactive species can be studied. These properties are, among other things, of importance for catalytic applications. The development of industrial catalysts is even nowadays still more an art than a science [11], thus fundamental data on active centers on the surface of the catalysts is definitely needed. Gas-phase studies can supply bond dissociation energies that might be of help for evaluating intermediates in catalytic cycles, and they further the understanding about the intrinsic properties of the metals. The last point can be studied quite conveniently by systematic variations. Variations of the metal ions in the reaction with a particular substrate can show the effects of additional d-electrons across a row of the periodic table while varying the chain length of the substrate or introducing additional substituents reveals preferential reaction pathways. In the following some of the insight that can be gained with these variations will be highlighted.

2. Instrumentation

Two different approaches have been used to study the reactions of transition-metal ions with selected substrates, based on either a multi-sector or a Fourier transform ion cyclotron resonance (FTICR) mass spectrometer. As

Ben S. Freiser (ed.), Organometallic Ion Chemistry, 123–155.

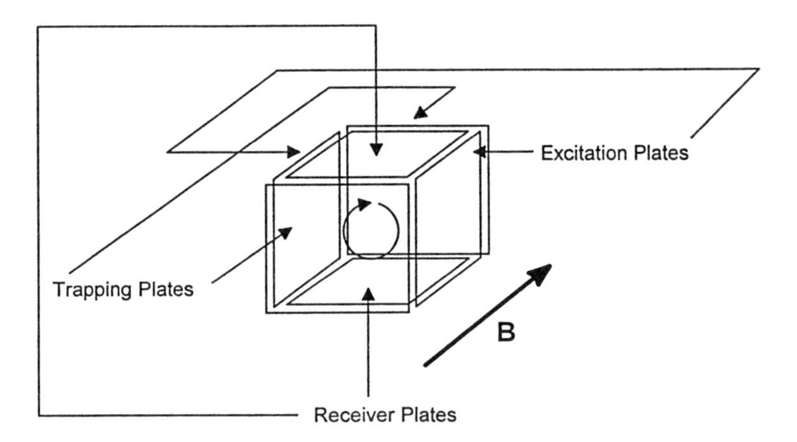

Fig. 1. Cubic FTICR cell with **B** indicating the magnetic field.

the techniques have been described extensively [12, 13], only a brief reference will be given here and some details concerning the spectrometers used.

FTICR instruments are ideally suited for the study of ion molecule reactions. Moving ions in a strong magnetic field **B** (1–7 Tesla) are forced on circular trajectories due to the Lorentz force, but they can still travel in the direction of the magnetic field vector or in the opposite direction. If this option is eliminated by the application of electric potentials on a pair of trapping plates (Figure 1 shows the case of a cubic FTICR cell located between the poles of an electromagnet or in the room-temperature bore of a superconducting magnet), the ions can be stored in the crossed fields. Their frequency of gyration, i.e. the number of rotations in the plane perpendicular to **B**, depends upon their mass-to-charge ratio and is thus used to obtain a mass spectrum. This cyclotron frequency lies in the radio-frequency range and depends on the field strength. To be detected however, the ions must be excited from their position in the middle of the cell to larger radii close to the cell boundaries. This is accomplished by irradiating the ions at their cyclotron frequencies, using the excitation plates and a short pulse consisting of the frequencies in question. In case of resonance they absorb energy which is converted to potential energy by enlarging their cyclotron radius. By using a broad range of radio frequencies, all ions can be simultaneously excited. Upon circulating close to the cell walls they induce image currents every time they pass the plates. The signal induced in the receiver plates is converted to a voltage, digitized and Fourier transformed. The Fourier transformation is necessary because ions of a different mass and hence different cyclotron frequency are simultaneously present in the cell. If ions of a single mass are to be ejected from the ion ensemble, a signal containing only their particular cyclotron frequency is applied to the excitation plates and the duration is chosen long enough so that the ions hit the cell plates where they are neutralized. If only one sort of ions shall be studied, all ions except for those of interest are ejected from the cell. Following isolation they are allowed to react with a substrate that is admitted to the cell either continuously or as a pulse of gas. After a certain reaction time, the products formed are detected or a certain product ion is again selected for further experiments. Structural information can be gained from collision-induced dissociation (CID) experiments [14]. These are performed by exciting the ions into a collision gas, e.g. Ar that is pulsed into the cell; due to their kinetic energy the ions undergo fragmentation.

The spectrometer used in the present case was a Bruker-Spectrospin CMS 47X, which is equipped with an external ion source [15]. External sources are advantageous because ionization techniques that require higher pressures can be used; inside the cell a fairly high vacuum is required as the signal in the receiver plates is damped by collisions with neutral molecules which then leads to broad signals. The instrument and further details of its operation have been previously described [16–18]. To this end, metal ions are generated by laser desorption/ionization [19] by focusing the beam of a Nd:YAG laser (Spectron Systems, 1064 nm) onto a high-purity rod of the desired transition metal which was affixed in the external ion source. The ions are extracted from the source and transferred into the analyzer cell by a system of electrostatic potentials and lenses. The ion source, transfer system, and ICR cell are differentially pumped by three turbomolecular pumps (Balzers TPU 330 for source and cell, respectively, and Balzers TPU 50 in the middle of the transfer system). After deceleration, the ions are trapped in the field of the superconducting magnet (Oxford Instruments) which has a maximum field strength of 7.05 T. The metal's most abundant isotope was isolated using FERETS [20], a combination of hard sweep pulses and soft single shots, and allowed to react with the substrate that was present with a constant pressure of $(1–3) \times 10^{-8}$ mbar; reaction times were typically 1–10 s. For CID experiments, argon was present as a buffer gas with a constant pressure of $(1–5) \times 10^{-7}$ mbar, as measured with an uncalibrated ionization gauge (Balzers IMG 070). All functions of the instrument are controlled by a Bruker Aspect 3000 minicomputer; broad band spectra over the whole frequency range were recorded with a fast ADC, digitized as 64K or 128K data points and zero filled [21] to 256K before Fourier transformation. Reaction products were unambiguously identified with high-resolution spectra over a narrow frequency range combined with mass analysis and their formation pathways by double-resonance and MS/MS techniques [22].

In the multi-sector approach, that has also been taken in many experiments, the decomposition pathways of organometallic complexes are studied. These complexes can be formed in the ion source of any tandem mass spectrometer that has at least two sectors. Usually, a volatile organometallic or coordination compound such as $Fe(CO)_5$ or $Cr(acac)_3$ is introduced together with the substrate of interest in a chemical ionization source. The mixture is ionized and often the many ion-molecule reactions that occur also lead to a 1:1 adduct complex of the metal ion and the substrate. This adduct complex is selected with the first or more of the sectors, and its metastable, i.e. unimolecular [23], or collision-induced decompositions [24] are monitored with the next sector in line. High-energy CID spectra are obtained by admitting a collision gas like He in a collision cell located in a field-free region between two sectors. They provide structural information about the complex while the metastable ion (MI) spectra reveal those fragmentation pathways that require the lowest activation energy. If a third or fourth sector is available, a product generated in the MI decompositions can be further characterized by selection with the next sector followed by CID in the next field-free region and scanning of the final sector. The starting complexes may also be preformed organometallic compounds [25] or fragments thereof, or may be generated by using a fast atom bombardment (FAB) source in which an inorganic salt such as $CuSO_4$ is bombarded with fast (7–9 keV) xenon atoms [26]. The liberated metal ions or clusters are then reacted in the ion source with the substrate to also form 1:1 adduct complexes.

The instrument that has been employed in Berlin used to be a VG ZAB-HF-3F mass spectrometer with a BEB configuration (B magnetic, E electric sector) and has been described before [27]. Part of the experiments have been performed with a later version of the instrument, where the second magnet was replaced with an AMD double focusing mass spectrometer so that altogether a BEBE configuration was achieved. The instrument is depicted in Figure 2 and described in [28]. Except for some MS^3 or MS^4 experiments, complexes were selected with the first two sectors to provide a better parent-ion resolution. For CID experiments, helium was introduced in one of the collision cells with a pressure of ca. 3×10^{-5} mbar, corresponding to 80% transmission or nearly single-collision conditions.

Both MS techniques are quite different in terms of the way that the potential energy surface for a system consisting of a metal ion M^+ and a substrate A-B is entered. Identical results from both approaches are therefore a priori not to be expected. Comparisons have shown, however, that more or less the same results are obtained if several restrictions are kept in mind [16, 17, 29–32]. Naturally, the first difference arises from the fact that M^+ is the primary ion in FTICR experiments and thus, ligand detachment from the collision complex $[M(A\text{-}B)^+]^*$ cannot be monitored, but is a process recorded by the sector approach. On the contrary, if intact adduct complexes $M(A\text{-}B)^+$ are formed in the FTICR, either by third-body association, energy redistribution, or by radiative stabilization, there is no equivalent in the sector instruments' MI or CID studies that rely on the decomposition of this very complex. The remaining product distributions are usually similar, except in cases where the activation energy for decomposition of $M(A\text{-}B)^+$ is much less than the binding energy gained upon complexation of M^+ with A-B. If this is the case, MI spectra sample those decomposition pathways where the available internal energy is just sufficient to make it over the respective activation barriers while in the corresponding FTICR experiment there is

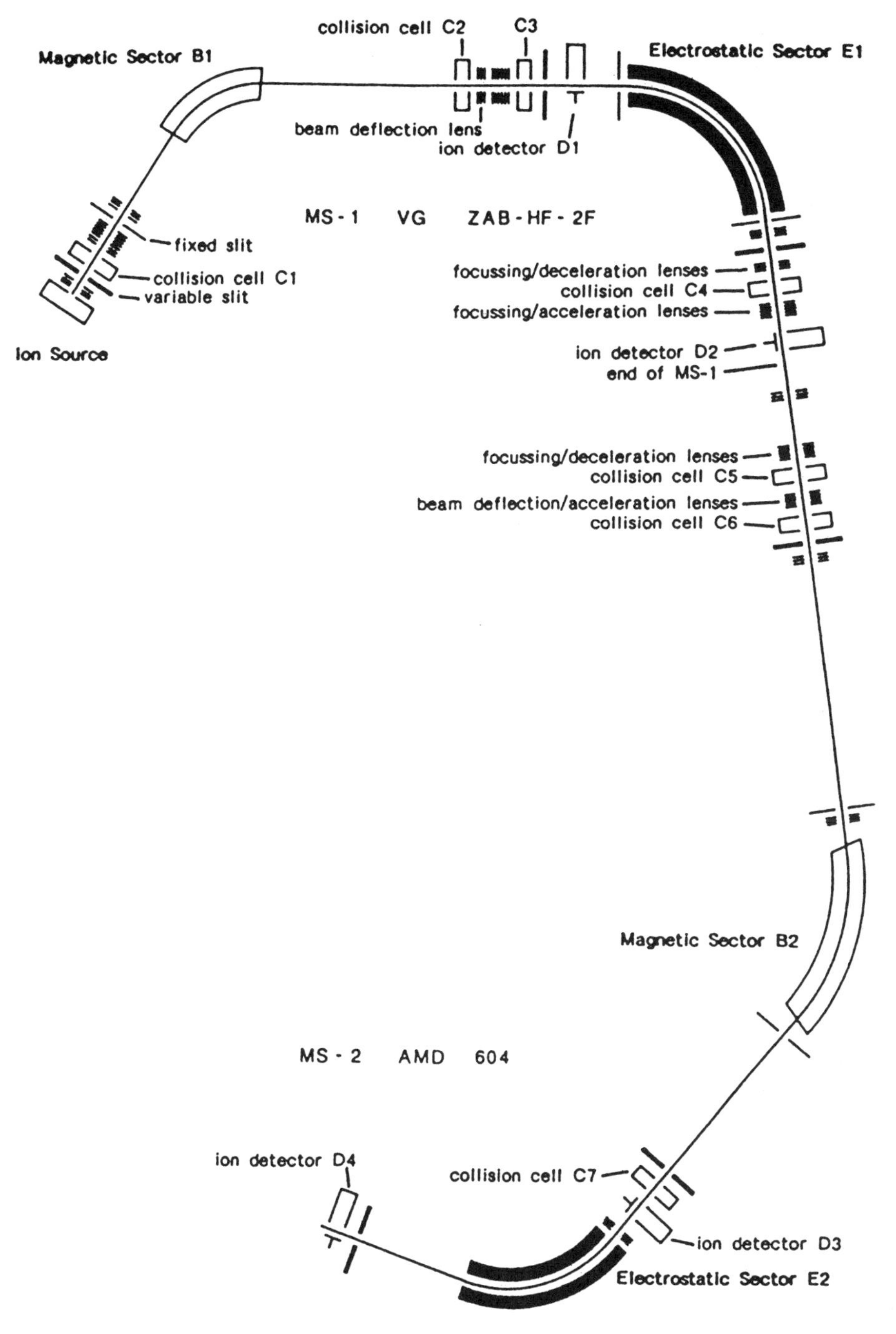

Fig. 2. Four-sector BEBE mass spectrometer at the TU Berlin.

plenty energy at hand so that also decomposition pathways with higher activation barriers can be used. But even if there is only one primary decomposition pathway, the product formed in an FTICR experiment may still have sufficient energy to decompose further, thus multiple losses are quite common in FTICR spectra. Very good agreement was found for the labeling distributions of individual products studied with both techniques. A caveat that should be kept in mind is that $M(A)(B)^+$ has the same mass as does $M(A\text{-}B)^+$, so it must be made sure that indeed the intact substrate is complexed and selected in sector studies. The fact that it is formed by ion/molecule reactions in the ion source does not guarantee its structure, as will be shown later.

3. Early discoveries: The remote functionalization mechanism

Most of the early work on gas-phase transition-metal ions had been done with hydrocarbons [1a, 2] as the selective activation of C-H or C-C bonds in alkanes represented the most challenging goal for researchers on catalysts, as it still does! As the systems studied grew larger, a multitude of reactions with little selectivity were observed. It was therefore decided to define the first step in the course of the bond activation by providing an anchor group for the metal ion. As nitriles are well-known from the condensed phase to be good ligands for transition-metal ions, they were selected. It had also already been reported that Cu^+ sequentially formed adduct complexes with acetonitrile to finally generate $Cu(CH_3CN)_3^+$ [33].

By forming complexes of the type $Fe(CH_3(CH_2)_nCN)^+$ in the ion source of the sector instrument and by studying their decomposition products, a remarkable chain-length effect was noted for different values of n [34]. Starting with $n = 2$, losses of H_2 and of C_2H_4 could be observed that increased in relation to the loss of the intact ligand with concomitant reformation of the bare Fe^+ ion, the so-called ligand detachment signal. By using labeled precursors, it became obvious that these two products were generated from the end of the alkyl chain and the mechanism depicted in Scheme 1 was proposed and termed *remote functionalization* with a view to some analogous mechanisms in enzymatic and biomimetic processes [35]. The metal ion in the adduct complex **1** is strongly held in position by the end-on coordination to the nitrile group. The inductively activated bonds in the vicinity of the functional group are therefore not within reach, and the metal ion can only undergo oxidative addition with C-H or C-C bonds from remote positions, available by folding back the chain. The metal ion selects a C-H bond and intermediate **2** is formed, which decomposes by either β-hydrogen shift or β-CC-cleavage to generate **3** or **4**, respectively. From **3**, reductive elimination of H_2 brings about formation of a complexed unsaturated nitrile, which can act as a chelating ligand. Alkene loss from **4** is combined with de-insertion to regenerate an intact nitrile complex, shortened with respect to **1**. That

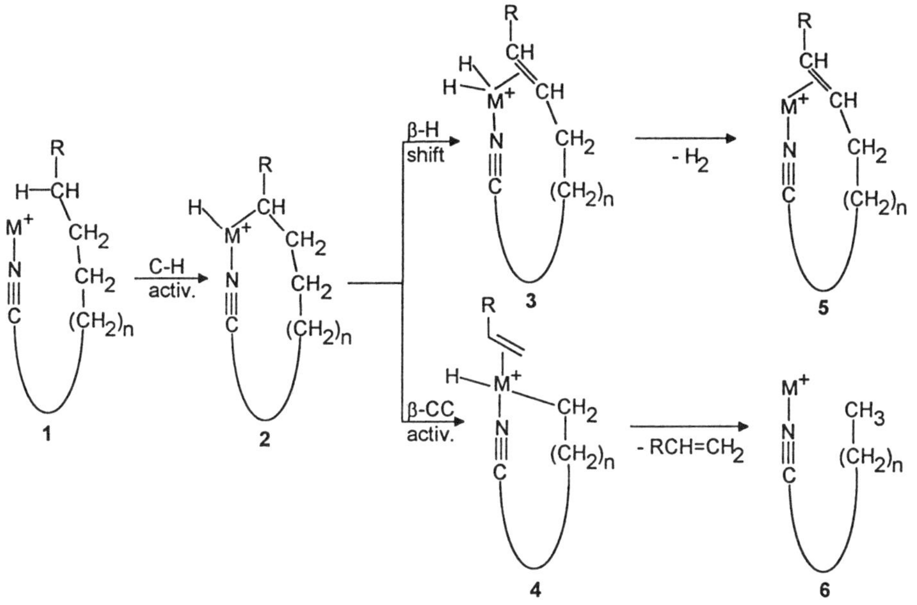

Scheme 1. Remote functionalization of a linear nitrile commencing at distant C-H bonds.

indeed **6** is formed and not a complex similar to **2**, with M-H and M-C bonds, could be shown by CID studies on the alkene loss product [34, 36] and by the labeling distributions of consecutive reactions, which are described below. In case of the reaction of Fe^+ with nitriles of an intermediate length [34], the selective activation of the terminal CH_3-group is noted and therefore only H_2 and C_2H_4 are observed as products. With longer nitriles the activation of internal CH_2-groups is observed as well, leading to the loss of H_2 from internal methylene groups and to the loss of propene and butene from the end of the chain [37, 38].

It was also discovered that the reaction was not limited to Fe^+; Co^+ and Ni^+ were found to react with nitriles under remote functionalization, too [37–39]. There were minor differences between the three ions concerning the preferred position of attack in the chain but the overall picture was very similar. Therefore it was somewhat surprising that Cu^+ behaved in a completely different manner [40], but the findings were in line with the already known chemistry of the four ions with alkanes, where Fe^+, Co^+, and Ni^+ were reactive and Cu^+ inert [1a, 2]. Nitrile complexes of Cu^+ also did not form any fragment ions in metastable ion spectra, hence the ion is unreactive with these activated substrates as well. Under collisional activation, however, products were observed, viz. $CuHCN^+$, $CuC_2H_2N^+$, and $CuC_3H_3N^+$ [40], but these are formed by processes which do not require any C-H or C-C insertion steps. $CuHCN^+$ is a product of the ion-dipole mechanism described below and the other two products are due to radical losses in the course of the CID process. Cleavage of the C-C bond of the nitrile complex generates $\cdot CH_2CN\text{-}Cu^+$ and a combined loss of R· and H· radicals from $R\text{-}CH_2\text{-}CH_2\text{-}CN\text{-}Cu^+$ complexes may lead to the acrylonitrile complexes $(CH_2 = CH\text{-}CN)Cu^+$. In ICR experiments Cr^+ was found to be inert towards heptanenitrile while Mn^+ showed a small amount of dehydrogenation [39].

4. Further variations of the theme: Limits and extensions of remote functionalization

As the reactions of Fe^+ with nitriles had shown that the metal ion had a preference for the activation of a C-H bond at $C_{(7)}$ and $C_{(8)}$, it was of interest whether the subsequent β-H shift could be replaced by a β-CH_3 shift if no hydrogen atoms are available at $C_{(8)}$. The reaction of 8,8-dimethylnonanenitrile with Fe^+ was chosen to test this possibility [41]. The main product, besides less intense losses of H_2 and C_4H_8, was indeed methane. ^{2}H-labeling showed that the hydrogen stemmed from the remote methylene groups from $C_{(6)}$ and $C_{(7)}$ and the alkene was formed by remote functionalization, too. Insertion into a C-H bond of one of the terminal CH_3-groups was followed by β-CC cleavage and loss of isobutene, a sequence analogous to Scheme 1 with $CH_2 = C(CH_3)_2$ in place of $RCH = CH_2$. The methane lost incorporated an intact methyl group and to roughly 10% a hydrogen from $C_{(7)}$ and to 90% from a terminal methyl group, i.e. from $C_{(9)}$. Although β-methyl shifts could not be ruled out, it was, however, considered more likely that initial C-C insertion was enforced upon the system by the *t*-butyl substitution and that the resulting intermediate underwent β-H shift from $C_{(7)}$ (10%) and $C_{(9)}$ (90%).

Replacing the $(CH_3)_3C$-group with a $(CH_3)_3Si$-group had a dramatic effect on the product distribution, with only hydrogen still being produced [42]. This is probably due to the fact that both methane losses as well as isobutene loss would generate a Si = C double bond, which is energetically much too demanding. In addition, the β-effect of silicon [43] may also be responsible for the high specificity of the reaction. Labeling showed unambiguously that dehydrogenation occurs only at the $C_{(7)}$ and $C_{(6)}$ positions next to the trimethylsilyl group.

In CID spectra of Fe^+ and Co^+ complexes of linear nitriles, formal losses of alkanes were observed [37], which were mechanistically described by a combined loss of hydrogen and an alkene (Scheme 2). By using unsaturated nitriles as starting compounds it was indeed found that alkene loss was the most prominent decomposition pathway already in the MI spectra [44]. Other products were of minor importance and mostly due to an metal-induced isomerization of the double bond. Labeling studies were supportive of the mechanism in Scheme 2 and the proposed isomerizations.

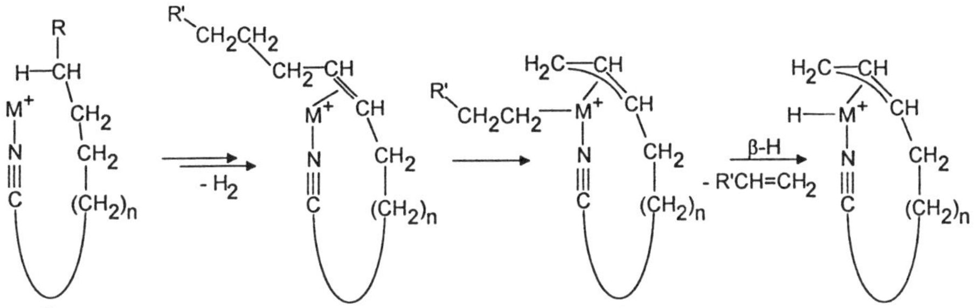

Scheme 2. Remote functionalization of a linear nitrile by dehydrogenation followed by allylic C-C activation and loss of an alkene. The combined loss of H_2 and $R'CH = CH_2$ is formally equal to an alkane loss.

In order to get more information about the rate determining effect for the remote functionalization in Scheme 1, the symmetric secondary nitrile $(C_4H_9)_2CHCN$ and several labeled isotopomers were studied [45]. Besides a small signal for the loss of methane, only H_2 and C_2H_4 loss were observed in the MI spectra, therefore enabling the determination of intramolecular kinetic isotope effects. It could be shown that the first step, oxidative addition of a C-H bond, was not rate determining, but for the other steps, intramolecular kinetic isotope effects were measured. Varying the chain length of the secondary nitriles, a remarkable feature was discovered with compounds having two long alkyl residues [46, 47]. Besides the losses of H_2 and C_nH_{2n} according to Scheme 1, alkane losses were already present in the metastable ion spectra. Labeling studies showed that they were definetely not due to the consecutive dehydrogenation/alkene detachment in Scheme 2. For instance, loss of "C_3H_8" was shifted to "$C_3H_4D_4$" for $(CD_3(CH_2)_5)_2CHCN$-Fe^+. Involvement of both ends of the chain could only be explained by a double remote functionalization as depicted in Scheme 3. Loss of propene-d_3 from the one end of the nitrile affords a shortened secondary nitrile whose longer chain is dehydrogenated in the $\omega/(\omega - 1)$-position. The alternative sequence, first dehydrogenation followed by alkene loss, could be excluded on the basis of MS/MS experiments. In the MI spectra, besides the "C_3H_8" loss, loss of H_2 and of C_3H_6 were both observed, but only the propene loss product further decomposed unimolecularly by H_2 loss if selected with the next sector of the instrument and studied in the next field-free region. The reason for only alkene loss followed by dehydrogenation being observed probably lies in the formation of the chelating structures that are produced by the dehydrogenation. Once chelation is established, the metal ion cannot swing around to the other chain anymore.

Quite understandable is the absence of remote functionalization for very short nitriles, and it is also reasonable to expect some mechanistic deviations for the first substrates in the row of linear nitriles who react with Fe^+,

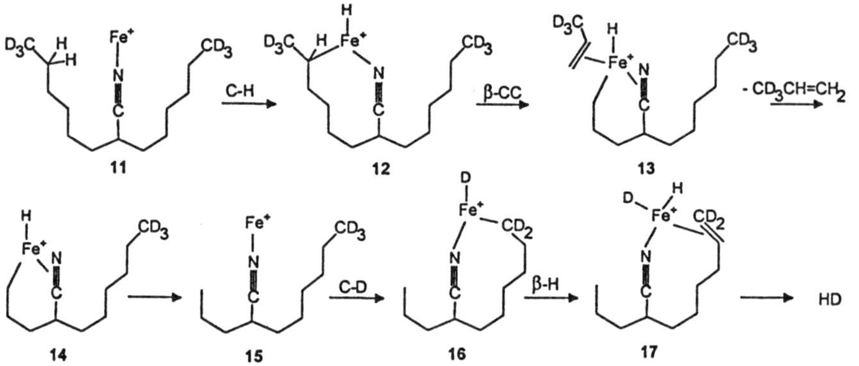

Scheme 3. Double remote functionalization leading to loss of propene and hydrogen from both chains.

Co^+, or Ni^+. The products formed from pentanenitrile and its ^{2}H-labeled isotopomers indeed revealed some unusual processes that were operative with these three metal ions [36, 48a]. For all three ions, loss of H_2 and C_2H_4 were major products, and for Co^+ and Ni^+, in addition, cleavage into propene and acetonitrile was also prominent. Dehydrogenation and ethene loss could well be described by the remote functionalization mechanism in Scheme 1, but it is also immediately obvious that intermediate **4** in this particular case should be considerably strained, with n being equal to 1. The dehydrogenation intermediate **3** is less strained and thus for all three ions, H_2 is exclusively formed by remote functionalization. In case of Fe^+, the activation of the $\omega/(\omega - 1)$-position was, however, preceded by a degenerate isomerization via the ferracyclobutane intermediate **21** (Scheme 4) [48b]. **21** was also involved before the loss of C_2H_4 from the encounter complex **18**, a process that for Fe^+ did not proceed by remote functionalization but by direct C-C activation leading via **22** to the $FeC_3H_5N^+$ isomer **24**. Structure **26** can be distinguished from the propionitrile complex **29**, formed from propionitrile and Fe^+ by third-body association, by its characteristic CID pattern as shown in Scheme 5. A third $FeC_3H_5N^+$ isomer **31** is formed in the reaction of ferracyclobutane **30**, itself obtained by decarbonylation of cyclobutanone, and acetonitrile [49]; upon CID it fragments to yield the alkylidene ion **32** [50]. For Co^+ and Ni^+, structures analogous to **21** were not observed, and furthermore their mode of ethene generation differed from each other and also from that of Fe^+. For Ni^+, the classic remote functionalization applied, probably via a side-on bonded intermediate **23**, and led to propionitrile complexes, as revealed by CID on **25**, that only afforded Ni^+. For Co^+ the encounter complex **19** to ca. 40% reacts by C-C activation, as did Fe^+, and to 60% via remote functionalization. Hence, two distinguishable $CoC_3H_5N^+$ isomers are generated, as could indeed be shown by performing CID with product ions formed from labeled pentanenitriles.

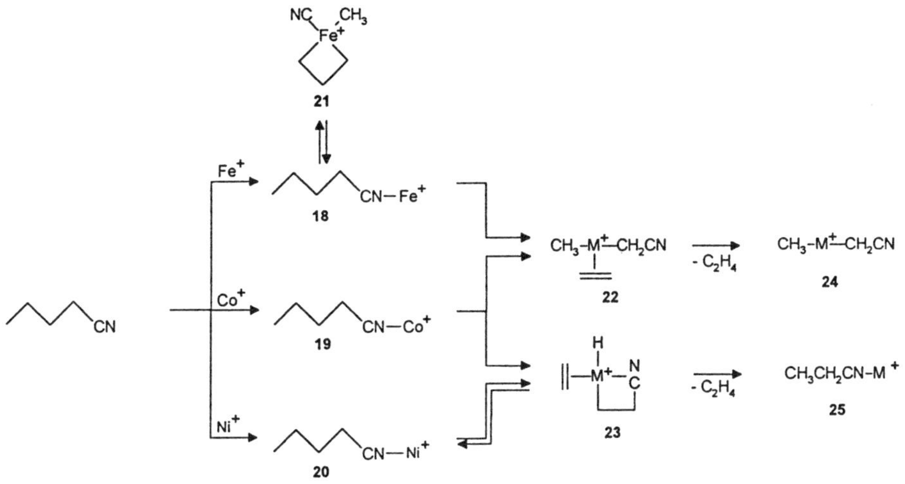

Scheme 4. Reaction of pentanenitrile with Fe^+, Co^+, and Ni^+, leading to loss of ethylene by initial C-C insertion or by remote functionalization.

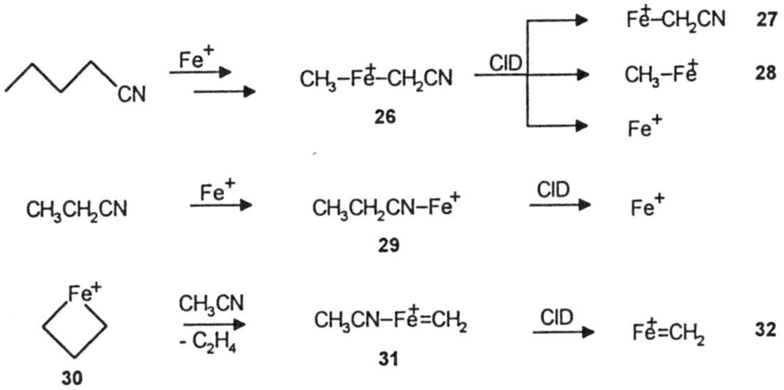

Scheme 5. Distinction of three $FeC_3H_5N^+$ isomers by collision induced dissociation.

For instance, $CD_3(CH_2)_3CN$ upon reaction with Co^+ loses 64% $C_2H_2D_2$ by remote functionalization and 36% C_2H_4 by activation of internal C-C bonds. Consequently, the $CoC_3H_4DN^+$ ion formed ($\equiv$ $DCH_2CH_2CN\text{-}Co^+$), upon CID exclusively yields Co^+ while the $CoC_3H_2D_3N^+$ isomer ($\equiv$ $CD_3\text{-}Co^+\text{-}CH_2CN$) gives rise to Co^+, $CoCD_3^+$, and $CoC_2H_2N^+$. Similar experiments were done with the other three ^{2}H-isotopomers labeled in the α-, β-, or γ-position [36].

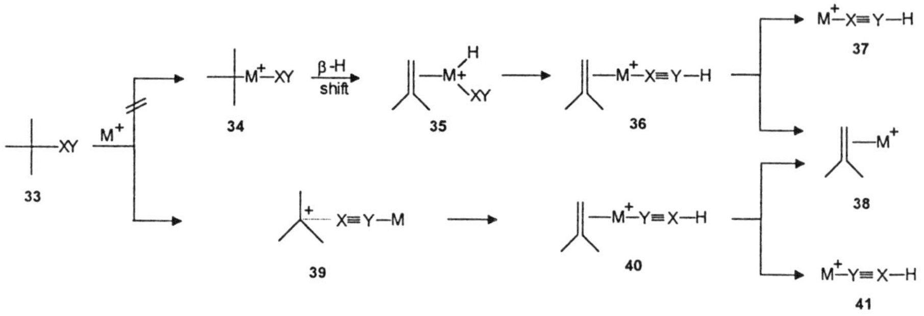

Scheme 6. The ion/dipole mechanism as applied to the reactions of *tert*-butyl cyanide and isocyanide with late transition-metal ions.

5. The ion/dipole mechanism

Reacting t-C_4H_9CN with Fe^+ could not be expected to lead to remote functionalization and indeed two new products were observed, $[Fe,H,C,N]^+$ by loss of C_4H_8 and the complementary $FeC_4H_8^+$ by loss of [H,C,N] [51]. Formally the same two products were generated upon reaction of Fe^+ with *tert*-butyl isocyanide, t-C_4H_9NC [51]. Other late transition metals behaved exactly in the same manner [52]. As a first guess, an insertion/β-hydrogen shift mechanism, as depicted in Scheme 6, could be used to explain their formation. Insertion into the weak C-XY bond in **33** followed by β-H shift and rearrangement to structure **36** could precede the competitive ligand loss therefrom. In that case, however, the $[Fe,H,C,N]^+$ ions formed would have been expected to possess structure **37**, which is not the case.

High-energy CID studies on the two ions, formed from *tert*-butyl cyanide and isocyanide, respectively, showed that they in fact had to be described by structure **41**. The two spectra were very similar, but differed in the occurrence of structure indicative FeC^+ signals for Fe-CNH^+ formed from t-BuNC and FeN^+ for Fe-NCH^+ formed from t-BuCN. To explain these structures, an ion/dipole mechanism [53] was proposed (Scheme 6). Coordination of the metal ion to the functional group XY in **33** induces the cleavage of the labile C-XY bond. And yet, the so formed ion/dipole complex **39** is unable to separate in the $C_4H_9^+$ ion and XYM because this would be endothermic. Therefore, protonation of the unblocked X atom in XYM leads to the intermediate **40**, whose competitive ligand losses produce $MHCN^+$ ions of the correct structure **41** and the isobutene complexes **38**.

The losses of the isomeric hydrogen cyanide and isocyanide neutrals are reflected in the branching ratios of **40** depending upon the structure of the precursor molecule **33**. If **40** is formed from t-BuCN it contains a HCN ligand, and it is found experimentally that this ligand is preferentially lost under retention of isobutene. If on the contrary **40** is formed from t-BuNC, it contains a HNC ligand. As hydrogen isocyanide is a much stronger ligand than hydrogen cyanide, it is consequently preferentially retained, and isobutene loss is the more abundant process in this case.

The intermediacy of the ion/dipole structure **39** could also be demonstrated. It should be stable for quite a long time as it is trapped on the reaction coordinate by a potential-energy barrier on one side and an entropic bottleneck on the other [53, 54]. The incipient carbenium ion therein is expected to undergo hydrogen scrambling, as is known from the reactions of free carbenium ions [55]. For *tert*-butyl systems, this obviously cannot be detected, with the isomerization being a degenerate one. As will be discussed in more detail below, the ion/dipole mechanism is not restricted to this particular group, but can also be found for other substrates. In particular, the reactions of Co^+, Ni^+, and Cu^+ with 2-methyl butanenitrile [17] or 2-ethyl butanenitrile [29] or of Fe^+ with 2,2-dimethyl butanenitrile [46] also follow, at least partially, the ion/dipole mechanism. In these cases not only could the structure of the $[M,H,C,N]^+$ ions be shown to be that of **41**, i.e. HCN-M^+, but scrambling of the hydrogen atoms in the alkyl residues could be demonstrated by using labeled nitriles as precursors. Thus the ion/dipole intermediates corresponding to **39** are stable enough for H/D scrambling taking place before the proton transfer to the neutral CNM part.

Another argument in favour of the ion/dipole mechanism is the trend observed in the reactions of Fe^+ through Cu^+ with 2-methyl butanenitrile and 2-ethyl butanenitrile [17, 29]. The ion/dipole mechanism is only one of three mechanisms operative in these systems, but its relative importance increases from Fe^+, where it is absent, to Cu^+, where it is the sole mechanism observed. This trend is due to the decreasing ability of the metal ions to undergo alternative reactions which require insertions into C-C or C-H bonds. For Cu^+ this ability is absent, for Cu^+ being a d^{10} metal ion, thus it has no other alternative but to react via the ion/dipole mechanism, which is devoid of oxidative addition steps. In other words, *the extent by which the ion/dipole mechanism is operative depends on the need for it.*

Some more arguments for the ion/dipole mechanism, such as the increasing amount of R^+ ions formed from RX with growing exothermicity of the X^- abstraction by M^+, will be presented along with the respective systems below. At this time a few words of comparison with the insertion/β-hydrogen shift mechanism and with the reactions of alkali and other main-group metal ions seem to be appropriate. While some of the arguments presented could still be disputed and could, albeit with a few assumptions, be explained with the help of the insertion/β-hydrogen shift mechanism, the collective evidence seems quite obvious and points to the presence of the ion/dipole intermediates. As discussed in more detail elsewhere [1b], there are more cases in the literature where the ion/dipole mechanism seems to be operative but where the reactivity has been described by "dissociative attachment" or "Lewis-acid chemistry". These are the reactions of main group ions with alcohols and alkyl halides. Li^+, Na^+, Mg^+, Al^+, Ga^+, and In^+ with RX (X = F, Cl,

Br, I, OH) afforded either adduct complexes $M(RX)^+$, X^- abstraction with concomitant R^+ formation, $M(HX)^+$ ions by loss of the respective alkene, or $M(alkene)^+$ complexes by loss of HX [56–63]. Clearly, these are also cases where insertion reactions are impossible for the metal ions and hence the ion/dipole mechanism represents the sole alternative. Mechanistically the observed reactivity previously could not be exactly described, with some unknown rearrangement assumed to be taking place between the initial encounter complex $RX\text{-}M^+$ and the immediate precursor for the ligand loss, $M(HX)(alkene)^+$. Steps along the lines of Scheme 6 will explain how this rearrangement takes place.

6. Further reaction of nitriles and isonitriles: Competition between different mechanisms

It has already been mentioned that secondary nitriles with two long chains such as $(n\text{-}C_4H_9)_2CHCN$ reacted mostly via remote functionalization. As is true for all the mechanisms presented here, upon variation of the chain length, deviations are bound to occur. Thus, studying the row of secondary nitriles R^1R^2CHCN with Fe^+, some interesting observations were made [46]. While the longer representatives ($R^1 = R^2 = n\text{-}C_5H_{11}$, $n\text{-}C_6H_{13}$) reacted exclusively by remote functionalization, as did the linear nitriles, the shortest secondary nitrile, $(CH_3)_2CHCN$, was unreactive in the sense that only ligand detachment, i.e. loss of the nitrile with concomitant formation of Fe^+, was noted in the MI spectra. The interesting aspect, however, was the intermediate region, where a new mechanism prevailed that led to methane loss (Figure 3). It was most prominent for 2-methyl butanenitrile ($C_2H_5(CH_3)CHCN$) and 2-ethyl butanenitrile ($(C_2H_5)_2CHCN$), so these substrates together with $(n\text{-}C_4H_9)_2CHCN$ were studied in more detail employing labeled compounds [17, 29, 46].

The labeling showed that one of the four hydrogen atoms lost as methane always came from a position β to the cyanide group and the other three originated from one of the terminal methyl groups. The most reasonable explanation for the incorporation of a β-hydrogen is the preceding activation of the C-CN bond, upon which a β-hydrogen shift to the metal ion is made possible (Scheme 7). Loss of HCN from the intermediate **44** formed this way seems to be unfavourable, though, since it is not observed. Instead, activation of a $C\text{-}CH_3$ bond followed by reductive elimination of methane is seen to occur. This mechanism also explains why CH_4 loss was most prominent for 2-methyl and 2-ethyl butanenitrile: in these cases the $C\text{-}CH_3$ bond to be activated is a relatively weak allylic one, and the mechanism leads to the formation of stable allyl complexes **46**! Another piece of evidence comes from the occasional observation of $CH_3\cdot$ radical losses in cases where CH_4 is lost too and which originate most likely from **45**.

The competition between the different mechanisms in Figure 3 can now

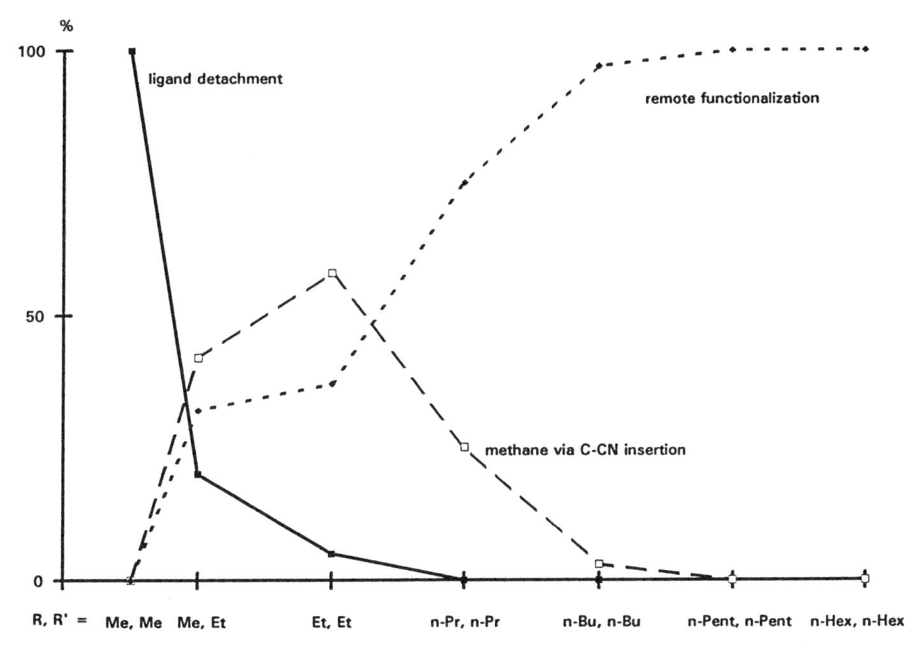

Fig. 3. Effect of the chain length of secondary nitriles RR′CH-CN reacting with Fe^+ according to three different mechanisms.

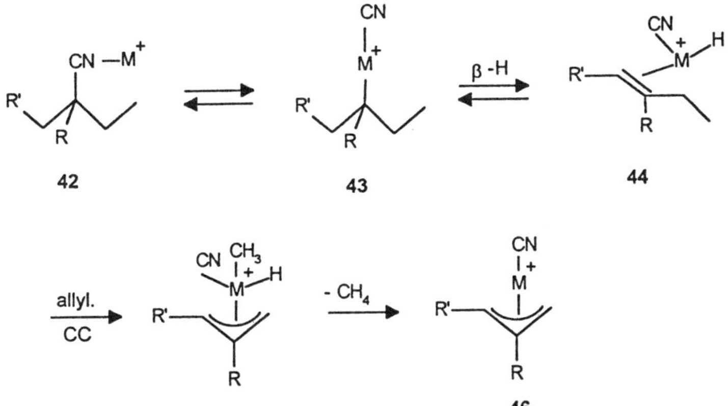

Scheme 7. Methane loss from small branched nitriles upon reaction with Fe^+ or Co^+ by initial C-CN insertion followed by β-H shift, C-CH_3 insertion, and reductive elimination.

be reasonably explained. For the smallest secondary nitrile, *i*-PrCN, neither remote functionalization nor the mechanism in Scheme 7 is possible, thus only ligand detachment is noted. For the immediate homologues, methane loss is most favourable due to the allylic position of the methyl group, but with increasing chain length, methane generation becomes more and more disadvantageous while on the contrary remote functionalization is facilitated by the decreasing strain in the intermediates **3** and **4** (Scheme 1). The mechanistic competition in case of the secondary nitriles reacting with Fe^+ is even topped by the reaction of this metal ion with 2,2-dimethyl butanenitrile, the reason being the ion/dipole mechanism which is possible for Fe^+ and tertiary nitriles. In the MI spectra of $Fe(C_2H_5C(CH_3)_2CN)^+$ hence products of three mechanisms feature: $Fe(HCN)^+$ and loss of HCN according to the ion/dipole mechanism, CH_4 and weak $CH_3\cdot$ loss signals formed according to Scheme 7 ($R = CH_3$, $R' = H$), and finally H_2 and C_2H_4 from the remote functionalization mechanism. The intensity of the latter two products is, however, relatively weak because the chain is still rather short [46]. For the higher homologues 2,2-dimethyl pentanenitrile and 2,2-dimethyl hexanenitrile the remote functionalization products again strongly gain importance [52].

Besides Fe^+, for 2-methyl and 2-ethyl butanenitrile the other first row transition-metal ions from Ti^+ through Zn^+ were studied as well [17, 29]. It has already been mentioned that in the row Fe^+, Co^+, Ni^+, Cu^+ the importance of the ion/dipole mechanism drastically increases. Fe^+ is unable to react with secondary nitriles according to Scheme 6 while for the closed shell Cu^+ ion this is the only mechanism possible. For Co^+ and Ni^+ the ion/dipole mechanism is possible also with secondary nitriles but other mechanisms compete, in particular the remote functionalization, which is seen to go through a maximum with these two metal ions. For Co^+, methane loss according to Scheme 7 is observed as well, but its intensity is smaller than in case of Fe^+ by one order of magnitude. The early transition metal ions Ti^+ and V^+ are very reactive and behave completely different from the late transition-metal ions. Their hallmark are C-H activations and usually many different products are formed. This is also true for the two secondary nitriles where under FTICR conditions single, double, and even triple dehydrogenations as well as combined losses of H_2 and other small neutrals were observed. In some cases these combined losses were slow enough so that it could be shown in double resonance experiments that dehydrogenation preceded the loss of the other neutral molecule. Due to the multitude of products formed, together with further complications arising from rapid hydrogen scrambling processes, the mechanism could not be determined despite of the labeling experiments. As the labeling distributions were different for Ti^+ and V^+, each metal ion must develop its own characteristic gas-phase chemistry. As was also found for the reaction with alkanes [1a, 2], Cr^+, Mn^+, and Zn^+ were unreactive with 2-methyl and 2-ethyl butanenitrile. While in MI spectra only ligand detachment was noted, exclusively adduct formation was seen in FTICR experiments. CID upon these adduct complexes led to three

Table 1.

	Cr^+	Mn^+	Fe^+	Co^+	Ni^+	Cu^+	Zn^+
Primary nitriles	−	−	−	−	−	−	−
Secondary nitriles	−	−	−	+	+	+	−
Tertiary nitriles	+	+	+	+	+	+	?

kinds of products, the most trivial being ligand detachment. More interesting was that the two products from the ion/dipole mechanism, $M(HCN)^+$ and HCN loss could be observed, which means that this particular mechanism can be enforced by the energy deposition during CID. In addition, acrylonitrile complexes were seen to be formed. This product was also seen in case of the high-energy CID spectra of $Cu(RCN)^+$ complexes [40]. It can be explained by loss of radicals during the CID process. Initial loss of a radical from the α-position, which has the weakest bonds in the molecule, is followed by another radical loss through cleavage of the β/γ-bond, so that acrylonitrile is generated.

Combining the results presented so far with some unpublished material [52], the ability of the late transition metal ions to react with different nitriles via the ion/dipole mechanism can be summarized as is done in Table 1. The classification does not exclude other mechanisms and the question mark represents a system that still has to be determined.

It has already been mentioned that *tert*-butyl isocyanide reacted with Fe^+ exclusively via the ion/dipole mechanism under formation of $Fe(HNC)^+$ and loss of HNC. For linear isonitriles RNC, the chemistry observed with Fe^+ [64] differed markedly from that of linear nitriles, where remote functionalization was most important. The isonitriles showed reactivity already starting with C_2H_5NC, where the two products of the ion/dipole mechanism were formed. Formation of $Fe(HNC)^+$ and loss of HNC were also very important for the next members of the row, but beginning with *n*-BuNC, the products of the remote functionalization mechanism began to appear in the spectra as well. For longer chain isonitriles the importance of the ion/dipole mechanism declined rapidly and mostly dehydrogenation by remote functionalization was observed.

Because of the inability of Cu^+ to react via remote functionalization, it was of interest to see what would happen if this ion was reacted with the row of isonitriles [65]. Indeed, signals from remote functionalization were mostly absent for all but the longest isonitriles studied. For R = *n*-Hex through *n*-Non, in the high-energy collisional activation spectra very weak signals for H_2 loss appeared, which could possibly be due to remote functionalization. Yet, this hypothesis was not followed in detail and thus must be regarded with caution. Instead, the products from the ion/dipole mechanism prevailed for all systems. Other products seen were loss of CuCN by formation of R^+ carbenium ions, a process that gained importance with increasing

R, and again radical losses were observed. Under collisional activation, cleavage of the α/β and β/γ bonds occurred and led to $\cdot CH_2NC\text{-}Cu^+$ and $\cdot CH_2CH_2NC\text{-}Cu^+$. Also formed were vinyl isocyanide complexes by the same mechanism that was discussed above for acrylonitrile complexes, involving two consecutive radical losses. The $Cu(HNC)^+$ ion was studied together with its $Cu(HCN)^+$ isomer, generated from *t*-BuNC, to verify its structure [66] and to see if the two ions could be neutralized. As expected, in high-energy CID experiments the two ions could be distinguished by their characteristic CuC^+ (for $HNC\text{-}Cu^+$) and CuN^+ (for $HCN\text{-}Cu^+$) ions. The two ions could also be successfully neutralized to HNC-Cu and HCN-Cu using the technique of neutralization/reionization mass spectrometry [NRMS, 67]. The NRMS spectra showed that they retained their structure upon neutralizing the fast ion beams with xenon gas and in the subsequent reionization step with oxygen, which is necessary for detection.

In a screening study, *n*-propyl, *n*-butyl, and *n*-pentyl isocyanide were studied with almost the whole d-block of transition-metal ions to reveal some trends of reactivity [65]. It could be shown that dehydrogenation was most prominent for early transition metal ions in the first row and, with the exception of group 11 and 12, for most of the second and third row ions, but it was also an important process for Fe^+ and to a lesser degree for Co^+. The ion/dipole mechanism was most important for the late transition-metal ions in all three rows, but it was observed for almost all metal ions to a certain amount. The intensity of the ligand detachment signals was taken as a sign of missing reactivity. According to this criterion, Mn^+ was the most unreactive ion in the first row and the same applies for the ions at the beginning and at the end of the second and third row. Altogether, these results are in line with the general knowledge about the reactivity of these ions with various substrates such as alkanes, alkenes or others [1a, 2].

7. Other monosubstituted substrates RX with unsaturated functional groups (X = CP, NCO, NCS)

To study the influence of other functional groups, some similar compounds of the type RX were investigated upon. Initially, phosphaalkynes [68] $R-C\equiv P$ were selected and studied with Fe^+ to compare them with the nitriles $R-C\equiv N$ as well as with 1-alkynes $R-C\equiv CH$ [69]. In the condensed phase, phosphaalkynes gained their name from their behaviour, which was more like to alkynes than to nitriles. The same was found in the gas-phase reactivity of *t*-BuCP and *t*-PentCP with Fe^+, which was dominated by "phosphapropargylic" insertions that led to alkene (C_2H_4) and alkane (CH_4) losses. While the corresponding nitriles (and isonitriles) under these circumstances reacted via the ion/dipole mechanism to form $Fe(HCN)^+$ ($Fe(HNC)^+$) and $Fe(alkene)^+$ by loss of HCN (HNC), in case of the alkynes also the propargylic bond was activated. Although the same products were formed, the product distributions were different for RCP vs. RCCH with regard to the intensities, mainly because of the occurrence of multiple losses already in the MI spectra.

Isocyanates R-N = C = O proved to be an interesting class of substrates with some unique characteristics bearing resemblance to condensed-phase chemistry, but there were also elements from well-known gas-phase processes. They were examined because there had been some reports about the reactions of *i*-PrNCO and *t*-BuNCO with a supported nickel catalyst [70]. Simple pyrolysis of both isocyanates required more than 900 K and led exclusively to HNCO and the respective alkenes. If nickel clusters on a carbon support were present, *i*-PrNCO at 400 K afforded CO and a surface bound nitrene *i*-Pr-N = [Surf], which at 500 K underwent methane loss to generate surface bound acetonitrile, which desorbed at a higher temperature. The transformation of *t*-BuNCO to HNCO and *i*-C_4H_8 was initiated on the catalyst already at 500 K, but no alkane loss or nitrile products could be detected, only NH_3 and N_2, which were believed to be decomposition products of *t*-Bu-N = [Surf].

The reaction of *i*-PrNCO with gaseous Ni^+ did, however, not lead to acetonitrile complexes but was found to be a clear-cut case of the ion/dipole mechanism [71–73]. The only products formed were $Ni(HNCO)^+$ by loss of C_3H_6 and $Ni(C_3H_6)^+$ by loss of HNCO. The same applied for all the ions from Cr^+ through Zn^+. Not only did all these "late" transition-metal ions form exclusively these two products, they also formed them in a similar ratio, with $M(HNCO)^+$ being strongly preferred. All ions formed $M(HNCO)^+$ as the sole isotopomer upon reaction with $(CH_3)_2CD$-NCO and revealed a kinetic isotope effect of k_H/k_D of around 1.5 ± 0.1 upon reaction with $(CD_3)(CH_3)CH$-NCO. Again, in principle two mechanisms can be considered to explain the formation of these products, the ion/dipole or the insertion/β-H shift mechanism. For some of the ions, for instance Zn^+, it is very unlikely that they should react by oxidative addition, as the latter mechanism would demand. Furthermore it is even more unlikely that they should do so with the same ease and isotope effect as some reactive ions such as Fe^+ or Co^+. It seems therefore obvious that the ion/dipole mechanism is operative. The absence of H/D scrambling is in line with other studies where a 2-propyl cation is generated as constituent of an ion/dipole complex of not too high energy [74] and must therefore not be taken as an argument against its operation.

Some further support for this hypothesis is gained from the reactions of Cr^+ through Zn^+ with *s*-BuNCO, *t*-BuNCO, *t*-PentNCO, and their ^{2}H-isotopomers. Besides $M(HNCO)^+$ and loss of HNCO only two further processes are observed. One is the subsequent decomposition of the initially formed alkene complexes from *s*-BuNCO and *t*-PentNCO, which is limited to Fe^+, Co^+, and Ni^+, and the other is the formation of carbenium ions by loss of neutral [MNCO]. The latter are obviously generated by dissociation of the ion/dipole intermediate and are an indication that there is sufficient energy available to fuel this otherwise endothermic reaction. It is known [53] that ion/dipole complexes are lying in an entropic well that is flanked by a loose and a tight transition state. The tight transition state precedes complex formation and the loose one determines further rearrangements and/or decompositions. At low energies the loose transition state is rate determining, and thus the complex has sufficient time to sample all processes that are energetically lower than the loose transition state, for instance hydrogen scrambling. At higher energies, the tight transition state becomes rate determining and the ion/dipole well is passed faster and faster, suppressing rearrangement processes. The observation of carbenium ions from the direct ion/dipole dissociation is a clear indication of the latter condition prevailing and explains the exclusive participation of the β-hydrogen atoms for the two products of the ion/dipole mechanism. Not all ions are, however, able to react via the ion/dipole mechanism with isocyanates; the alkali ions Na^+ and K^+ were only observed to form adduct complexes at a very slow rate [73].

An interesting observation was made for the secondary reactions of Cr^+ through Zn^+ with the secondary and tertiary isocyanates [73, 75]. The primary ions formed, $M(HNCO)^+$ and $M(C_nH_{2n})^+$, reacted with a further isocyanate molecule to $MC_{n+1}H_{2n+1}NO^+$ and, occasionally, to $M(HNCO)_2^+$ and $M(C_nH_{2n})_2^+$. With labeled substrates it could be shown that, depending on the metal ion and the particular isocyanate, the $MC_{n+1}H_{2n+1}NO^+$ ions were either of the $M(HNCO)(C_nH_{2n})^+$ or of the $M(C_nH_{2n+1}NCO)^+$ structure. The two isomers differed in their CID spectra; while the latter ones predominantly formed M^+ upon CID, the former also gave rise to $M(HNCO)^+$ and $M(C_nH_{2n})^+$. With secondary isocyanates, the relatively unreactive ions Cr^+, Mn^+, and Zn^+ only formed intact adduct complexes, while Fe^+ through Cu^+ formed the "dissociated" $M(HNCO)(C_nH_{2n})^+$ products. With tertiary isocyanates, Cr^+ also belonged to that latter group. As mentioned in the instrumental section, ion/molecule reactions like the ones studied here are used, albeit at a higher pressure, to form adduct complexes in ion sources, thus caution is advisable that these are indeed obtained, and not any "dissociated" isomers.

Quite a different picture was seen upon reaction of the early transition-metal ions Ti^+ and V^+ with isocyanates [71–73]. Although upon reaction with isopropyl isocyanate a small amount of $M(HNCO)^+$ ions was noted for both reagents too, the main product in each case was $M(CH_3CN)^+$ by loss of CO/CH_4. These acetonitrile complexes were characterized by CID. In addition, combined loss of CO and H_2 as well as formation of MNH^+ ions was seen for both metal ions in smaller intensities. The observation of CO losses made it very likely that a mechanism similar to the condensed phase was operative, being initiated by decarbonylation. The results from the labeled isocyanates were supportive of this conclusion as they showed that the methane loss en route to the acetonitrile complexes proceeded as a formal 1,1-elimination from the α-position. Scheme 8 features two possible scenarios for the formation of $CH_3CN\text{-}M^+$ and $CD_3CN\text{-}M^+$ from d_3-labeled *i*-PrNCO

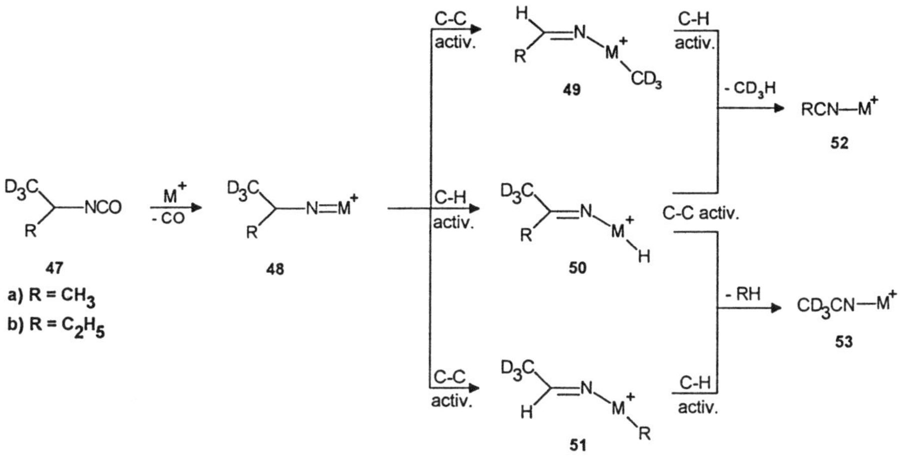

Scheme 8. Formation of nitrile complexes from secondary isocyanates upon reaction with Ti^+ or V^+ as depicted for d_3-labeled isopropyl and *sec*-butyl isocyanate.

(**47a**). If C-H activation precedes C-C activation, or vice versa, could unfortunately not be distinguished from the data at hand.

Very similar observations were made for *sec*-butyl isocyanate, which produced CH_3CN-M^+ by loss of CO/C_2H_6 together with C_2H_5CN-M^+ by loss of CO/CH_4 for $M^+ = Ti^+$ and V^+ [73]. These products were also formed in a 1,1-elimination, as can be seen in Scheme 8 for $C_2H_5CH(CD_3)NCO$ (**47b**). Nitrile complexes were, however, absent for the tertiary isocyanates *t*-BuNCO and *t*-PentNCO. The main product for Ti^+ and V^+ was always $M(HNCO)^+$, quite similar to the reactions of late transition-metal ions. Yet in addition to this ion, there was a multitude of other products in smaller intensities, but the mechanisms that led to their formation could not even with the help of several labeled *tert*-pentyl isocyanates be established. It thus seems that there is a strong similarity to the condensed phase studies, where *t*-BuNCO also did not form nitriles but partly decomposed and otherwise only formed HNCO and *i*-C_4H_8. Since nitrene formation should not be very much different between secondary and tertiary isocyanates, one of the following steps has to be impossible for tertiary substrates. It can be speculated that this could be the first step, which would imply that C-H activation precedes C-C activation for secondary substrates. Although there is no further evidence for this assumption, it is probably more likely than the conclusion that the second step, namely a *second* C-C activation, is impossible for tertiary isocyanates.

It still remains to be explained why Ti^+ and V^+ do react via nitrene complexes to form nitrile complexes while the late transition-metal ions are obviously unable to do so. The stability of the final product cannot be the determining factor as nitrile complexes are in fact very stable for Cr^+ through

Zn^+, quite in contrast to Ti^+ or V^+ [1, 5]. The reactions of the tertiary isocyanates show that Ti^+ and V^+ are also able to form $M(HNCO)^+$ ions, so these too cannot be the reason for the different behaviour. It thus seems to be the stability of the initially formed nitrene complexes that is vital for the outcome of the reactions. Nitrene complexes are quite common products from isocyanates in condensed-phase reactions of transition-metal complexes [76] and they are also involved in the Ni_x/C_∞-catalyzed acetonitrile formation from *i*-PrNCO [70]. There is not much data available for the gas phase, but it is known that the bond dissociation energies of Ti^+ and V^+ to the simplest nitrene are indeed quite strong: $D°(HN = Ti^+) = (466 \pm 12)$ kJ mol^{-1} [77] and $D°(HN = V^+) = (415 \pm 15)$ kJ mol^{-1} [78]. This contrasts with the much weaker value for Fe^+: $D°(HN = Fe^+) = (230 \pm 60)$ kJ mol^{-1} [79]. Furthermore, Cr^+ through Cu^+ are unreactive with ammonia [19, 33, 79, 80] while Sc^+ through V^+ exothermically dehydrogenate NH_3 to form $HN = M^+$ nitrene complexes [77–79, 81]; in the absence of any barriers this could be interpreted as $D°(HN = M^+) < 390$ kJ mol^{-1} for $M = Cr - Cu$. Thus the gas-phase studies show that for nitrile production from isocyanates a high nitrene bond dissociation energy is necessary; as it seems likely that this applies for the Ni_x/C_∞-catalyzed reaction too, other catalysts with similar properties could also be effective. For instance, CH_3NCO is adsorbed non-dissociatively on a Cu(100) single crystal surface, but on Pt(110) or a potassium-doped Cu(100) surface, dissociation into CO and CH_3N = [Surf] is observed [82]. For Fe(111) and Fe(100) it has been estimated that NH is adsorbed by about 420 kJ mol^{-1} [83], hence with a similar bond dissociation energy as obtains to Ti^+ or V^+.

Simply changing one atom in the functional group can make quite a difference, as will be demonstrated with the reactions of isothiocyanates, R-N = C = S [84, 85]. The smaller members of the row of alkyl isothiocyanates behave unexceptional with bare Fe^+ [84], the main product for ethyl and propyl isothiocyanate being $Fe(HNCS)^+$ accompanied by a much smaller HNCS loss and some minor products, among which H_2S loss is noteworthy. But already for butyl isothiocyanate, $Fe(HNCS)^+$ and HNCS loss were significantly suppressed in favour of H_2S loss, which was the main product, dehydrogenation, and ethylene loss. The occurrence of H_2 and C_2H_4 loss with increasing chain length clearly points to the remote functionalization mechanism, and indeed ^2H labeling showed that dehydrogenation involved the $\omega/(\omega - 1)$-position. For the ethylene loss the remote functionalization mechanism is hence very likely too, but could not unambiguously be proven due to the presence of other losses ([H,N,C], $C_2H_5\cdot$), which for the labeled compounds could not be resolved from $C_2H_xD_{4-x}$ in the MI spectra. The labeling further showed that all four positions in butyl isothiocyanate contributed to the $Fe(HNCS)^+$ ions and to the HNCS loss. It can therefore be concluded that their formation is due to the ion/dipole mechanism.

The most interesting product, however, is the H_2S loss. This remarkable combination of an atom of the functional group with hydrogen atoms from

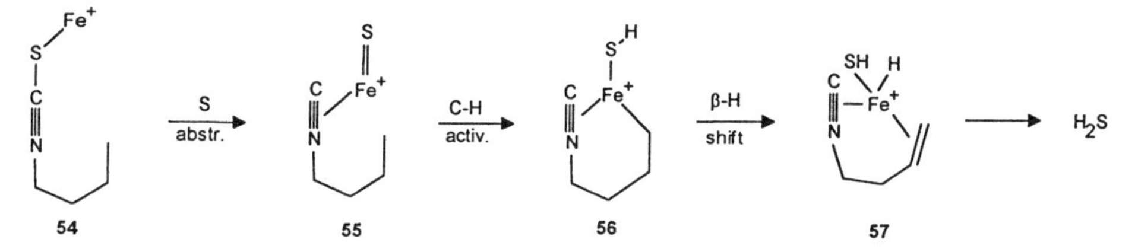

Scheme 9. Dehydrosulfurization of butyl isothiocyanate by remote functionalization after initial sulfur abstraction.

the chain greatly gained importance upon going from propyl to butyl isothiocyanate. It was therefore suspected that the hydrogen atoms were provided by the $\omega/(\omega - 1)$-position, which is not within reach for the smaller isothiocyanates. This could indeed be supported by the labeling results which revealed that ca. 90% of the hydrogen atoms were provided by the terminal positions. The mechanism in Scheme 9 accounts for these findings. Initially sulfur abstraction can be assumed since it is known that the C-S bond is relatively weak. Moreover, sulfur abstraction by Fe^+ has been noted before [86], and it is also known from solution studies of isothiocyanates with transition-metal complexes [87]. The following steps are analogous to the remote functionalization of bare metal ions, only with a ligated metal ion reacting in this case. The C-H activation may proceed stepwise with an initial hydrogen transfer to the metal ion or as a concerted addition of the C-H bond across the FeS group. Subsequently β-H shift and reductive elimination lead to a chelating butenyl isocyanide complex. That a ligated metal ion is indeed able to react by remote functionalization was also shown in reactions of FeO^+ with ketones, where a water molecule lost incorporated the metal-bound oxygen and remote hydrogen atoms from a ketone chain [88].

To get more information about this interesting system, for which a transition from the ion/dipole mechanism (for EtNCS and PrNCS) to two kinds of remote functionalization reactions (H_2/C_2H_4 and H_2S) is observed, Ti^+ through Zn^+ were studied under FTICR conditions with BuNCS and four of its ^{2}H-labeled isotopomers [85]. A shift from the remote functionalization products to those of the ion/dipole mechanism was noted for Fe^+, demonstrating, as discussed in the instrumental section, that MI spectra sample slightly cooler products than FTICR spectra. $M(HNCS)^+$ was also the main product for most of the other metal ions, with all positions of the alkyl chain contributing to its formation. As discussed above for the isocyanates, the overall similarity for different metal ions, combined with the H/D scrambling observed, strongly supports the assumption of the ion/dipole mechanism being responsible again. For Fe^+, Co^+, and Ni^+, besides HNCS loss, subsequent dehydrogenation of the resulting $M(C_4H_8)^+$ products to $M(C_4H_6)^+$ ions was active. For Ti^+, sulfur abstraction to TiS^+ was a prominent pathway; this process has been noted for other sulfur-containing molecules before [6a,

86a]. Cr^+ showed an unusual reactivity with many different products being formed. The most important process was dehydrogenation, which unspecifically involved all positions of the substrate. H_2S loss was also seen but differed in its labeling distribution from the Fe^+ induced H_2S loss. The only specific reaction for Cr^+ was propene loss which occurred by insertion into the α-CC bond with subsequent β-H shift. The same process is seen for Mn^+ and Fe^+. The manganese ion, besides losing propene and unspecifically some H_2S, mainly affords C_2H_4, which is formed in a highly specific manner from the α/β-position. Zn^+ mainly underwent charge transfer with butyl isothiocyanate, which is in accord with the ionization potentials of Zn (9.394 eV [89a]) versus BuNCS (9.02 eV [89b]).

8. Primary amines and alcohols

Amines and alcohols are different from the substrates discussed so far since they can also be dehydrogenated involving their functional group. This is indeed observed in some cases, mainly for the smallest members of the alkyl series [1], but will not be discussed here. We will again be concerned with somewhat larger substrates where mechanistic competition is possible. We will also restrict the discussion largely on Fe^+, Co^+, and Ni^+, the metal ions for which there is the most material available. No special mention is made on the hydride abstraction occasionally noted for the latter two ions, which specifically involves the α-position of the amines.

Propylamine has been studied with these metal ions in an ICR [90, 91] and in the Berlin BEB three-sector instrument [92, 93]. The latter studies employed ^2H-labeled substrates and hence were able to revise some of the mechanisms proposed in the earlier ICR studies, which were formulated based on chemical intuition. For all three ions, dehydrogenation was observed, but while for Fe^+ and Co^+ this reaction proceeded as remote functionalization of the $\omega/(\omega-1)$-positions, for Ni^+ the reaction proved to be unselective with contributions from all positions, including the NH_2 hydrogens. The main product, respectively, was ethylene, but only in case of Co^+ and Ni^+, the alkene was generated by remote functionalization too. For Fe^+ a special mechanism applied, which also led to loss of NH_3 and propene (Scheme 10). Ethylene originated from the α- and β-position, and the reaction was initiated by C-C activation. A path from **58** to **62** could be excluded from the labeling results of the other two products, NH_3 and C_3H_6. Both neutrals incorporated the hydrogen atoms of the α- and β-position in a statistical manner, which can only be interpreted if a preceding intermediate renders them indistinguishable. This intermediate, **60**, is also the immediate precursor for the ethylene loss. That indeed ethylene insertion in the CH_3-Fe^+ bond in **61** is possible, was shown in an FTICR study, where propene loss was noted following reaction of **61** with C_2H_4 [94].

Butylamine reacted with Fe^+ mainly by dehydrogenation, which was due

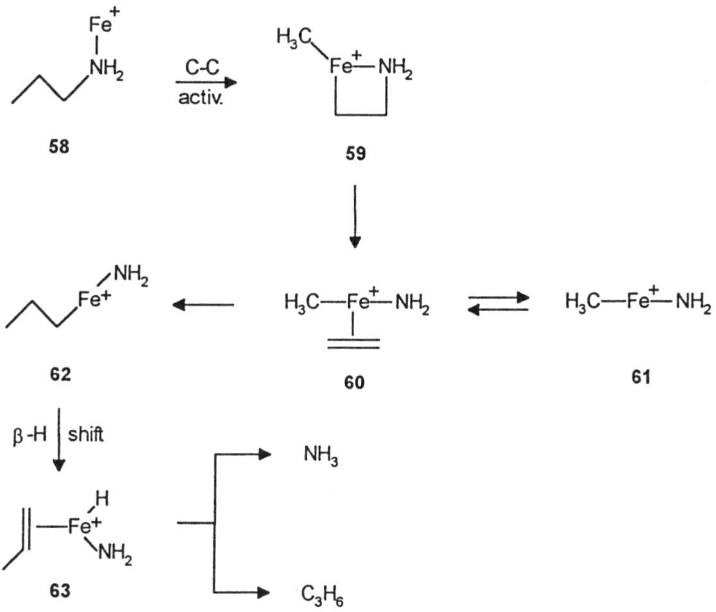

Scheme 10. Reaction of Fe^+ with propylamine, initiated by C-C activation and giving rise to loss of ethene, propene, and ammonia.

to remote functionalization, and to a small amount of ethylene [95]. In case of Co^+, eight different products were noted for butylamine and six for isobutylamine under ICR conditions [90]. Isobutylamine has also been studied with Fe^+, Co^+, and Ni^+ in a sector instrument employing labeled compounds [96]. Under these conditions less products were formed. As already observed for propylamine, dehydrogenation was specific for Fe^+ and Co^+ and proceeded by remote functionalization while Ni^+ reacted by extensive scrambling or involving unspecifically all positions of the substrate. Methane, which was the most important product for Fe^+ and Co^+, was formed by C-C activation, which afforded **65**, and subsequent β-H shift (Scheme 11). Propene, the major product for Ni^+, in case of both Co^+ and Ni^+ was generated via **68** by C-H activation followed by C-C cleavage. For Fe^+ this process only accounted for roughly 17% of the propene loss, the remainder being formed via **65** by C-N cleavage, producing **61** and propene that incorporated C_α and C_γ. Neopentylamine showed a very similar dependence on the metal ion employed [96]. Again, for Fe^+ and Co^+ methane was the most important product, and for all three metal ions it was formed as depicted in Scheme 11 (R = CH_3). Isobutene was the only other significant product and in case of Co^+ and Ni^+ was formed specifically via **68**. For Fe^+ again two processes were operative, and in this case both contributed to about 50% to the isobutene formation.

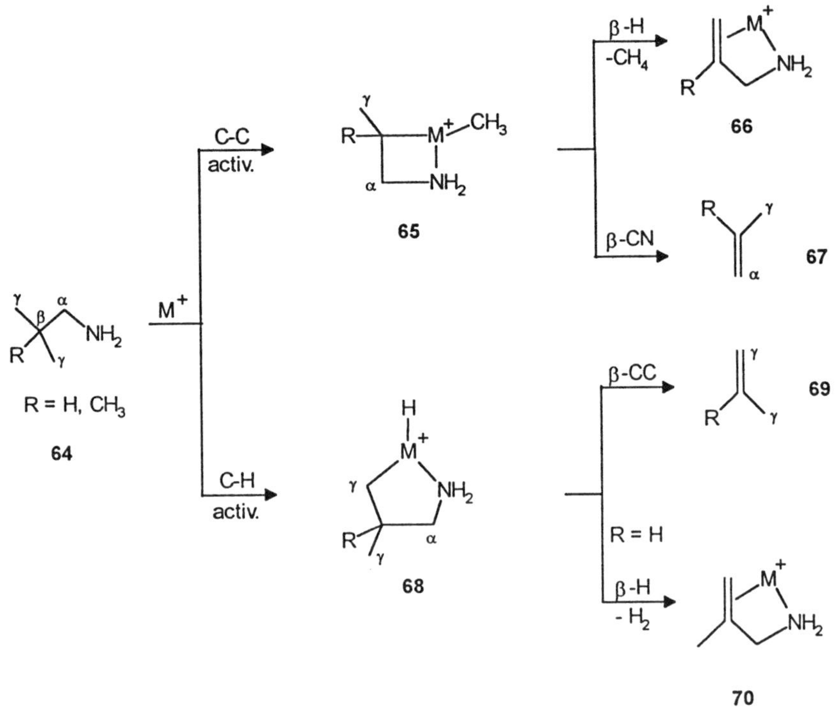

Scheme 11. Reaction of $Fe^+ - Ni^+$ with isobutylamine (R = H) and neopentylamine (R = CH_3).

Insertion of Fe^+ into the terminal C-CH_3 bond is quite common in gas-phase reactions of amines. For example, *tert*-pentylamine with Fe^+ [94, 97] or Co^+ [52] mainly afforded CH_3-M^+-NH_2 (**61**) by loss of isobutene. Other examples may be found in the reactions of Fe^+ and Co^+ with the isomeric substrates 2-ethyl-1-butylamine and 2,2-dimethyl-1-butylamine [95, 98]. Although in each case the main product was hydrogen, a product from remote functionalization as revealed by ^2H labeling, C_4H_8 loss was also present. For 2-ethyl-1-butylamine this corresponded to isobutene while for 2,2-dimethyl-1-butylamine it was 1-butene that was produced. Both neutrals for either of the two metal ions arose from initial C-CH_3 insertion followed by C-C cleavage. The similarity with Scheme 10 and the upper part of Scheme 11 is obvious.

The ion/dipole mechanism is also observed for amines. For instance, Fe^+ and Co^+ produced with *tert*-butylamine under FTICR conditions mainly MNH_3^+ and $M(C_4H_8)^+$ together with some methane [99]. For Ni^+ and Cu^+ these products from the ion/dipole mechanism were also observed, but there were significant amounts of methyl abstraction affording $(CH_3)_2C = NH_2^+$ that dominated the spectra. In MI spectra of $M(t\text{-}BuNH_2)^+$ complexes (M =

Fe – Ni) the main product was methane loss, indicating that this process has the lowest activation barrier [99]. The ion/dipole mechanism was also operative for Fe^+ or Co^+ reacting with *tert*-pentylamine [52, 94], but since there are other alternatives, in particular the isobutene loss by C-CH_3 insertion, its significance was greatly reduced.

The discussion of the alcohol reactions will be similarly restricted as for the amines. For more detailed information the reader is referred to the comprehensive reviews [1]. We will begin with propanol as this molecule gives another example of the ion/dipole mechanism being operative. In metastable ion studies, propanol with Fe^+ exclusively afforded loss of H_2O [93, 100]. For Co^+ and Ni^+, other products, including propene loss, were observed too, but dehydration was still the most important reaction [93]. For all three ions this reaction proved to be unspecific, with contributions from all positions of the molecule, including the OH group. These scrambling reaction are a hint to ion/dipole intermediates being involved. Co^+ and Ni^+ also formed C_2H_4 and CH_3OH, but only in case of Co^+ this reaction was specific and proceeded in analogy to Scheme 10. C-CH_3 activation is followed by β-cleavage, but in addition to the ethylene loss observed for propylamine, reductive elimination of methanol is also operative for propanol reacting with Co^+. For isopropanol, cleavage into H_2O and propene is even more important than for *n*-propanol. Very likely this results from the weaker C-OH bond that facilitates the initial OH abstraction necessary for the ion/dipole mechanism. The latter's operation is concluded from the similarity in the behaviour of otherwise quite different metal ions reacting with isopropanol to afford significant amounts of $M(C_3H_6)^+$ and $M(H_2O)^+$. Not only Fe^+ [57b, 101], Co^+ [57b], and Ni^+ [57b] have been reported to produce these ions, but also Cr^+ [101a, 102], Cu^+ [103], Au^+ [103b], and Mo^+ [101a], making it unlikely that the insertion/β-H shift mechanism applies.

As already observed for several other systems, the ion/dipole mechanism is strongly suppressed as soon as other mechanisms become feasible. For the alcohols this can be seen in the reactions of Fe^+ and Co^+ with some labeled *n*-pentanols [104]. While H_2O was the only reaction product from propanol, Fe^+ with pentanol showed less than 5% dehydration, but instead H_2 and C_2H_4 were formed by remote functionalization. Quite similarly, Co^+ with pentanol produced less than 2% H_2O, and the main products were those of the remote functionalization mechanism, namely H_2 and C_2H_4 originating from the $\omega/(\omega-1)$-position. Another feature of the remote functionalization mechanism can be seen in the alcohol system by varying the chain length. If one proceeds to *n*-hexanol, the main product for Co^+ becomes propene [105]. As could be demonstrated by ^2H labeling, this product is also formed by remote functionalization through activation of the $(\omega-1)$ C-H bond followed by β-CC cleavage (c.f. Scheme 1). Ethylene is still formed along with the propene, originating from the $\omega/(\omega-1)$-position, and hydrogen is formed via both C-H activation intermediates. A somewhat unusual reaction in the Co^+/*n*-HexOH system is the combined loss of C_2H_4 (from the end of the end of the chain) and H_2, but further details of the mechanism could not be resolved [105].

Acknowledgements

The work described in this chapter would have been impossible without the support of my colleagues and coworkers. In particular the continuous help of Prof. Helmut Schwarz is gratefully acknowledged, other names may be found in the references. Financial support came from the *Fonds der Chemischen Industrie* and the *Stiftung Volkswagenwerk*.

References

1. (a) K. Eller and H. Schwarz, *Chem. Rev.* **91**, 1121, (1991); (b) K. Eller, *Coord. Chem. Rev.* **126**, 93 (1993); *Coord. Chem. Rev.* **129**, 247 (1994).
2. J. Allison, *Prog. Inorg. Chem.* **34**, 627 (1986).
3. R. R. Squires, *Chem. Rev.* **87**, 623 (1987).
4. D. K. MacMillan and M. L. Gross, in *Gas Phase Inorganic Chemistry*, edited by D. H. Russell (Plenum Press, New York, 1989), p. 369.
5. (a) G. Czekay, T. Drewello, K. Eller, C. B. Lebrilla, T. Prüsse, C. Schulze, N. Steinrück, D. Sülze, T. Weiske, and H. Schwarz, in *Organometallics in Organic Synthesis 2*, edited by H. Werner and G. Erker (Springer-Verlag, Berlin, 1989), p. 203; (b) H. Schwarz, *Acc. Chem. Res.* **22**, 282 (1989); (c) K. Eller and H. Schwarz, *Chimia* **43**, 371 (1989); (d) K. Eller, S. Karrass, and H. Schwarz, *Ber. Bunsenges. Phys. Chem.* **94**, 1201 (1990).
6. (a) R. C. Burnier, G. D. Byrd, T. J. Carlin, M. B. Wiese, R. B. Codym, and B. S. Freiser, in *Ion Cyclotron Resonance Spectrometry II, Lecture Notes in Chemistry*, Vol. 31, edited by H. Hartmann and K.-P. Wanczek (Springer-Verlag, Berlin, 1982), p. 98; (b) B. S. Freiser, *Talanta* **32**, 697 (1985); (c) B. S. Freiser, *Anal. Chim. Acta* **178**, 137 (1985); (d) S. W. Buckner and B. S. Freiser, *Polyhedron* **7**, 1583 (1988); (e) B. S. Freiser, *Chemtracts-Anal. Phys. Chem.* **1**, 65 (1989); (f) B. S. Freiser, in *Bonding Energetics in Organometallic Compounds, ACS Symposium Series* 428, edited by T. J. Marks (American Chemical Society, Washington, DC, 1990), p. 55; (g) R. R. Weller, T. J. MacMahon, and B. S. Freiser, in *Lasers in Mass Spectrometry*, edited by D. M. Lubman (Oxford, New York, 1990), p. 249; (h) L. M. Roth and B. S. Freiser, *Mass Spectrom. Rev.* **10**, 303 (1991).
7. (a) P. B. Armentrout, in *Structure/Reactivity and Thermochemistry of Ions*, edited by P. Ausloos and S.G. Lias (D. Reidel, Dordrecht, 1987), p. 97; (b) P. B. Armentrout and R. Georgiadis, *Polyhedron* **7**, 1573 (1988); (c) P. B. Armentrout and J. L. Beauchamp, *Acc. Chem. Res.* **22**, 315 (1989); (d) P. B. Armentrout, in *Gas Phase Inorganic Chemistry*, edited by D. H. Russell (Plenum Press, New York, 1989), p. 1; (e) P. B. Armentrout, in *Selective Hydrocarbon Activation*, edited by J. A. Davies, P. L. Watson, A. Greenberg, and J. F. Liebman (VCH Verlagsgesellschaft, Weinheim, 1990), p. 467; (f) P. B. Armentrout, *Int. Rev. Phys. Chem.* **9**, 115 (1990); (g) P. B. Armentrout, in *Bonding Energetics in Organometallic Compounds, ACS Symposium Series* 428, edited by T. J. Marks (American Chemical Society, Washington, DC, 1990), p. 18; (h) P. B. Armentrout, *Ann. Rev. Phys. Chem.* **41**, 313 (1990); (i) P. B. Armentrout, *Science* **251**, 175 (1991); (j) P. B. Armentrout and L. S. Sunderlin, in *Transition Metal Hydrides*, edited by A. Dedieu (VCH Publishers, New York, 1992), p. 1; (k) P. B. Armentrout and D. E. Clemmer, in *Energetics of Organometallic Species*, edited by J. A. Martinho Simões (Kluwer, Dordrecht, 1992), p.

321; (l) P. B. Armentrout, in *Isotope Effects in Gas-Phase Chemistry*, *ACS Symposium Series* 502, edited by J. A. Kaye (American Chemical Society, Washington, DC, 1992), p. 194; (m) P. B. Armentrout, in *Gas-Phase Metal Reactions*, edited by A. Fontijn (Elsevier, Amsterdam, 1992), p. 301; (n) P. B. Armentrout, in *Advances in Gas Phase Ion Chemistry*, Vol. 1, edited by N. G. Adams and L. M. Babcock (JAI Press, Greenwich, CT, 1992), p. 83.

8. (a) J. L. Beauchamp, A. E. Stevens, and R. R. Corderman, *Pure Appl. Chem.* **51**, 967 (1979); (b) J. L. Beauchamp, in *High-Energy Processes in Organometallic Chemistry*, *ACS Symposium Series* 333, edited by K. S. Suslick (American Chemical Society, Washington, DC, 1987), p. 11; (c) P. A. M. van Koppen, M. T. Bowers, J. L. Beauchamp, and D. V. Dearden, in *Bonding Energetics in Organometallic Compounds*, *ACS Symposium Series* 428, edited by T. J. Marks (American Chemical Society, Washington, DC, 1990), p. 34; (d) J. L. Beauchamp and P. A. M. van Koppen, in *Energetics of Organometallic Species*, edited by J. A. Martinho Simões (Kluwer, Dordrecht, 1992), p. 287.
9. (a) J. Allison, A. Mavridis, and J. F. Harrison, *Polyhedron* **7**, 1559 (1988); (b) A. Mavridis, K. Kunze, J. F. Harrison, and J. Allison, in *Bonding Energetics in Organometallic Compounds*, *ACS Symposium Series* 428, edited by T. J. Marks (American Chemical Society, Washington, DC, 1990), p. 263.
10. (a) J. C. Weisshaar, in *State-Selected and State-to-State Ion-Molecule Reaction Dynamics*, Part 1, edited by C.-Y. Ng and M. Baer (Wiley, New York, 1992), p. 213; (b) J. C. Weisshaar, in *Gas-Phase Metal Reactions*, edited by A. Fontijn (Elsevier, Amsterdam, 1992), p. 253; (c) J. C. Weisshaar, *Acc. Chem. Res.* **26**, 213 (1993).
11. (a) R. Schlögl, *Angew. Chem.* **105**, 402 (1993); (b) Comment: *Appl. Catal. A* **104**, N3 (1993).
12. Some recent reviews on FTICR: (a) M. V. Buchanan, *Fourier Transform Mass Spectrometry*, *ACS Symposium Series* 359 (American Chemical Society, Washington, DC, 1987); (b) B. S. Freiser, in *Techniques for the Study of Ion-Molecule Reactions*, edited by J. M. Farrar and W. H. Saunders, Jr. (Wiley-Interscience, New York, 1988), p. 61; (c) M. P. Chiarelli and M. L. Gross, in *Analytical Applications of Spectroscopy*, edited by C. S. Creaser and A. M. Davies (Royal Society of Chemistry, London, 1988), p. 263; (d) K. P. Wanczek, *Int. J. Mass Spectrom. Ion Processes* **95**, 1 (1989); (e) C. L. Wilkins, A. K. Chowdhury, L. M. Nuwaysir, and M. L. Coates, *Mass Spectrom. Rev.* **8**, 67 (1989); (f) C. D. Hanson, E. L. Kerley, and D. H. Russell, in *Treatise on Analytical Chemistry*, Vol. 11, edited by J. D. Winefordner, M. M. Bursey, and I. M. Kolthoff (Wiley-Interscience, New York, 1989), p. 117; (g) A. G. Marshall and F. R. Verdun, *Fourier Transform in NMR, Optical, and Mass Spectrometry* (Elsevier, Amsterdam, 1990), Chapter 7; (h) A. G. Marshall and P. B. Grosshans, *Anal. Chem.* **63**, 215A (1991); (i) A. G. Marshall and L. Schweikhard, *Int. J. Mass Spectrom. Ion Processes* **118/119**, 37 (1992).
13. For multi-sector instruments, see: K. L. Busch, G. L. Glish, and S. A. McLuckey, *Mass Spectrometry/Mass Spectrometry: Techniques and Applications of Tandem Mass Spectrometry* (VCH Verlagsgesellschaft, Weinheim, 1988).
14. (a) R. B. Cody and B. S. Freiser, *Int. J. Mass Spectrom. Ion Phys.* **41**, 199 (1982); (b) R. B. Cody, R. C. Burnier, and B. S. Freiser, *Anal. Chem.* **54**, 96 (1982); (c) R. C. Burnier, R. B. Cody, and B. S. Freiser, *J. Am. Chem. Soc.* **104**, 7436 (1982); (d) R. B. Cody, *Analusis* **16**(6), XXX (1988).
15. (a) P. Kofel, M. Allemann, H. P. Kellerhans, and K. P. Wanczek, *Int. J. Mass Spectrom. Ion Processes* **65**, 97 (1985); (b) P. Kofel, M. Allemann, H. P. Kellerhans, and K. P. Wanczek, *Adv. Mass Spectrom.* **10**, 885 (1985); (c) P. Kofel, Ph.D. Thesis, Universität Bremen (1987).
16. K. Eller and H. Schwarz, *Int. J. Mass Spectrom. Ion Processes* **93**, 243 (1989).
17. K. Eller, W. Zummack, and H. Schwarz, *J. Am. Chem. Soc.* 112, 621 (1990).
18. K. Eller, Ph.D. Thesis, Technische Universität Berlin, D83 (1991).
19. R. B. Cody, R. C. Burnier, W. D. Reents, Jr., T. J. Carlin, D. A. McCrery, R. K. Lengel, and B. S. Freiser, *Int. J. Mass Spectrom. Ion Phys.* **33**, 37 (1980).

20. R. A. Forbes, F. H. Laukien, and J. Wronka, *Int. J. Mass Spectrom. Ion Processes* **83**, 23 (1988).
21. (a) G. Horlick and W. K. Yuen, *Anal. Chem.* **48**, 1643 (1976); (b) M. B. Comisarow and J. D. Melka, *Anal. Chem.* 51, 2198 (1979); (c) C. Giancaspro and M. B. Comisarow, *Appl. Spectrosc.* **37**, 153 (1983).
22. (a) M. B. Comisarow, V. Grassi, and G. Parisod, *Chem. Phys. Lett.* **57**, 413 (1978); (b) R. T. McIver, Jr. and W. D. Bowers, in *Tandem Mass Spectrometry*, edited by F. W. McLafferty (Wiley-Interscience, New York, 1983), p. 287.
23. R. G. Cooks, J. H. Beynon, R. M. Caprioli, and G. R. Lester, *Metastable Ions* (Elsevier, Amsterdam, 1973).
24. (a) K. Levsen and H. Schwarz, *Angew. Chem.* **88**, 589 (1976); *Angew. Chem. Int. Ed. Engl.* 15, 509 (1976); (b) R. G. Cooks, *Collision Spectroscopy* (Plenum Press, New York, 1978); (c) K. Levsen and H. Schwarz, *Mass Spectrom. Rev.* **2**, 77 (1983); (d) R. N. Hayes and M. L. Gross, in *Methods in Enzymology*, Vol. 193, edited by J. A. McCloskey (Academic Press, San Diego, CA, 1990), p. 237; (e) J. Bordas-Nagy and K. R. Jennings, *Int. J. Mass Spectrom. Ion Processes* **100**, 105 (1990).
25. See e.g.: D. Schröder, J. Müller, and H. Schwarz, *Organometallics* **12**, 1972 (1993).
26. (a) R. B. Freas and J. E. Campana, *J. Am. Chem. Soc.* 107, 6202 (1985); (b) T. Drewello, K. Eckhart, C. B. Lebrilla, and H. Schwarz, *Int. J. Mass Spectrom. Ion Processes* **76**, R1 (1987).
27. (a) T. Weiske, Ph.D. Thesis, Technische Universität Berlin, D83 (1985); (b) J. K. Terlouw, T. Weiske, H. Schwarz, and J. Holmes, *Org. Mass Spectrom.* **21**, 665 (1986).
28. R. Srinivas, D. Sülze, T. Weiske, and H. Schwarz, *Int. J. Mass Spectrom. Ion Processes* **107**, 369 (1991).
29. K. Eller, S. Karrass, and H. Schwarz, *Organometallics* 11, 1637 (1992).
30. (a) R. B. Freas and D. P. Ridge, *J. Am. Chem. Soc.* 102, 7129 (1980); (b) B. S. Larsen and D. P. Ridge, *J. Am. Chem. Soc.* **106**, 1912 (1984).
31. M. A. Hanratty, J. L. Beauchamp, A. J. Illies, and M. T. Bowers, *J. Am. Chem. Soc.* **107**, 1788 (1985).
32. D. K. MacMillan, R. N. Hayes, D. A. Peake, and M. L. Gross, *J. Am. Chem. Soc.* **114**, 7801 (1992).
33. R. C. Burnier, T. J. Carlin, W. D. Reents, Jr., R. B. Cody, R. K. Lengel, and B. S. Freiser, *J. Am. Chem. Soc.* **101**, 7127 (1979).
34. C. B. Lebrilla, C. Schulze, and H. Schwarz, *J. Am. Chem. Soc.* **109**, 98 (1987).
35. (a) R. Breslow, *Chem. Soc. Rev.* **1**, 553 (1972); (b) R. Breslow, *Acc. Chem. Res.* **13**, 170 (1980).
36. K. Eller, W. Zummack, and H. Schwarz, *Int. J. Mass Spectrom. Ion Processes* **100**, 803 (1990).
37. C. B. Lebrilla, T. Drewello, and H. Schwarz, *J. Am. Chem. Soc.* **109**, 5639 (1987).
38. C. B. Lebrilla, T. Drewello, and H. Schwarz, *Int. J. Mass Spectrom. Ion Processes* **79**, 287 (1987).
39. R. M. Stepnowski and J. Allison, *Organometallics* **7**, 2097 (1988).
40. C. B. Lebrilla, T. Drewello, and H. Schwarz, *Organometallics* **6**, 2450 (1987).
41. T. Prüsse, C. B. Lebrilla, T. Drewello, and H. Schwarz, *J. Am. Chem. Soc.* **110**, 5986 (1988).
42. A. Hässelbarth, T. Prüsse, and H. Schwarz, *Chem. Ber.* **123**, 209 (1990).
43. (a) S. Patai and Z. Rappoport, *The Chemistry of Organic Silicon Compounds* (Wiley-Interscience, New York, 1989); (b) J. B. Lambert, *Tetrahedron* **46**, 2677 (1990).
44. T. Prüsse, T. Drewello, C. B. Lebrilla, and H. Schwarz, *J. Am. Chem. Soc.* **111**, 2857 (1989).
45. G. Czekay, T. Drewello, and H. Schwarz, *J. Am. Chem. Soc.* **111**, 4561 (1989).
46. G. Czekay, T. Drewello, K. Eller, W. Zummack, and H. Schwarz, *Organometallics* **8**, 2439 (1989).

47. G. Czekay, K. Eller, D. Schröder, and H. Schwarz, *Angew. Chem.* **101**, 1306 (1989); *Angew. Chem. Int. Ed. Engl.* **28**, 1277 (1989).
48. (a) K. Eller, W. Zummack, H. Schwarz, L. M. Roth, and B. S. Freiser, *J. Am. Chem. Soc.* **113**, 833 (1991); (b) The overall reaction of Fe^+ with pentanenitrite proceeds at about 1/3 of the collision rate: D. Stöckigt, S. Sen, and H. Schwarz, *Chem. Ber.* **126**, 2553 (1993).
49. Structure **30** is possibly in equilibrium with $(C_2H_4)Fe^+ = CH_2$: (a) D. Stöckigt, Diploma Thesis, Technische Universität Berlin (1991); See however: (b) P. A. M. van Koppen, D. B. Jacobson, A. Illies, M. T. Bowers, M. Hanratty, and J. L. Beauchamp, *J. Am. Chem. Soc.* **111**, 1991 (1989); (c) R. H. Schultz and P. B. Armentrout, *Organometallics* **11**, 828 (1992).
50. D. B. Jacobson and B. S. Freiser, *Organometallics* **3**, 513 (1984).
51. K. Eller and H. Schwarz, *Organometallics* **8**, 1820 (1989).
52. K. Eller and H. Schwarz (unpublished results).
53. For ion/dipole complexes in the reactions of organic ions see: (a) T. H. Morton, *Tetrahedron* **38**, 3195 (1982); (b) D. J. McAdoo, *Mass Spectrom. Rev.* **7**, 363 (1988); (c) R. D. Bowen, *Acc. Chem. Res.* **24**, 364 (1991); See also: (d) S. Hammerum, *J. Chem. Soc., Chem. Commun.*, 858 (1988); (e) N. Heinrich and H. Schwarz, in *Ion and Cluster Ion Spectroscopy and Structure*, edited by J. P. Maier (Elsevier, Amsterdam, 1989), p. 329.
54. (a) P. Longevialle and R. Botter, *J. Chem. Soc., Chem. Commun.*, 823 (1980); (b) P. Longevialle and R. Botter, *Int. J. Mass Spectrom. Ion Phys.* **47**, 179 (1983); (c) P. Longevialle and R. Botter, *Org. Mass Spectrom.* **18**, 1 (1983); (d) T. H. Morton, *Org. Mass Spectrom.* 27, 353 (1992).
55. (a) D. Bethell and V. Gold, *Carbonium Ions* (Academic Press, London, 1967); (b) J. L. Fry and G. J. Karabatsos, in *Carbonium Ions*, Vol. II, edited by G. A. Olah and P. v. R. Schleyer (Wiley-Interscience, New York, 1970), p. 521; (c) J. T. Keating and P. S. Skell, in *Carbonium Ions*, Vol. II, edited by G. A. Olah and P. v. R. Schleyer (Wiley-Interscience, New York, 1970), p. 573; (d) G. A. Olah and J. A. Olah, in *Carbonium Ions*, Vol. II, edited by G. A. Olah and P. v. R. Schleyer (Wiley-Interscience, New York, 1970), p. 715; (e) R. E. Leone, J. C. Barborak, and P. v. R. Schleyer, in *Carbonium Ions*, Vol. IV, edited by G. A. Olah and P. v. R. Schleyer (Wiley-Interscience, New York, 1973), p. 1837; (f) M. Saunders, P. Vogel, E. L. Hagen, and J. Rosenfeld, *Acc. Chem. Res.* **6**, 53 (1973).
56. (a) R. D. Wieting, R. H. Staley, and J. L. Beauchamp, *J. Am. Chem. Soc.* **97**, 924 (1975); (b) R. H. Staley, R. D. Wieting, and J. L. Beauchamp, *J. Am. Chem. Soc.* **99**, 5964 (1977).
57. (a) J. Allison and D. P. Ridge, *J. Organomet. Chem.* **99**, C11 (1975); (b) J. Allison and D. P. Ridge, *J. Am. Chem. Soc.* **101**, 4998 (1979).
58. R. V. Hodges and J. L. Beauchamp, *Anal. Chem.* **48**, 825 (1976).
59. (a) W. R. Creasy and J. M. Farrar, *J. Phys. Chem.* **89**, 3952 (1985); (b) W. R. Creasy and J. M. Farrar, *J. Chem. Phys.* **85**, 162 (1986); (c) W. R. Creasy and J. M. Farrar, *J. Chem. Phys.* **87**, 5280 (1987).
60. R. V. Hodges, P. B. Armentrout, and J. L. Beauchamp, *Int. J. Mass Spectrom. Ion Phys.* **29**, 375 (1979).
61. A. K. Chowdhury and C. L. Wilkins, *Int. J. Mass Spectrom. Ion Processes* **82**, 163 (1988).
62. J. S. Uppal and R. H. Staley, *J. Am. Chem. Soc.* **104**, 1229 (1982).
63. L. Operti, E. C. Tews, and B. S. Freiser, *J. Am. Chem. Soc.* **110**, 3847 (1988).
64. K. Eller, C. B. Lebrilla, T. Drewello, and H. Schwarz, *J. Am. Chem. Soc.* **110**, 3068 (1988).
65. K. Eller and H. Schwarz, *Chem. Ber.* **123**, 201 (1990).
66. K. Eller, D. Sülze, and H. Schwarz, *Chem. Phys. Lett.* **154**, 443 (1989).
67. For reviews on NRMS, see: (a) C. Wesdemiotis and F. W. McLafferty, *Chem. Rev.* **87**, 485 (1987); (b) S. F. Selgren and G. I. Gellene, *Anal. Instrum.* **17**, 131 (1988); (c) J. L.

Holmes, *Mass Spectrom. Rev.* **8**, 513 (1989); (d) H. Schwarz, *Pure Appl. Chem.* **61**, 685 (1989); (e) F. W. McLafferty, *Science* **247**, 925 (1990).
68. Reviews: (a) M. Regitz and P. Binger, *Angew. Chem.* **100**, 1541 (1988); *Angew. Chem. Int. Ed. Engl.* **27**, 1484 (1988); (b) M. Regitz and P. Binger, *Nachr. Chem. Tech. Lab.* **37**, 896 (1989); (c) M. Regitz, *Chem. Rev.* **90**, 191 (1990); (d) L. N. Markovski and V. D. Romanenko, *Tetrahedron* **45**, 6019 (1989); (e) M. Regitz, in *Heteroatom Chemistry*, edited by E. Block (VCH Publishers, New York, 1990), p. 295.
69. K. Eller, T. Drewello, W. Zummack, T. Allspach, U. Annen, M. Regitz, and H. Schwarz, *J. Am. Chem. Soc.* **111**, 4228 (1989).
70. (a) H. Bock and O. Breuer, *Angew. Chem.* **99**, 492 (1987); *Angew. Chem. Int. Ed. Engl.* **26**, 461 (1987); (b) H. Bock, *Polyhedron* **7**, 2429 (1988).
71. K. Eller and H. Schwarz, *Inorg. Chem.* **29**, 3250 (1990).
72. K. Eller and H. Schwarz, *Ber. Bunsenges. Phys. Chem.* **94**, 1339 (1990).
73. K. Eller, D. Schröder, and H. Schwarz, *J. Am. Chem. Soc.* **114**, 6173 (1992).
74. (a) R. D. Bowen, B. J. Stapleton, and D. H. Williams, *J. Chem. Soc., Chem. Commun.*, 24 (1978); (b) D. H. Williams, B. J. Stapleton, and R. D. Bowen, *Tetrahedron Lett.*, 2919 (1978); (c) R. D. Bowen and D. H. Williams, *Int. J. Mass Spectrom. Ion Phys.* **29**, 47 (1979); (d) R. D. Bowen, *J. Chem. Soc., Perkin Trans. 2*, 1219 (1980); (e) R. D. Bowen and D. H. Williams, *J. Am. Chem. Soc.* **102**, 2752 (1980); (f) H. Schwarz and D. Stahl, *Int. J. Mass Spectrom. Ion Phys.* **36**, 285 (1980); (g) R. D. Bowen, *J. Chem. Soc., Perkin Trans. 2*, 919 (1989); (h) E. L. Chronister and T. H. Morton, *J. Am. Chem. Soc.* **112**, 133 (1990); (i) R. D. Bowen, A. W. Colburn, and P. J. Derrick, *J. Chem. Soc., Perkin Trans. 2*, 147 (1991).
75. K. Eller and H. Schwarz, *Int. J. Mass Spectrom. Ion Processes* **108**, 87 (1991).
76. See e.g.: (a) S. Cenini and G. La Monica, *Inorg. Chim. Acta* **18**, 279 (1976); (b) P. Braunstein and D. Nobel, *Chem. Rev.* **89**, 1927 (1989); (c) I. S. Kolomnikov, Y. D. Koreshkov, T. S. Lobeeva, and M. E. Volpin, *J. Chem. Soc., Chem. Commun.*, 1432 (1970); (d) Further references may be found in ref. [73].
77. D. E. Clemmer, L. S. Sunderlin, and P. B. Armentrout, *J. Phys. Chem.* **94**, 3008 (1990).
78. D. E. Clemmer, L. S. Sunderlin, and P. B. Armentrout, *J. Phys. Chem.* **94**, 208 (1990).
79. S. W. Buckner, J. R. Gord, and B. S. Freiser, *J. Am. Chem. Soc.* **110**, 6606 (1988).
80. (a) P. M. Holland and A. W. Castleman, Jr., *J. Am. Chem. Soc.* **102**, 6174 (1980); (b) P. M. Holland and A. W. Castleman, Jr., *J. Chem. Phys.* **76**, 4195 (1982); (c) D. Wittneben, Ph.D. Thesis, Universität Bielefeld (1991).
81. (a) B. C. Guo, K. P. Kerns, and A. W. Castleman, Jr., *J. Phys. Chem.* **96**, 4879 (1992); (b) $D°(HN = Sc^+)$ has been studied theoretically in ref. [9b].
82. (a) M. Surman, F. Solymosi, R. D. Diehl, P. Hofmann, and D. A. King, *Surf. Sci.* **146**, 144 (1985); (b) D. Lackey, M. Surman, and D. A. King, *Surf. Sci.* **162**, 388 (1985). See also: (c) M. Surman, F. Solymosi, R. D. Diehl, P. Hofmann, and D. A. King, *Surf. Sci.* **146**, 135 (1984).
83. M. Grunze, F. Bozso, G. Ertl, and M. Weiss, *Appl. Surf. Sci.* **1**, 241 (1978).
84. K. Eller, S. Akkök, and H. Schwarz, *Helv. Chim. Acta* **73**, 229 (1990).
85. K. Eller, S. Akkök, and H. Schwarz, *Helv. Chim. Acta* **74**, 1609 (1991).
86. (a) T. J. Carlin, M. B. Wise, and B. S. Freiser, *Inorg. Chem.* **20**, 2743 (1981); (b) T. C. Jackson, T. J. Carlin, and B. S. Freiser, *Int. J. Mass Spectrom. Ion Processes* **72**, 169 (1986); (c) R. L. Hettich, T. C. Jackson, E. M. Stanko, and B. S. Freiser, *J. Am. Chem. Soc.* **108**, 5086 (1986); (d) J. R. Gord and B. S. Freiser, *Anal. Chim. Acta* **225**, 11 (1989); (e) T. J. MacMahon, T. C. Jackson, and B. S. Freiser, *J. Am. Chem. Soc.* **111**, 421 (1989); (f) W. J. Gwathney, L. Lin, C. Kutal, and I. J. Amster, *Org. Mass Spectrom.* **27**, 840 (1992).
87. (a) T. A. Manuel, *Inorg. Chem.* **3**, 1703 (1964); (b) R. O. Harris, J. Powell, A. Walker, and P. V. Yaneff, *J. Organomet. Chem.* **141**, 217 (1977); (c) W. P. Fehlhammer and A. Mayr, *J. Organomet. Chem.* **191**, 153 (1980); (d) H. Werner, S. Lotz, and B. Heiser, *J. Organomet. Chem.* **209**, 197 (1981); (e) J. Fortune, A. R. Manning, and F. S. Stephens,

J. Chem. Soc., Chem. Commun., 1071 (1983); (f) J. C. Bryan, S. J. Geib, A. L. Rheingold, and J. M. Mayer, *J. Am. Chem. Soc.* **109**, 2826 (1987); (g) F.-M. Su, J. C. Bryan, S. Jang, and J. M. Mayer, *Polyhedron* **8**, 1261 (1989); (h) G. R. Lee and N. J. Cooper, *Organometallics* **8**, 1538 (1989); (i) E. P. Cullen, J. Fortune, A. R. Manning, P. McArdle, D. Cunningham, and F. S. Stephens, *Organometallics* **9**, 1443 (1990); (j) I. R. Beaumont, M. J. Begley, S. Harrison, and A. H. Wright, *J. Chem. Soc., Chem. Commun.*, 1713 (1990); (k) U. Riaz, O. Curnow, and M. D. Curtis, *J. Am. Chem. Soc.* **113**, 1416 (1991); (l) M. D. Curtis, *Appl. Organomet. Chem.* **6**, 429 (1992).

88. D. Schröder, K. Eller, and H. Schwarz, *Helv. Chim. Acta* **73**, 380 (1990).
89. (a) D. R. Lide (Ed.), *CRC Handbook of Chemistry and Physics*, 72nd Ed. (CRC Press, Boca Raton, FL, 1991); (b) R. D. Levin and S. G. Lias, *Ionization Potential and Appearance Potential Measurements 1971–1981* (National Bureau of Standards, Washington, 1982).
90. B. D. Radecki and J. Allison, *J. Am. Chem. Soc.* **106**, 946 (1984).
91. S. J. Babinec and J. Allison, *J. Am. Chem. Soc.* **106**, 7718 (1984).
92. S. Karraß, K. Eller, C. Schulze, and H. Schwarz, *Angew. Chem.* **101**, 634 (1989); *Angew. Chem. Int. Ed. Engl.* **28**, 607 (1989).
93. S. Karrass, T. Prüsse, K. Eller, and H. Schwarz, *J. Am. Chem. Soc.* **111**, 9018 (1989).
94. S. Karrass, D. Stöckigt, D. Schröder, and H. Schwarz, *Organometallics* **12**, 1449 (1993).
95. S. Karrass and H. Schwarz, *Helv. Chim. Acta* **72**, 633 (1989).
96. S. Karraß, K. Eller, and H. Schwarz, *Chem. Ber.* **123**, 939 (1990).
97. K. Eller, D. Stöckigt, S. Karrass, and H. Schwarz (unpublished results).
98. S. Karrass and H. Schwarz, *Organometallics* **9**, 2034 (1990).
99. K. Eller, S. Karraß, and H. Schwarz (unpublished results).
100. T. Prüsse, Ph.D. Thesis, Technische Universität Berlin, D83 (1991).
101. (a) S. Huang, R. W. Holman, and M. L. Gross, *Organometallics* **5**, 1857 (1986); (b) S. Karraß, D. Schröder, and H. Schwarz, *Chem. Ber.* **125**, 751 (1992).
102. H. Kang and J. L. Beauchamp, *J. Am. Chem. Soc.* **108**, 7502 (1986).
103. (a) R. W. Jones and R. H. Staley, *J. Phys. Chem.* **86**, 1669 (1982); (b) D. A. Weil and C. L. Wilkins, *J. Am. Chem. Soc.* **107**, 7316 (1985).
104. T. Prüsse and H. Schwarz, *Organometallics* **8**, 2856 (1989).
105. T. Prüsse, J. Allison, and H. Schwarz, *Int. J. Mass Spectrom. Ion Processes* **107**, 553 (1991).

5. Electronic state effects in sigma bond activation by first row transition metal ions: the ion chromatography technique

PETRA A.M. VAN KOPPEN, PAUL R. KEMPER and
MICHAEL T. BOWERS
Department of Chemistry, University of California, Santa Barbara, CA 93106, USA

1. Introduction

The ability of transition metal ions to activate C—H and C—C bonds of small saturated hydrocarbons has been attributed to the high density of low-lying excited states available to the transition metal center. The details of the interactions of these metal ion states with the reactant molecule, however, are not well understood and the desire to understand these details has stimulated significant experimental and theoretical interest. On the experimental side, gas-phase transition metal ion chemistry has been explored extensively [1–3] and recently state specific studies have been carried out [4–17]. Theoretically, ab initio electronic structure calculations provided insight regarding the nature of the bonding and trends in bond energies [18–25]. In addition, ab initio potential energy surfaces allow chemists to explore the interaction of the low-lying excited states of metal ions with reactant molecules [26]. Both theory and experiment indicate the reactivity of a given state of the metal ion will depend on its interaction with other nearby electronic states. Surface crossings are commonplace and "spin-forbidden" reactions are often observed for transition metal ions (spin is generally conserved for bimolecular reactions involving light elements, but only the total angular momentum must be rigorously conserved for heavier elements) [27]. Because reactions involving atomic transition metal ions are inherently complex, quantitative state specific experimental studies along with theoretical calculations are necessary to understand their reaction energetics and mechanisms.

The valence electron configurations for first-row ground and low-lying excited atomic transition metal ion states are either $3d^n$ or $4s3d^{n-1}$. The 4s and 3d orbitals are similar in energy giving rise to an abundance of low-lying electronic states for first-row transition metal ions. Radiative lifetimes of many excited states of metal ions are long (on the order of seconds) owing to parity forbidden transitions [28, 29]. Consequently, once these long lived excited electronic states are produced, their reactivity can be studied. The electronic state configurations and corresponding energies for first row transition metal ions, Sc^+ through Zn^+ are given in Table I [30]. The ground and

Ben S. Freiser (ed.), Organometallic Ion Chemistry, 157–196.

Table I. Electronic states of first row transition metal ions

Ion	State	Configuration[a]	Energy[b]	Boltzmann distribution at 2200 K
Sc^+	a^3D	$3d(^2D)4s^a$	0.013	0.886
	a^1D	$3d(^2D)4s$	0.315	0.060
	a^3F	$3d^2$	0.608	0.054
	a^1D	$3d^2$	1.357	<0.001
Ti^+	a^4F	$3d^2(^3F)4s$	0.028	0.626
	b^4F	$3d^3$	0.135	0.356
	a^2F	$3d^2(^3F)4s$	0.593	0.016
	a^2D	$3d^2(^1D)4s$	1.082	<0.001
	a^2G	$3d^3$	1.124	0.001
V^+	a^5D	$3d^4$	0.026	0.806
	a^5F	$3d^3(^4F)4s$	0.363	0.191
	a^3F	$3d^3(^4F)4s$	1.104	0.002
	a^3P	$3d^4$	1.452	<0.001
Cr^+	a^6S	$3d^5$	0.000	0.998
	a^6D	$3d^4(^5D)4s$	1.522	0.002
	a^4D	$3d^4(^5D)4s$	2.458	≪0.001
	a^4G	$3d^5$	2.543	
Mn^+	a^7S	$3d^5(^6S)4s$	0.0	0.999
	a^5S	$3d^5(^6S)4s$	1.175	0.001
	a^5D	$3d^6$	1.808	<0.001
Fe^+	a^6D	$3d^6(^6D)4s$	0.052	0.793
	a^4F	$3d^7$	0.300	0.204
	a^4D	$3d^6(^5D)4s$	1.032	0.003
	a^4P	$3d^7$	1.688	≪0.001
Co^+	a^3F	$3d^8$	0.086	0.851
	a^5F	$3d^7(^4F)4s$	0.515	0.148
	b^3F	$3d^7(^4F)4s$	1.298	0.001
	a^1D	$3d^8$	1.445	<0.001
	a^3P	$3d^8$	1.655	
Ni^+	a^2D	$3d^9$	0.075	0.990
	a^4F	$3d^8(^3F)4s$	1.160	0.009
	a^2F	$3d^8(^3F)4s$	1.757	<0.001
	b^2D	$3d^8(^1D)4s$	2.899	
Cu^+	a^1S	$3d^{10}$	0.00	1.000
	a^3D	$4s3d^9$	2.808	<0.001
Zn^+	^2S	$4s3d^{10}$	0.000	1.000
	^2D	$4s^23d^9$	7.909	

[a]The state indicated in parenthesis refers to the coupling of the 3d electrons. [b]Ref. 30, in eV; average over J levels.

low-lying excited states usually differ in either their electron configuration or spin multiplicity, allowing both of these electronic state effects to be studied.

We have taken two approaches to further our understanding of sigma bond activation by transition metal centers. First, we have measured electronic state specific rate constants for first row transition metal ions reacting with propane [8] and methyl iodide [9] at thermal energy. Second, dissociation energies for H_2 loss from $Co(H_2)_n^+$, $V(H_2)_n^+$ and $Sc(H_2)_n^+$ (n = 1–7) and CH_4 and C_2H_6 loss from $Co(CH_4)_{1,2}^+$ and $Co(C_2H_6)_{1,2}^+$, have been determined via temperature-dependent equilibrium measurements [31–34]. The interaction of a transition-metal ion with H_2 is the simplest system to study experimentally and lends itself well to theoretical calculations. It is our goal to extend what we have learned from these simple systems to larger more complex systems.

We have recently developed a new "chromatographic" technique [35] to characterize populations of electronically excited metal ions. The $3d^n$ and $4s3d^{n-1}$ valence electron configurations of atomic transition metal ions exhibit large differences in mobility (typically 40%). Because the 4s orbital is larger than the 3d orbital, the $4s3d^{n-1}$ configurations have a more repulsive interaction with the bath gas than the $3d^n$ configurations. This gives rise to the differences in mobility and to a spatial and temporal spread of the metal ions in different electronic configurations as they drift through a buffer gas under the influence of a weak electric field. If a mass selected ion beam is pulsed into the reaction cell containing a buffer gas, ions with different electronic configurations separate as they drift through the cell and are observed at different times in the arrival time distribution (ATD). The excited and ground state populations can be determined by integrating the corresponding peak areas. Reaction rate constants are measured as a function of the percent ground and excited state to obtain the state selected rate constants for reaction (by extrapolation to 100% ground and excited state M^+). In the equilibrium experiments, excited state deactivation by collisions with the H_2 or alkane reactant gas (which is also the buffer gas in these experiments), usually leads to essentially 100% ground state metal ions. This is not true in experiments with He or Ne buffer gases, however, when essentially no deactivation of $4s3d^{n-1}$ excited states to the ground state occurs.

In this chapter we will review how we determine the electronic state population of transition metal ions using the ion chromatography (IC) technique. We will also indicate under what conditions we observe electronic state deactivation. The reactivity of ground and excited states of Co^+ with CH_3I will be discussed, including quantitative determination of rate constants for all the observed products: adduct formation as well as CH_3 and I elimination channels. By modeling the reaction efficiencies for the elimination channels on the ground state surface with statistical phase space theory, the Co^+—CH_3 and Co^+—I bond energies were determined. Finally we will discuss the measured dissociation energies for H_2 loss from $Co(H_2)_n^+$, $V(H_2)_n^+$ and $Sc(H_2)_n^+$ (n = 1–7) and the implications of these results and those of ab initio calculations on the mechanisms for sigma bond activation in these and more complex systems.

2. Experimental

2.1. State specific metal ion generation

Atomic metal ions generated by electron ionization (EI) [35] surface ionization (SI) [6, 36] or laser vaporization [37, 38], are produced in a mixture of ground and excited electronic states. With EI, the number of excited states is often large and varies strongly with electron energy. The limited data available indicate that laser vaporization also produces a significant fraction of electronically excited ions as well. The electronic state population of ions formed by SI, is presumed to follow a Boltzmann distribution at the temperature of the ionizing surface. A detailed discussion of the validity of the Boltzmann distribution of states has been presented [36a]. As will be shown in the next section, the IC technique provides direct evidence that this presumption is correct. Thus, only with SI and Resonant Enhanced Multiphoton Ionization (REMPI) [4, 39–41] are the electronic state distributions well characterized.

Several investigators [6, 11, 12] have determined the presence of excited states and their populations by the observance of an endothermic reaction at thermal energy. This gives a lower bound for the energy of one or more excited states, but little or no information as to the ground and excited state populations (due to the unknown reactivities of the states) nor does it identify the excited state(s) involved. Elkind and Armentrout have reported a large number of experiments [6, 13, 28, 42–46] aimed at characterizing the production (by SI, EI and EI combined with a high pressure drift cell) and reactivity of ground and excited state atomic metal ions. By determining the reaction cross section as a function of kinetic energy for ions formed by SI and EI, state specific cross sections can be obtained. The states which can be formed by SI are limited to rather low energy ($\lesssim$0.5 eV) [6] and to investigate higher energy states, ions must be formed by EI. In these experiments, reactive states are detected by their thresholds in the cross section vs. kinetic energy curve (excitation function); however, the contribution of a state to the excitation function depends on the state's population and reactivity, both of which are unknown. Furthermore, any non-reactive states present do not appear, except as an overall reduction in cross section. Populations of low energy states in EI experiments can be estimated by comparing EI and SI data. Weisshaar has used REMPI to form essentially pure populations of vanadium [16, 39–41] and iron [17] ground and excited state ions, even to the point of producing a particular J level. The drawback to the REMPI technique is its complexity and difficulty. While less specific than REMPI, the IC technique provides a simple method of producing and characterizing populations of electronically excited metal ions.

It was shown a number of years ago that ground and excited electronic states C^+ [47, 48] and O^+ [49] have significantly different mobilities (5–20%). We have found that different electronic states for a given transition metal atomic ion often have mobilities as much as 50% different [35]. The valence electronic configurations in these ions are either $3d^n$ or $4s3d^{n-1}$ [30]. States with different configurations exhibit the largest mobility differences; however, mobilities of electronic states with high and low spin $4s3d^{n-1}$ configurations can also be significantly different [35]. When mobilities of electronic states are significantly different, the spatial and temporal distributions of the states separate as they drift through the buffer gas. Consequently, if all of the electronic states enter the drift cell at the same point in time, they will arrive at the exit slit, and at the detector, at different times. Because the different interactions of the states with the buffer gas causes the temporal separation, we refer to this method as the "Ion Chromatography" technique.

The IC technique provides, in many cases, an absolute determination of ground and excited state populations which allows investigation of state specific reactions. The metal ion electronic state distribution produced by SI and EI is discussed below. The main limitation of the method is the difficulty in the assignment of electronic states under EI conditions. EI produces electronic states over a wide range in energy and different states with the same electronic configuration are not resolved. However, by varying the ionizing energy from threshold to 50 eV, state specific rate constants have been obtained even when the electronic configurations are not resolved chromatographically [8]. In addition to the determination of ground and excited state populations, the mobilities of these ions are interesting in their own right. The dependence of mobility on temperature or field strength (effective temperature) [50–52] can be used to determine the interaction potential between the ion and buffer gas [48, 52, 53].

In the following sections we will describe the experimental method and the physical basis for the temporal resolution we observe between different electronic states of the same ion. These techniques will be applied to the first-row transition metal series.

2.2. The instrument

A detailed description of the apparatus and experimental technique has appeared elsewhere [54], and only an overview is given here. A schematic of the reaction cell and IC experiment is given in Figure 1. Both a traditional EI and SI source are used to study state specific transition metal ion chemistry. The EI source forms atomic metal ions by high energy electron impact on volatile metal containing compounds. The SI source is similar to one developed by Armentrout and coworkers [14, 55] and forms ions by thermal heating on a very hot filament (2200 K ± 200 K). Our version allows both standard EI and SI on the same sample. Having both sources simultaneously available is very useful, as will be evident in the experiments described.

The chromatography cell (or reaction cell) is the key part of the experiment. In a typical experiment, atomic metal ions are mass selected in a double focusing mass spectrometer, decelerated to a kinetic energy of 2 to

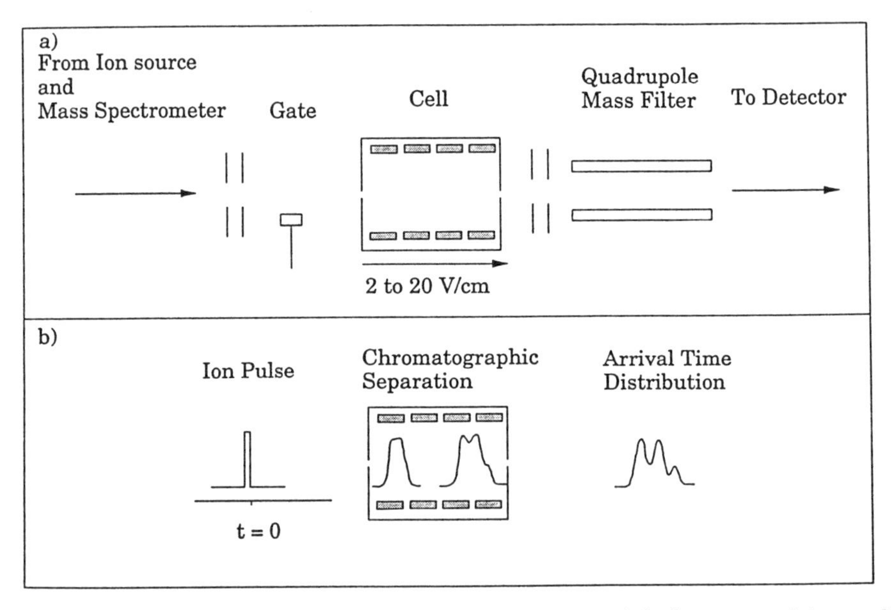

Fig. 1. Ion chromatography experiment. (a) Essential elements of the instrument. A beam of mass-selected ions is focused on the 0.5 mm entrance hole of the IC cell, which contains 2 to 5 Torr of He. A 1- to 10-μs pulse is injected, at low energy (2 or 3 eV, lab frame), and drifts through the cell under the influence of a small electric field. The ions exit the cell, are mass analyzed, and detected (b) Spatial and temporal ion beam profile for a model system with three components. A pulse of ions is formed at t = 0 and chromatographically separated in the cell, and subsequently detected.

3 eV and focused into the reaction cell containing 2 to 5 Torr of He and a small amount of reactant gas ($\sim 1 \times 10^{-5}$ to 1×10^{-3} Torr). Cell entrance and exit holes are variable from 0.5 to 1.0 mm and the drift length is 4 cm. The ions are quickly thermalized by collisions with He as they enter the cell [54]. The ions drift through the cell under the influence of a weak electric field (typically 1 to 10 V/cm) with drift times $t_d = 100$ to 3000 μs and typically experience 5×14^4 to 2×10^5 collisions with He. Ions which exit the cell are quadrupole mass analyzed and collected. The cell temperature is variable from ~80 K to 600 K.

2.3. Ion mobility measurements

The basic definition of ion mobility is given in equation (1) [52], where K is the ion mobility, v_d is the ion drift velocity and E is the electric field to which the ion is

$$v_d = KE \qquad (1)$$

subjected. To measure mobilities, the mass selected ions are pulsed into the drift cell (pulse width ~1–3 μs). The pulse simultaneously triggers either a multi-channel scalar (MCS) scan or a time-to-pulse-height converter (TPHC) ramp. Ions which exit the cell are then collected as a function of time, giving an arrival time distribution (ATD). The time it takes an ion to traverse the cell depends on its mobility, which in turn depends inversely on the collision cross section of the species with He gas: the larger the cross section, the smaller the mobility, and the longer the time. Because the mobility is inversely proportional to the cross section, the time the ion spends in the reaction cell, t_d is directly proportional to the collision cross section of that ion with the bath gas as shown in equation (2),

$$t_d = \frac{z}{v_d} = \frac{z}{KE} = \frac{z\sigma}{AE} \tag{2}$$

where z is the cell length, σ the collision cross section [56], and A a constant term that incorporates known properties of the system (mass, charge, and temperature) [57], Thus, two types of ions, type (1) and type (2), having the same mass but different collision cross sections with the bath gas will exit the cell at different times as shown in equation (3).

$$t_d(1) = \frac{\sigma(1)}{\sigma(2)} t_d(2) \tag{3a}$$

$$t_d(1) - t_d(2) = \frac{z}{AE}[\sigma(1) - \sigma(2)] \tag{3b}$$

As we have noted, these effects can be very large for different electronic states in a given atomic transition metal ion.

One factor that affects the utility of the device is resolution. It is straightforward to show [35] that the resolution is proportional to $(V/T)^{1/2}$ equation (4), where Δt_d is the time width of an ion packet of a specific species due

$$\text{Resolution} = \frac{t_d}{\Delta t_d}\left[\frac{qV}{8kT}\right]^{1/2} \tag{4}$$

to collisional broadening in the cell, q is the ion charge, V is the voltage drop across the cell, k is the Boltzmann constant, and T is the temperature. In the ideal case, $\Delta t_d \ll [t_d(1) - t_d(2)]$ so that abundances of ions having different collision cross sections can be clearly resolved. Because the resolution increases (i.e., Δt_d decreases) in proportion to $(V/T)^{1/2}$, higher values of V and lower values of T are desirable. We have, in fact, shown [35] dramatic increases in resolution for transition metal ions in reducing T from 300 to 150 K, confirming that aspect of equation (4). Further reductions in T are not useful, however, because the different mobilities begin to converge to a single value (the Langevin mobility) [52]. Thus, although the resolution increases, there is less to resolve.

The other path to higher resolution (increasing V) requires a simultaneous increase in bath gas pressure to avoid "translational heating" of the ions and loss of resolution. Our current instrument has an effective upper limit of ~5 Torr (due to pumping speed restrictions). We do observe significant improvement in resolution in going from 2 to 5 Torr, as expected. In reality, to increase resolution by increasing the pressure reduces signal intensity and the current experiments reflect a balance of the intensity and resolution effects. Helium is used as the bath gas because it provides the greatest IC peak separation. This is due to the very weak M^+-He interaction ($D_0^o \lesssim 1$ kcal/mol for $4s3d^{n-1}$ states and ~3 kcal/mol for $3d^n$ states [53]. The greatest difference in mobility occurs when one state has a binding energy $\gg$kT (where the mobility (K) = Langevin mobility (K_L)) and the other state has a binding energy $\cong$kT (where $K \cong 1.4(K_L)$) [52]. Mobility measurements in H_2 for example, show only one peak (where $K = K_L$) because the binding energies of all the states are $\gg$kT.

Mobilities are determined in our experiment by plotting the arrival times of the various peak centers as a function of 1/V [35]. This gives a linear plot with a slope inversely proportional to the ion mobility (K), as defined in equation (5).

$$K = v_d/E = \frac{z/t_d}{V/z} = z^2 \cdot \left[\frac{t_d}{1/V}\right]^{-1} \tag{5}$$

The intercept of the t_d vs. 1/V plot is the time spent in the quadrupole. These mobilities are converted to reduced mobilities (K_0) to remove the first order effects of helium density (N) [51, 52],

$$K_0 = K \cdot \frac{P}{760} \cdot \frac{273.15}{T} \tag{6}$$

where P is the He pressure in Torr.

Our method of analysis, using the slope of the arrival time vs. 1/V plot, makes three main assumptions: (1) K is independent of the drift field, (2) the center of the ATD peak accurately represents the drift time of the gaussian ion packet, and (3) the ion packet does not penetrate significantly into the cell before thermalization. Regarding (1), it has been shown that the effect of the ion drift velocity on K can be represented by a change in an "effective temperature" [50–52]

$$T_{eff} = T + \frac{M_B \cdot v_d^2}{3k} \tag{7}$$

where T is the thermodynamic temperature, M_B is the buffer gas mass and k is Boltzmann's constant. At the highest E/N values used in our experiments ($E/N \leqslant 8 \times 10^{-17}$ cm^2V (=8 Td) at 305 K; $\leqslant$4Td at 160 K) T_{eff} is calculated to be 348 and 172 K (respectively), about a 10% increase. It was necessary to use these high E/N values to resolve the ATD peaks. Changes in thermodynamic temperature (T) from 150–305 K produced changes in mobility of

<10% (see results). The observed variation in K_0 with effective temperature for most ions is even smaller [50]. We conclude that the increased ion energy due to the drift field has no significant effect ($\lesssim$2%) on our mobility results under almost all experimental conditions we use.

Our second approximation (regarding the observed peak center) arises from the time dependent attenuation of the gaussian ion packet as it exits the cell [52, 54]. The effect is to shift the apparent drift time to a slightly shorter value than the drift time of the ion packet center. The effect is accentuated at long drift times and low He pressures where the time width of the packet increases. For the relatively high pressures and drift velocities used in these experiments, it can be shown [54] that this perturbation is truly minor ($\lesssim$1%); and our second assumption is well justified.

Our final assumption (3) concerns the starting point of the ion packet in the cell. The excess ion kinetic energy remaining at the cell entrance must be dissipated in collisions with helium. This can result in penetration into the cell before thermalization, producing an effectively shorter cell length. This effect can be roughly calculated, and at the pressures and ion energies used in the present experiments, the worst case penetration is $\leqslant$3% of the cell length [54]. There are reasons to believe that it is probably much less [54].

As a final point, we note that all our mobility measurements agree very well (±2%) with previous high quality measurements [50, 51].

3. Arrival time distribution (ATDs)

3.1. Metal ion electronic state distributions produced by SI and EI

The mobilities of atomic transition metal ions in He can be used to separate electronic states with different electron configurations ($4s3d^{n-1}$ and $3d^n$) and, in many cases, high and low spin states within the $4s3d^{n-1}$ configuration. The large difference in mobility between configurations is due to the repulsion between the 4s electron and the filled He 1s electron cloud. This repulsion partly cancels the long range charge induced dipole (Langevin) attraction [52]. The result is a reduced attraction between the metal ion and helium which leads to an increase in mobility for the states with a $4s3d^{n-1}$ configuration. The states with $3d^n$ configuration had in all cases mobilities very close to the Langevin mobility [35]. These mobilities showed little temperature dependence. States with $4s3d^{n-1}$ configurations had mobilities 40–50% higher than Langevin. There was considerable variation in these mobilities from metal to metal and a large temperature dependence was usually observed [53].

The ATD (in He) for V^+ formed by SI of $VOCl_3$ on a heated filament

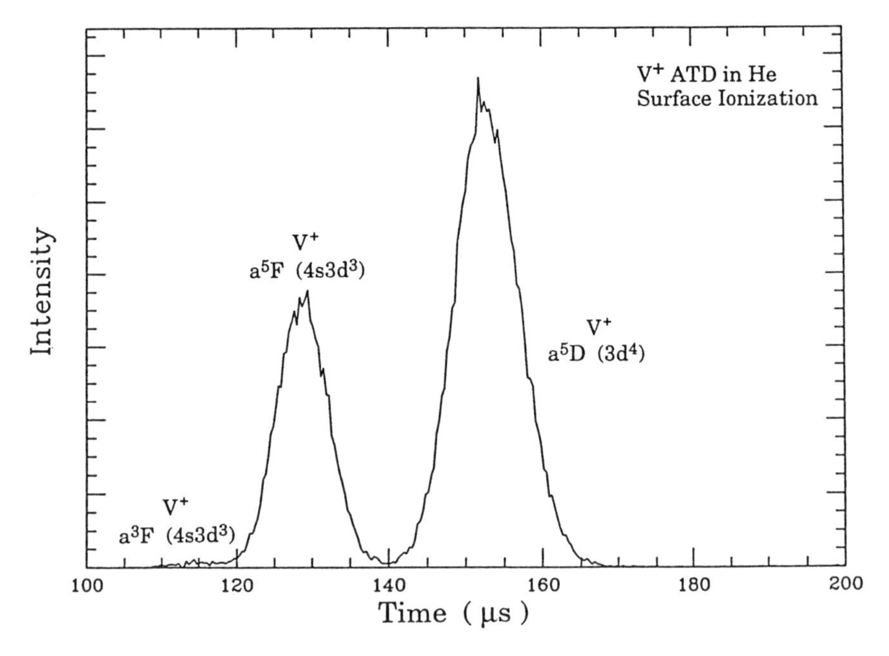

Fig. 2. Arrival time distribution (ATD) of V^+ ions that have been chromatographically separated in the IC cell. The V^+ ions were formed by surface ionization of $VOCl_3$ The three peaks are assigned as shown [35] and their relative intensities correspond to those expected for a Boltzmann distribution at the filament temperature of ~2200 K. Note that the peaks are baseline resolved indicating essentially no deactivation is occurring.

(2200 ± 200 K) is shown in Figure 2. The Boltzmann distribution at 2200 K is 80.6% ^5D ($3d^4$) ground state V^+, 19.1% ^5F ($4s3d^3$) and only 2% ^3F ($4s3d^{n-1}$) (Table I). The peaks are baseline resolved and correspond to $3d^n$ and $4s3d^{n-1}$ electronic configurations as indicated. Note that the high and low spin states of the $4s3d^3$ configuration are chromatographically separated in the ATD although with only 2% of the ^3F ($4s3d^3$) state this is not clear until we look at the ATD of V+ produced by EI shown in Figure 3. The absence of V^+ ions with arrival times intermediate between those of excited and ground states indicates deactivation does not occur (or occurs extremely slowly) from the excited states to the ground state while the ion traverses the reaction cell. Because deactivation does not occur, the integrated ATD peak areas equal the populations of the electronic configurations. The integrated peak areas follow a Boltzmann distribution of electronic state population for V^+ as expected for ions formed by SI at 2200 K (Table I).

Under EI conditions, the assignment of electronic states is less straightforward, because different states with the same configuration ($3d^n$, high spin $4s3d^{n-1}$ and low spin $4s3d^{n-1}$) are not resolved. For example, the ATD for V^+, formed by EI, consists of three peaks as shown in Figure 3. These

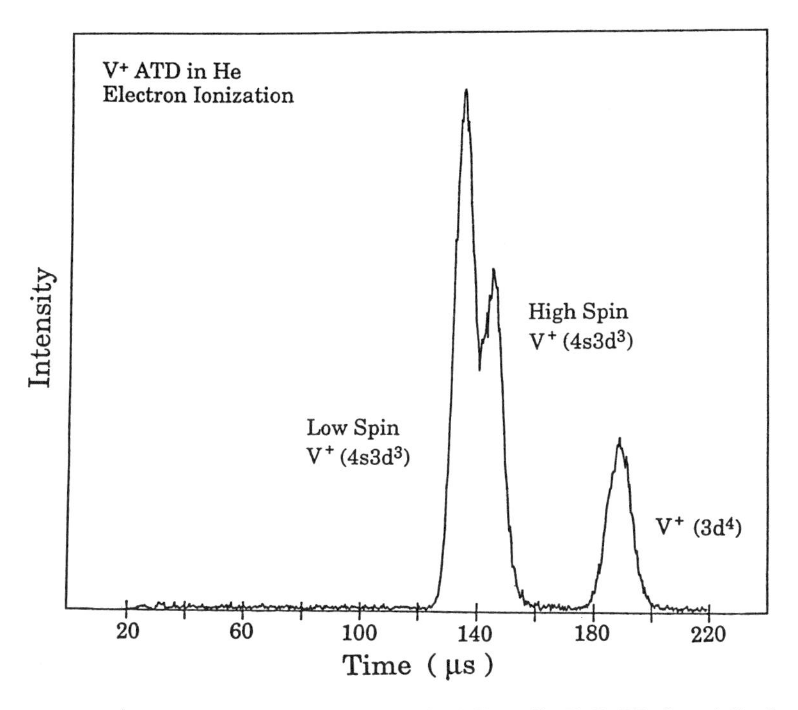

Fig. 3. ATD of V^+ formed by electron ionization of $VOCl_3$ at 40 eV. In EI a broad distribution of electronic states can be formed. While each of the three resolved features in the figure can be assigned to specific electron configurations, each configuration could have more than one electronic state present. The populations of principal electronic states of V^+ formed by EI have been published [35].

correspond to the $3d^n$ and $4s3d^{n-1}$ (high and low spin) states. Note the very different intensity pattern from the SI results of Figure 2. In order to quantitatively determine the populations of the major states present in V^+ formed by EI, the chromatography results must be coupled with previous work [14, 28]. The populations of principal electronic states of first-row transition metal ions formed by electron ionization have been published [35].

3.2. Excited state deactivation

Using the IC technique, we can obtain information on the collisional deactivation of electronically excited states to the ground state. In recent studies of first row transition metal ions (Sc^+ through Zn^+), collisional deactivation was observed only for Sc^+, Ti^+, Mn^+, and Fe^+ in He [35]. In each of these cases an excited state of $3d^n$ configuration was observed to deactivate to a $4s3d^{n-1}$ ground state. Only in the Sc^+ case did excited $4s3d^{n-1}$ states collisionally deactivate in He at thermal energy. Loh *et al.* [38] first proposed a mechanism to explain this effect. The ground state $4s3d^{n-1}$ potential energy

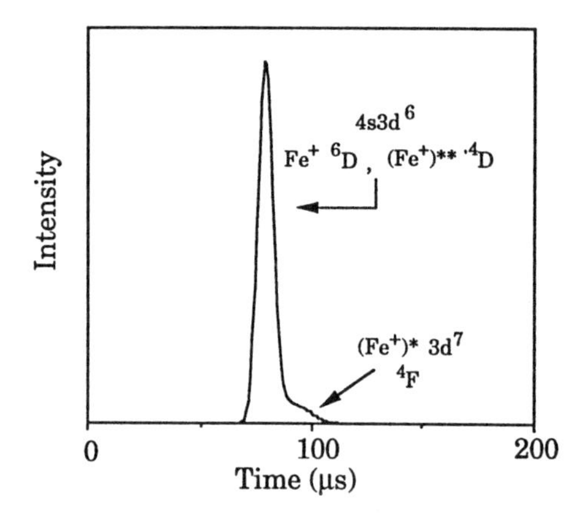

Fig. 4. ATD of Fe^+ formed by electron ionization of $Fe(CO)_5$ at 50 eV. The ^4D excited and ^6D ground states have the same $4s3d^6$ electronic configuration with identical arrival time distributions. The ^4F state, however, has a $3d^7$ electronic configuration and an ATD distinct from the ^6D and ^4D states. Collisional deactivation of the ^4F state to the ^6D state gives rise to the tail at long times in the ATD. Without deactivation two baseline resolved peaks would have been observed in the ATD.

curve will become repulsive at larger internuclear distance than the $3d^n$ excited configuration because of the repulsion between the 4s electron and the filled 1s shell of He. Thus, the two potential energy curves cross at long range, providing a means for deactivation in the collision. Such a crossing does not occur for ions with $3d^n$ ground states and $4s3d^{n-1}$ excited states.

Collisional deactivation is clearly observed in the arrival time distribution for Fe^+ shown in Figure 4. Unlike the ATD for V^+, where baseline resolved peaks corresponding to ground and excited states of V^+ are observed, the Fe^+ ATD consists of a single peak with a slight tail. The main peak is composed of the ^6D ($4s3d^6$) ground state and the ^4D ($4s3d^6$) second excited state, while the tail contains the ^4F ($3d^7$) first excited state. The ^4F state would have been base-line resolved from the ^6D, ^4D states in the absence of collisional deactivation (see Figure 2 for V^+). However, collisional deactivation of the ^4F to the ^6D state does occur giving rise to Fe^+ arrival times intermediate between that of the ^4F and ^6D states. The fraction of the $4s3d^6$ and $3d^7$ electronic state configurations as well as the rate of deactivation of the ^4F state to the ^6D state are accurately and uniquely determined [8] by modeling the experimental ATDs with the known transport properties of the ground and excited states of Fe^+. In this instance the deactivation rate constant in He is 9.5×10^{-13} cm^3/sec.

Deactivation of Co^+ and Ni^+ $4s3d^{n-1}$ excited states does occur in Ne. The rate is slow ($k_d \sim 10^{-16}$ cm^3/sec) [53b]. Deactivation in Ar has not been measured, but should be somewhat faster. It seems likely, however, that many collisions ($\gg$1000) with Ar will be necessary to deactivate these isolated excited states. Deactivation in diatomic and polyatomic bath gases is considerably faster. Nonetheless, complete deactivation of the Co^+ ^5F and ^3F ($4s3d^7$) excited states required more than 20,000 collisions with H_2 in our equilibrium experiment [32].

Both the ^3D ground and ^1D first excited state of Sc^+ have a 4s3d configuration, yet the ^1D state is efficiently deactivated to the ^3D ground state. In this instance it appears both of these states cross a higher lying $3d^2$ state (probably the ^3F state, see Table I) allowing deactivation to occur. The mechanism for deactivation in this interesting system is not yet well understood, however.

4. State selected rate constant measurements

4.1. Varying the electronic state population to obtain rate constants

The electronic state population of transition metal ions can be changed in a completely known and simple way by varying formation conditions. For example, under EI conditions, the electronic state population is a strong function of both the ionizing electron energy and the precursor used. SI is a relatively mild ionization technique producing states limited to low energies and varying only slightly over the range of temperature of the filament (1800–2500 K). Combining the EI and SI methods, the electronic state population can span a very broad range (in the case of Co^+, from 97% ground state to 36% ground state, see Figure 5). Consequently, this method provides an attractive alternative to traditional experimental methods for exploration of electronic state specific reactivity of transition metal ions with neutral molecules. The possibility exists of forming highly excited states (with EI) which are not resolved in the IC spectra. However, no long lived excited states above the b^3F ($4s3d^7$) second excited state of Co^+ are observed under multi-collision conditions. The a^1D and a^3P ($3d^8$) states of Co^+, which lie only 0.147 and 0.357 eV above the b^3F ($4s3d^7$) state respectively, should rapidly collisionally deactivate to the b^3F state. Because deactivation is not observed in the ATD for Co^+, the a^3P and higher lying excited states are assumed not to be present ($\leq 2\%$) under the high pressure conditions of our experiment. (Deactivation of the ^3P state would appear as reverse tailing from the slow to fast peak in the ATD.)

In the past two years we have applied the IC method to a number of systems in order to study state-specific transition metal ion chemistry [8, 9]. Here we will use the reactions of Co^+ ions with methyl iodide as an example of the method [9]. Previous studies [58–61] of atomic metal ions with alkyl halides have been reported but the effects of electronic excited states on the mechanism and energetics of Co^+ reacting with CH_3I were not completely understood. Thus, by measuring the rate constants as a function of tempera-

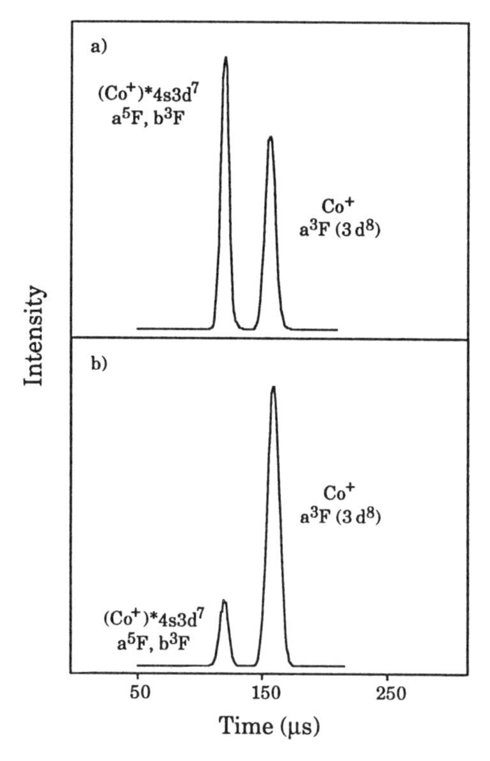

Fig. 5. ATDs of Co^+ ions chromatographically separated by the IC cell. The ions are formed (a) by electron ionization of $Co(CO)_3NO$ at 75 eV, which yields 48% ground state and 52% excited state, and (b) by electron ionization of $CoCp(CO)_2$ at 50 eV, which forms 83% ground state and 17% excited state.

ture for the a^3F, a^5F and b^3F states of Co^+ reacting with CH_3I, and by modeling the reaction efficiencies for the elimination channels on the ground state surface using statistical phase space theory, insight into the mechanism and energetics of these reactions was obtained.

The Co^+ ion has an a^3F ($3d^8$) ground state and two relatively low-lying metastable excited states: a^5F ($4s3d^7$, +0.43 eV) and b^3F ($4s3d^7$, +1.20 eV) [30]. The short time peak in Figure 5 corresponds to the ^5F and ^3F excited states [8], and it is apparent they are not resolved (separation of the high and low spin $4s3d^{n-1}$ states is best for the early transition metal ions and decreases with ion size). However, using a number of techniques, including surface ionization which formed only the lower energy ^5F state, it was possible to show their reactivity with methyl iodide is similar.

The three observed reaction channels for Co^+ reacting with methyl iodide, are shown in the following reaction scheme.

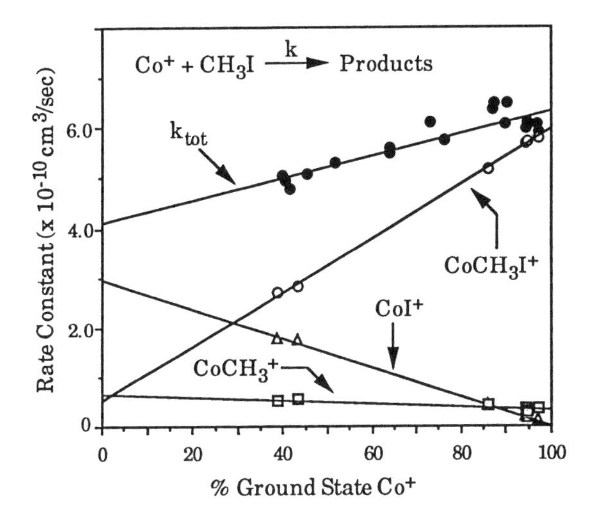

Fig. 6. The total and differential rate coefficients for Co^+ reacting with CH_3I, as a function of % ground state Co^+ (300 K, 1.75 Torr of He, 10^{-5} Torr of CH_3I). The linear least squares fit to the experimental data points is used to extrapolate the rates of reaction corresponding to 100% ground state and 100% excited state Co^+.

$$Co^+ + CH_3I \underset{k_b}{\overset{k_f}{\rightleftarrows}} (CoCH_3I^+)^* \xrightarrow{k_s(He)} CoCH_3I^+$$

$$(CoCH_3I^+)^* \xrightarrow{k'} CoCH_3^+ + I$$

$$(CoCH_3I^+)^* \xrightarrow{k'} CoI^+ + CH_3$$

By varying the electronic state composition of the reactant Co^+ ions, it was possible to determine state specific rate constants for all three product channels. The results are given in Figure 6. Under our experimental conditions (10^{-5} Torr of CH_3I in 1.75 Torr of He), adduct formation is the dominant product for the a^3F ($3d^8$) ground state of Co^+, with only small amounts of elimination products observed. A strong positive temperature dependence was observed for both elimination channels and a negative temperature dependence was observed for adduct formation.

The a^5F and b^3F, $4s3d^7$ excited states show greatly reduced clustering and enhanced elimination. This result is consistent with the expectation that the 4s electron will rather strongly destabilize the $CoCH_3I^+$ electrostatic complex. The excited state reactions also differ from those of the ground state in that a negative temperature dependence was observed for both elimination channels and that the $CoI^+ + CH_3$ channel is enhanced over the $CoCH_3^+ + I$ channel.

On the ground state surface, Co^+ and CH_3I react to form the $Co^+ \cdots ICH_3$ electrostatic complex. Oxidative addition of the cobalt ion to the methyl iodide bond yields the $I—Co^+—CH_3$ inserted intermediate [62]

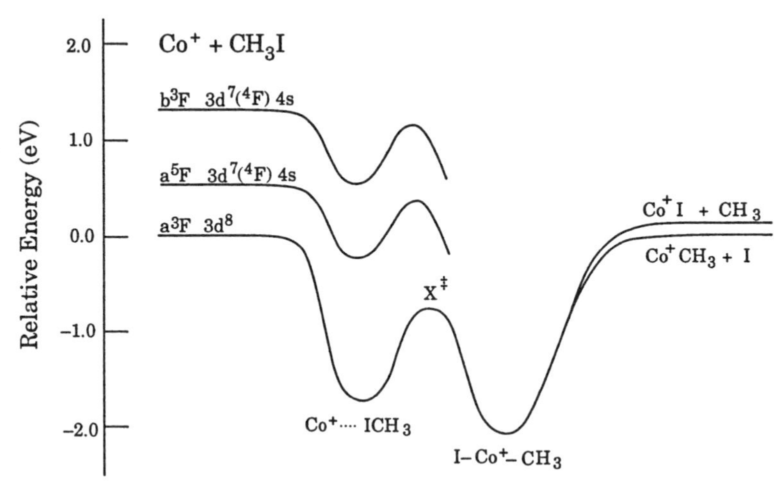

Fig. 7. Schematic reaction coordinate diagram for the a^3F, a^5F and b^3F states of Co^+ reacting with CH_3I. The symbol $X^‡$ respresents the insertion transition state.

which can dissociate to form the $CoI^+ + CH_3$ or $CoCH_3^+ + I$ products if sufficient energy is available. The transition state for I—C bond activation, denoted as $X^‡$ in Figure 7, must be located well below the reactant and product energies for the following reasons. The transition state for C—H bond activation for Co^+ reacting with propane was determined to be located 0.11 eV below the asymptotic energy of the reactants [63]. Because methyl iodide has a permanent dipole moment, it is probably bound more strongly to Co^+ than propane which does not have a permanent dipole moment. In addition, the I—C bond strength is substantially less than the C—H bond strength. Hence, with a deeper electrostatic well and a lower bond activation energy, the oxidative addition transition state, $X^‡$, for the $Co^+ + CH_3I$ system must be well below that for the $Co^+ + C_3H_8$ system and therefore well below the reactant and product energies as shown in Figure 7. The rate limiting transition states for both elimination channels are thus the orbiting transition states in the exit channels. This deduction is consistent with the positive temperature dependence observed for these reactions.

If the I—C bond activation transition state were rate limiting for ground state Co^+ reacting with CH_3I, a negative temperature dependence would be observed for the elimination channels because the density of states at the orbiting transition state (in the entrance channel) would increase much more quickly with temperature than the density of states at the tight transition state (which ultimately leads to products). In this case, competition between the tight and orbiting transition states would increasingly favor dissociation back to reactants with increasing temperature in contrast to what is observed experimentally.

Both methyl radical and iodine atom elimination channels were determined to be endothermic on the ground state surface by modeling the efficiencies of these reactions using statistical phase space theory (assuming the rate limiting transition state is the orbiting transition state in the exit channel). This result is consistent with the strong positive temperature dependence observed experimentally for these reactions. The bond energies determined for Co^+—CH_3 and Co^+—I were $D_0^o = 53.3 \pm 2$ and 50.6 ± 2 kcal/mole respectively, in good agreement with literature values for the Co^+—CH_3 bond energy [19, 59]. No other values for the Co^+—I bond energy have been reported.

For the a^5F, b^3F ($4s3d^7$) excited states of Co^+ reacting with CH_3I, the observed negative temperature dependence of the rate constants for both elimination channels implies that the C—I bond activation transition state is rate limiting and that the reactions must be exothermic. The greater the exothermicity, the greater the stabilization of the C—I bond activation transition state and hence the greater the efficiencies for the elimination reactions. The enhancement of the $CoI^+ + CH_3$ channel over the $CoCH_3 + I$ channel for both the a^5F and b^3F excited states of Co^+ was reproduced in the phase space calculations. It is due to the greater rotational densities of states of the $CoI^+ + CH_3$ products.

The a^3F ($3d^8$) ground and the b^3F ($3d^7(^4F)4s$) second excited state of Co^+ can conserve spin to form a triplet I—Co^+—CH_3 intermediate. However, with the 4s orbital occupied in the b^3F state of Co^+, one electron is forced into an antibonding orbital of the inserted intermediate, as shown for sigma bond activation of H_2 in Figure 8. This results in destabilization of the complex formed [7]. Reaction could still occur, however, by mixing of the diabatic triplet surfaces (correlating to the b^3F and a^3F states) at the I—C bond activation transition state to produce a triplet I—Co^+—CH_3 intermediate on the ground state surface. For the a^5F ($3d^7(^4F)4s$) excited state of Co^+ mixing of the diabatic quintet and triplet surfaces (correlating to the a^5F and a^3F states), could also produce a triplet I—Co^+—CH_3 intermediate on the ground state surface.

4.2. Product ion arrival time distributions

Reaction rate constants can also be obtained directly from the product ion ATDs. In these experiments, a pulse of reactant ions is admitted into the cell as above, however the quadrupole mass filter following the cell is set to transmit the product ion of interest rather than the reactant ion. The resulting product ATD shows components due to *both* reactant ion configurations (if they are both present and reactive). When the reactant ion electronic state distribution is known, their contributions to the product ion ATDs gives an indication of the reactivity of the different states. Quantitative reaction rate constants can be obtained by theoretically modeling the experimental product ion ATDs for both the ground and excited state reactions from the known transport properties of these ions in He.

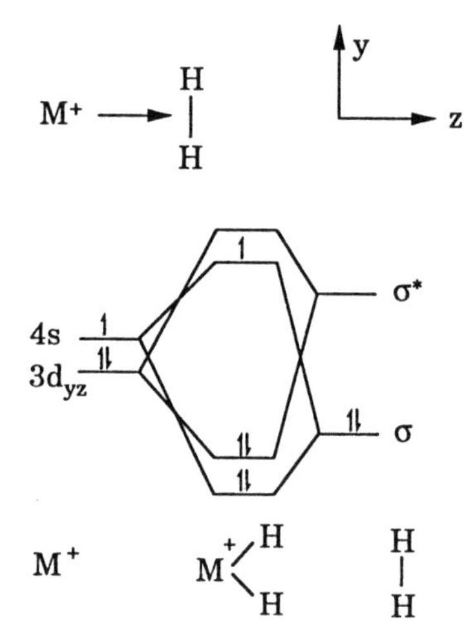

Fig. 8. Simple molecular orbital diagram for H_2 approaching M^+ in C_{2v} symmetry (along the z axis in the yz plane). The orbital occupancy of the metal ion shown indicates sigma bond activation is less favorable for 4s occupied states because it forces one electron into an anti-bonding orbital of the H—M^+—H complex. However, sd hybridization reduces the anti-bonding character of this orbital and stabilizes the inserted molecule.

The product ATD for $CoCH_3^+$ formed in the reaction of ground and excited state Co^+ reacting with CH_3I, is shown in Figure 9a and will be used as an example. The observed peak is broad and consists of two components. The onsets of these components correlate exactly to the ATDs of ground state Co^+ and excited state $(Co^+)^*$ as indicated by the dashed arrows (see Figure 9b). We interpret this result as follows. The arrival time for a particular $CoCH_3^+$ product ion is a function of the position at which it was formed in the cell. If Co^+ reacts with CH_3I at the exit end of the cell, the arrival time of the product $CoCH_3^+$ will correspond to the arrival time of Co^+. The longest $CoCH_3^+$ arrival times correspond to the formation of $CoCH_3^+$ at the entrance of the cell ($CoCH_3^+$ is larger and drifts more slowly through the helium in the cell than Co^+). The longest arrival time for the $CoCH_3^+$ product is the same for both excited and ground state Co^+ because the decrease in mobility due to the CH_3 ligand is much greater than the difference in mobility due to the different electronic state configurations. In other words, $CoCH_3^+$ products formed in the reaction of ground and excited states of Co^+ with CH_3I have about the same mobility. Hence the excited state $(Co^+)^*$ contribution to the $CoCH_3^+$ ATD goes all the way across the distribution as shown by the shaded area in Figure 9a [8]. Ions which react at intermediate positions in the cell have intermediate arrival times giving rise to the broad peak we observe.

The observed shapes of the product ATDs are complicated. However, the ATDs for both the ground and excited states reacting to form the

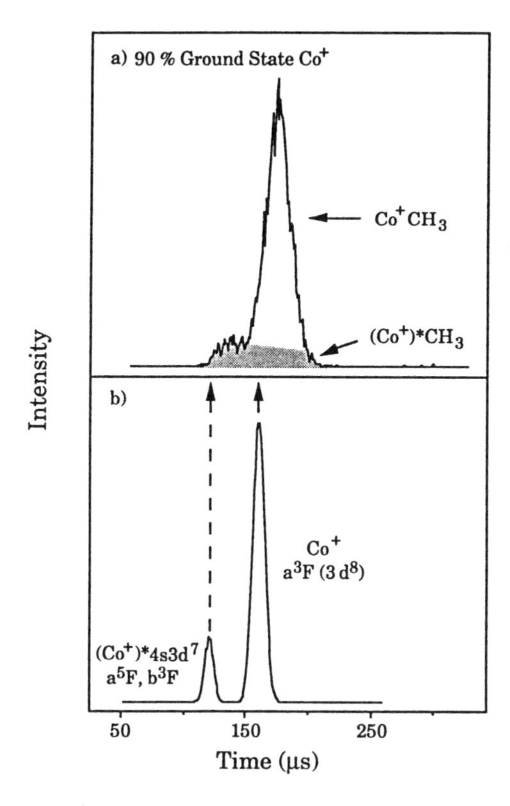

Fig. 9. (a) ATD for $CoCH_3^+$ product ions formed in the reaction of ground and excited states of Co^+ with CH_3I. The data shown is for a Co^+ ion beam containing 90% ground state Co^+ and 10% excited state $(Co^+)^*$. This beam is injected into the IC cell where it reacts with CH_3I to form $CoCH_3^+$. The arrival time of the $CoCH_3^+$ product ion is a function of the position at which the reaction took place in the cell. The contribution due to excited state reaction corresponds to the shaded area and is denoted as $(Co^+)^*CH_3$ (this notation does not imply excited state products are formed). (b) The ATD of unreacted Co^+ showing the presence of both ground and excited state Co^+ Note the onset of the features in the upper panel exactly correspond to the arrival times of ground and excited state Co^+ in the lower panel allowing an unambiguous assignment to be made.

$CoCH_3^+$ can be calculated from known transport properties. The resulting peaks are expected to be broad and to decrease in intensity at long times. Based on this expectation, as well as experimentally observed peak shapes where a single state dominates the reaction, we approximate the peak shape for excited state $(Co^+)^*$ reacting to form the $CoCH_3^+$ product ion by the shaded area in Figure 9a (denoted as $(Co^+)^*CH_3$ in Figure 9). Because in this experiment the initial Co^+ beam consisted of 90% ground state Co^+ and 10% excited state $(Co^+)^*$, it appears from the relative areas in Figure 9a, that ground state Co^+ forms the $CoCH_3^+$ product ion about as efficiently as the excited state does at 300 K. These qualitative observations are in very good agreement with the results shown in Figure 6, where the rate constant for the $CoCH_3^+$ channel is essentially independent of the electronic state population. Quantitative reaction rate constants will be obtained by fitting the product ion ATDs for both the ground and excited state reactions in the near future.

5. Ligand binding energies

5.1. Overview

A further application of the IC method is measurement of equilibrium constants to obtain ligand binding energies. We have reported bond dissociation energies and entropies for Co^+ and V^+ with up to seven H_2 ligands [31–33] as well as for $Sc^+(H_2)_{1-4}$ [34], $Na^+(H_2)_{1,2}$ and $K^+(H_2)_{1,2}$ [64]. High level electronic structure calculations have been done on $Na^+(H_2)_{1,2}$ [65–68], $Sc^+(H_2)$ [26], $V^+(H_2)_n$ [69] and $Co^+(H_2)_n$ [21, 65, 69] that give insight into the bonding. For the closed shell alkali metal ions, the charge quadrupole and charge induced-dipole electrostatic interactions dominate their bonding. In the case of transition metal ions, donation of charge from the ligand to the metal ion and back donation from the metal ion to the ligand results in a much stronger interaction. Thus, a comparison of the bonding of H_2 with transition metal ions and alkali metal ions gives insight into the electrostatic and covalent or dative contributions to the bonding.

There are several factors which influence the binding energy of transition metal ions with H_2. These include the loss of d-d exchange, the $3d^n$ to $4s3d^{n-1}$ promotion energy, the cost of hybridization, repulsion between the metal ion and the ligand, extent of electron donation and back donation and finally mixing of low lying electronic state surfaces. A comparison of the bonding of H_2 with Co^+, V^+ and Sc^+ will be made giving insight into specific details regarding the nature of d-orbital participation in the bonding. First, a description of the experiment is presented.

5.2. The equilibrium experiment

Results are obtained by measuring equilibrium constants as a function of temperature for the various clustering reactions:

$$M^+(H_2)_{n-1} + H_2 \leftrightarrows M^+(H_2)_n \tag{8}$$

Bond dissociation energies (BDE) are then extracted from statistical mechanical fitting of the data.

After the temperature and H_2 pressure are observed to be stable within the reaction cell, product/parent ion ratios are measured as a function of reaction time. This time is varied by changing the drift voltage across the cell. As the drift time is increased, the product/parent ion ratios eventually become constant, indicating that equilibrium has been reached. With the smaller clusters ($M^+(H_2)_{1,2}$) this usually occurred at the shortest accessible drift times (about 150 μs). The larger clusters required more time and in all cases the reactions were probed out to 3.5 ms to ensure equilibrium had been established and to insure that the drift-field is not significantly perturbing the ion thermal kinetic energy. For selected experiments, the pressure of H_2 was varied by a factor of two with no significant change observed in ΔG^0_T.

The ion ratios are converted to equilibrium constants using equation (9),

$$K_p^0 = \frac{M^+(H_2)_n \ 760}{M^+(H_2)_{n-1} P_{H_2}} \tag{9}$$

where P_{H_2} is the hydrogen pressure in Torr and $M^+(H_2)_n$ and $M^+(H_2)_{n-1}$ are the measured intensities of the cluster ions of interest. The standard free energy change is calculated using equation (9),

$$\Delta G_T^0 = -RT \ln K_p^0 \tag{10}$$

where R is the gas constant, and T is the temperature. Free energies are measured at a variety of temperatures until a satisfactory range of data has been collected.

A number of potential sources of error are present in these experiments, including pressure and temperature inaccuracies; mass discrimination and resolution; and the presence of electronically excited metal ions. The effect of these factors on our dissociation energies and entropies are discussed in detail elsewhere [32, 53b]. The net result is that the bond dissociation energies are essentially unaffected by these uncertainties. The entropies are affected to a small extent by any mass discrimination in the quadrupole mass analyzer, but in these experiments such mass discrimination should be very small because the parent/product mass difference is only two AMU. In essentially all cases, we are able to measure the absolute association entropy to within 2 cal/mol K.

Two separate effects lead to non-linear ΔG^0 vs. T plots: lack of mass resolution (leading to overlapping of peaks) and the presence of M^+ excited electronic states. Care is taken to baseline resolve the different product masses and to avoid overlap of adjacent peaks. Further, excited electronic states with $4s3d^{n-1}$ configurations are much more difficult to deactivate than $3d^n$ excited states, as discussed earlier. The presence of long lived $4s3d^7$ excited states created a problem in the $Co^+ + H_2$ experiments [32], making it difficult to accurately measure K_p for addition of the first H_2 ligand.

Vanadium also has a d^n ground state. The effect of the excited ^5F and ^3F ($4s3d^3$) electronic states was not a problem in the V^+ experiments [33], however, because these excited V^+ states are effectively quenched in collisions with H_2. This quenching was demonstrated in two ways. First, equilibrium was attained rapidly in the $V^+ + H_2 \leftrightarrows V^+(H_2)$ reaction, showing that the residual amounts of excited V^+ ^5F ($4s3d^3$) and ^3F ($4s3d^3$) were far smaller than the true equilibrium amount of V^+ ^5D ($3d^4$) ground state after about 10^4 collisions with H_2. This result implies that the quenching rate constant for excited $4s3d^3$ electronic states is $k > 5 \times 10^{-14}$ cm^3 s^{-1}. In contrast, the Co^+ ions required roughly ten times more collisions for equilibrium to be established [32]. The second check on the deactivation of excited V^+ was made by collecting the arrival time distributions of V^+ in pure He, and in He with small amounts of D_2 added. The ATD in He of V^+ formed by SI resolves the three lowest V^+ electronic states in the ratio expected for SI (Figure 2). As small amounts of H_2 (or D_2) are added to the He bath gas, the peaks corresponding to the $4s3d^3$ and $3d^4$ states are no longer base line resolved because deactivation of the excited states begins to take place (Figure 10a). This process continues with increasing D_2 pressure until complete coalescence of the two peaks has occurred, producing 100% a ^5D ($3d^4$) ground state V^+. In Figure 10b it is shown that the arrival time distribution of $V(D_2)^+$ can essentially be superimposed on the ATD for ground state V^+. We therefore conclude that excited electronic states of V^+ are rapidly quenched and that we are observing clustering of ground state V^+.

For Sc^+, rapid deactivation of both the ^3F ($3d^2$) and the ^3D (4s3d) excited states to the ^3D (4s3d) ground state is observed in H_2 as well as in He. We are therefore observing clustering of ground state Sc^+. As we will see, however, Sc^+ has a different, but interesting problem that makes measurement of the first equilibrium constant challenging.

5.3. Results and discussion

The binding energies derived for the H_2 clustering reactions for V^+, Co^+, Sc^+, Na^+ and K^+ are summarized in Table II. For Na^+ and K^+, the relatively small bond energies for both the first and second H_2 ligands reflects the rather weak electrostatic interactions which are responsible for the bonding in these closed shell alkali metal ions. The $K^+(H_2)_{1,2}$ binding energies are smaller than Na^+ $(H_2)_{1,2}$ because of the larger ionic radius of K^+ relative to Na^+. Several theoretical investigations have been done on the $Na^+(H_2)$ system [64–68]. The best theoretical value for the binding energy, $D_e(Na^+ - H_2)$, is 3.1 kcal/mol, in good agreement with the experimentally derived D_e of 3.45 ± 0.3 kcal/mol (including both experimental and zero point uncertainties) [64]. All the theoretical studies indicate that the H_2 approaches side on (to maximize the charge-quadrupole interaction) and the Na^+ and the H_2 ligands are largely unperturbed resulting in the fairly weak interactions observed and calculated.

For V^+ and Co^+, the derived binding energies with H_2 are 10.2 ± 0.5 and 18.2 ± 1.0 kcal/mol [31–33], respectively, which are substantially larger than those for Na^+ and K^+ (2.45 and 1.45 kcal/mole, respectively) [64]. Unlike the $K^+(H_2)$ and $Na^+(H_2)$ systems, metal ligand charge transfer and deshielding on the metal (i.e., ligand induced sd hybridization) occurs in the $V^+(H_2)$ and $Co^+(H_2)$ clusters [21, 69, 70]. The hybridization results in electron

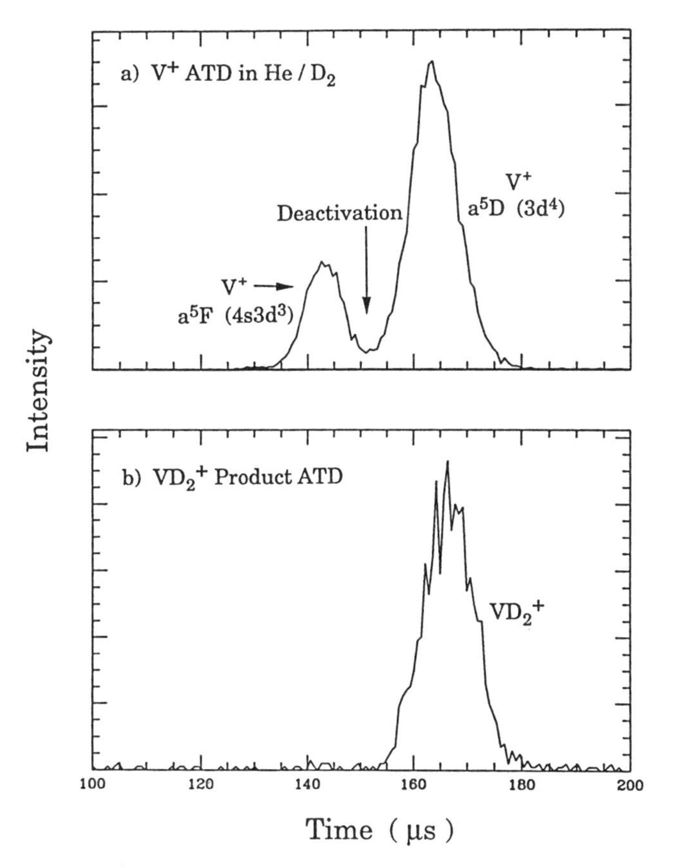

Fig. 10. (a) ATD of V^+ ions chromatographically separated by the IC cell containing 3 Torr of He and 6×10^{-4} Torr D_2. The V^+ ions are formed by SI of $VOCl_3$. Note the filling in between the peaks, indicating that deactivation has begun. Note also the fraction of the a^5F ($4s3d^3$) excited state has decreased from that shown in the ATD of V^+ in pure He (see Figure 2). This deactivation process continues with increasing D_2 partial pressure until the peaks completely coalesce. (b) ATD of the VD_2^+ product ion under the same conditions as panel (a). Note that the ATDs of the VD_2^+ peak can be nearly superimposed on the V^+ a^5d ($3d^4$) ground state peak, indicating that ground state V^+ is the exclusive source of VD_2^+.

density directed away from the internuclear axis, reducing Pauli repulsion and allowing a closer approach of the H_2 ligand to the metal center. The resulting bond length is shorter than expected for an unhybridized metal ion, increasing the bond strength due to a larger electrostatic interaction. This increased bond strength more than offsets the energy necessary to hybridize the dσ orbital.

The binding energies for molecular H_2 to V^+, $V^+(H_2)_n$ (n = 1–7), show a pairwise behavior similar to that observed for $Co^+(H_2)_n$ clusters, except in two ways. One, the first and second H_2 ligands are much more weakly

Table II. Binding energies ($-\Delta H_0^o$ in kcal/mol) for: $M^+ (H_2)_{n-1} + H_2 \rightleftarrows M^+ (H_2)_n$

n	Na^{+a}	K^{+a}	Sc^{+b}	V^{+c}	Co^{+d}
1	2.45 ± 0.2	1.45 ± 0.2	5.5 ± 0.3	10.2 ± 0.5	18.2 ± 1.0
2	2.25 ± 0.2	1.35 ± 0.4	6.9 ± 0.3	10.7 ± 0.5	17.0 ± 0.7
3			5.4 ± 0.3	8.8 ± 0.4	9.6 ± 0.5
4			5.0 ± 0.4	9.0 ± 0.4	9.6 ± 0.6
5				4.2 ± 0.5	4.3 ± 0.7
6				9.6 ± 0.5	4.0 ± 0.7
7				< 2.5	0.8 ± 0.5

[a]See Ref 64. [b]For Sc^+—H_2, Sc^+ inserts into the H—H bond with an average Sc^+—H bond energy of 54.4 kcal/mol (Ref. 34). [c]See Ref 33. [d]See Ref. 32.

bound (10.2 and 10.7 kcal/mol for V^+ versus 17.0 and 18.2 kcal/mol for the Co^+ clusters). Because the radial extent [71] of the ^5D ($3d^4$) ground state of V^+ is greater than the ^3F ($3d^8$) ground state of Co^+, the electrostatic interaction for Co^+ and H_2 will be greater than for V^+ and H_2. In addition, calculations indicate Co^+ back-donates charge to the σ^* orbital of H_2 from its filled $d\pi$ orbitals whereas V^+, with its half filled $d\pi$ orbitals, is unable to back-donate charge because of the associated loss in d-d exchange energy [69].

The second difference in the Co^+ and V^+ systems is that the sixth H_2 ligand, in the V^+ system, has a much larger bond dissociation energy and much more negative ΔS of association than expected. As originally suggested by Armentrout [72], these changes are almost certainly due to a transition from a quintet to a triplet surface. Calculations by Maitre and Bauschlicher [73] support this conclusion and show that the quintet to triplet promotion energy on the metal is overcome by the two-electron donation from V^+ into the antibonding orbitals of four H_2 ligands in the $V(H_2)_6^+$ cluster.

The Sc^+/H_2 system is somewhat different. Tolbert and Beauchamp [74] suggested that the H—Sc^+—H inserted molecule might be formed via the exothermic reaction (11),

$$Sc^+ + CH_2O \rightarrow Sc^+ \begin{matrix} \diagup H \\ \diagdown H \end{matrix} + CO \qquad (11)$$

although they had no proof of the structure of the product. Subsequent theoretical papers [24, 26] indicated the H—Sc^+—H is stable but formation of the inserted intermediate from ground state Sc^+ ^3D (3d4s) reacting with H_2 required surface crossings with excited states and predicted a substantial energy barrier to formation. These authors considered the electrostatic Sc^+—H_2 complex but did not find a minimum on the potential energy surface for this species. Sunderlin et al. [75] measure the Sc^+—H and HSc^+—H bond energies to be 56.3 ± 1.2 and 55.4 ± 2.3 kcal/mol, respectively, which implies that ground state Sc^+ can insert into H_2 exothermically, ΔH_0^o =

-8.8 ± 2.6 kcal/mol. Perhaps the most definitive proof that ground state Sc^+ inserts into H_2 comes from our equilibrium measurements [34]. In contrast to the other systems we studied, the attainment of equilibrium was quite difficult to achieve, requiring 10^5 to 10^6 collisions at 300 K. Taken by itself this result is consistent with either a barrier of 10 kcal/mol, an inefficient surface crossing or both. It is definitely at odds with simple formation of an electrostatically bound complex between Sc^+ and H_2. Analysis of the rate data for approach to equilibrium indicates insertion of Sc^+ into H_2 is very slow ($k = 3\text{–}13 \times 10^{-17}$ cm^3/sec) [34]. Further, this rate constant has a negative temperature dependence which is incompatible with a simple insertion energy barrier [34].

To understand the mechanism by which ground state Sc^+ inserts into the H—H bond we must consider the interaction of the ^3D (4s3d), ^1D (4s3d), ^3F ($3d^2$) and ^1D ($3d^2$) states of Sc^+ with H_2. Ab initio calculations [24, 26] indicate that the H_2 approaches Sc^+ in C_{2v} symmetry (along the z axis in the yz plane, as shown in Figure 8) and the term components of the lowest Sc^+ electronic states mix. Below 2.5 eV there are seven electronic states of Sc^+ giving rise to 65 components all of which could potentially contribute to the reaction with H_2. However, symmetry and spin considerations severely restricts the number of components which correlate to the H—Sc^+—H molecule. The two hydrogen atoms are bound to two equivalent sd hybrid orbitals on Sc^+ and H—Sc^+—H has a $^1A_1$ ($a_1^2b_2^2$) ground state configuration. The σ bonding orbital of H_2 and the 4s orbital of Sc^+ interact to form the a_1 bonding and antibonding molecular orbitals. The σ^* antibonding orbital of H_2 must interact with a d orbital of Sc^+ with the same symmetry to form the b_2 bonding and antibonding molecular orbitals. These symmetry requirements are met only by the doubly occupied $3d_{yz}$ component of the ^1D ($3d^2$) state of Sc^+. Consequently, in order for Sc^+ to insert into the H_2 bond without an activation energy barrier, the 4s orbital on Sc^+ must be unoccupied, to accept electron density from the H_2 σ orbital, and the d orbital on Sc^+ must be doubly occupied and of the right symmetry, to donate into the σ^* orbital of H_2.

A simplified schematic reaction coordinate diagram for Sc^+ reacting with H_2 is shown in Figure 11. The correlation from the ^1D ($3d^2$) state of Sc^+ to H—Sc^+—H is shown. The high spin ^3D (4s3d) and ^3F ($3d^2$) states and the low spin ^1D (4s3d) state of Sc^+ produce repulsive potentials with H_2 (following the electrostatically bound complex). There is a strongly avoided crossing between the singlet surfaces evolving from the ^1D (4s3d) and ^1D ($3d^2$) states of Sc^+ and a spin-orbit coupled crossing between the triplet and singlet surfaces which may allow the ^3D (4s3d) ground state of Sc^+ to react adiabatically to form H—Sc^+—H. However, this may not be feasible energetically. Theory predicts an insertion barrier of approximately 19.0 kcal/mol [26]. An alternative reaction mechanism, involving the uninserted $Sc(H_2)^+_{1-3}$ clusters has been proposed [34]. Due to the relatively strong electrostatic interaction between the ^3F ($3d^2$) state of Sc^+ and successive H_2

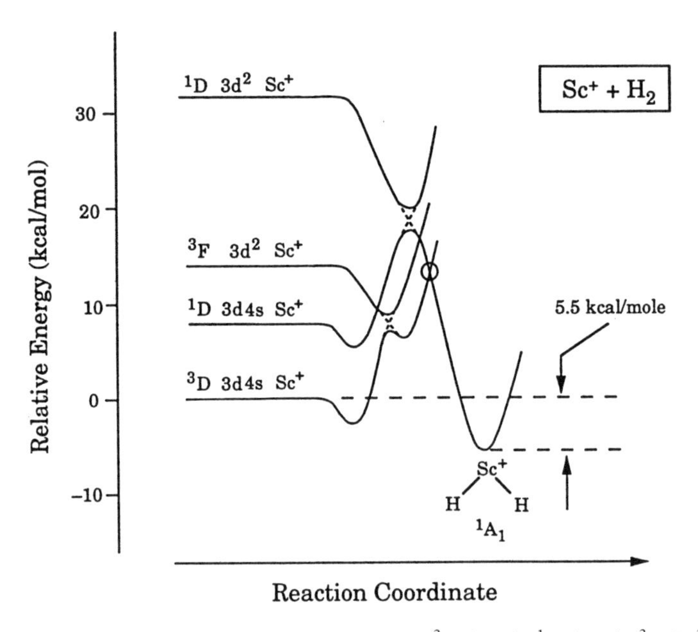

Fig. 11. Schematic reaction coordinate diagrams for the ^{3}D (3d4s), ^{1}D (3d4s), ^{3}F ($3d^2$) and ^{1}D ($3d^2$) states of Sc^+ reacting with H_2. The ^{1}D ($3d^2$) state of Sc^+ diabatically correlates to the inserted $^1A_1$ H—Sc^+—H molecule. The strongly avoided crossings between the singlet surfaces (^{1}D ($3d^2$) and ^{1}D (3d4s)) and between the triplet surfaces (^{3}D and ^{3}F)) are indicated by dashed lines and the weakly avoided crossing due to spin orbit coupling between the triplet ground and singlet excited state surfaces is circled.

ligands, the energy of the weakly avoided crossing between the triplet and singlet surface is significantly decreased [34]. The addition of the third H_2 ligand appears to provide sufficient stabilization energy to bring the insertion barrier below the asymptotic energy of the reactants (the ^{3}D (4s3d) ground state of $Sc^+ + H_2$). The proposed "cluster mediated" mechanism for sigma bond activation of H_2 by Sc^+ is consistent with all the data, including the negative temperature dependence of the rate constant for insertion.

The equilibrium experiments [34] indicate that the reaction of ground state Sc^+ with H_2 to form H—Sc^+—H is exothermic by 5.5 ± 0.3 kcal/mol (equation (12)),

$$Sc^+ + H_2 \rightarrow Sc^+\!\!\begin{array}{l} \diagup H \\ \diagdown H \end{array} \qquad \Delta H^o_0 = -5.5 \text{ kcal/mol} \qquad (12)$$

with an average Sc^+—H bond energy of 54.4 kcal/mol. Reaction (12) is not observed under single collision conditions. However, Tolbert and Beauchamp [74] predicted reaction (12) to be exothermic by $\sim 4 \pm 7$ kcal/mol based on the thermochemistry of reaction (11), where ScH_2^+ is observed and as noted earlier Sunderlin et al. [75] find this reaction to be exothermic by 8.8 ± 2.6 kcal/mol. These results are in reasonable agreement with the 5.5 ± 0.3 kcal/mol exothermicity determined in the equilibrium experiment. The theoretically predicted [26] heat of reaction of +1.8 kcal/mol, is also in reasonable agreement with our results considering the complexity of the potential energy surfaces involved.

In the reaction of Co^+ with H_2 the ground state of Co^+ has the right electron configuration to insert into the H—H bond without a barrier. In the H—Sc^+—H molecule, the hydrogen atoms are bound to two equivalent sd hybrid orbitals. However, the cost of sd hybridization is too large for formation of the H—Co^+—H molecule to be favorable energetically [76]. As a result, Co^+ binds to molecular H_2 but does not insert into the H—H bond. The sd hybrid orbitals are formed more easily for early transition metal ions because the difference in size between the 4s and 3d orbitals is smaller for early transition metal ions.

For the V^+ system, the interaction of the ^5D ($3d^4$), ^5F ($4s3d^3$), ^3F ($4s3d^3$) and ^3P ($3d^4$) states with H_2 must be considered [4, 28]. The ^5D ($3d^4$) ground state can cluster with H_2 but cannot diabatically insert into the H_2 bond because the necessary doubly occupied d orbital is absent. The high spin ^5F ($3d^3(^4F)4s$) first excited state and the low spin ^3F ($3d^3(^4F)4s$) second excited state will also be repulsive to insertion into the H_2 bond because the d orbitals are singly occupied in both states (note that the only difference between the ^5F and ^3F states is the spin coupling between the $3d^3(^4F)$ and the 4s electron). Only the ^3P ($3d^4$) excited state of V^+ has the right spin and electron configuration to diabatically correlate to an inserted product. A strongly avoided crossing between the triplet surfaces and spin orbit coupling between the quintet and triplet surfaces could lead to adiabatic reactions for the ^5F ($4s3d^3$) and ^3F ($4s3d^3$) excited states and the ^5D ($3d^4$) ground state of V^+ with H_2 to form H—V^+—H. However, as noted, ground state V^+ only clusters with molecular hydrogen and is not observed to insert into the H—H bond, most likely because the cost of promotion, including the loss in d-d exchange energy, is too great.

6. Sigma bond activation in small alkanes

6.1. Overview

Studies of atomic transition metal ions reacting with simple alkanes indicate that the chemistry is quite diverse. There are some trends, however, including the observed increase in reactivity with increasing alkane size. For first row transition metal ions only Sc^+ and Ti^+ react with CH_4 by inserting into the C—H bond and react with C_2H_6 to eliminate H_2 at thermal energy. Even though H_2 elimination is exothermic for several other first row transition metal ions, including Co^+, Fe^+ and Ni^+ reacting with C_2H_6, we believe these

Table III. Experimental and theoretical M^+-ligand dissociation energies (D_0^o in kcal/mol)

$M^+—L$	Theory[a]		Experiment	
			Equilibrium measurements[b]	Threshold collisional activation[c]
	D_e	D_0^o	D_0^o	D_0^o
$Co^+—H_2$	18 3	17.4 ± 1	18.2 ± 1.0	–
$Co^+—CH_4$	21.4	22.7 ± 2	22.5 ± 0.7	21.4 ± 1.4
$(CH_4)Co^+—CH_4$	23.1	24.4 ± 2	24.8 ± 0.8	23.3 ± 1.2
$Co^+—C_2H_6$	25.0	26.4 ± 2	28.0 ± 1.6	23.4 ± 1.2
$Co^+—C_3H_8$	29.3 ± 0.4	30.8 ± 2	–	30.9 ± 1.4

[a]Ref. 21. [b]Ref 31. [c]Ref. 78.

reactions are not observed because the barrier associated with initial C—H bond activation is above the reactant energy [77]. Recent experimental [31, 78] and theoretical studies [21] indicate Co^+-alkane bond energies increase with ligand size, ranging from 22.9 kcal/mol for $Co^+—CH_4$ to 30.8 kcal/mol for $Co^+—C_3H_8$ as summarized in Table III. C—H bond activation is believed to occur for propane because electrostatic stabilization of the initial $Co(C_3H_8)^+$ cluster is sufficient to bring the insertion barrier below the $Co^+ + C_3H_8$ asymptotic energy.

In their theoretical study of molecular complexes of Co^+ with H_2, CH_4, C_2H_6 and C_3H_8, Perry et al. [21] not only obtained bond energies for these species (Table III) but also gained considerable insight into the nature of the bonding. As discussed earlier for $Co^+—H_2$, insertion into the H_2 bond is endothermic and is therefore not observed. H_2 is bound to Co^+ in C_{2v} symmetry (along the z axis in the yz plane). H_2 donates electron density into the empty 4s orbital of Co^+ and back-bonding involving charge transfer from the doubly occupied d_{yz} orbital of Co^+ into the empty H_2 antibonding orbital is observed.

The binding of ground state Co^+ to CH_4 is superficially similar to $Co^+—H_2$ in that two C—H bonds coordinate to Co^+ in C_{2v} symmetry (along the z axis in the yz plane). However, the $Co^+—H_2$ and $Co^+—CH_4$ bonds are fundamentally different. First, CH_4 donates electron density from both C—H bonds into the empty 4s and the singly occupied $3d_{yz}$ orbitals of Co^+. No back-bonding is observed. Back-bonding is more difficult for CH_4 than for H_2 because the C—H antibonding orbitals are much higher in energy relative to the H—H antibonding orbital. Second, the dσ orbital ($d_{y^2-z^2}$) is doubly occupied. Extensive sd hybridization, involving the nonbonding doubly occupied $d_{y^2-z^2}$ orbital, polarizes this orbital along the y axis to reduce its repulsion to the ligand and consequently polarizes the empty 4s orbital along the z axis increasing its ability to accept charge from the ligand. This sd hybridization also deshields the metal nucleus along the z axis and therefore increases the electrostatic interaction [79].

For ethane and propane, the bonding is similar to that with CH_4 and coordination to C—H bonds is preferred over coordination to C—C bonds. Coordination of four primary C—H bonds of propane to Co^+ was calculated to be somewhat more stable than the coordination of two secondary C—H bonds to Co^+, probably because of a greater charge-induced polarization for the η^4 geometry [21]. However, calculations indicate that the different orientations are remarkably similar in energy indicating that the metal ion can approach various reaction sites. These results are consistent with the experimental observation of both initial primary and secondary C—H bond activation.

For ground state first row transition metal ions reacting with propane, the observed branching ratios for adduct formation, CH_4 and H_2 elimination were found to vary widely. Under multicollision conditions [8, 80, 81], adduct formation dominates for V^+ through Zn^+. CH_4 and H_2 elimination are observed for Fe^+, Co^+ and Ni^+ whereas only dehydrogenation and no demethanation is observed for V^+. For Sc^+ and Ti^+, H_2 elimination dominates over both CH_4 elimination and adduct formation. Exclusive adduct formation is observed for Cr^+, Mn^+, Cu^+ and Zn^+.

The observed reactivity of ground state M^+ with propane can be qualitatively explained by conservation of electron spin for C—H bond activation [4, 6, 7]. For first row transition metal ions, only Co^+, Ni^+ and Cu^+ can conserve spin on the ground state surface to activate C—H bonds. However, the ^1S ($3d^{10}$) ground state of Cu^+ does not react at thermal energy because the filled d shell prevents the formation of two sigma bonds. For Sc^+, Ti^+, V^+ and Fe^+ ions, C—H bond activation is observed because their ground states interact with low lying excited states that can conserve spin in the reaction. For Cr^+ and Mn^+, the excited states which can conserve spin are too high in energy relative to the ground state (see Table I), and hence they do not react at thermal energy. We will discuss the effect of spin and other aspects of sigma bond activation in the following sections.

6.2. Reactions of Fe^+, Co^+ and Ni^+ with propane

For Fe^+, Co^+ and Ni^+ reacting with propane, the total cross sections for H_2 and CH_4 elimination are observed to be small, less than 15% of the collision limit, and to decrease with the extent of deuteration [63, 82]. The inefficiency of the reaction indicates that a rate limiting transition state exists along the reaction coordinate. The decrease in cross section with extent of deuteration suggests that this rate limiting transition state is associated with C—H bond activation. Kinetic energy release distributions for H_2 and CH_4 loss from nascent $M(C_3H_8)^+$ complexes for Fe, Co and Ni were measured. For H_2 loss the distribution is bimodal. Studies with propane-2,2-d_2 and propane-1,1,1,3,3,3-d_6 indicate that both initial primary and initial secondary C—H bond activation are involved which gives rise to the observed bimodality [63]. The kinetic energy release distribution for demethanation is statistical. Modeling the relative reaction cross sections for C_3H_8 and C_3D_8 and the kinetic energy release distribution for demethanation using statistical phase space theory allows for accurate determination of the energy of the C-H bond activation transition state. For Fe^+, Co^+ and Ni^+ reacting with propane the C—H bond activation transition states are located only a few kcal/mol ($\leq$2.5 kcal/mol) below the reactant energy [63, 82].

Using the IC technique, state specific rate constants were measured for ground and low lying excited states of Fe^+, Co^+ and Ni^+ reacting with propane [8, 81]. At 1.75 Torr of He pressure, adduct formation is the dominant product for the a^3F ($3d^8$) ground state of Co^+, with only small amounts of elimination products observed. The Co^+ a^5F ($3d^7(^4F)4s$) and b^3F ($3d^7(^4F)4s$) excited states react with similar rate constants and show greatly reduced clustering (due to the repulsive 4s electron) and enhanced elimination channels.

The a^3F ($3d^8$) ground and the b^3F ($3d^7(^4F)4s$) excited state of Co^+ can conserve spin to form a triplet H—Co^+—C_3H_7) intermediate whereas the a^5F ($3d^7(^4F)4s$) excited state of Co^+ presumably involves mixing of the diabatic quintet and triplet surfaces (correlating to the a^5F and b^3F states), to produce a triplet H—Co^+—C_3H_7 intermediate.

For the ^6D ($4s3d^6$) ground state of Fe^+, adduct formation is dominant, despite the 4s electron, owing to a crossing from the $Fe^+(^6D, 4s3d^6)C_3H_8$ ground-state surface to the low lying $Fe^+(^4F, 3d^7)C_3H_8$ first excited-state surface where the adduct is more strongly bound [83]. This mechanism also allows the ^6D ($4s3d^6$) ground state of Fe^+ to react adiabatically with propane to form the quartet H—Fe^+—C_3H_7 intermediate. The Fe^+ ^4D ($4s3d^6$) second excited state and the Co^+ a^5F and b^3F ($4s3d^7$) excited states react similarly. With all of these states, adduct formation is strongly reduced relative to the ground state and elimination reactions enhanced. As noted above, the Fe^+ ^4F ($3d^7$) first excited state is observed to collisionally deactivate to the ^6D ($4s3d^6$) ground state. It also showed the largest H_2 and CH_4 elimination rates. Presumably this is due to an efficient coupling of the electronic energy into the reaction. These results are consistent with electronic state specific reactivity studies of the Fe^+/propane system of Schultz, Armentrout and coworkers [6, 15] and Hanton, Weisshaar and coworkers [4, 17].

The ground state for Ni^+ is a ^2D ($3d^9$) and thus all the excited states must have the 4s orbital occupied (up to 6.4 eV, above this energy 4p orbitals are also occupied). The ^2D ($3d^9$) ground state and the $4s3d^8$ excited states of Ni^+ are baseline resolved in the Ni^+ arrival time distribution, indicating that deactivation does not occur. The observed reactivity for the $4s3d^8$ and $3d^9$ states of Ni^+ reacting with propane is similar to the $4s3d^7$ and $3d^8$ states of Co^+ reacting with propane. At 1.75 Torr of He pressure, adduct formation is the dominant product of the ^2D ($3d^9$) ground state of Ni^+, with only small amounts of elimination products observed. The $4s3d^8$ excited states show greatly reduced clustering (due to the repulsive 4s electron) and enhanced elimination channels. Now, it is possible that some highly excited Ni^+ states are present because, unlike the electronic states of Co^+ where both $3d^8$ and $4s3d^7$ excited electronic states are present (providing a means for deactivation for at least some of the excited states), all the excited states of Ni^+ (below 6.4 eV) have the 4s orbital occupied. As a result, if excited states of Ni^+ are produced, they may not collisionally deactivate. Even though the distribution of excited electronic states of Ni^+ is not known in our experiment, the fact that we observe H_2 and CH_4 elimination for $4s3d^8$ excited states of Ni^+ reacting with propane indicates that sigma bond activation is possible for 4s occupied states even in the absence of a higher lying $3d^n$ state. States with $4s3d^{n-1}$ configurations are generally repulsive to sigma bond activation [6, 7, 80] and thermal energy reactions are believed to occur only through the interaction with an accessible $3d^n$ state (via surface crossings) [4, 26], which in this case must be the ground state.

The initial interaction with propane is more repulsive for $4s3d^{n-1}$ states relative to $3d^n$ states and this could lead to a C—H bond activation transition state above the reactant energy for $4s3d^{n-1}$ states. However, mixing of diabatic surfaces (correlating to ground and low lying excited electronic states of Ni^+) at the C—H bond activation transition state counters this effect. The increase in exothermicity for electronically excited $4s3d^{n-1}$ states of M^+ reacting with propane presumably lowers the C—H bond activation transition state.

All of the data for ground state Fe^+, Co^+ and Ni^+ reacting with propane can be explained by initial C—H bond activation, without the need to invoke initial C—C bond activation. Threshold collisional activation data for $Fe(C_3H_8)^+$ indicate that the C—C bond activation transition state is located $\sim$8 kcal/mol above the C—H bond activation transition state [84]. These results can be rationalized by using results of Low and Goddard [85] who determined relative transition state energies for reductive H—H, C—H and C—C coupling to eliminate H_2, CH_4 and C_2H_6 from the corresponding Pt and Pd complexes. Due to the directionality of the sp^3 hybrid orbital on carbon as opposed to the spherically symmetric s orbital on hydrogen, the transition state for C—C reductive coupling was calculated to be 10 kcal/mol higher than that for C—H coupling, which in turn was 10 kcal/mol higher than the H—H coupling transition state. Blomberg et al. [23] extended these calculations to include both first and second row transition metals, Fe, Co, Ni, Rh, and Pd. For all these metals, the transition state for C—C bond activation in ethane was found to be 14–20 kcal/mol higher than C—H bond activation in methane. For Fe^+, Co^+ and Ni^+ reacting with propane, we determined that the transition state for initial C—H bond activation is only a few kcal/mol below the reactant energy. Thus, if the initial C—C bond activation transition state is 10–20 kcal/mol higher in energy than C—H bond activation, it will be located $\sim$8–18 kcal/mol above the reactant energy. Such a high transition state energy would explain why initial C—C bond activation would not occur at low energy.

6.3. Reactions of V^+ with propane

State selected studies using REMPI [16] and IC [86] indicate that while both the ^5D ($3d^4$) ground state and ^5F ($3d^3(^4F)4s$) excited state of V^+ react slowly with propane to eliminate H_2 at thermal energy, the ^3F ($3d^3(^4F)4s$) excited state of V^+ is observed to eliminate H_2 at least two orders of magnitude faster than the ground state and to eliminate a small amount of CH_4 ($\sim$1%) as well. As Aristov and Armentrout [14] have pointed out, the ^3P ($3d^4$) excited state of V^+, has the right symmetry and electronic configuration to insert into the C—H bond without an insertion barrier. As noted in the $V(H_2)^+$ discussion, the d orbitals in the ^5D ($3d^4$) ground state and the ^5F ($3d^3(^4F)4s$) and ^3F ($3d^3(^4F)4s$) excited states of V^+ are all singly occupied and cannot diabatically insert. This is mitigated, however, by a strongly avoided crossing between the triplet surfaces and spin orbit coupling between the quintet and triplet surfaces which allow the ^3F ($4s3d^3$) and ^5D ($3d^4$) states of V^+ to react with C_3H_8 to eliminate H_2 (see Figure 12).

The increase in reactivity observed for the ^3F ($4s3d^3$) excited state of V^+ relative to the ground state is most likely due to two effects. First, the strongly avoided crossing between the triplet surfaces is expected to be more efficient than the weakly avoided crossing between the quintet and triplet surfaces. Second, the available energy for the triplet state is larger than that for the quintet states.

Although the elimination of CH_4 in the reaction of ground state $V^+ + C_3H_8$ is exothermic, this channel is not observed at thermal energies. In fact, a threshold for CH_4 elimination is observed in the ion beam studies of ground state V^+ reacting with propane [87]. We can explain these results as follows. H_2 and CH_4 loss are significantly less exothermic for V^+ reacting with propane (3 and 10 kcal/mol, respectively) than for Fe^+, Co^+ and Ni^+ (Table IV). Because the transition state for C—H reductive coupling was calculated to be 10 to 20 kcal/mol higher in energy than the H—H reductive elimination transition state (due to the directionality of the sp^3 hybrid orbital on carbon [23, 85] it appears the transition state for C—H reductive coupling is above the threshold for the reactants, preventing CH_4 elimination at thermal energy. By fitting the above reaction cross section data for CH_4 elimination the energy of this transition state was determined to be 18 kcal/mol above the reactant energy, as shown in Figure 12. This relatively large barrier along the reaction coordinate for CH_4 elimination is consistent with the small CH_4/H_2 branching ratio observed for the ^3F excited state of V^+ reacting with propane.

6.4. Reactions of Sc^+ and Ti^+ with propane

Under multicollision conditions, ground state Sc^+ ^3D ($4s3d$) and Ti^+ ^4F ($4s3d^2$) are the only transition metal ions in the first row where the elimination reactions dominate over adduct formation. The ^3D ($3d4s$) ground state

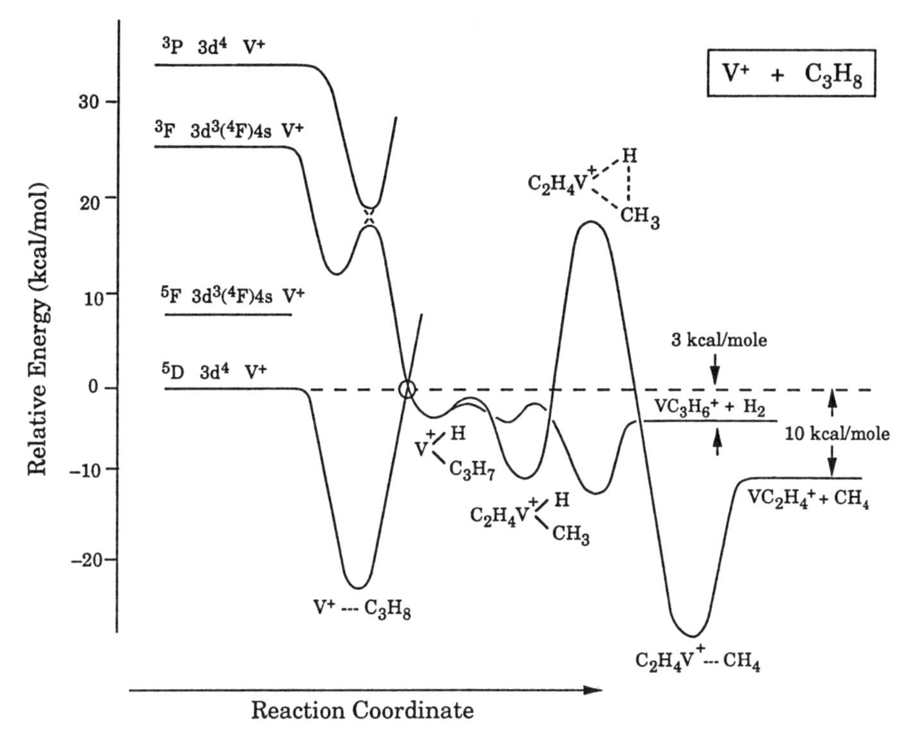

Fig. 12. Schematic reaction coordinate diagrams for the ^5D ($3d^4$), ^5F ($3d^3$ (^4F)4s), ^3F ($3d^3$(^4F)4s) and ^3P ($3d^4$) states of V^+ reacting with C_3H_8. The ^3P ($3d^4$) state of V^+ diabatically correlates to the inserted triplet H—V^+—C_3H_7 intermediate. The strongly avoided crossing between the triplet surfaces is indicated by dashed lines and the weakly avoided crossing due to spin orbit coupling between the quintet ground and triplet excited state surfaces is circled.

and the ^1D (3d4s) and ^3F ($3d^2$) excited states of Sc^+ all have singly occupied d orbitals and will be repulsive to sigma bond activation. However, the d_{yz}^2 component of the ^1D ($3d^2$) state diabatically correlates to the inserted singlet H—Sc^+—C_3H_7 intermediate. Spin orbit coupling of the triplet and singlet surfaces allows ground state Sc^+ to insert into the C—H bond. Thus, the potential energy surface for Sc^+ reacting with propane is similar to Sc^+

Table IV. Reaction thermochemistry[a]

		$-\Delta H_{rxn}$ (kcal/mole)							
		Sc	Ti	V	Cr	Mn	Fe	Co	Ni
$M^+ + C_3H_8$	→ $MC_3H_6^+ + H_2$	9	≥5	3	0.3	7	9	17	15
	→ $MC_2H_4^+ + CH_4$	17	≥14	10	8	15	17	24	26

[a]See Refs. 20, 36b, 89, 90.

reacting with H_2 (see Figure 11). However, in the "cluster mediated" mechanism proposed for Sc^+ reacting with H_2, the addition of three H_2 ligands is required to provide sufficient stabilization energy to bring the weakly avoided crossing between the triplet and singlet surface below the asymptotic energy of the reactants (the ^3D (4s3d) ground state of $Sc^+ + H_2$). In the case of propane, both H_2 and CH_4 elimination channels are observed at thermal energy under single collision conditions [88]. This implies that only a single propane ligand is required to bring the insertion barrier below the ground state reactant energy, most likely because Sc^+ is much more strongly bound to propane than to hydrogen.

For Ti^+, the ^4F ($3d^2(^3F)4s$) ground state and the ^4F ($3d^3$) and ^2F ($3d^2(^3F)4s$) excited states all have singly occupied d orbitals and cannot diabatically insert into the C—H bonds of propane. The ^2D ($3d^2(^1D)4s$) and ^2G ($3d^3$) states do have at least one component with the correct electronic configuration to activate the C—H bond. These states are sufficiently low in energy to be accessible to the ground state which can adiabatically react by spin orbit coupling between the quartet and doublet surfaces to form the doublet $H—Ti^+—C_3H_7$ intermediate. The stability of the $Ti^+—C_3H_8$ and $H—Ti^+—C_3H_7$ complexes are apparently sufficient to bring the energy of the quartet/doublet surface crossing below the asymptotic energy of the Ti^+/C_3H_8 ground state reactants.

The key difference for early transition metal ions is that the s and d orbital radii are approximately equal and can efficiently form sd hybrid orbitals. As was observed for $H—Sc^+—H$, where the hydrogen atoms are bound to two equivalent hybrid orbitals, calculations indicate [19] the carbon and hydrogen in the $H—Sc^+—C_3H_7$ and the $H—Ti^+—C_3H_7$ complexes are bound to equivalent sd hybrid orbitals. As a result, the C—H bond activation transition state energy is significantly lowered and elimination is favored over adduct formation.

7. Summary and conclusions

The application of IC to transition metal ions provides a unique opportunity to carry out electronic state specific reactivity studies. Due to the difference in mobility of the $3d^n$ and $4s3d^{n-1}$ configurations of atomic transition metal ions, the electronic state populations can be characterized. States with different electronic configurations exhibit the largest mobility differences; however, mobilities of high and low spin $4s3d^{n-1}$ figurations can also be significantly different, especially in the first half of the transition metal series, as was shown for ^5F and ^3F ($4s3d^3$) states of V^+ (Figures 2 and 3). The IC technique provides, in many cases, an absolute determination of ground and excited state populations which allows investigation of state specific reactions.

To understand the diverse chemistry of more complex systems, we first applied the IC technique to simple systems. Even for the simplest systems, the interaction of first row transition metal ions with H_2 we observe large differences in the binding energy and reactivity. For example, the binding energy for Co^+—H_2 (18.2 kcal/mol) is much larger than for V^+—H_2 (10.2 kcal/mol) while ground state Sc^+ actually inserts into the H—H bond. These investigations prompted high level ab initio calculations which yielded considerable insight into the bonding energetics and mechanism for sigma bond activation.

A number of factors are seen to affect the metal-ligand interaction. We can conveniently examine these factors by dividing the $M^+ + L$ reaction into five steps: 1) the initial electrostatic complex, 2) the barrier to the bond insertion, 3) rearrangement of the inserted complex, 4) the barrier to reductive coupling, and 5) the orbiting transition state leading to products. In the following, we consider these steps (and the factors which affect them) in order.

The electrostatic complex

First, the depth of the initial electrostatic well greatly affects the reaction by controlling the energy of subsequent transition states relative to the reactant energy. As noted, the increase in reactivity from C_2H_6 to C_3H_8 to C_4H_{10} is primarily due to the increase in the depth of the electrostatic well as the size of the alkane increases. Many experimental investigations, including the equilibrium studies and the associated theoretical calculations, point to four major effects which determine this well depth. First are the purely electrostatic factors, including ion size, ligand polarizability and the presence of permanent dipole or quadrupole moments of the ligand. Second is the ability of the metal ion to back donate charge to the ligand. As discussed for Co^+—H_2 this occurs via a filled $d\pi \rightarrow \sigma^*$ transfer and stabilizes the metal by increasing the d-d exchange energy. Third is the presence of a 4s electron on M^+. If the large 4s orbital is occupied, metal ligand repulsion greatly increases and the metal ions ability to accept electron density from the ligand decreases, both of which reduce the M^+—L binding energy. Fourth, the initial M^+—L attraction can be increased by $d\sigma$—4s hybridization on M^+. This removes electron density along the bond axis and decreases M^+—L repulsion. These four factors largely determine the M^+—L binding energy which influences subsequent reactivity.

The initial insertion barrier

In order for a reaction to proceed beyond the initial well, a bond insertion barrier must be overcome. Again, a number of factors are important. The first factor involves the type of bond being broken. In thermal energy reactions of M^+ with small alkanes, C—H bond insertion is greatly favored because C—C bond activation requires 8–10 kcal/mol more energy. In reactions with larger alkanes, the initial well depth may be large enough to allow

C—C bond activation. As discussed by Armentrout [6, 13], Weisshaar [4] and coworkers, simple molecular orbital arguments can be used to understand the requirements for H—H, C—H and C—C bond activation. The most favorable conditions for sigma bond activation by first row transition metal ions requires that 1) the 4s orbital on M^+ is unoccupied to facilitate the metals ability to accept electron density from the sigma bond being broken, and 2) a doubly occupied 3d orbital of the appropriate symmetry on M^+ is present to donate electron density into the antibonding orbital of the sigma bond being broken.

Sigma bond activation is less favorable with the 4s orbital occupied for two reasons. First, the electrostatic stabilization is less when the 4s orbital is occupied thereby increasing the energy of the insertion transition state. Second, occupation of the 4s orbital forces one electron into an antibonding orbital of the inserted intermediate. However, mixing of diabatic surfaces (correlating to states with $3d^n$ and $4s3d^{n-1}$ electronic configurations) can lead to sigma bond activation for 4s occupied states.

Finally, conservation of electron spin for sigma bond activation is an important factor which must be taken into consideration. Some first row transition metal ions can conserve spin on the ground state surface during activation of sigma bonds but for others an interaction with a sufficiently low lying excited state that can conserve spin is required to observe insertion.

Rearrangement of the inserted complex

The third step of the reaction with alkanes involves either a β—H or β—alkyl shift. For Fe^+ Co^+ and Ni^+ reacting with propane, initial C—H bond activation rather than the β—H or β—CH_3 transfer was determined to be the rate limiting transition state for H_2 and CH_4 elimination. However, the energy of the transition state associated with the β—H shift in V^+/C_3H_8 must be close to the reactant energy to explain the observed H/D isotope effects [87]. Thus, this step could play a significant role in some systems.

Reductive elimination

In the reaction of metal ions with small saturated hydrocarbons, the fourth step, reductive elimination of molecular hydrogen or alkanes (involving H—H or C—H coupling), could in principle be rate limiting and could determine the product distribution. The factors influencing these transition state energies and the product branching ratio include the reaction exothermicity, the stability of the inserted complex and the difference in activation energy for C—H and H—H coupling (due to the directionality of the sp^3 hybrid orbital on carbon versus the spherically symmetric s orbital on hydrogen). As the reaction exothermicity increases, the energy of the reductive elimination transition state decreases due to the greater product stability. Increasing the inserted complex stability has the same effect. The increase in the transition state energy associated with C—H relative to H—H reductive coupling has been predicted theoretically and is seen in the V^+/C_3H_8 system where H_2 but not CH_4 is eliminated at thermal energy, even though the CH_4 channel is more exothermic.

THE ORBITING TRANSITION STATE

Finally, the orbiting transition states leading to products can strongly influence the observed product distribution because these are the last transition states experienced as the system dissociates. This effect was seen in the $Co^+ + CH_3I$ reaction where the greater rotational density of states enhances the $Co^+—I + CH_3$ products relative to $Co^+—CH_3 + I$ as the available energy increases.

Acknowledgements

The support of the National Science Foundation under grant CHE91–19752 is gratefully acknowledged. In addition, we gratefully acknowledge useful discussions with Peter Armentrout, Philippe Maitre and Jason Perry.

References

1. For a recent review, see K. Eller and H. Schwartz, *Chem. Rev.* **91**, 1121 (1991).
2. D. H. Russel (ed.), *Gas Phase Inorganic Chemistry*, (Plenum Press, New York, 1989).
3. T. J. Mark (Ed.), *Bonding Energetics in Organometallic Compounds*, *ACS Symposium Series*, Vol. 428 (American Chemical Society, Washington DC, 1990).
4. J. C. Weisshaar, in *Advances in Chemical Physics*, Vol. 82, edited By C. Ng, (Wiley-Interscience, New York, 1992 and references therein).
5. J. C. Weisshaar, *Acc. Chem. Res.* **26**, 213 1993.
6. P. B. Armentrout, *Annu. Rev. Phys. Chem.* **41**, 313 (1990) (and references therein).
7. P. B. Armentrout, *Science* **251**, 175 (1991).
8. P. A. M. van Koppen, P. R. Kemper and M. T. Bowers, *J. Am. Chem. Soc.* **114**, 1083 (1992); P. A. M. van Koppen, P. R. Kemper and M. T. Bowers, *J. Am. Chem. Soc.* **114**, 10941 (1992).
9. P. A. M. van Koppen, P. R. Kemper and M. T. Bowers, *J. Am. Chem. Soc.* **115**, 5616 (1993).
10. M. T. Bowers, P. R. Kemper, G. von Helden, G. and P. A. M. van Koppen, *Science* **260**, 1446 (1993).
11. F. Strobel and D. P. Ridge, *J. Phys. Chem.* **93**, 3635 (1989).
12. J. V. B. Oriedo and D. H. Russell, *J. Phys. Chem.* **96**, 5314 (1992).
13. J. L. Elkind and P. B. Armentrout, *J. Phys. Chem.* **91**, 2037 (1987).
14. N. Aristov and P. B. Armentrout, *J. Phys. Chem.* **91**, 6178 (1987).
15. R. H. Schultz, J. L. Elkind and P. B. Armentrout, *J. Am. Chem. Soc.* **110**, 411 (1988); R. H. Schultz, Thesis, University of Utah (1990).
16. L. Sanders, S. D. Hanton and J. C. Weisshaar, *J. Chem. Phys.* **92**, 3498 (1990).
17. S. D. Hanton, R. J. Noll and J. C. Weisshaar, *J. Chem. Phys.* **96**, 5176 (1992).

18. C. W. Bauschlicher, Jr., S. R. Langhoff and H. Partridge, in *Modern Electronic Structure Theory*, D. R. Yarkony (Ed.), World Scientific Pub. Co., London (1995).
19. C. W. Bauschlicher, Jr. and S. R. Langhoff, *Int. Rev. Phys. Chem.* **9**, 149 (1990).
20. M. Sodupe, C. W. Bauschlicher, Jr. and S. R. Langhoff, *J. Phys. Chem.* **96**, 2118 (1992).
21. J. K. Perry, G. Ohanessian and W. A. Goddard, *J. Phys. Chem.* **97**, 5238 (1993).
22. E. A. Carter and W. A. Goddard, *J. Phys. Chem.* **92**, 6679 (1988).
23. M. R. A. Blomberg, P. E. M. Siegbahn, U. Nagashima and J. Wennerberg, *J. Am. Chem. Soc.* **113**, 424 (1991).
24. A. E. Alvarado-Swaisgood and J. F. Harrison, *J. Phys. Chem.* **89**, 5198 (1985).
25. D. G. Musaev, K. Morokuma, N. Koga, K. A. Nguyen, M. S. Gordon and T. R. Cundari, *J. Phys. Chem.* **97**, 11435 (1993).
26. A. K. Rappe and T. H. Upton, *J. Chem. Phys.* **85**, 4400 (1986).
27. G. Herzberg, *Atomic Spectra and Atomic Structure*, Dover, New York (1944).
28. J. L. Elkind and P. B. Armentrout, *J. Phys. Chem.* **89**, 5626 (1985).
29. R. H. Garstang, *Mont. Not. R. Astron. Soc.* **124**, 321 (1962).
30. (a) C. E. Moore, *Atomic Energy Levels*, U.S. National Bureau of Standards: Washington, DC, Circ. 467 (1952); (b) J. Sugar and C. Corliss, *J. Phys. Chem. Ref. Data*, **10**,197, 1097 (1981); (c) ibid., **11**, 135 (1982).
31. P. R. Kemper, J. E. Bushnell, P. A. M. van Koppen and M. T. Bowers, *J. Phys. Chem.* **97**, 1810 (1993).
32. P. R. Kemper, J. Bushnell, G. von Helden and M. T. Bowers, *J. Phys. Chem.* **97**, 52 (1993).
33. J. E. Bushnell, P. R. Kemper and M. T. Bowers, *J. Phys. Chem.* **97**, 11628 (1993).
34. J. E. Bushnell, P. R. Kemper and M. T. Bowers, *J. Am. Chem. Soc.* **116**, 9710 (1994).
35. (a) P. R. Kemper and M. T. Bowers, *J. Phys. Chem.* **95**, 5134 (1991); (b) P. R. Kemper and M. T. Bowers, *J. Am. Chem. Soc.*, **112**, 3231 (1990).
36. (a) L. S. Sunderlin and P. B. Armentrout, *J. Phys. Chem.*, **92**, 1209 (1988); (b) L. S. Sunderlin and P. B. Armentrout, *Int. J. Mass Spectrom. Ion Proc.* **94**, 149 (1989).
37. R. B. Cody, R. C. Burnier, W. D. Reents, Jr., T. J. Carlin, D. A. McCrery, R. K. Lengel and Freiser, *Int. J. Mass Spectrom. Ion Phys.* **33**, 37 (1980).
38. S. K. Loh, E. R. Fisher, Li Lian, R. H. Schultz and P. B. Armentrout, *J. Phys. Chem.* **93**, 3159 (1989).
39. L. Sanders, A. D. Sappy and J. C. Weisshaar, *J. Phys. Chem.* 1**85**, 6952 (1986).
40. L. Sanders, A. D. Sappy and J. C. Weisshaar, *J. Phys. Chem.* **91**, 5145 (1987).
41. S. Hanton, L. Sanders and J. C. Weisshaar, *J. Phys. Chem.* **93**, 1963 (1989).
42. J. L. Elkind and P. B. Armentrout, *J. Chem. Phys.* **84**, 4862 (1986).
43. J. L. Elkind and P. B. Armentrout, *J. Chem. Phys.* **90**, 5736 (1986).
44. J. L. Elkind and P. B. Armentrout, *J. Chem. Phys.* **90**, 6576 (1986).
45. J. L. Elkind and P. B. Armentrout, *J. Chem. Phys.* **86**, 1868 (1987).
46. J. L. Elkind and P. B. Armentrout, *Int. J. Mass. Spectrom. Ion Proc.* **83**, 259 (1988).
47. N. D. Twiddy, A. Mohebati and Tichy, *Int. J. Mass Spectrom. Ion Proc.* **74**, 251 (1986).
48. S. T. Grice, P. W. Harland, R. Maclagan and R. W. Simpson, *Int. J. Mass Spectrom Ion Proc.* **87**, 181 (1989).
49. B. R. Rowe, D. W. Fahey, F. C. Fehsenfeld and D. L. Albritton, *J. Chem. Phys.* **73**, 194 (1980).
50. (a) H. W. Ellis, et al. Transport Properties of Gaseous Ions Over a Wide Range, *At. Nucl. Data Tables* **17**, 177 (1976); **22**, 179 (1978); **31**, 113 (1984); (b) W. Lindinger and D. C. Albritton, *J. Chem. Phys.* **62**, 3517 (1975).
51. M. McFarland, D. L. Albritton, F. C. Fehsenfeld, E. E. Ferguson and A. L. Schmeltekopf, *J. Chem. Phys.* **59**, 6610 (1976).
52. E. W. McDaniel and E. A. Mason, *The Mobility and Diffusion of Ions in Gases*, John Wiley, New York (1973).
53. (a) G. von Helden, P. R. Kemper, M-T. Hsu and M.-T. Bowers, *J. Chem. Phys.* **96**, 6591 (1992); (b) P. R. Kemper, M.-T. Hsu and M. T. Bowers, *J. Phys. Chem.* **95**, 10,600 (1991).

54. P. R. Kemper and M. T. Bowers, *J. Am. Soc. Mass Spectrom.* **1**, 197 (1990).
55. N. Aristov and P. B. Armentrout, *J. Am. Chem. Soc.* **108**, 1806 (1986).
56. The quantity σ is really the collision cross section only for "hard sphere" collisions of large ions with He. For smaller species, such as transition metal ions, σ is the collision integral which requires consideration of the details of the interaction potential.
57. Equation (2) holds for heavy ions colliding with light neutrals for relatively low values of E. These conditions are fulfilled for all systems reported in this chapter.
58. J. Allison and D. P. Ridge, *J. Am. Chem. Soc.* **101**, 4998 (1979).
59. E. R. Fisher, L. S. Sunderlin and P. B. Armentrout, *J. Phys. Chem.* **93**, 7375 (1989).
60. E. R. Fisher, R. H. Schultz and P. B. Armentrout, *J. Phys. Chem.* **93**, 7382 (1989).
61. R. Georgiadis, E. R. Fisher and P. B. Armentrout, *J. Am. Chem. Soc.* **111**, 4251 (1989).
62. Allison and Ridge [58] obtained strong evidence for oxidative addition to the methyl iodide bond for Fe^+ reacting with CH_3I. They found that $FeCH_3I^+$ reacts with CD_3I to form $FeICD_3I^+ + CH_3$ exclusively; no $FeICH_3I^+ + CD_3$ was observed. Exclusive CH_3 elimination indicates a symmetric $Fe(CH_3I)(CD_3I)^+$ complex is not formed and the Fe^+ ion must insert into the I—C bond to form the I—Fe^+—CH_3 intermediate which reacts with CD_3I eliminating CH_3. Other first row transition metal ions, including Co^+ are expected to activate methyl iodide in a similar fashion.
63. P. A. M. van Koppen, J. Brodbelt-Lustig, M. T. Bowers, D. V. Dearden, J. L. Beauchamp, E. R. Fisher and P. B. Armentrout, *J. Am. Chem. Soc.* **113**, 2359 (1991); Ibid. **112**, 5663 (1990).
64. J. E. Bushnell, P. R. Kemper and M. T. Bowers, *J. Phys. Chem.* **98**, 2044 (1994).
65. C. W. Bauschlicher, Jr. H. Partridge and S. R. Langhoff, *J. Phys. Chem.* **96**, 2475 (1992).
66. L. A. Curtiss and J. A. Pople, *J. Phys. Chem.* **92**, 894 (1988).
67. D. A. Dixon, J. L. Gole, and A. J. Komornicki, *J. Phys. Chem.* **92**, 1378 (1988).
68. M. F. Falcetta, J. L. Pazun, M. J. Dorko, D. Kitchen and P. E. Siska, *J. Phys. Chem.* **97**, 1011 (1993).
69. P. Maitre and C. W. Bauschlicher Jr., *J. Phys. Chem.* **97**, 11912 (1993).
70. C. W. Bauschlicher Jr., H. Partridge and S. R. Langhoff, *Chem. Phys. Lett.* **165**, 272 (1990).
71. L. A. Barnes, M. Rosi and C. W. Bauschlicher Jr., *J. Chem. Phys.* **93**, 609 (1990).
72. P. B. Armentrout (private communication).
73. P. Maitre and C. W. Bauschlicher Jr. (private communication).
74. M. A. Tolbert and J. L. Beauchamp, *J. Am. Chem. Soc.* **106**, 8117 (1984).
75. L. S. Sunderlin, N. Aristov and P. B. Armentrout, *J. Am. Chem. Soc.* **109**, 78 (1987). See P. B. Armentrout, in this book for updated bond energies.
76. J. K. Perry, W. A. Goddard and G. Ohanessian, *J. Chem. Phys.* **97**, 7560 (1992).
77. R. H. Schultz and P. B. Armentrout, *J. Phys. Chem.*, **96**, 1662 (1992).
78. C. Haynes and P. B. Armentrout (work in progress).
79. M. R. A. Blomberg, P. E. M. Siegbahn and M. J. Svensson, *J. Phys. Chem.* **95**, 4313 (1991).
80. R. Tonkyn, M. Ronan and J. C. Weisshaar, *J. Phys. Chem.*, **92**, 92 (1988).
81. P. A. M. van Koppen, P. R. Kemper and M. T. Bowers, State specific studies for Ni^+ reacting with propane (unpublished).
82. P. A. M. van Koppen, M. T. Bowers, E. R. Fisher and P. B. Armentrout, *J. Am. Chem. Soc.* **116**, 3780 (1994).
83. J. K. Perry, California Institute of Technology, Thesis (1993).
84. R. H. Schultz and P. B. Armentrout, *J. Am. Chem. Soc.* **113**, 729 (1991).
85. J. J. Low and W. A. Goddard, *J. Am. Chem. Soc.* **106**, 8321 (1984).
86. P. A. M. van Koppen, P. R. Kemper and M. T. Bowers (unpublished results).
87. P. A. M. van Koppen, M. T. Bowers, C. Haynes and P. B. Armentrout (work in progress).
88. L. S. Sunderlin and P. B. Armentrout, *Organometallics* **9**, 1248 (1990).
89. (a) P. A. M. van Koppen, D. B. Jacobson, A. J. Illies, M. T. Bowers, M. A. Hanratty and J. L. Beauchamp, *J. Am. Chem. Soc.*, **111**, 1991 (1989); (b) M. A. Hanratty, J. L.

Beauchamp, A. J. Illies, P. A. M. van Koppen and M. T. Bowers, *J. Am. Chem. Soc.* **110**, 1 (1988); (c) P. A. M. van Koppen, M. T. Bowers, J. L. Beauchamp and D. V. Dearden, in *Bonding Energetics in Organometallic Compounds*, T. J. Mark Ed., ACS Symposium series 428; American Chemical Society: Washington DC (1990).

90. R. H. Schultz and P. B. Armentrout (work in progress).

6. Development of a fourier-transform ion cyclotron resonance (FTICR) mass spectrometry method for studies of metal ion excited states

DAVID H. RUSSELL, J.V.B. ORIEDO and TOURADJ SOLOUKI
Department of Chemistry, Texas A&M University, College Station, Texas 77843, U.S.A.

1. Introduction

Studies of ion-molecule reaction chemistry of gas-phase transition metal ions have rapidly expanded and the field of gas phase organometallic chemistry has emerged [1]. Eller and Schwarz have compiled an amazingly comprehensive review of this entire field that covers the period 1973 to 1992 [2]. The pioneering gas-phase ion chemistry studies emphasized the type of reactions, product ion distribution, and speculation on reaction mechanisms; however, the most important contribution of much of this work is to the understanding of reaction energetics and bond energies to metal centers [3]. In the last few years, several groups have placed considerable attention on the specific electronic state(s) of the reacting metal ion (M^+) and how the reactivity of M^+ might change if different excited states are formed by the ionizing process.

The effect of long-lived metastable electronic states on the reaction chemistry of atomic transition metal ions was first demonstrated for the Cr^+ system [4]. Freas and Ridge noted that Cr^+ formed by electron impact ionization of $Cr(CO)_6$ reacted with CH_4 at thermal energies to yield $CrCH_2^+$ (reaction (1)). Reaction (1) is endothermic for ground state Cr^+ by 45 kcal/mol [5]. Beauchamp and co-workers later observed that Cr^+ formed by surface ionization (which yields primarily ground state ions) of $CrCl_3$ reacted with CH_4 under high energy conditions to form CrH^+, whereas Cr^+ formed by electron impact ionization of $Cr(CO)_6$ reacts with CH_4 by reaction (1) [5]. In addition, the rate of product ion formation for reaction (1) increases as the kinetic

$$Cr^+ + CH_4 \rightarrow CrCH_2 + H_2 \tag{1}$$

energy of the reactant Cr^+ ion increases suggesting that the reaction is exothermic. Thus, Beauchamp concluded that reaction (1) must involve an electronically excited state of Cr^+, viz. electronic states with energies of ~2.5 eV above the $^6S(3d^5)$ ground state Cr^+ such as the $^4D(4s^13d^4)$, $^4G(3d^5)$, and $^4P(3d^5)$ states.

Ridge and coworkers re-examined the reactions of metastable electronic

Ben S. Freiser (ed.), Organometallic Ion Chemistry, 197–228.

states of Cr^+ with CH_4 [6]. These studies involved kinetics of reactions of Cr^+ with $Cr(CO)_6$ and collision-induced dissociation of $CrCH_4^+$. Because reaction (1) does not follow pseudo-first-order-kinetics Ridge concluded that Cr^+ ions are formed in two different electronic states, viz. the ground (25%) and excited states (75%). Further studies that illustrate the presence of long-lived electronically excited Cr^+ ions were reported by Huang and Gross [7]. For example, the decay of Cr^+ ions formed by multiphoton ionization of $Cr(CO)_6$ follow pseudo-first-order kinetics, whereas Cr^+ formed by electron impact ionization show composite decay curves. Thus, they concluded that Cr^+ formed by MPI corresponded to a single state of Cr^+, probably the ground state. In related studies Ridge showed that approximately 22% of the Mn^+ ions produced by electron impact ionization of $Mn_2(CO)_{10}$ are formed in an electronically excited state(s) having radiative lifetimes of greater than 5.8 ± 0.7 s [8]. Ionization methods such as laser ablation [9], surface ionization [10], particle bombardment and multiphoton ionization [7] are used to generate M^+ ions, and, with the exception of multiphoton ionization and surface ionization, all the methods form ion populations composed of ground and excited state species.

Armentrout has studied a number of state-specific gas phase ion-molecule reactions of first row transition metal ions [11]. These studies show that the ground state $Cr^+[^6S(3d^5)]$ reacts endothermically with CH_4 to form CrH^+, $CrCH_2^+$ and $CrCH_3^+$ ionic products, whereas the $^4D(4s^13d^4)$ and $^4G(3d^5)$ excited states react with CH_4 more efficiently to form the same ions. The first excited state $^6D(4s^13d^4)$ is relatively unreactive [12]. Similarly, results from state-specific reactions of Fe^+ with H_2, HD and D_2 show that the reactivity of the ground state is different from that of the excited states [11b]. For instance, ground state Fe^+ $[^6D(4s^1\,3d^6)]$ is relatively unreactive with H_2, whereas the first excited state $Fe^+[^4F(3d^7)]$ (0.25 eV above ground state) reacts efficiently to form FeH^+ ionic product. Studies by Weisshaar show that the reaction of Fe^+ with C_3H_8 is affected by the spin-orbit level of the reactant Fe^+ ions [13]. That is, the $^4F_{5/2}(3d^7)$ state is twice as reactive as the $^4F_{9/2}(3d^7)$ and $^4F_{7/2}(3d^7)$ states.

2. Experimental methods for probing metastable electronic states of transition metal ions

Several experimental methods for determining the abundances of long-lived, metastable electronic states of transition metal ions, denoted $(M^+)^*$, have been developed. Weisshaar and coworkers have developed a state-selective resonance-enhanced multiphoton ionization angle-resolved time-of-flight photoelectron spectrometry method to generate ions formed in specific states [13], and this method can be used to study the chemistry of state selected $(M^+)^*$ species. The experiment is rather complex and only a limited number of $(M^+)^*$ species have been generated, however, in terms of probing state-selected chemistry this is the most definitive approach.

The majority of $(M^+)^*$ ion chemistry studies utilize ion beam methods such as those developed by Armentrout's research group [14]. In these studies Armentrout has used reaction cross-sections to compare the reactivity of atomic transition metal ions formed by surface ionization (SI) (yields predominantly ground state species, denoted $(M^+)^o$) and electron impact ionization (EI) which forms both $(M^+)^*$ and $(M^+)^o$. Absolute reaction cross-sections are measured as a function of the reactant M^+ ions translational energy (reaction probability at a given kinetic energy). The energies of the reactant $(M^+)^*$ ions are detected by observing shifts in the reaction cross-section as a function of translational energy of the reactant ion. The limitation of this method as a probe for $(M^+)^*$ is that the translational energy behavior for ion-molecule reactions is different for exothermic and endothermic reactions. To date the method has not been developed to the point that the energies of the reacting states can be measured; however, as more reactions of specific $(M^+)^*$ species are determined it should be possible to expand the scope of this method.

Bowers and Kemper have developed an electronic state chromatography (ESC) method based on ion mobility in a buffer gas (such as helium) that measures the relative abundances of d^n and $d^{n-1}s^1$ electron configurations of M^+ [15]. The time it takes for a mass selected ion to transverse the collision cell containing a buffer gas depends upon the electronic configuration of the ion. Metal ions with different electronic configurations have different mobilities due to the different collision cross-sections of these states. The $3d^n$ states have low mobilities because they undergo capture collisions with the buffer gas, whereas the $4s^13d^{n-1}$ states have high mobilities due to repulsive interaction with the buffer gas. Thus M^+ ions which have different electron configuration (e.g., $3d^n$ versus $4s^13d^{n-1}$) can be separated and probed in terms of reaction chemistry. This method also provides a means for probing the dynamics of collisional relaxation [16] (see also Chapter 5 of this text). Clearly, the ESC method does not resolve electronic states within a particular configuration and the signal intensities of the chromatogram reflect the abundance of the M^+ electron configuration rather than specific electronic states.

3. FTICR method for measuring abundances and collisional relaxation of metastable electronic states of metal ions

The methods developed by Weisshaar, Armentrout, and Bowers and Kemper can be used to measure abundances and to study the ion chemistry of $(M^+)^*$ and $(M^+)^o$ species, but there would be advantages associated with a comparable method using Fourier-transform ion cyclotron resonance (FTICR) mass spectrometry. The most striking advantage of the FTICR method would be the ability to extend the timeframe over which the products of ion-molecule reactions are sampled. That is, both Armentrout's and Bowers and Kemper's method sample the reaction products formed on the 10^{-6} s to 10^{-4} s timescale, whereas FTICR samples reaction products formed on the 10^{-3} to 10^{2} s timescale. Another advantage of an FTICR experiment performed by using two-section ion cell technology would be that ions could be formed in one region, partitioned to another ion cell to perform ion chemistry or structural probes (e.g., collision-induced dissociation, photodissociation, etc.), and partitioned back to the original ion cell for further studies (ion chemistry or structural probe).

Although there are distinct advantages for the FTICR experimental method and such methods would provide complimentary data to that obtained by other experimental probes, a probe for "reading" the energies and abundances of ions formed by the initial ionization method has not been developed. Therefore, in 1989 we began a series of studies to ascertain whether an FTICR experimental method for determining the energies and abundances of $(M^+)^*$ could be developed. We decided that the simplest probe for measuring the energies and abundances of $(M^+)^*$ would be charge-exchange ion-molecule reactions of systems having well-defined thermochemical requirements. Although it may seem a trivial matter to select neutral systems that undergo charge-exchange ion-molecule reactions with M^+ species, competing reactions are undesirable. For example, if the M^+ ions undergo condensation reactions with the neutral reagent, the rate of depletion of M^+ is a composite of all reaction channels; therefore, it is desirable to use a reagent neutral that reacts with $(M^+)^*$ to form a single reaction product or series of reaction products. Ideally, the reagent neutral would not react with $(M^+)^o$. For these reasons we chose charge-exchange ion-molecule reactions of M^+ with the corresponding neutral metal carbonyl ($M(CO)_y$).

3.1. Instrumentation and experimental method

The instrument (Figure 1) used for these studies has been described in detail elsewhere [17]. Briefly, the instrument consists of a home-built, differentially pumped two-section ICR cell and vacuum system, an Extrel FTMS 2001 computer and electronics system, and a 3-T (15-cm bore) superconducting Oxford magnet. The two vacuum regions are equipped with separate sample inlet systems that can handle up to three reagent gases. Gaseous reagents can also be introduced by variable leak valves or pulsed valves and semi-volatile solid samples can be introduced by a heated direct insertion probe. Ions can be formed by electron impact ionization, multiphoton ionization (Nd:YAG or excimer lasers) and by using laser ablation.

The charge-exchange ion molecule reaction chemistry can be performed using either single-section or two-section ion cell instruments [17]. Fe^+ is initially formed by electron impact ionization (electron beam energies from threshold to 100 eV) of $Fe(CO)_5$ in region 1 of the two-section cell (Figure

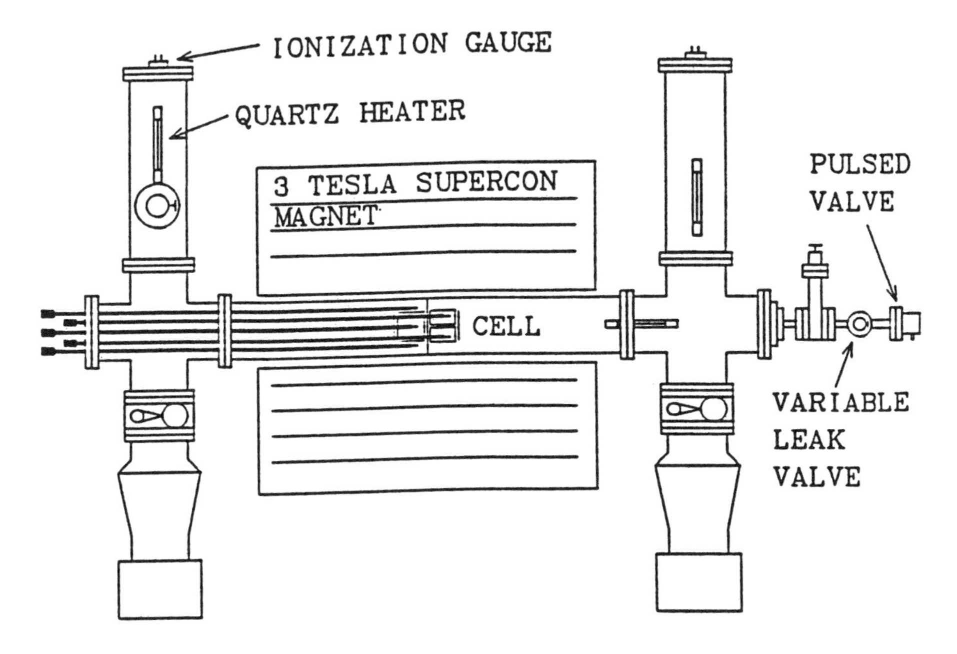

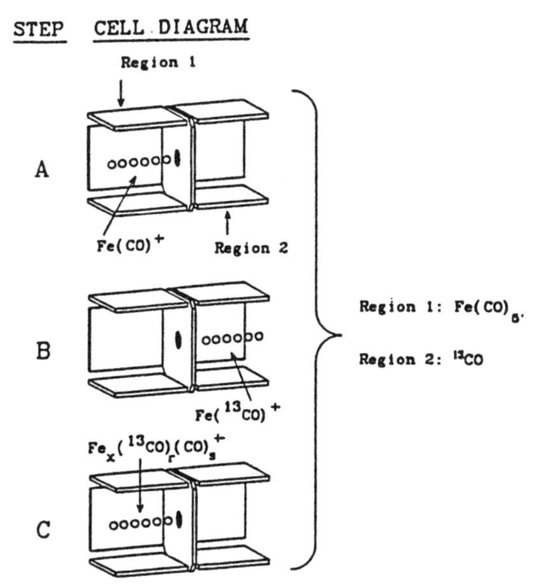

Fig. 1. Schematic drawings of the TAMU FTMS-1000 two-section cell Fourier-transform ion cyclotron resonance mass spectrometer.

2a). In the single-section cell mode the reactant Fe^+ ion is isolated by using ion-ejection techniques. Following the isolation step (Figure 2b), Fe^+ is allowed to react with the neutral $Fe(CO)_5$ for a specified period of time. Experiments involving the two-section ion cell are performed in a similar manner. $Fe(CO)_5$ is introduced into both region 1 and 2 at a static pressure of 2×10^{-7} Torr. The reactant Fe^+ is produced in region 1 and region 2 is used as the reaction chamber. The Fe^+ ion is isolated by using ion ejection methods described above. Region 2 is quenched of all ions formed during the ionization event prior to partitioning the mass-selected Fe^+ ion (from region 1) to region 2. The Fe^+ ion is then allowed to react with the neutral $Fe(CO)_5$. The reaction time between Fe^+ and $Fe(CO)_5$ can be varied from a few milliseconds to several seconds. In our experiments we keep the reaction time relatively short (typically 50 ms) to minimize the chances of sampling reaction products formed by multiple collisions. Multiple collisions are undesirable because the initial collisions may result in collisional relaxation of the $(Fe^+)^*$ and the abundances of the $Fe(CO)_y^+$ charge-exchange product ions do not reflect the true distribution of energies of the initially formed ion population. In addition, the $Fe(CO)_y^+$ charge-exchange product ions are reactive with $Fe(CO)_5$, thus the abundance of these ions is depleted if long reaction times are used.

The two-section ion cell experiment can also be used to study collisional relaxation of $(Fe^+)^*$. This experiment requires the use of the two-section ion cell, and the methods are similar to those used to synthesize $Fe(^{13}CO)^+$ [17]. The experiment is performed by admitting $Fe(CO)_5$ to region 1 and the desired collisional relaxation reagent gas to region 2. Fe^+ is formed in region 1, mass-selected, and partitioned to region 2 where it undergoes collisions with the reagent gas. Following a specified collisional relaxation time, Fe^+ is partitioned back to region 2. The energies of the $(Fe^+)^*$ are then determined by examining the $Fe(CO)_y^+$ charge-exchange product ions (Figure 2c,d). The abundances and energies of the $Fe(CO)_y^+$ charge-exchange product ions are compared with those for ions not subjected to collisional relaxation to determine which ions have reacted.

3.2. Charge-exchange ion-molecule reaction chemistry of the $Fe+/Fe(Co)_5$ system

The charge-exchange ion-molecule reaction chemistry method that we have developed provides an excellent method for measuring the abundances and bracketing the energies of high-lying metastable states of M^+. To date the most detailed studies have been performed on Fe^+, thus most of this chapter will deal with the $Fe^+/Fe(CO)_5$ system. The term charge exchange is used to indicate both dissociative and non-dissociative charge-exchange reactions of Fe^+ with the neutral $Fe(CO)_5$ (reaction (2)). Charge exchange of $(Fe^+)^*$ with $Fe(CO)_5$ to form $Fe(CO)_5^+$ is endothermic for ground-state $(Fe^+)^o$, but reaction (2) is exothermic for $(Fe^+)^*$. Likewise, dissociative charge-exchange

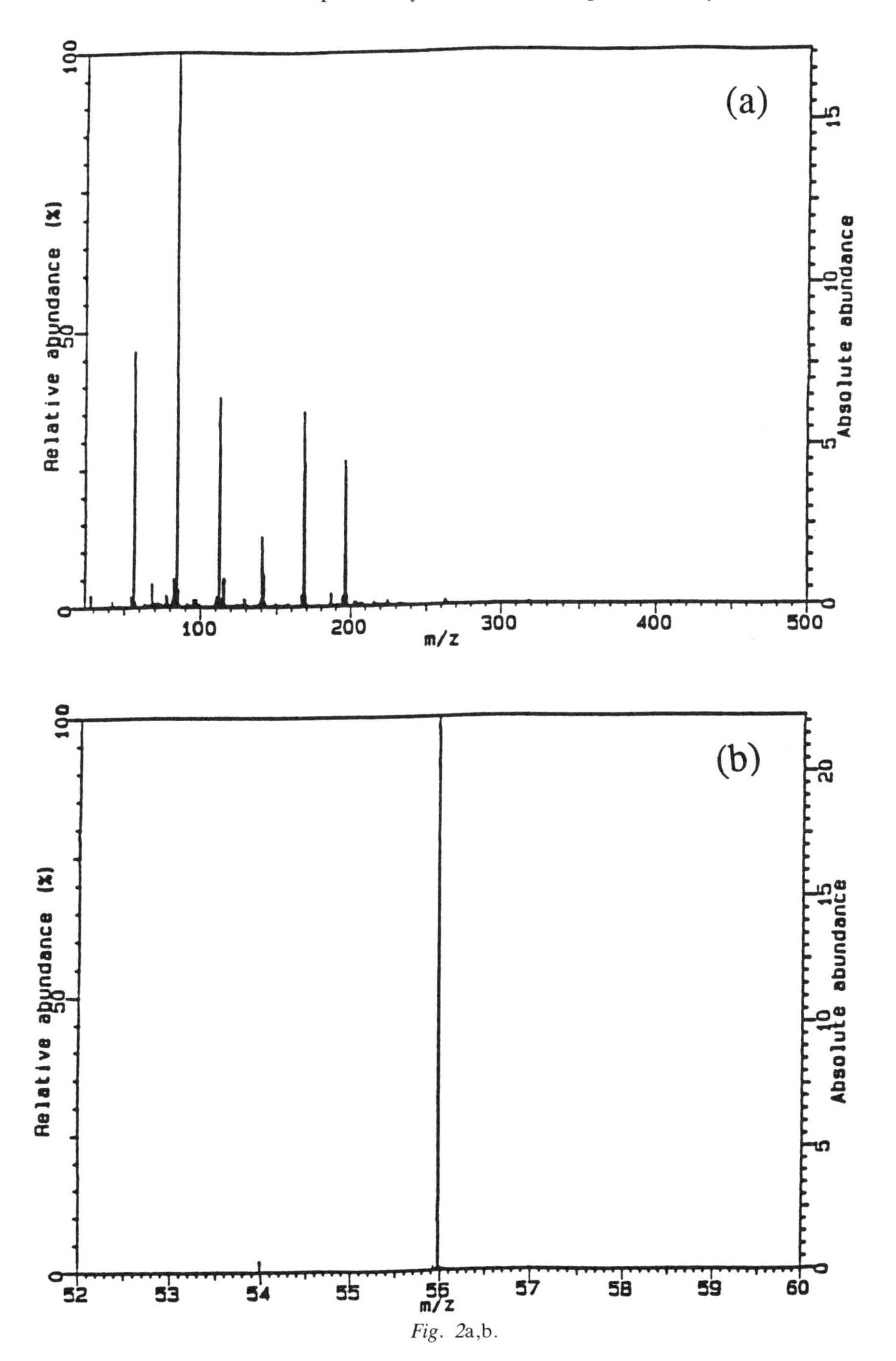

*Fig. 2*a,b.

Fig. 2. 70 eV mass spectrum of $Fe(CO)_5$. (a) Before the reactant ion is isolated; (b) after the reactant ion (Fe^+) is isolated using ion-ejection coupled with ion transfer method.

Charge-transfer Reaction Product (ions) before (a) and after (b) Collisional Relaxation of $Fe^{+\bullet}$:

$Fe^{+\bullet}$ + $Fe(CO)_5$ -----------------------------> $Fe(CO)_y^+$ + (5-y)CO

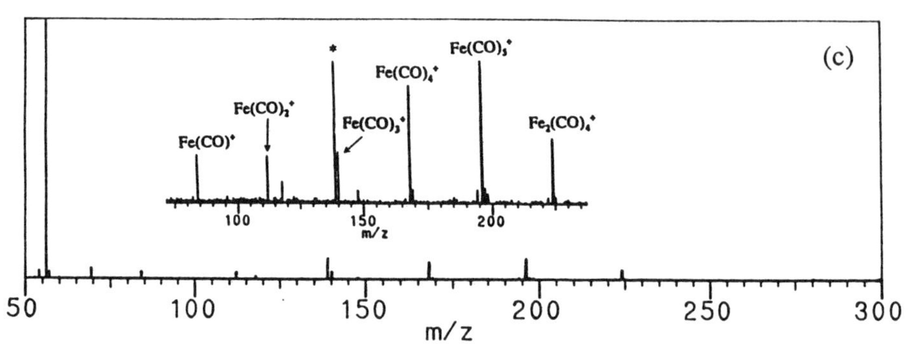

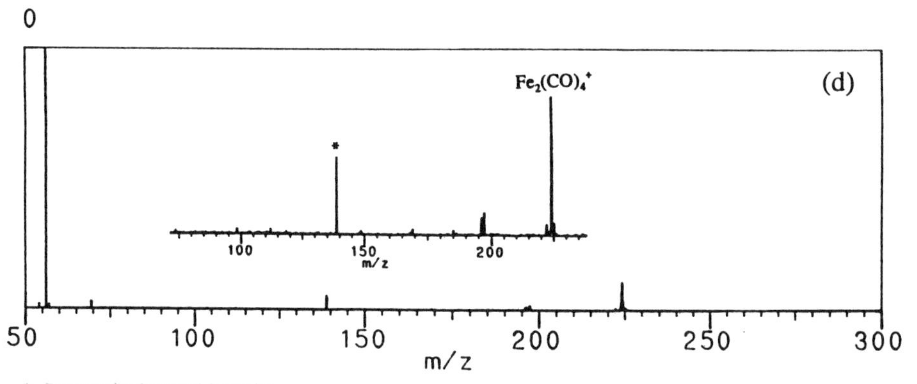

* Denotes background impurity.

Fig. 2c,d.

reactions of $(Fe^+)^*$ with $Fe(CO)_5$ to form $Fe(CO)_y^+$ y = 1–4, are exothermic for more highly excited $(Fe^+)^*$ ions. Reents *et al.* have also used the occurance of charge-exchange between Fe^+ and $Fe(CO)_5$ to implicate long-lived $(Fe^+)^*$ ions by electron impact ionization of the precursor [6].

$$(Fe^+)^* + Fe(CO)_5 \rightarrow Fe(CO)_y^+ + 5 - y\,CO \qquad (2)$$

Clustering reactions of M^+ with neutral $M(CO)_n$ have also been reported [18, 19]. Recent studies have shown that the ionic cluster fragments formed by the reaction of Fe^+ with $Fe(CO)_5$ (reaction (3)) depend upon the internal energy of the reactant Fe^+ ions. That is, $(Fe^+)^*$ reacts with $Fe(CO)_5$ to form $Fe_2(CO)_2^+$ and $Fe_2(CO)_3^+$ ionic cluster fragments, and $(Fe^+)^o$ [formed by collisional relaxation of Fe+ formed by 70-eV EI of $Fe(CO)_5$, see section 6] reacts with $Fe(CO)_5$ to form only the $Fe_2(CO)_4{}^+$ ionic cluster fragments. Thus, the gas-phase ion-molecule reactions of M^+ are strongly influenced by the presence of long-lived excited-state ions and the specificity of the product

Table I. Ionization energies of selected electronic states of Fe^+, the appearance energies of the charge transfer product ions, and the relative abundances of excited electronic states of Fe^+ [70–50 eV EI of $Fe(CO)_5$] estimated from the charge transfer ion-molecule reaction chemistry data

State	Configuration	Ionization energy (eV)	Appearance energy (eV)	Charge transfer product	Excited states relative abundance %
^2G	$4s^13d^6$	11.64			10
			11.53	$Fe(CO)^+$	
^2F	$4s^13d^6$	11.26			
^2H	$4s^13d^6$	11.12			
^2P	$4s^13d^6$	11.07			10
^4G	$4s^13d^6$	11.02			
^6S	$4s^23d^5$	10.76			
^4F	$4s^13d^6$	10 68	10.81	$Fe(C0)_2{}^+$	
^4H	$4s^13d^6$	10.51			
^4P	$4s^13d^6$	10.45			
^2D	$3d^7$	10.42			10
^2H	$3d^7$	10.39			
^2P	$3d^7$	10.17			
			9.87	$Fe(C0)_3^+$	
^2G	$3d^7$	9.86			
^4P	$3d^7$	9.56			30
^4D	$4s^13d^6$	8.90			
			8.77	$Fe(C0)_4^+$	
^4F	$3d^7$	8.17			40
			7.98	$Fe(CO)_5^+$	
^6D	$4s^13d^6$	7.92			

ions formed by reactions of excited-state ions is not limited to reactions with small diatomics or other small molecules [11].

$$Fe^+ + Fe(CO)_5 \rightarrow Fe_2(CO)_y^+ + 5 - y\,CO \qquad (3)$$

Table I contains data for the $(Fe^+)^*$ states that lie within ca. 4 eV of the $(Fe^+)^o$ [20], the electron configurations and energies of the $(Fe^+)^*$ states, the appearance energies of the $Fe(CO)_y^+$ product ions, and the abundances of $(Fe^+)^*$ determined by charge-exchange ion-molecule reactions for Fe^+ produced by 70 eV ionization of $Fe(CO)_5$. In our first paper describing the charge-exchange reaction chemistry of $(Fe^+)^*$ we adopted appearance energy values reported by Distefano because these values seemed to be the most widely accepted [21]. Subsequently, Ng and Armentrout reported appearance energies for the $Fe(CO)_y^+$ ions (y = 1–5) that differ from those reported by Distefano. For example, Ng and co-workers obtained an appearance energy of 10.84 ± 0.04 eV and 10.88 ± 0.05 eV for $Fe(CO)_2^+$ by using photoionization and photoelectron-photoion coincidence (PEPICO) spectroscopy, respectively [22]. On the basis of sequential bond energies of $Fe(CO)_y^+$ (y = 1–5) (determined by collision-induced dissociation) Armentrout and co-workers calculated an appearance energy value for $Fe(CO)_2^+$ as 10.85 ± 0.16 eV [23]. Distefano's original value for $Fe(CO)_2^+$ was 10.68 ± 0.1 eV. The appearance energy values that we adopted are the average of the reported appearance energies, 10.81 ± 0.2 eV. The appearance energies for the y = 1, 3–5 fragment ions are also the average of the available experimental data. The uncertainty in these average values were determined by taking the square root of the sum of the squares of the individual measurements.

The extent of electronic excitation of Fe^+ can be estimated from the data of Table I, and these data can be used to bracket the energy levels of the states giving rise to the charge-transfer product ions. Because $Fe(CO)_5{}^+$ has an appearance energy of 7.98 eV, the lowest lying state that can give rise to this product ion is the $[^4F(3d^7)]$ $(Fe^+)^*$ at 8.17 eV. The energy required to form $Fe(CO)^+$ is approximately 3.61 eV above $(Fe^+)^o$, thus, the $Fe(CO)^+$ ionic product must arise from the reaction of $(Fe^+)^*$ with energies greater than or equal to the $^2G(4s^13d^6)$. Similarly, the $Fe(CO)_4{}^+$ ionic product can be attributed to the $^4D(4s^13d^6)$ $(Fe^+)^*$ ion or states of higher energy.

On the basis of charge-exchange ion-molecule reaction chemistry data contained in Table I we estimate that $(Fe^+)^*$ with energies of up to about 4.0 eV are populated when Fe^+ is produced by 20–70–eV EI of $Fe(CO)_5$. That is, the $^2G(4s^13d^6)$ state is the lower bound of the highest populated electronic states. We find no evidence to suggest that $(Fe^+)^*$ ions with sufficient energy to form Fe^+ by charge-exchange with $Fe(CO)_5$ are formed by ionization of $Fe(CO)_5$. For example, reacting $^{54}Fe^+$ with $Fe(CO)_5$ does not produce any detectable $^{56}Fe^+$. From the data contained in Table I we estimate that the first excited state $[^4F(3d^7)]$ accounts for approxiamtely 40% of the $(Fe^+)^*$ formed by 70 eV EI of $Fe(CO)_5$. States from the $^4D(4s^13d^6)$ to the $^2G(3d^7)$ are estimated at 30%, and states between $^2P(3d^7)$ and $^2F(4s^13d^6)$ account for an additional 20%. Approximately 10% of the $(Fe^+)^*$ are composed of $^2G(4s^13d^6)$ and possibly higher lying states. On the basis of the abundances of charge-exchange product ions and ionic cluster fragments $[Fe_2(CO)_y^+$ (y = 3 and 4) we estimate that approximately 35% of Fe^+ ions produced by 70 eV electron impact of $Fe(CO)_5$ are in ground electronic state.

3.3. Basic assumptions and limitation of charge-exchange ion-molecule reactions to probe energies of reactant ions

The use of charge-exchange ion-molecule reactions to measure abundances and energies of $(M^+)^*$ requires that certain fundamental assumptions be made. Of key importance is whether the $(M^+)^*$ ions react to form the lowest energy product ions or react via channels that result in the lowest energy difference (ΔE) between the reactant $(M^+)^*$ states and ionic products. For example, high-lying $(Fe^+)^*$ ions, ^2G $(4s^13d^6)$ with an ionization energy of 11.64 eV, could react with $Fe(CO)_5$ to form $Fe(CO)_y^+$ ions where y ranges from 1 to 5, i.e., all possible charge-exchange product ions are formed. In an earlier paper we examined this question and argued that (Fe^+) reacts to form charge-transfer product ions that are controlled by ΔE [24]. Specifically, the high-lying ^2G state only reacts to form the $Fe(CO)^+$ ion. The strongest experimental evidence supporting this view comes from collisional relaxation studies of $(Fe^+)^*$ states which show that the high-lying $(Fe^+)^*$ states that give rise to the $Fe(CO)_2^+$ charge-transfer product ion are not depleted in reactions with NO [25]. That is, the high-lying $(Fe^+)^*$ state(s) that are unreactive with NO react with $Fe(CO)_5$ to form exclusively $Fe(CO)_2^+$. Clearly these $(Fe^+)^*$ states have sufficient energy to yield $Fe(CO)_y^+$ (y = 3–5) but react via a single reaction channel. Another observation consistent with a high degree of selectivity for charge-exchange reactions is that $(Fe^+)^*$ formed at near-threshold ionizing energy reacts with $Fe(CO)_5$ to form a greater amount of $Fe(CO)_y^+$ (y = 1–2) relative to $Fe(CO)_y^+$ (y = 3–4). This observation implicates two points: (i) at low ionizing beam energy highlying $(Fe^+)^*$ ions are formed in greater abundance than are the low-energy $(Fe^+)^*$ ions, and (ii) the high-lying $(Fe^+)^*$ states react to form high-energy product ions, e.g., y = 1–2. More detailed studies related to all these conclusions are still underway.

The rationale we use for explaining the selectivity of the charge-exchange reaction is to consider product ion formation as a two-step process, reaction (4).

$$(Fe^+)^* + Fe(CO)_5 \rightarrow Fe^o + [(Fe(CO)_5)^+]^* \rightarrow Fe(CO)_y^+ + 5\text{–}y\ CO \tag{4}$$

The fragment ions formed by dissociation of $[(Fe(CO)_5)^+]^*$ are determined by the internal energy content of the ion; however, if two product ions are energetically accessible, formation of the lower energy product ion requires that a greater amount of the excess internal energy be partitioned as translational energy or retained as internal energy of the ionic and neutral fragments. Although this explanation is somewhat speculative, the arguments are consistent with results obtained by laser-ion beam photodissociation studies for the $Fe(CO)_y^+$ species [26]. For example, we observed that the only photofragment ion of $[Fe(CO)_4^+]^*$, formed by photoexcitation at 458–514.5 nm, is $Fe(CO)_2^+$. Clearly, the $[(Fe(CO)_4)^+]^*$ ion formed by photoexcitation at these wavelengths has sufficient energy to produce $Fe(CO)_3^+$, but the excess energy (ca. 0.81 eV) must be partitioned as energy of translation or internal energy of the photofragment ion and neutral. The problem in making arguments about the dissociation dynamics of $[Fe(CO)_5)^+]^*$ on the basis of photodissociation is that it is difficult to differentiate single photon and multiphoton processes. For example, $[Fe(CO)_3{}^+]^*$ formed by photoexcitation of $Fe(CO)_y^+$ yields $Fe(CO)_y^+$ (y = 1 and 2) in a ratio of 1/1, but there is clear evidence in these experiments that some photofragment ions are formed by multiphoton processes. Armentrout suggests that spin conservation plays an important role in the dissociation reactions of $Fe(CO)_y^+$ species, and it is quite possible that spin conservation is responsible for differences in the photofragment ions sampled by FTICR and laser-ion beam photodissociation experiments [27]. Armentrout suggest that because the $Fe(CO)_y^+$ (y = 1–2) ions have quartet ground states the $Fe(CO)_3^+$ ions efficiently dissociate along an adiabatic pathway. The dissociation reactions of $[(Fe(CO)_5)^+]^*$ formed by charge-exchange may also be subject to constraints of spin conservation effects.

We also compared the abundances of $(Fe^+)^*$ by using other transition metal carbonyls, possibly opening new reaction channels less hindered by spin conservation, different ionizing beam energy, and collisional relaxation studies. For example, Fe^+ (from $Fe(CO)_5$) was reacted with $Cr(CO)_6$ and the abundance of $Cr(CO)_y^+$ was used to bracket the energies of the $(Fe^+)^*$ ions. The results from these studies correlate very well (within experimental error ±5%) with the data obtained from reactions of $(Fe^+)^*$ with $Fe(CO)_5$.

Another key question is whether $Fe(CO)_y^+$ product ions are formed by charge-transfer or ligand-exchange ion-molecule reactions. Armentrout and co-workers examined the reactions of translationally excited $(Fe^+)^o$ $[^6D(4s^13d^6)]$ ion with $Fe(CO)_5$ and concluded that the charge transfer and ionic cluster formation were the two main reaction channels [23]. To ensure that $Fe(CO)_y^+$ ions are indeed charge-transfer and not ligand-exchange product ions, reactions of $^{54}Fe^+$ with $Fe(CO)_5$ were examined. $^{54}Fe^+$ was isolated (see Experimental section) from $^{56}Fe^+$ and $^{57}Fe^+$ isotopes following 70–25 eV electron impact ionization of $Fe(CO)_5$ and allowed to react with $Fe(CO)_5$ for 100 ms. Results from these studies show that $^{54}Fe^+$ [70–25 eV EI of $Fe(CO)_5$] reacts with $Fe(CO)_5$ to produce $^{56}Fe(CO)_y^+$ (y = 1–5) charge-transfer ionic products. $^{54}Fe^+$ also reacts with $Fe(CO)_5$ to produce $^{54}Fe\,^{56}Fe(CO)_y^+$ (y = 2–4) ionic cluster fragments. It is apparent from these results that $Fe(CO)_y^+$ are indeed charge-transfer product ions.

3.4. Comparisons of abundances and energies of $(Fe^+)^*$ determined by other methods

Armentrout and Bowers have independently reported estimates for the abundances of $(Fe^+)^*$ produced by electron impact of $Fe(CO)_5$. Table II compares the abundances of $(Fe^+)^*$ and $(Fe^+)^o$ determined by charge-exchange ion-molecule reaction chemistry, electronic state chromatography (ESC) [15a], and Armentrout's method (SI) based on comparison of ion-molecule reaction cross section Fe^+ produced by EI and SI [28]. There are some differences between the excited states estimated from the charge-exchange ion-molecule reaction chemistry data and the ion mobility data. These differences can be explained by examining various assumptions made by each model. First, electronic state chromatography (ESC) does not resolve electronic states within a particular electron configuration so that the relative population of the $(Fe^+)^*$ ions is actually a measure of the relative abundances of particular electron configurations of the metal ion population. Second, Bowers' esti-

Table II. Relative abundance of excited states of atomic transition metal ions produced by electron impact ionization of the respective neutral precursor. The data summarizes relative abundances determined by charge-exchange ion-molecule reaction chemistry and previous studies

Ion	This work	Previous studies		
		Amentrout *et al.*	Bowers *et al.*	Ridge *et al.*
$Cr^+[Cr(CO)_6]$	80%	80%	54% [a]	75%
$Fe^+[Fe(CO)_5]$	65%	60%	–	–
$Co^+[Co(CO)_3(NO)]$	75%	–	64%	–
$Co^+[Co(CO)_3(NO)]$ 30 eV EI	68%	–	62%	–

[a] 50 eV, 3 Torr He.

mated fraction of each ionic state is based on prior knowledge of the extent of electronic excitation. The $(Fe^+)^*$ ions estimated by Bowers were based on the premise that only electronic states below 2 eV are populated when Fe^+ ions are formed by 30–50-eV EI of $Fe(CO)_5$. On the basis of the charge-exchange reactions of Fe^+ with $Fe(CO)_5$ it is apparent that electronic states with energies greater than 2 eV are populated by 30–70 eV EI of $Fe(CO)_5$. A considerable number of these states have $4s^13d^6$ electronic configuration. Because the ESC data do not resolve electronic states within a particular electron configuration, it is difficult to explicitly characterize the relative population of these states (states higher than 2 eV).

4. Relative abundances of metastable electronic states for other metal systems and by other ionization methods

4.1. Comparisons of Cr^+ (from $Cr(CO)_6$, Co^+ ($Co(CO)_3NO$) with Fe^+ ($Fe(CO)_5$)

We have compared the relative abundances of $(M^+)^o$ and $(M^+)^*$ ions formed by ionization of $Cr(CO)_6$, $Co(CO)_3NO$ with that for $Fe(CO)_5$. The electronic states, electron configurations, and ionization energies for Cr^+ and Co^+ are contained in Tables III–V, respectively. These tables also contain the thermochemical data for the charge-exchange product ions and the abundances of the $(M^+)^*$ and $(M^+)^o$ ions. Figure 3 compares the relative abundances of $(M^+)^*$ and $(M^+)^o$ for each of the three metal systems. In each case the abundance of $(M^+)^o$ is less than that for the $(M^+)^*$. Cr^+ and Co^+ are similar to Fe^+ in that in each case there is an appreciable abundance of high-lying excited states, but for both species the most abundant excited state is the lowest lying state, e.g., ^6D $(4s^13d^{4)}$ $(Cr^+)^*$ and ^5F $(4s^13d^7)$ $(Co^+)^*$. In the case of Cr^+, as was observed for Fe^+, the measured amounts of $(M^+)^*$ and $(M^+)^o$ do not vary depending on the charge-exchange reagent used for

Table III. Ionization energies of selected electronic states of Cr^+, the appearance energies of the charge transfer product ions, and the relative abundances of excited electronic states of Cr^+ [70–50 eV EI of $Cr(CO)_6$] estimated from the charge transfer reaction chemistry data [reactions of Cr^+ with $Fe(CO)_5$]

State	Configuration	Ionization energy (eV)	Appearance energy (eV)	Charge transfer product	Excited states relative abundance (%)
^2F	$3d^5$	10.78			10
			10.81	$Fe(C0)_2^+$	
^2D	$3d_5$	10.65			
^4F	$3d^44s^1$	10.62			
^4H	$3d^44s^1$	10.51			
^2I	$3d^5$	10.50			10
^4P	$3d^44s^1$	10.48			
^4D	$3d^5$	9.87	9.87	$Fe(C0)_3^+$	
^4P	$3d^5$	9.47			
^4G	$3d^5$	9.31			20
^4D	$3d^44s^1$	9.19			
			8.77	$Fe(CO)_4^+$	
^6D	$3d^44s^1$	8.25			60
			7.98	$Fe(CO)_5^+$	
^6S	$3d^5$	6.77			

Table IV. Ionization energies of selected electronic states of Cr^+, the appearance energies of the charge transfer product ions, and the relative abundances of excited electronic states of Cr^+ [70–50 eV EI of $Cr(CO)_6$] estimated from the charge transfer reaction chemistry data [reactions of Cr^+ with $Cr(CO)_6$]

State	Configuration	Ionization energy (eV0	Appearance energy (eV)	Charge transfer product	Excited states relative abundance (%)
^2F	$3d^5$	10.78			10
^2D	$3d^5$	10.65			
^4F	$3d^44s^1$	10.62	10.62	$Cr(CO)_3^+$	
^4H	$3d^44s^1$	10.51			
^2I	$3d^5$	10.50			10
^4P	$3d^44s^1$	10.48			
				$Cr(CO)_4^+$	
^4D	$3d^5$	9.87			
^4P	$3d^5$	9.47			
^4G	$3d^5$	9.31			15
^4D	$3d^44s^1$	9.19			
			9.17	$Cr(CO)_5^+$	
^6D	$3d^44s^1$	8.25			65
			8.14	$Cr(CO)_6^+$	
^6S	$3d^5$	6.77			

Table V. Ionization energies of selected electronic states of Co^+, the appearance energies of the charge transfer product ions, and the relative abundances of excited electronic states of Co^+ [70–50 eV EI of $Co(CO)_3(NO)$] estimated from the charge transfer ion-molecule reaction chemistry data

State	Configuration	Ionization energy (eV0	Appearance energy (eV)	Charge transfer product	Excited states relative abundance (%)
^1H	$4s^13d^7$	11.65			10
			11.53	$Fe(CO)^+$	
^1P	$4s^13d^7$	11.28			
^3D	$4s^13d^7$	11.27			12
^3H	$4s^13d^7$	11.22			
^1G	$4s^13d^7$	10.98			
^3P	$4s^13d^7(^2P)$	10.95			
^3P	$4s^13d^7(^4P)$	10.84			
			10.81	$Fe(CO)_2^+$	
^3G	$4s^13d^7$	10.54			
^1G	$3d^8$	10.24			8
^5P	$4s^14d^7$	10.06			
			9.87	$Fe(CO)_3^+$	
^3P	$3d^8$	9.50			
^1D	$3d^8$	9.30			23
^3F	$4s^13d^7$	9.08			
			8.77	$Fe(CO)_4^+$	
^5F	$4s^13d^7$	8.27			47
			7.98	$Fe(CO)_5^+$	
^3F	$3d^8$	7.86			

bracketing the energies of the excited states (e.g., $Fe(CO)_5$ or $Cr(CO)_6$). The abundances of $(Co^+)^*$ were determined by reaction with $Fe(CO)_5$ because Co^+ is quite reactive with $Co(CO)_3NO$ and the abundances of $Co(CO)_y^+$ as well as $Co(CO)_yNO^+$ complicate data analysis.

One of the more interesting aspects of the data for Cr^+ and Co^+ systems is the dependence of the distribution of charge-transfer product ions on the ionizing energy (Figures 4 and 5). In the case of Cr^+ the abundance of the $Fe(CO)_y^+$ ions varies smoothly as the ionizing energy is dropped below approximately 40 eV, whereas in the case of Co^+ the abundance of $Fe(CO)_y^+$ y = 1–3 drops off smoothly but the y = 4–5 product ions remain relatively constant down to approximately 20 eV. We previously reported that lowering the ionizing energy used to ionize $Fe(CO)_5$ down to approximately 20 eV did not decrease the relative abundance of the high-lying $(Fe^+)^*$ ions, but it appears that the high-lying $(Cr^+)^*$ and $(Co^+)^*$ ions are less populated at low ionizing energy. Note also that in the case of Cr^+ the abundance of the excited states that give rise to $Fe(CO)_2^+$, and to a lesser extent $Fe(CO)_3^+$, increases in abundance between 40–60 eV ionizing energy, but such behavior is not observed for Co^+. More detailed studies on these systems are underway.

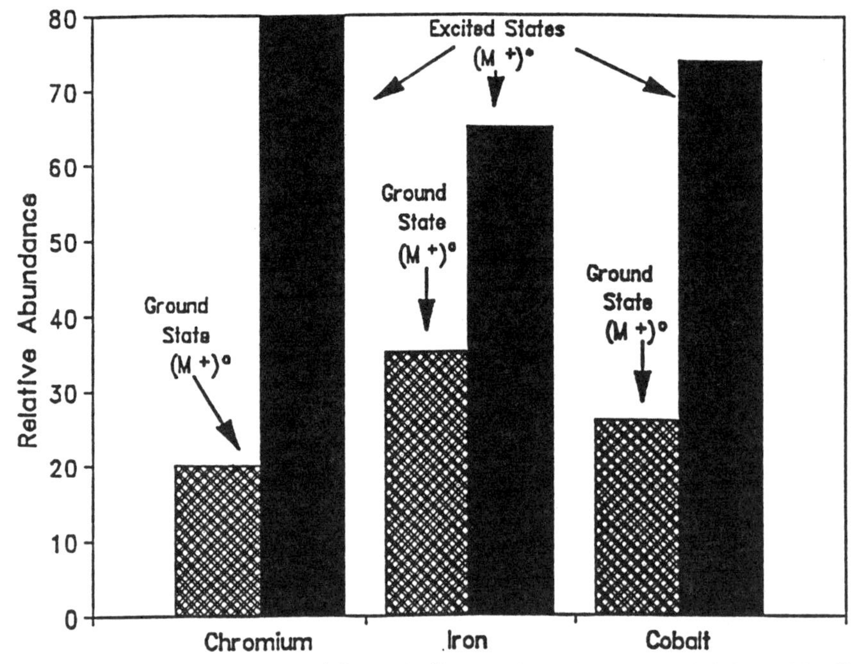

Fig. 3. Relative abundance of $(M^+)^o$ and $(M^+)^*$ ions (M = Cr, Fe, and Co) formed by 50 to 70 eV electron impact of $Cr(CO)_6$, $Fe(Co)_5$, and $Co(CO)_3(NO)$, respectively.

4.2. Fe^+ FORMED BY COLLISION-INDUCED DISSOCIATION

Several laboratories have used collision-induced dissociation as a method for producing metal ions [29] (anions, ligated metal ions, and ionic clusters) and Armentrout [23] has used CID to determine M—CO bond energies for the $M(CO)_y^+$ ions. Clearly, such methods can result in the formation of excited state ions and it is important to know the relative abundance of $(M^+)^o$ and $(M^+)^*$, especially for thermochemical determinations. The only systems we have examined to date are the Fe^+ ions formed by CID of $Fe(CO)^+$ and $Fe(CO)_2^+$. The data for these systems are contained in Tables VI and VII. Fe^+ is the primary CID product ion of $Fe(CO)^+$, but Fe^+ is primarily formed by consecutive reactions involving loss of CO from $Fe(CO)_2^+$. Note, however, that in both cases the most abundant ion is the first excited state $(Fe^+)^*$, but appreciable amounts of higher-lying excited states are also formed.

One of the most important pitfalls associated with CID for forming M^+ ions is that the precursor ion, $M(CO)_y^+$, is translationally excited, thus M^+ may be formed with excess kinetic energy. Under these conditions it is important to know whether the charge-exchange reactivity of M^+ is due to the presence of $(M^+)^*$ ions or simply translationally excited $(M^+)^o$ ions. In our studies we performed the CID by using off-resonance rf excitation to

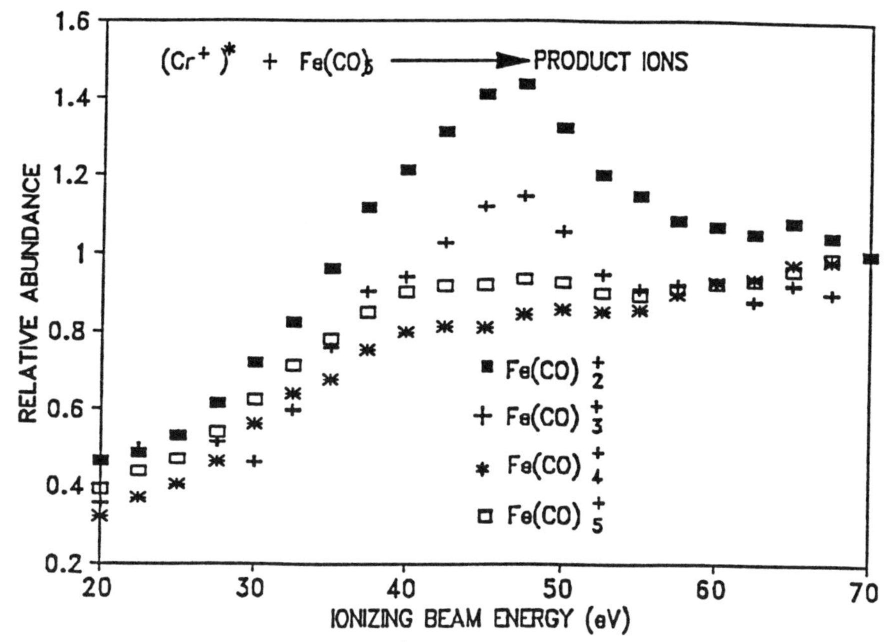

Fig. 4. The relative abundance of $Fe(CO)_y^+$ charge-exchange product ions as a function of the ionizing electron beam energy used to produce the Cr^+ ions.

minimize the translational energy imparted to the ion, thus the CID product ions are formed by multiple collisions between the precursor ion and CID reagent gas [30]. In addition, the experiments were performed by using the two-section ion cell and the upper limit of the translational energy of the ion is determined by the radius of the aperature of the conductance limit separating the two ion cells [31]. Even though we have taken every precaution to minimize the translational energy of the CID product ions, it should be recognized that we cannot be absolutely certain that the energies of the $(Fe^+)^*$ measured in this experiment are not skewed to higher values. On the other hand, we have translationally excited $(Fe^+)^o$ ions (formed by collisional relaxation of the Fe^+ ions formed by EI of $Fe(CO)_5$) and we do detect significant amounts of charge-exchange product ions, thus we conclude that 1–5 eV translationally excited $(Fe^+)^o$ do not react with $Fe(CO)_5$ by endothermic charge-exchange ion-molecule reactions. This result is also consistent with our studies of the trapping voltage on the abundance of $Fe(CO)_y^+$ charge-exchange product ions [25].

4.3. Fe^+ AND Co^+ FORMED BY LASER ABLATION

Laser ablation is frequently used to generate M^+ ions for ion-molecule reaction chemistry studies [32], but very little has been done to characterize

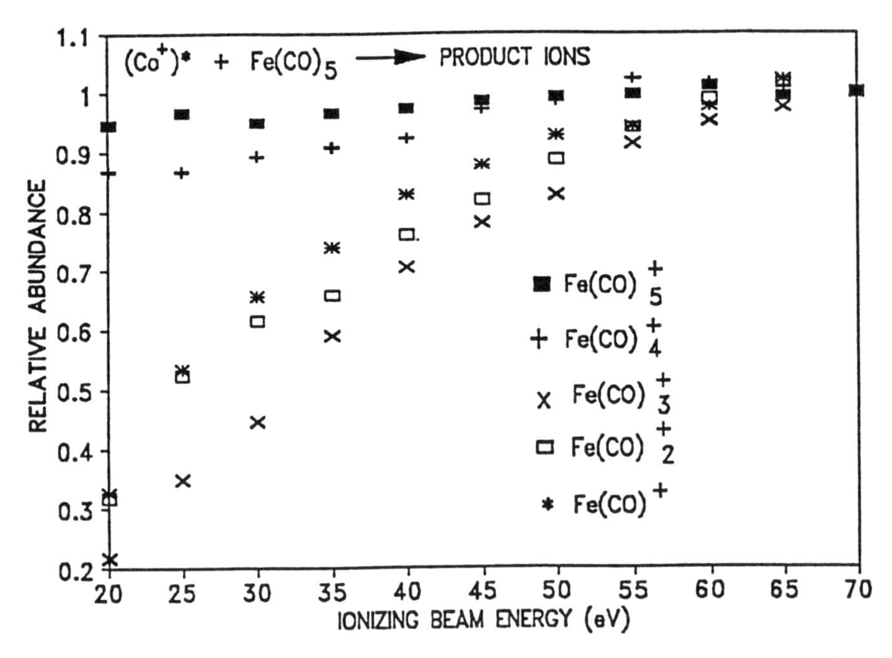

Fig. 5. Plot of the relative abundance of $Fe(CO)_y^+$ charge-transfer product ions as a function of the ionizing beam energy used to the reactant Co^+ ions.

Table VI. Extent and distribution of electronic states of Fe^+ formed by off-resonance CID of $Fe(CO)^+$ [produced by EI of $Fe(CO)_5$

State	Configuration	Ionization energy (eV)	Excited states relative abundance (%)
^4H	$4s^13d^6$	10.51	
^4P	$4s^13d^6$	10.45	
^2D	$3d^7$	10.42	5
^2H	$3d^7$	10 39	
^2P	$3d^7$	10.17	
^2G	$3d^7$	9.86	
^4P	$3d^7$	9.56	20
^4D	$4s^13d^6$	8.90	
^4F	$3d^7$	8.17	65
^6D	$4s^13d^6$	7.92	10

the energies of the ions produced. The charge-exchange reaction chemistry can be used to probe such systems. To date we have only examined the Fe^+ ions generate by laser ablation (337 nm) of 304 stainless steel. Figure 6 compares the abundances of the $Fe(CO)_y^+$ charge-exchange product ions observed for $(Fe^+)^*$ formed by EI and laser ablation (337 nm) of 304 stainless steel.

Table VII. Extent and distribution of electronic states of Fe^+ formed by off-resonance CID of $Fe(CO)^+$ [produced by EI of $Fe(CO)_5$]

State	Configuration	Ionization energy (eV)	Excited states relative abundance (%)
^4H	$4s^13d^6$	10.51	
^4P	$4s^13d^6$	10.45	
^2D	$3d^7$	10.42	15
^2H	$3d^7$	10 39	
^2P	$3d^7$	10.17	
^2G	$3d^7$	9.86	
^4P	$3d^7$	9.56	20
^4D	$4s^13d^6$	8.90	
^4F	$3d^7$	8.17	40
^6D	$4s^13d^6$	7.92	25

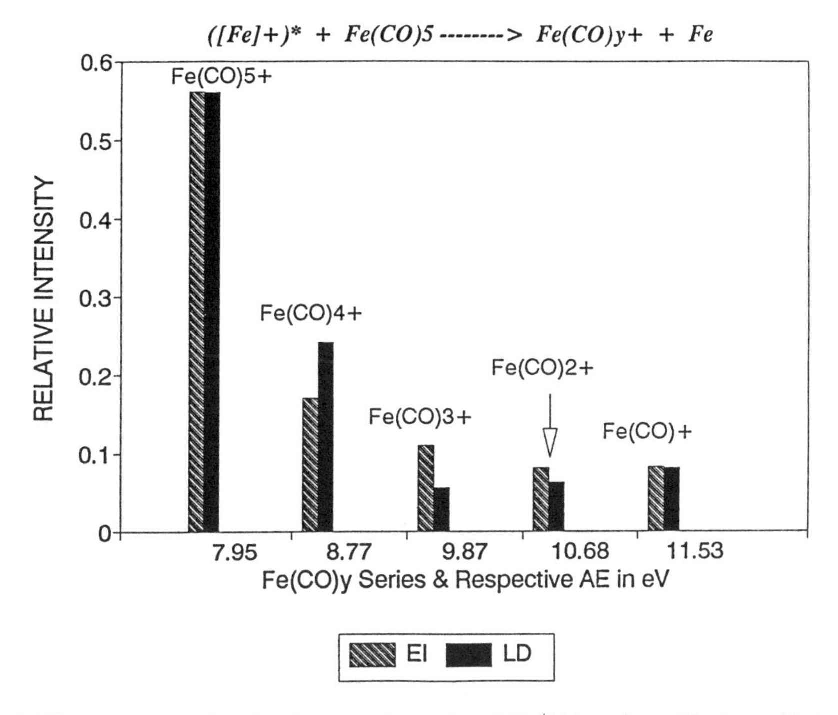

Fig. 6. Histogram comparing abundances and energies of $(Fe^+)^*$ ions formed by laser ablation (337 nm) of 304 stainless steel and electron impact ionization (70 eV) of $Fe(CO)_5$.

5. Collisional relaxation studies of Fe^+

A primary advantage of the two-section ion cell ICR charge-exchange ion-molecule reaction chemistry method for measuring abundances of metastable electronic states of metal ions is that it can be easily combined with studies

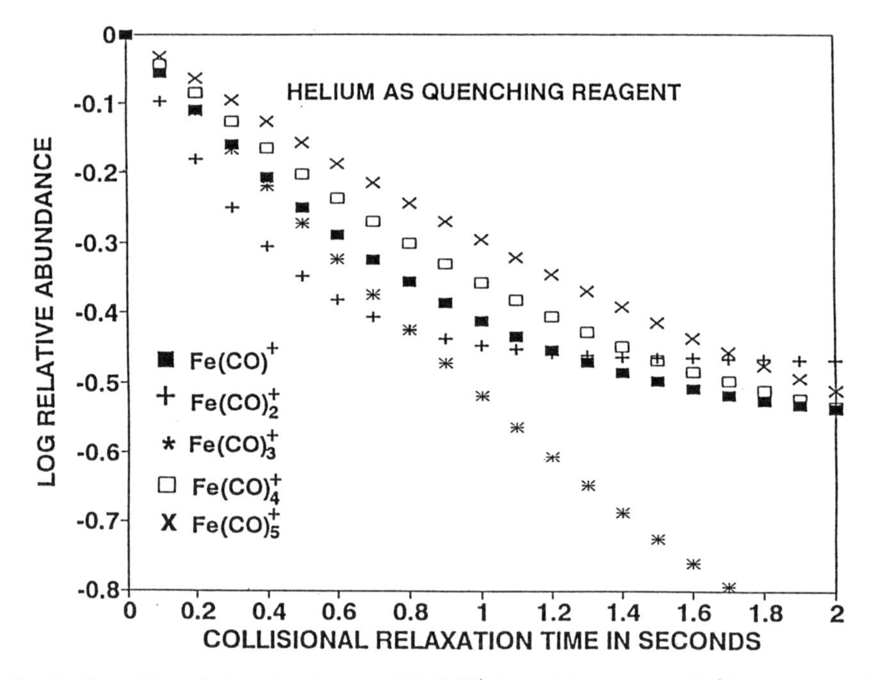

Fig. 7. Plot of log relative abundances of $Fe(CO)_y^+$ formed by reaction $(Fe^+)^*$ with $Fe(CO)_5$ following collisional relaxation with He (2×10^{-7} Torr).

of collision relaxation of the metal ions. For instance, metal ions can be formed in region 1, partitioned to region 2 where the ions react with some reagent gas, the ions are then partitioned back to region 1 and allowed to react with $Fe(CO)_5$. Because the reaction time of the ion in region 2 and the pressure of the reagent gas in region 2 are defined, the reaction rates for the collisional relaxation process can be determined with good (±10%) accuracy. As with all kinetic studies performed by ICR methods the greatest source of error is the absolute pressure measurement for the reagent gas and uncertainities associated with the kinetic energies of the reacting ion.

Figures 7–9 contain $\log_{10}$ plots of the abundance of the $Fe(CO)_y^+$ charge-exchange product ions as a function of reaction time with helium, argon, and krypton. There are several very striking features about this data. For each reagent gas the $(Fe^+)^*$ ions that yield $Fe(CO)_3^+$ show a single reactive component that decays linearly. All the other charge-exchange product ions decay in a non-linear fashion and several of these show reactive and non-reactive ion populations. For example, the $Fe(CO)_2^+$ product ion abundance levels off after approximately 1 s reaction time leaving an unreactive or very slowly reacting $(Fe^+)^*$ ion population. We previously observed an unreactive form of $(Fe^+)^*$ with NO and suggested the electron configuration of the ^6S state (s^2d^5) as the reason for the lack of reactivity. In addition, we argued

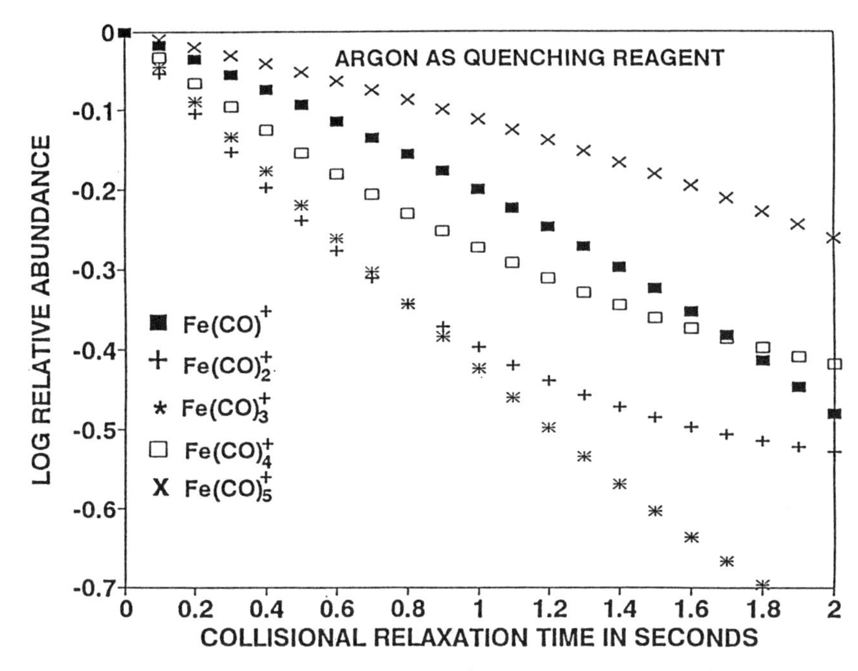

Fig. 8. Plot of log relative abundances of $Fe(CO)_y^+$ formed by reaction $(Fe^+)^*$ with $Fe(CO)_5$ following collisional relaxation with Ar (2×10^{-7} Torr).

that s^1d^7 states that lie higher in energy than the ^6S state react via an electron transfer mechanism to yield the ^6S state (Scheme 1). Assuming this explanation is correct, then it would appear that with the rare gases the s^1d^7 to s^2 d^5 change in electron configuration does not occur. This gives further validation to our suggestion that the electron configuration changing reaction observed with NO is due to the unpaired electron of NO.

It is also interesting to compare the rates of collisional relaxation of the various $(Fe^+)^*$ ions with the rare gas atoms. Helium and argon relax $[^4F(3d^7)]$ $(Fe^+)^*$ at approximately the same rate, and this rate is approximately twice that for relaxation by Kr. The general trend for the collisional relaxation of the high-lying states is in the reverse order. For example, Kr relaxes electronic states with energies greater than 10.81 eV (states that form $Fe(CO)^+$ and $Fe(CO)_2^+$) approximately 7 times faster than He and 2 times faster than Ar. The relative rates (Kr > Ar > He) suggest that collisional relaxation of $(Fe^+)^*$ by rare gas atoms is not simply a physical process of converting electronic energy into energy of translation. Because the higher-lying $(Fe^+)^*$ states consist mainly of $4s^13d^6$ states and the first excited state has a $3d^7$ configuration, it appears that Ar and Kr can interact with the s^1d^6 configuration on a more attractive potential surface than He.

The most plausible mechanism for collsional relaxation of $(Fe^+)^*$ by the

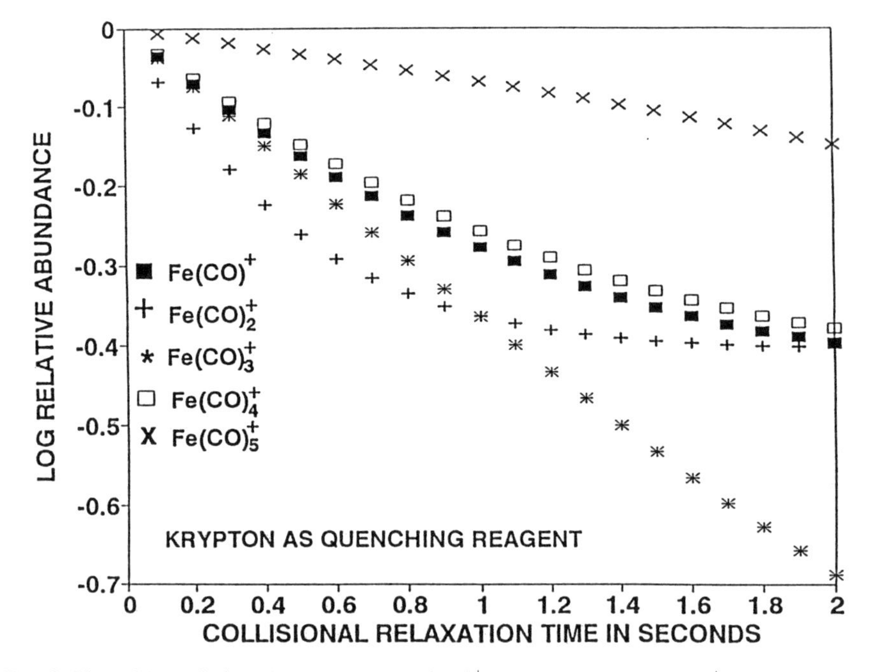

Fig. 9. Plot of log relative abundances of $Fe(CO)_y^+$ formed by reaction $(Fe^+)^*$ with $Fe(CO)_5$ following collisional relaxation with Kr (2×10^{-7} Torr).

rare gas atoms is the curve-crossing mechanism proposed by Armentrout [14a, 33]. According to a curve crossing mechanism collisional relaxation of $(Fe^+)^*$ is likely to occur when the potential energy curves between $(Fe^+)^*$ states and lower-lying states cross. Thus, high-lying states should be less efficiently quenched to the ground state under thermal collision conditions and the larger rare gas atoms should be more efficient quenchers than are lighter rare gas atoms. The experimental data for relaxation of $(Fe^+)^*$ by He, Ar, and Kr is in general agreement with a curve-crossing mechanism. The fact that the trend in the relaxation data for the $^4F(3d^7)$ $(Fe^+)^*$ is in the reverse order, He > Ar > Kr, is probably related to the small energy differences between these states and the promotion energy required for the $3d^7$ to $4s^13d^6$ reaction.

Collisional relaxation studies of metastable electronic states of Fe^+ have shown that $Fe_2(CO)_4{}^+$ ionic cluster fragment is exclusively formed by ground-state Fe^+. For example, Figure 10 contains a plot of the relative abundance of $Fe_2(CO)_4^+$ as a function of the collisional relaxation time for $(Fe^+)^*$ with He buffer gas. The relative abundance of the y = 2–3 ionic cluster fragments (not shown in the figure) decreases with increasing relaxation time and the abundance of the ionic cluster fragment increases. At the very longest relaxation time virtually all the ionic cluster fragments formed correspond

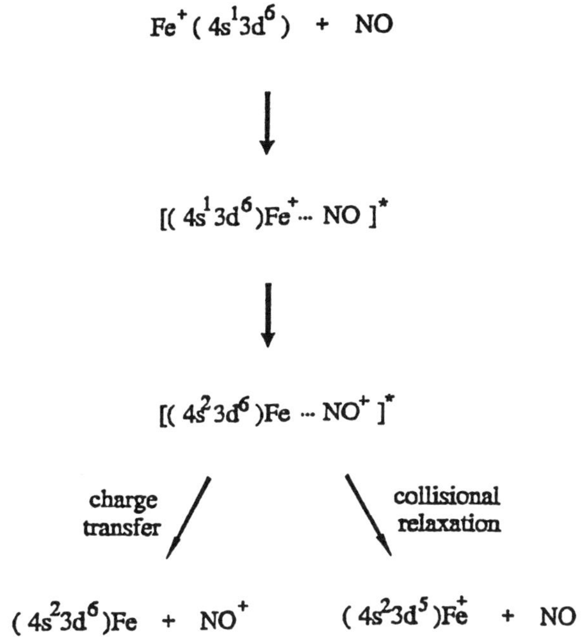

$$Fe^{+}(4s^{1}3d^{6}) + NO$$

$$\downarrow$$

$$[(4s^{1}3d^{6})Fe^{+} \cdots NO]^{*}$$

$$\downarrow$$

$$[(4s^{2}3d^{6})Fe \cdots NO^{+}]^{*}$$

charge transfer ↙ ↘ collisional relaxation

$$(4s^{2}3d^{6})Fe + NO^{+} \qquad (4s^{2}3d^{5})Fe^{+} + NO$$

Scheme 1.

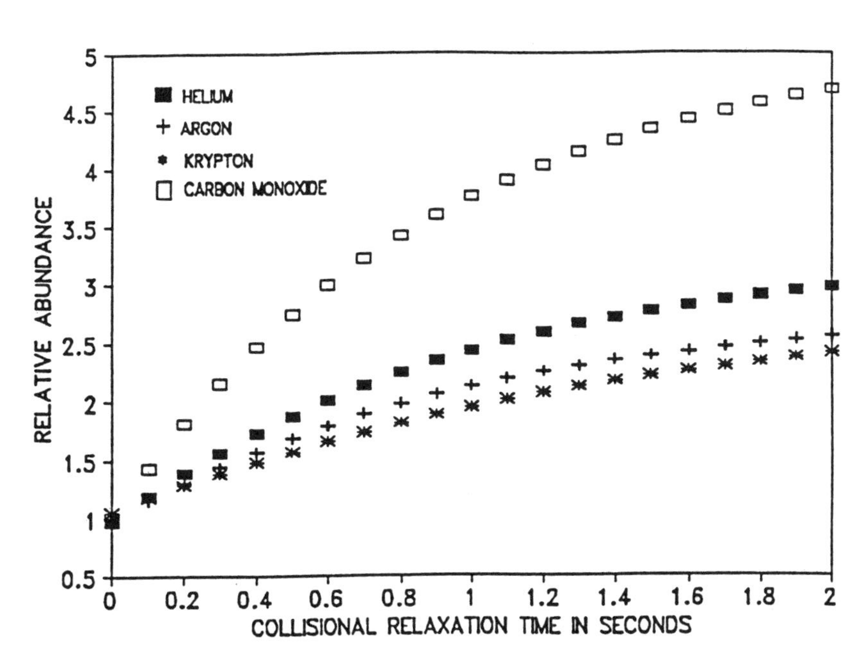

Fig. 10. Plot of the relative abundances of $Fe_2(CO)_4^+$ as a function of collisional relaxation time of $(Fe^+)^*$ using He, Ar, Kr, and CO reagent gases.

to $Fe_2(CO)_4^+$. We interpret this data as evidence that the y = 2–3 ionic cluster fragments are formed exclusively by reaction of $(Fe^+)^*$ ions, whereas the ionic cluster fragment is formed by reaction of $(Fe^+)^o$ ions.

The collisional relaxation results obtained from the FTICR experiment appear inconsistent with results from ion-beam studies. For example, Armentrout suggested that $(M^+)^*$ ions survive many thousands of collisions without being relaxed to $(M^+)^o$, but the results from the FTICR experiment show that most of the $(Fe^+)^*$ ions are relaxed to $(Fe^+)^o$ after 3–10 collisions with the bath gas. There are two factors that must be considered when comparing the data from the two very different experimental methods: (i) the ion beam experiment measures the abundance of $(M^+)^o$ as a function of the number of collisions between $(M^+)^*$ and the collision gas, and (ii) the ions are sampled on the 10^{-5} s timescale. The FTICR experiment directly measures of the number of $(M^+)^*$ ions that react to form the charge-exchange product ions, and the experiment is performed on a much longer timescale (10^{-3} s to 10^2 s). The numbers of collisions required to relax the $(M^+)^*$ species quoted by Armentrout is for relaxing the entire ion population. The data in Figures 7–9 clearly show the same trends, e.g., some of the $(Fe^+)^*$ ions do not completely decay following 3–10 collisions. Thus, some excited states may undergo efficient collisional relaxation, whereas other states relax with low efficiency. In the specific case of $(Fe^+)^*$, Armentrout found a specific ion that only undergoes efficient collisional relaxation with O_2 [34]. This was a low abundance state (ca. 5%), and this is probably the same $(Fe^+)^*$ species that we found to be unreactive with NO, e.g., the ^6S state.

6. Reactions of $(M^+)^*$ and $(M^+)^o$ ions with polar organic molecules

One of the long range objectives of our metal ion chemistry research is to develop methods for structure determination of complex organic (denoted org) molecules. For example, several years ago we showed that the dissociation reactions of organo-alkali metal ions, $[org + Na]^+$ as a specific example, differed significantly from the dissociation reactions of the protonated molecule, $[org + H]^+$ [35]. We attributed these differences to the "structure" of the $[org + Na]^+$ and $[org + H]^+$ ions. Specifically, the attachment site(s) and binding energies of the cations are different [36, 37]. Numerous other laboratories have reported similar results and it is generally agreed that organo-metal ion complexes provide unique insight into ion structure [38, 39]. One way to extend this methodology is by using metal ions that can bind at specific sites of the polar molecule and that have different binding energies toward various functional groups. That is, different metals will have different affinities for C, N, or O containing functional groups and the relative binding energies of these atoms to the M^+ species will be different [40]. In addition, transition metal-organic bonds are more covalent than alkali metal ion-organic bonds and the energetics and mechanisms of reaction are quite different [41, 42].

Another factor to consider in the case of transition metal ions is the relative reactivities of the different $(M^+)^*$ species. For example, the different electron configurations of the various $(M^+)^*$ species should alter the reactivities toward different functional groups [43]. Clearly, the reactivities of the various $(M^+)^*$ species may differ, and the total energy of the system will be different; the energy of $[org + (M^+)^*]$ will be greater than that for $[org + (M^+)^o]$ by an amount equal to the energy for the $(M^+)^* \rightarrow (M^+)^o$ transition. There are several examples that illustrate the effect of electron configuration on the reactivity of Fe^+ and Co^+. For instance, the $3d^n$ state of Co^+ and Fe^+ react with propane more efficiently than the $4s^1 3d^{n-1}$ states to form the collisionally stablized $[M^+—C_3H_8]$ adduct ion [44]. Similarly, the $3d^8$ states of Co^+ are more reactive (with CH_3I) than the $4s^1 3d^7$ states that react to form $CoCH_3I$, $CoCH_3{}^+$, and CoI^+ [45]. Tonkyn and Weisshaar also observed that M^+ ions (Sc through Zn, formed by laser ablation) with d^n ground or low energy states react more efficiently to form the adduct ion [46]. The results are explained by a repulsive interaction for the occupied 4s orbital.

We have performed studies of Cr^+ and Fe^+ with N-methylformamide, N,N-dimethylformamide, N,N-diethylformamide, N,N-dimethylacetamide, and N-ethylacetamide. The initial objectives of this work are to compare the reactivities of the two metal ions with polar organic molecules, identify product ions that are unique to $(M^+)^*$ and $(M^+)^o$ ions, and evaluate the utility of the ion-molecule reaction chemistry for structural determination of complex molecules. Cr^+ and Fe^+ were chosen for the initial studies because we know the energies and abundances of $(M^+)^*$ and $(M^+)^o$, both ions can be generated in high abundance, and the chemistry of the two ions are quite different (see section VII.B of reference 2). The formamides and acetamides were chosen because the neutral molecules are relatively volatile, easily introduced into the ICR vacuum system, and the molecules are structurally similar to amino acids and peptides. The second phase of our studies will be to study the ion-molecule reaction chemistry of these same metal ions with amino acids and peptides.

The M^+ ions react with the amides by four pathways (Scheme 2). The M^+ ions react by (i) charge-exchange, (ii) collisional relaxation, (iii) to form adduct ions of the type $[M^+—amide]$, and (iv) to form $[M^+—amide\ fragment]$, denoted $[M^+—F_i]$. The detailed molecular beam studies reported previously by Sonnenfroh and Farrar would not pickup the charge-exchange and collisional relaxation reaction channels [42]. The major product observed for the charge-exchange ion-molecule reactions is the protonated amide ion, $[amide + H]^+$, formed by reaction of the amide radical cations (formed by charge-exchange) with amide neutral or fragments of the amide radical cation. The fragment ions are formed by dissociation of the amide radical cation and/or dissociation of the $[amide + H]^+$ ion. Because these are secondary

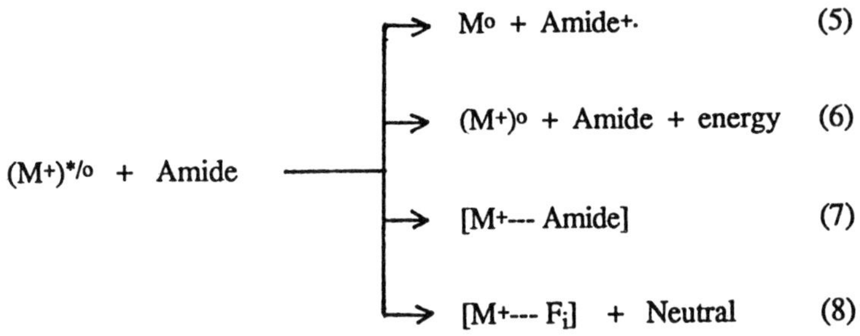

$$(M^+)^{*/o} + \text{Amide} \longrightarrow \begin{cases} M^o + \text{Amide}^{+\cdot} & (5) \\ (M^+)^o + \text{Amide} + \text{energy} & (6) \\ [M^+\text{---}\,\text{Amide}] & (7) \\ [M^+\text{---}\,F_i] + \text{Neutral} & (8) \end{cases}$$

Scheme 2.

Table VIII. Selected products of the reaction of Fe^+ < % > and Cr^+ with formamides and acetamides

Reactant	Fe^+		CR^+	
	% loss of CH_3	Major product	% loss of CH_3	Major product
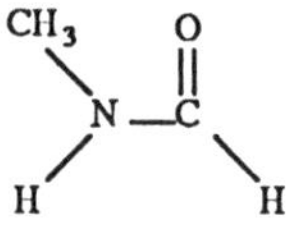 $CH_3(H)N{-}C(=O){-}H$	100	loss of CH_3	0	Cr^+[org}
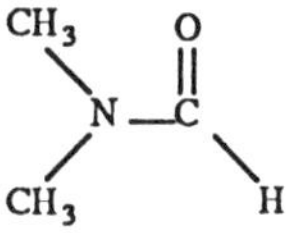 $(CH_3)_2N{-}C(=O){-}H$	100	loss of CH_3	0	Cr^+[org}
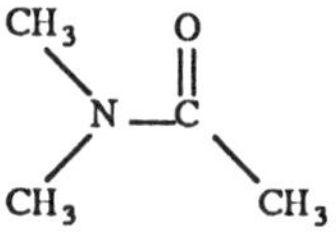 $(CH_3)_2N{-}C(=O){-}CH_3$	100	loss of CH_3	0	Cr^+[org}
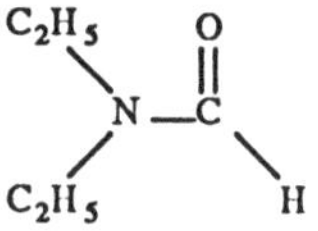 $(C_2H_5)_2N{-}C(=O){-}H$	25	$Fe^{=}[org{-}H_2]$	0	Cr^+[org}
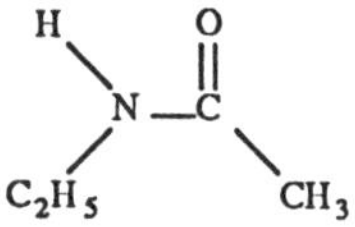 $H(C_2H_5)N{-}C(=O){-}CH_3$	100	loss of CH_3	–	–

ion-molecule products no further details on these reaction will be presented in this report.

Fe^+ reacts with the formamides and acetamides by charge-exchange, dissociative and nondissociative, and to form metal containing product ions, viz.

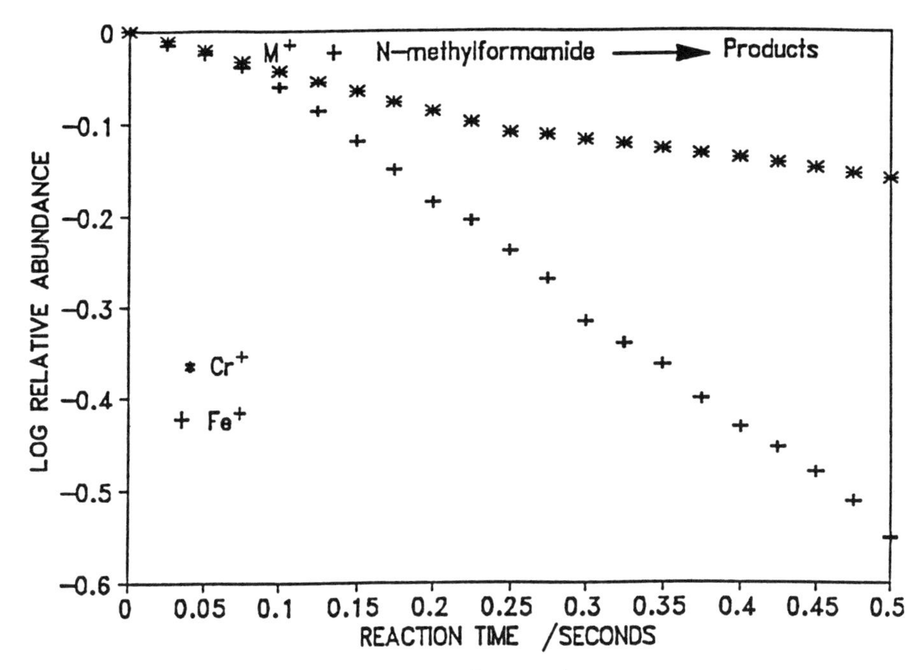

Fig. 11. Plots of log relative abundances of Cr^+ and Fe^+ as a function of reaction time N-methylformamide. Compares the relative reactivities of Cr^+ and Fe^+.

[Fe^+—amide] and [Fe^+—F_i], where F_i involves either loss of CO or CH_3 (see Table VIII). Similar types of reactions are observed for Fe^+ ions reacting with carbonyl compounds, and these product ions arise as a result of C—C bond insertion by the metal ion [47]. Conversely, Cr^+ reacts with the amides by charge-exchange, both dissociative and non-dissociative, and to form [Cr^+—amide] adduct ions (Table VIII). Dehydrogenation products of the [Cr^+—amide] adduct ions, commonly observed reaction products of Cr^+ [48], are also observed but, because these product ions do not provide structural information about the reacting neutral, these reactions will not be discussed further in this report [49]. The charge-exchange reactions of Fe^+ and Cr^+ are endothermic for $(M^+)^o$ and must involve $(M^+)^*$ species.

Fe^+ is much more reactive with these compounds than is Cr^+ (see Figure 11) and it appears that the reaction chemistry of these metals is controlled by a single functional group as opposed to synergistic effects of the amide group. Sonnenfroh and Farrar have examined detailed energetics of the Fe^+ and Cr^+ acetaldehyde reaction and suggested that the initial site of interaction is the oxygen atom followed by OC—C insertion [46]. In Figure 11 the decay of Cr^+ is biexponential but the rate of decay is similar for both portions of the curve. On the other hand, the curvature in the decay curve for Fe^+ suggest that initially reacting ions react slowly and the rate accelerates with

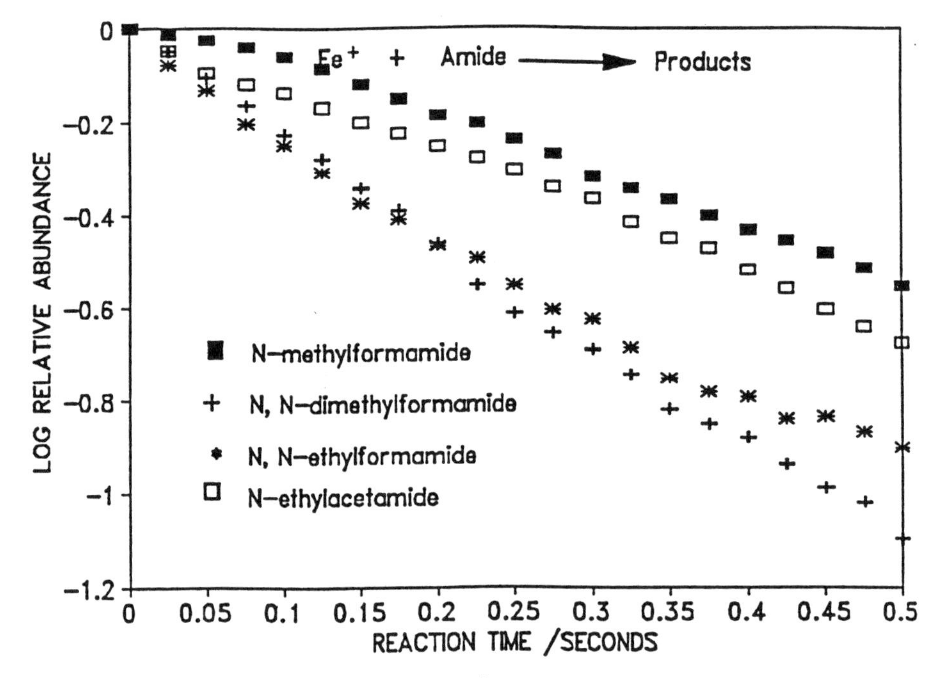

Fig. 12. Plots of log relative abundance of Fe^+ as a function of reaction time with various amides. The reactant Fe^+ was produced by 70 eV EI of $Fe(CO)_5$.

time. Such decay curves suggest that high-energy $(Fe^+)^*$ ions undergo collisional relaxation to a lower-lying $(Fe^+)^*$ ion or $(Fe^+)^o$, a slow process, and the resulting Fe^+ ions are more reactive. Similar composite decay curves are observed in Figures 12 and 13 where we have compared the decay curves for Fe^+ and Cr^+ reacting with N-methylformamide, N,N-dimethylformamide, N,N-diethylformamide, and N-ethylacetamide.

7. Future directions

The rich chemistry of metal ions with organic molecules assures that this field will continue to develop, and many new observations on specific reactions of long-lived excited state metal ions will be reported. At the current time, in the area of developing metal ion structural probes of complex molecules there appear to be few examples of truly novel chemistry for Cr^+ and Fe^+, both metal ions react to form product ions similar to those reported for mono-functional group organic molecules. Specifically, Fe^+ ions react to form product ions that can be explained by bond insertion reactions, and Cr^+ ions react to form adduct ions. In addition, Amster and coworkers have studied the reaction chemistry of metal ions with laser desorbed neutral

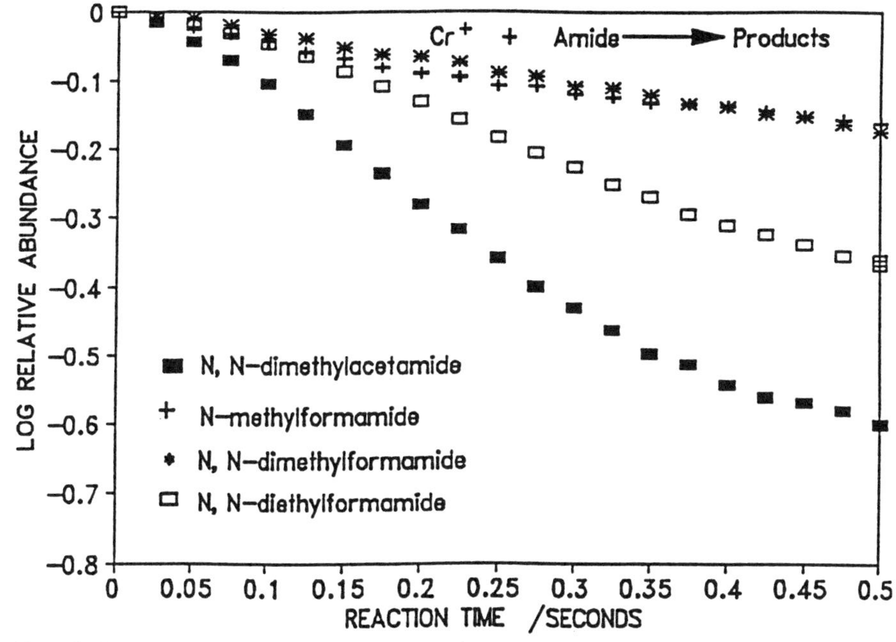

Fig. 13. Plots of log relative abundance of Cr^+ as a function of reaction time with various amides. The reactant Cr^+ was produced by 70 eV EI of $Cr(CO)_6$.

peptides and the fragmentation reactions of these systems are basically similar to those observed by tandem mass spectrometry of the $[org + H]^+$ ions [50]. We find that the reaction chemistry of the $(M^+)^*$ ions is dominated by charge-exchange and collisional relaxation (to form $(M^+)^o$), and these results are not inconsistent with the types of reaction products observed by Amster. The large cross-section for the charge-exchange reaction channel is of interest for probing gas-phase ion chemistry of complex molecules because this reaction provides a convenient method for producing radical cations of peptides and other molecules of biological interest. There is considerable interest in investigating the fragmentation reactions of odd-electron species, especially methods that might yield information complementary to multiphoton ionization.

The more interesting reaction chemistry of metal ions with polar organic molecules will probably involve systems such as Co^+ and Ni^+ because the chemical reactions of these metal ions is due to Lewis acid character [51]. It will also be interesting to compare the reactivities of Co^+ and Ni^+ with metals such as Cr^+ and Fe^+ toward polar organic molecules and model peptides. Furthermore, it will be interesting to compare the reactions of Co^+ and Ni^+ $(M^+)^*$ and $(M^+)^o$ species. The ICR based experiments will be especially useful for these studies because the chemistry of $(M^+)^*$ and $(M^+)^o$ can be identified and a complete picture of the system can be obtained.

Acknowledgements

This work was supported by the U.S. Department of Energy, Division of Chemical Sciences, Office of Basic Energy Sciences, the National Science Foundation, and the Robert A. Welch Foundation.

References

1. J. Allison, *Prog. Inorg. Chem.* **34**, 627 (1986); (b) S. W. Buckner and B. S. Freiser, *Polyhedron* **74**, 1583 (1988); (c) D. H. Russell (Ed.), *Gas Phase Inorganic Chemistry* (Plenum Press, New York, 1989).
2. K. Eller and H. Schwarz, *Chem. Rev.* **91**, 1121 (1991).
3. J. A. Simoes and J. L. Beauchamp, *Chem. Rev.* **90**, 629 (1990).
4. R. B. Freas and D. P. Ridge, *J. Am. Chem. Soc.* **102**, 7129 (1980).
5. L. F. Halle, P. B. Armentrout, and J. L. Beauchamp, *J. Am. Chem. Soc.* **103**, 962 (1981).
6. W. D. Reents, Jr., F. Strobel, R. B. Freas, J. Wronka, and D. P. Ridge, *J. Phys. Chem.* **89**, 5666 (1985).
7. S. K. Huang and M. L. Gross, *J. Phys. Chem.* **89**, 4422 (1985).
8. F. Strobel and D. P. Ridge, *J. Phys. Chem.* **93**, 3635 (1989).
9. R. B. Cody, R. C. Burnier, W. D. Reents, T. J. Carlin, R. K. McCrery, R. K. Lengal, and B. S. Freiser, *Int. J. Mass Spectrom. Ion Phys.*, **33**, 37 (1980).
10. P. B. Armentrout and J. L. Beauchamp, *J. Chem. Phys.* **48**, 315 (1980); (b) R. Georgiadis and P. B. Armentrout, *J. Am. Chem. Soc.* **108**, 2119 (1986).
11. P. B. Armentrout, *Annu. Rev. Phys. Chem.* **41**, 313 (1990); (b) P. B. Armentrout, in *Gas Phase Inorganic Chemistry*, edited by D. H. Russell (Plenum Publ. Co., New York, 1989), pp. 1–42.
12. R. Georgiadis and P. B. Armentrout, *J. Phys. Chem.* **92**, 7067 (1988).
13. S. D. Hanton, R. J. Noll, and J. C. Weisshaar, *J. Phys. Chem.* **94**, 5655 (1990).
14. S. K. Loh, D. A. Hales, L. Lian, and P. B. Armentrout, *J. Chem. Phys.* **90**, 5466 (1989); (b) P. B. Armentrout, *Comments At. Mol. Phys.* **22**, 1336 (1988); (c) K. Ervin and P. B. Armentrout, *J. Chem. Phys.* **83**, 166 (1985).
15. P. R. Kemper, G. von Helden, and M. T. Bowers, *J. Phys. Chem.* **95**, 5134 (1991); (b) P. R. Kemper, P. van Koppen, and M. T. Bowers, *Science* **260**, 1446 (1993).
16. P. R. Kemper, M. Hsu, and M. T. Bowers, *J. Phys. Chem.* **95**, 10600 (1991); (b) G. V. Helden, P. R. Kemper, M. Hsu, and M. T. Bowers, *J. Chem. Phys.* **96**, 6591 (1992).
17. E. L. Kerley and D. H. Russell, *J. Am. Chem. Soc.* **112**, 5959 (1990).
18. M. S. Foster; J. L. Beauchamp, *J. Am. Chem. Soc.* **93,** 4924–4926 (1971); (b) M. S. Foster; J. L. Beauchamp, *J. Am. Chem. Soc.* **97,** 4808–4814 (1975); (c) M. S. Foster and J. L. Beauchamp, *J. Am. Chem. Soc.* **75,** 4814–4817 (1975); (d) W. K. Mechstroth and D. P. Ridge, *Int. J. Mass Spectrom. Ion Proc.* **61,** 149–152 (1984); (e) W. K. Meckstroth, R. B. Freas, W. D. Reents, Jr., and D. P. Ridge, *Inorg. Chem.* **24,** 3139–3146 (1985); (f) D. A. Freden and D. H. Russell, *J. Am. Chem. Soc.* 107, 3762–3768 (1985); (g) D. A. Fredeen and D. H. Russell, *J. Am. Chem. Soc.* **108**, 1860 (1986); (h) D. A. Fredeen and D. H. Russell, *J. Am. Chem. Soc.* **109**, 3903 (1987).
19. See also, D. P. Ridge and W. K. Mechstroth in *Gas Phase Inorganic Chemistry* , edited by D. H. Russell (Plenum Publ. Co., New York, 1989), pp. 93–116; (b) D. H. Russell, D. A. Fredeen, and R. E. Tecklenburg, in *Gas Phase Inorganic Chemistry*, edited by D. H. Russell (Plenum Publ. Co., New York, 1989), pp. 117–136.
20. Data taken from C. E. Moore, *Atomic Energy Levels*, (National Bureau of Standards, Washington, DC, 1952); (b) J. Sugar and C. J. Corlis, *J. Phys. Chem. Ref. Data* **6**, 317 (1977); (c) R. H. Garstang, *Mon. Not. R. Astrom. Soc.* **124**, 321 (1962).

21. G. J. Distefano, *Res. Natl. Bur. Stands., A, Phys. Chem.* **74**, 233 (1970); (b) H. M. Rosenstock, K. Draxl, B. W. Steiner, and J. T. Herron, *J. Phys. Chem. Ref. Data* **6**, 1-157 (1977).
22. K. Norwood, K. Ali, G. D. Flesch, and C. Y. Ng, *J. Am. Chem. Soc.* **112**, 7502 (1990).
23. R. H. Schultz, K. C. Crellin, and P. B. Armentrout, *J. Am. Chem. Soc.* **113**, 8590 (1991).
24. J. V. B. Oriedo and D. H. Russell, *J. Phys. Chem.* **96**, 5314 (1992).
25. J. V. B. Oriedo and D. H. Russell, *J. Am. Chem. Soc.* **115**, 8376 (1993).
26. R. E. Tecklenberg, Jr., D. L. Bricker, and D. H. Russell, *Organomet.* **7**, 2506 (1988).
27. C. J. Cassady and B. S. Freiser, *J. Am. Chem. Soc.* **106**, 6176 (1984).
28. J. L. Elkind and P. B. Armentrout, *J. Phys. Chem.* **90**, 5736 (1986).
29. D. B. Jacobson and B. S. Freiser, *J. Am. Chem. Soc.* **106**, 4623 (1984); (b) L. Sallans, K. R. Lane, R. R. Squires, and B. S. Freiser, *J. Am. Chem. Soc.* **107**, 4379 (1985); (c) D. B. Jacobson and B. S. Freiser, *J. Am. Chem. Soc.* **108**, 27 (1986); (d) R. H. Forbes, L. M. Lech, and B. S. Freiser, *Int. J. Mass Spectrom. Ion Proc.* **77**, 107 (1987).
30. J. W. Gauthier, T. R. Trautman, and D. B. Jacobson, *Anal. Chim. Acta* **246**, 211 (1991).
31. E. L. Kerley, C. D. Hanson, M. E. Castro, and D. H. Russell, *Anal. Chem.* **61**, 2528 (1989).
32. G. D. Byrd and B. S. Freiser, *J. Am. Chem. Soc.* **104**, 594 (1982); (b) R. C. Burnier, G. D. Byrd, and B. S. Freiser, *J. Am. Chem. Soc.* **103**, 4360 (1981).
33. S. K. Loh, E. R. Fisher, L. Lian, R. H. Schultz, and P. B. Armentrout, *J. Phys. Chem.* **93**, 3159 (1989); (b) G. von Helden, P. R. Kemper, M.-T. Hsu, and M. T. Bowers, *J. Chem. Phys.* **96**, 6591 (1992).
34. P. B. Armentrout (private communication).
35. L. M. Mallis and D. H. Russell, *Anal. Chem.* **58**, 1076 (1986); (b) D. H. Russell, E. S. McGlohon, and L. M. Mallis, *Anal. Chem.* **60**, 1818 (1988).
36. L. M. Mallis and D. H. Russell, *Int. J. Mass Spectrom. Ion Proc.* **78**, 147 (1987).
37. F. Jensen, *J. Am. Chem. Soc.* **114**, 9533 (1992).
38. D. Renner and G. Spiteller, *Biol. Mass Spectrom.* **15**, 75 (1988); (b) R. P. Grese, R. L. Cerny, and M. L. Gross, *J. Am. Chem. Soc.* **111**, 2835 (1989); (c) R. P. Grese and M. L. Gross, *J. Am. Chem. Soc.* **112**, 5098 (1990); (d) J. A. Leary, T. D. Williams, and G. Bott, *Rapid Commun. Mass Spectrom.* **3**, 192 (1989); (e) L. M. Teesch, R. C. Orlando, and J. Adams, *J. Am. Chem. Soc.* **113**, 3668 (1991).
39. For reviews on the subject of gas-phase ion chemistry of organoaıkali metal ion complexes see: J. Adams, *Org. Mass Spectrom.* **27**, 913 (1993); (b) J. Adams, in *Experimental Mass Spectrometry*, edited by D. H. Russell (Plenum Publ. Co., New York, 1994), pp. 39–70.
40. S. Karrass, K. Eller, and H. Schwarz, *Chem. Ber.* **123**, 939 (1990).
41. R. C. Burnier, G. D. Byrd, and B. S. Freiser, *J. Am. Chem. Soc.* **103**, 4360 (1981); (b) J. Allison and D. P. Ridge, *J. Am. Chem. Soc.* **101**, 4998 (1979).
42. D. M. Sonnenfroh and J. M. Farrar, *J. Am. Chem. Soc.* **108**, 3521 (1986).
43. A. Bjarnason, J. W. Taylor, J. A. Kinsinger, R. B. Cody, and D. A. Weil, *Anal. Chem.* **61**, 1889 (1989); (b) A. Bjarnason and J. W. Taylor, *Organomet.* **8**, 2020 (1989).
44. P. A. M. van Koppen, P. R. Kemper, and M. T. Bowers, *J. Am. Chem. Soc.* **114**, 10941 (1992).
45. P. A. M. van Koppen, P. R. Kemper, and M. T. Bowers, *J. Am. Chem. Soc.* **115**, 5616 (1993).
46. R. Tonkyn and J. C. Weisshaar, *J. Phys. Chem.* **90**, 2305 (1986).
47. (a) R. C. Burnier, G. D. Boyd, and B. S. Freiser, *J. Am. Chem. Soc.* **103**, 4360 (1981); (b) B. S. Larsen and D. P. Ridge, *J. Am. Chem. Soc.* **106**, 1912 (1984); (c) L. F. Halle, W. E. Crowe, P. B. Armentrout, and J. L. Beauchamp, *Organomet.* **3**, 1694 (1984); (d) C. J. Cassidy and B. S. Freiser, *J. Am. Chem. Soc.* **107**, 1566 (1985); (e) M. A. Tolbert and J. L. Beauchamp, *J. Phys. Chem.* **90**, 5015 (1986); (f) D. M. Sonnenfroh and J. M. Farrar, *J. Am. Chem. Soc.* **108**, 3521 (1986); (g) M. A. Hanratty, J. L. Beauchamp, A. J. Illies, P. van Koppen, and M. T. Bowers, *J. Am. Chem. Soc.* **110**, 1 (1988).

48. B. D. Radecki and J. Allison, *J. Am. Chem. Soc.* **106**, 946 (1984); (b) W. L. Grady and M. M. Bursey, *Int. J. Mass Spectrom. Ion Proc.* **52**, 247 (1983).
49. Details of these reactions will be presented in T. Solouki, J. V. B. Oriedo, and D. H. Russell (manuscript in preparation for *J. Am. Chem. Soc.*).
50. J. P. Speir, G. S. Gorman, and I. J. Amster, *J. Am. Soc. Mass Spectrom.* **4**, 106 (1993).
51. P. C. Staire, *J. Am. Chem. Soc.* **104**, 4044 (1982); (b) K. Eller and H. Schwarz, *Chimia* **43**, 371 (1989).

7. Gas-phase organometallic ion photochemistry

Y.A. RANASINGHE, I.B. SURJASASMITA and BEN S. FREISER
H.C. Brown Laboratory of Chemistry, Purdue University, West Lafayette, in 47907, *U.S.A.*

1. Introduction

The combination of Fourier transform ion cyclotron resonance mass spectrometry (FTICRMS) [1–7] with various light sources has proven to be a powerful means of studying the photochemistry of a wide variety of metal-containing ions. In our laboratory, pulsed laser ionization [8, 9] is used to generate metal ions which are then stored in the ion trap of the FTICRMS where they are permitted to react with background gases [10, 11]. Isolation of the desired ion is readily achieved, followed by light irradiation for variable periods on the order of 1 to 20 s. Light absorption is measured indirectly by monitoring some change in ion intensity. Photodissociation is the most commonly observed process, whereby an ion dissociates following light absorption [12–14]. An alternative process in which the reactivity of an ion is altered following light absorption offers a potentially powerful method for obtaining information at longer wavelengths, at and below the ion dissociation energy [15, 16]. Similarly, multiphoton visible [17, 18] and infrared [19–22] absorption processes take advantage of the long trapping times and relatively slow radiative relaxation rates to effect some change in ion intensity. With these methods, one overcomes the burden of obtaining absorption information from species at 10^{-17} M. In this chapter we focus on some recent results from our laboratory on simple singly and doubly charged metal-ligand ions in the UV-VIS and in the infrared.

The ion storage capability of the ICR spectrometer makes it ideally suited for the study of the photochemistry of ions in the gas phase. This method was demonstrated and exploited in the early seventies primarily by Dunbar [12], Beauchamp [23, 24] and Brauman [25] and its applications have continued to grow to yield a substantial literature on a wide variety of positively and negatively charged ions from simple molecules to clusters and polymers, from organic and inorganic systems to those of biological interest, and so forth [26–32]. Other techniques and instruments have also been brought to bear on this problem including rf quadrupole traps [33, 34], tandem quadrupole instruments [35], drift tubes [36], time-of-flight instruments [37–39], and

Ben S. Freiser (ed.), Organometallic Ion Chemistry, 229–258.

ion beams [40, 41]. FTICRMS, like ICR, has the advantage of long ion storage times but, in addition, it greatly extends the variety of ions that can be studied, such as those synthesized through fairly complex multistep processes [10]. Furthermore, FTICRMS has the advantage of being able to monitor the product ions and precursor ions simultaneously.

2. Experimental

The fundamental principles, instrumentation, and applications of ICR and FTICRMS have been discussed in detail in a number of excellent recent reviews [1–7]. The work detailed in this paper was performed on two different instruments in our laboratory. The first is a prototype Extrel FTMS-1000 Fourier transform mass spectrometer equipped with a 6 in diffusion pump, a 15 in electromagnet maintained at about 1 T, and a 2 in cubic cell which has mesh transmit plates to permit irradiation of the ions. A Quanta-Ray Nd:YAG pulsed laser beam (1064 nm fundamental) is focused onto one of several metal targets embedded in one of the mesh transmit plates to generate metal ions by laser ablation [8]. A 2.5 kW Hg-Xe arc lamp is used either in conjunction with a Schoeffel 0.25 m monochrometer operating at 10 nm resolution or, alternatively, with cut-off filters. The filters are used primarily for obtaining photodissociation thresholds for thermochemical studies, while the monochrometer provides absorption information.

The second instrument is an Extrel FTMS-2000 Fourier transform mass spectrometer equipped with a differentially pumped dual-cell in a superconducting magnet maintained at about 3T. The dual-cell configuration permits ions to be transferred back and forth from one cell to the other through a 2 mm hole, while maintaining different background gas environments [42, 43]. For example, the ion of interest can be synthesized in one cell by an ion-molecule reaction sequence involving various reagent gases and then transferred to the other cell where a low background pressure is maintained. In this way the ion can be more efficiently trapped while being irradiated by light, since it is less likely to react away or be scattered out of the cell.

Associated with the FTMS-2000 instrument is a Spectra Physics argon-ion dye laser combination, a Synrad CW CO_2 laser (10.6 μ), and a Lumonix XeCl excimer laser (available through the Purdue laser facility). A ZnSe window is used to permit irradiation with the infrared laser, while a sapphire window is used for the visible lasers. Ions are generated by laser ablation from metal targets, mounted in a 4–way cross external to the magnetic field, using a second pulsed Nd:YAG laser. Alternatively, the YAG laser can be focused onto a target in a Smalley mini-supersonic source [44] mounted on an adjoining 6–way cross. This relatively new addition to our instrument permits the generation of a wide variety of cluster ions.

The inlet system of the FTMS-1000 consists of three Varian precision leak valves and two General Valve Series 9 pulsed solenoid valves. The FTMS-

2000 is equipped with two leak valves and two pulsed valves on both of the two differentially pumped regions. The pulsed valves allow the introduction of reagents at selected times during the pulse sequence [45].

2.1. Photodissociation

Photodissociation, process (1), will be observed depending on 3

$$AB^+ + h\nu \rightarrow A^+ + B \tag{1}$$

factors: the ion must first absorb a photon, second, the photon energy must be greater than the A^+-B bond energy and, third, the quantum yield for dissociation must be non-zero. The onset or threshold for photodissociation, namely the longest wavelength at which dissociation occurs, will be attributable either to the absorption characteristics of the ion or the enthalpy for the bond energy, $D(A^+\text{-}B)$. The former is referred to as a spectroscopic threshold while the latter is a thermodynamic threshold. Organic ions tend to have a low density of electronic states in the UV-VIS leading to spectroscopic thresholds, which generally provide only upper limits on $D(A^+\text{-}B)$ [12]. In contrast, our studies have amply shown that metal-containing complexes tend to have a high density of low-lying electronic states leading to very broad absorption and, in most cases, thermodynamic thresholds providing reasonable estimates of $D(A^+\text{-}B)$ [11]. The flip-side of this, however, is that assigning specific excited states from photodissociation spectra of organometallic ions may be difficult, if not impossible. Nevertheless, isomer differentiation based on the spectral variations represents a powerful structural tool [12, 14]. While, like CID, ion rearrangement may occur following photon absorption and prior to dissociation, the wavelength dependence will be a fingerprint of the original ground state structure.

The studies in our laboratory have thus far focused on five general categories of metal containing ions:

1. ML^+ (L = O, S, CH_2, C_4H_6, C_4H_8, C_6H_6, CH_3OH, etc.)
2. LM^+L' (both L = L′ and L ≠ L′: L = C_2H_4, C_3H_6, C_4H_8, C_6H_6, etc.)
3. MFe^+ (M = Sc, Ti, Cr, Fe, Co, Ni, Cu, Nb, Ta, Mg)
4. $MFeC_6H_6{}^+$ (M = Nb, V, Co, Sc)
5. ML^{2+} (M = Y, La: L = C_2H_2, C_2H_4, C_3H_6)

Although the structural information provided by wavelength dependent photodissociation spectra has been key in a few cases, and certainly interesting, by far the major focus of this work to date has been to obtain absolute metal ligand bond energies from photodissociation thresholds. A few representative recent examples are discussed here.

2.1.1. $MC_2H_2^+$ (*M* = *Sc*, *Y*, *La*)

This study was prompted by our interest in group 3 metal ion chemistry and by the substantial discrepancies (~20 kcal/mol) between recent theoretical computations for $MC_2H_2^+$(Sc,Y) and ion beam results. For example, theoreti-

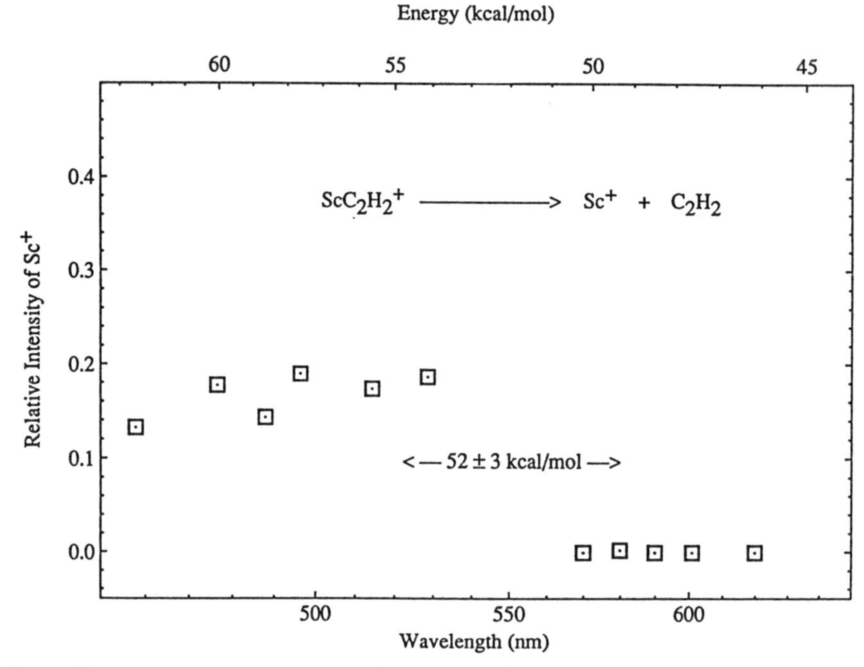

Fig. 1. Photoappearance spectrum of Sc^+ from $ScC_2H_2^+$ as a function of wavelength. The bond energy threshold is assigned as $D(Sc^+\text{-}C_2H_2) = 52 \pm 3$ kcal/mol. Figure copied from reference 51 with permission.

cal computations by Bauschlicher and coworkers yielded $D(M^+\text{-}C_2H_2) = 53$ kcal/mol [48], while Armentrout and coworkers using the ion beam method reported $D(Sc^+\text{-}C_2H_2) = 77.9 \pm 2.3$ kcal/mol and $D(Y^+\text{-}C_2H_2) = 70.8 \pm 2.3$ kcal/mol [49]. Such discrepancies are unusual considering that there is generally excellent agreement between theory and experiment, within 5 kcal/mol or so.

In our experiment $MC_2H_2^+$ species were generated readily by reacting M^+ with C_2H_4, reaction (2). Following its formation,

$$M^+ + C_2H_4 \rightarrow MC_2H_2^+ + H_2 \tag{2}$$

$MC_2H_2^+$ was isolated by swept double-resonance ejection techniques [50] and photodissociated using selected wavelengths from the argon-ion laser and argon-ion pumped dye laser. Photodissociation proceeds by complete ligand loss, exclusively, reaction (3).

$$MC_2H_2^+ + h\nu \rightarrow M^+ + C_2H_2 \tag{3}$$

Figure 1 shows representative data obtained on the photoappearance of Sc^+ from irradiation of $ScC_2H_2^+$ as a function of wavelength. Before and after each irradiation, a blank spectrum was taken in the absence of light to identify any background reactions. The photoappearance data are blank subtracted and normalized for the number of photons.

As seen in Figure 1, the product ion intensity is approximately constant for wavelengths up to 528 nm (the longest wavelength obtainable with substantial intensity using the argon ion laser). Irradiation with the dye laser at 570 nm and above yields no photodissociation products, indicating a threshold has been reached. In order to insure that the dye laser was focused properly, other ions were generated and observed to undergo photodissociation under otherwise identical conditions. Given this, the threshold was bracketed between 528 nm and 570 nm which corresponds to a bond energy of $D(Sc^+\text{-}C_2H_2) = 52$ kcal/mol in excellent agreement with theory. A ± 3 kcal/mol error bracket is assigned and is believed to be conservative.

Similar results were obtained for the Y and La analogues and the values were corroborated by ion-molecule bracketing methods [51]. The high values for the ion-beam experiment were believed to be due to trace amounts of ethylene impurity in samples of ethane. While these metal ions undergo reaction (1) with C_2H_4 to generate $MC_2H_2^+$, the only exothermic reaction with ethane yields $MC_2H_4^+$. Thus, trace amounts of C_2H_4 in a C_2H_6 sample would yield spuriously low $MC_2H_2^+$ thresholds in the ion beam experiment. The ion beam results have since been reinterpreted taking this into account. The new values, $D(Sc^+\text{-}C_2H_2) = 57.3 \pm 4.6$ kcal/mol [52] and $D(Y^+\text{-}C_2H_2) = 63 \pm 7$ kcal/mol [49], now are in closer accord with the photodissociation and theoretical values.

2.1.2. *Doubly charged ions*

Although the majority of gas-phase metal ion studies has involved singly charged species, the number of studies on doubly charged ions has been growing [53]. Interestingly, despite the fact that the second ionization energies of metals are higher than the first ionization energies of most organic molecules, a variety of reactions, in addition to simple charge transfer, has been observed. These reactions have primarily involved early transition metals, which have relatively low second ionization energies, and have yielded a variety of doubly charged metal ligand species, providing the opportunity to compare the reactivity and thermochemistry of the singly charged ions to their doubly charged counterparts.

Photodissociation has proven to be a particularly effective method for obtaining quantitative metal-ligand bond energies for these species. In addition a surprising array of products are often observed from simple systems, as exemplified by the Y^{2+} systems discussed below.

Y^{2+} (2nd IE 12.24 eV) reacts with small linear alkanes via dehydrogenation, demethanation, charge and hydride transfer reactions (Table I) [54]. Reaction with ethane (IE 11.52 eV) produces $YC_2H_4{}^{2+}$ in 98% abundance. $YC_2H_4{}^{2+}$ gives three photoproducts according to reactions (4)–(6) [55]. Also shown below are the photoappearance thresholds which, for reaction (4), yield a direct measure of $D(Y^{2+}\text{-}C_2H_4)$.

Table I. Primary product intensity ratios for the reactions of La^{2+} and Y^{2+} with small alkanes

Reagent	Products	La^{2+} [a]	Y^{2+} [b]
CH_4		NR	NR
C_2H_6	$MH^+ + C_2H_5^+$	NR	2
	$MC_2H_4^{2+} + H_2$		98
C_3H_8	$M^+ + C_3H_8^+$		3
	$MH^+ + C_3H_7^+$		7
	$MCH_3^+ + C_2H_5^+$		35
	$C_3H_6^{2+} + H_2$	81	19
	$MC_2H_4^{2+} + CH_4$	19	36
C_4H_{10}	$MC_4H_6^{2+} + 2H_2$	11	11
	$MCH_3^+ + C_3H_7^+$	8	28
	$MC_4H_8^{2+} + H_2$	38	3
	$MC_2H_4^{2+} + C_2H_6$	34	34
	$MH^+ + C_4H_9^+$	3	3
	$MC_3H_6^{2+} + CH_4$	6	6
	$MC_2H_5^+ + C_2H_5^+$		5
	$M^+ + C_4H_{10}^+$		10

[a] Ref. 47. [b] Ref. 55.

		Threshold (kcal/mol) arc lamp	laser	
$YC_2H_4^{2+} \xrightarrow{h\nu}$	$Y^{2+} + C_2H_4$	50±5	49±3	(4)
	$YC_2H_2^{2+} + H_2$	48±5	46±3	(5)
(<2%)	$YC_2H_3^{2+} + H$			(6)

The arc lamp data used to obtain the appearance thresholds for reactions (4) and (5) are presented in Figures 2 and 3. The thermochemical values calculated using these thresholds are listed in Table II. The photochemistry of $YC_2H_4^{2+}$ is similar to that of $LaC_2H_4^{2+}$ [47] and the bond energies calculated are comparable (Table III). H loss, reaction (6), at < 2% intensity, was too low for a threshold to be obtained. Charge separation was not observed even though formation of Y^+ and $C_2H_4^+$ is a lower energy process. Collision-induced dissociation of $MC_2H_4^{2+}$ (M = Y, La) yields complete ligand loss as the primary pathway, indicating an intact ligand structure. Theoretical calculations predict a symmetrical C_{2v} structure with the Y^{2+} binding predominantly electrostatically to the C-C bond midpoints [48, 56]. However, an asymmetric C_s structure having only a small energy difference from the symmetric one could not be completely ruled out. The $D(Y^{2+}-C_2H_2)$ bond energy calculated from the arc lamp threshold for H_2 loss in reaction (5) is 43 ± 7 kcal/mol. Direct photodissociation data or CID data could not be obtained due to the insufficient intensity of $YC_2H_2^{2+}$ to perform those experiments successfully. Theory predicts a symmetric C_{2v} structure as in the case of $YC_2H_4^{2+}$ [48, 56]. Figure 4 gives the wavelength dependence

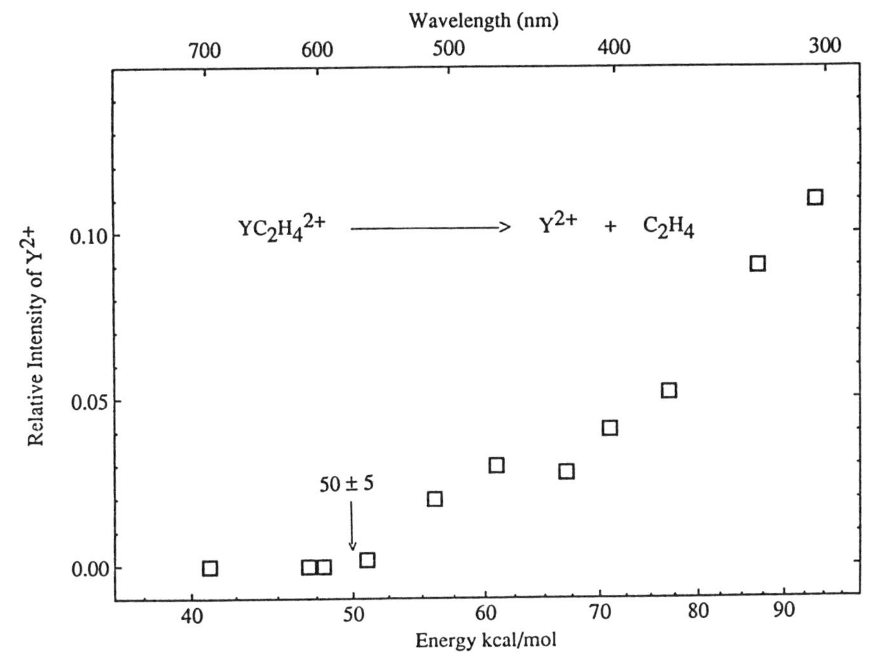

Fig. 2. Photoappearance spectrum of Y^{2+} from $YC_2H_4^{2+}$ as a function of wavelength. The bond energy threshold is assigned as $D(Y^{2+}{-}C_2H_4) = 50 \pm 5$ kcal/mol.

of the photoproducts from reactions (4) and (5), obtained using the argon ion and the dye lasers. The photoappearance intensities obtained from each laser cannot be compared directly, due to the different conditions used in each case. The thresholds compare favorably with the arc lamp dissociation results yielding $D(Y^{2+}\text{-}C_2H_4) = 49 \pm 3$ kcal/mol. It is also interesting to note that there is a cross-over of intensities at ~550 nm. The shorter wavelengths predominantly give Y^{2+}, while the longer wavelengths predominantly give $YC_2H_2^{2+}$. This is consistent with the determined threshold for $YC_2H_2^{2+}$ being at a lower energy than that for Y^{2+}.

When Y^{2+} reacts with propane, it produces $YC_3H_6^{2+}$ in ~19% abundance [54]. In analogy to $LaC_3H_6^{2+}$ [47], $YC_3H_6^{2+}$ yields a variety of photoproducts upon irradiation, as represented by reactions (7)–(12). The bond energies calculated from the thresholds using the supplemental thermochemical data [57] are: $D(Y^{2+}\text{-}C_3H_6) = 52 \pm 5$ kcal/mol, $D(Y^{2+}\text{-}C_2H_2) = 41 \pm 7$ kcal/-mol, $D(Y^{2+}\text{-}C_3H_3) = 131 \pm 7$ kcal/mol, $D(Y^{2+}\text{-}C_3H_4) = 44 \pm 7$ ($CH_2 = C = CH_2$) or 45 ± 7 ($CH_3\text{-}C \equiv CH$) kcal/mol, $D(Y^{2+}\text{-}C_3H_5) = 90 \pm 7$ kcal-/mol, and $D(Y^{2+}\text{-}C_3H_2) = 119 \pm 8$ kcal/mol (propadienylidine) or 126 ± 8 kcal/mol (propargylene) [58]. Figures 5 and 6 show the appearance thresholds for reactions (8) and (11).

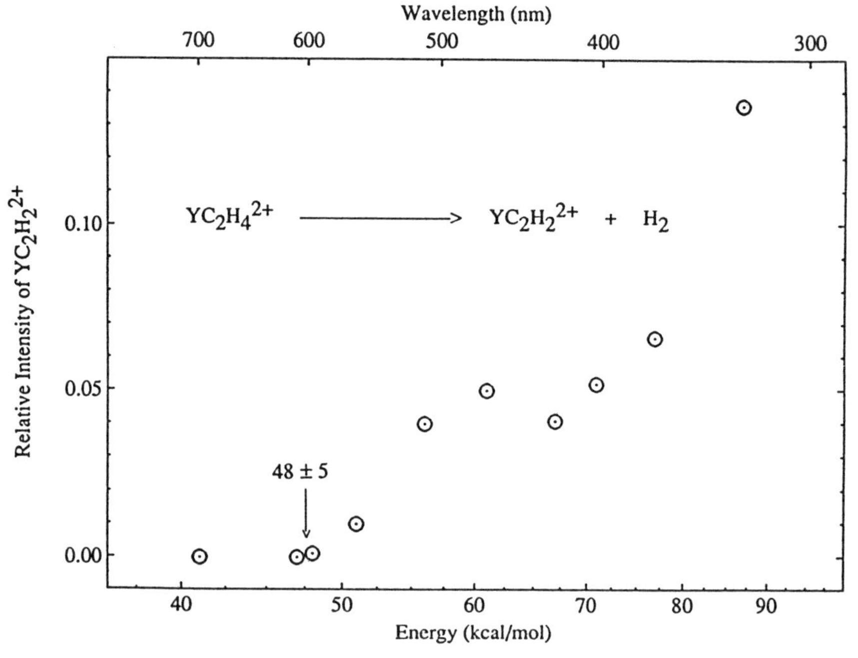

Fig. 3. Photoappearance spectrum of $YC_2H_2^{2+}$ from $YC_2H_4^{2+}$ as a function of wavelength. The threshold for the process is assigned as 48 ± 5 kcal/mol.

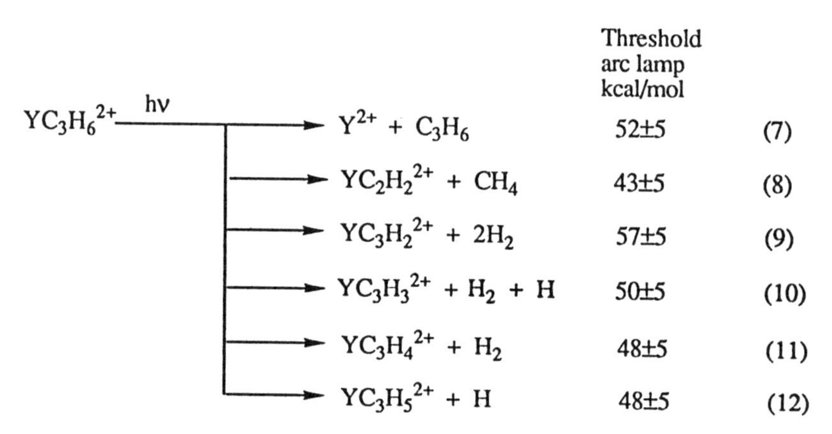

The loss of radicals is observed for these doubly charged ions, but no charge separation. The complete ligand loss (reaction (7)) is not the lowest energy threshold, in contrast to $LaC_3H_6^{2+}$ [47]. In CID experiments, in addition to complete ligand loss, charge separation reactions are observed, as well as other equally competitive fragmentation pathways.

It is gratifying to note that the bond energy $D(Y^{2+}\text{-}C_2H_2)$ calculated by using the thresholds from the photodissociation of $YC_2H_4^{2+}$ is in excellent

Table II. Thermochemical information derived from photodissociation thresholds[a]

Photodissociation of $YC_2H_4^+$	kcal/mol
$\Delta H_f(YC_2H_4^{2+})$	493 ± 5
$\Delta H_f(YC_2H_2^{2+})$	541 ± 7
$D°(Y^{2+}\text{-}C_2H_4)$	50 ± 5
$D°(Y^{2+}\text{-}C_2H_2)$	43 ± 7
Photodissociation of $YC_3H_6^{2+}$	
$\Delta H_f(YC_3H_6^{2+})$	483 ± 5
$\Delta H_f(YC_2H_2^{2+})$	544 ± 7
$\Delta H_f(YC_3H_2^{2+})$	540 ± 7
$\Delta H_f(YC_3H_3^{2+})$	481 ± 7
$\Delta H_f(YC_3H_4^{2+})$	531 ± 7
$\Delta H_f(YC_3H_5^{2+})$	479 ± 7
$D°(Y^{2+}\text{-}C_3H_3)$	131 ± 7
$D°(Y^{2+}\text{-}C_3H_4)$	
$CH_3\text{-}C \equiv CH$	44 ± 7
$CH_2 = C = CH_2$	45 ± 7
$D°(Y^{2+}\text{-}C_3H_5)$	90 ± 7
$D°(Y^{2+}\text{-}C_2H_2)$	41 ± 7
$D°(Y^{2+}\text{-}C_3H_6)$	52 ± 5
$D°(Y^+ - C_3H_2)$	
$CH_2 = C = C:$	119 ± 8
$:CH\text{-}C \equiv C\text{-}H$	126 ± 8

[a] Heats of formation are obtained from photodissociation thresholds and using the supplementary data in reference 57.

Table III. Theoretical and experimental bond energies obtained for $M^{n+} - L$, n = 1, 2 and L = C_2H_2, C_2H_4, C_3H_6

Bond	La		Y	
	Theory (kcal/mol)	Experiment (kcal/mol)	Theory (kcal/mol)	Experiment (kcal/mol)
n = 1				
$D°(M^+\text{-}C_2H_2)$	66[a]	52 ± 3[b]	53[a]	52 ± 3[b]
$D°(M^+\text{-}C_2H_4)$	46[c]	43 ± 10[d]	33[c]	>33[e] <54 ± 5
$D°(M^+\text{-}C_3H_6)$	51[c]	51 ± 10[d]	43[c]	>30[e] <54 ± 5[e]
n = 2				
$D°(M^{2+}\text{-}C_2H_2)$	44[c]	38 ± 10[d]	47[c]	42 ± 8[e]
$D°(M^{2+}\text{-}C_2H_4)$	41[c]	42 ± 5[d]	47[c]	50 ± 5[e]
$D°(M^{2+}\text{-}C_3H_6)$	48[c]	42 ± 5[d]	54[c]	52 ± 5[e]

[a] Ref. 78. [b] Ref. 51. [c] Refs. 48, 56. [d] Ref. 47. [e] Ref. 55.

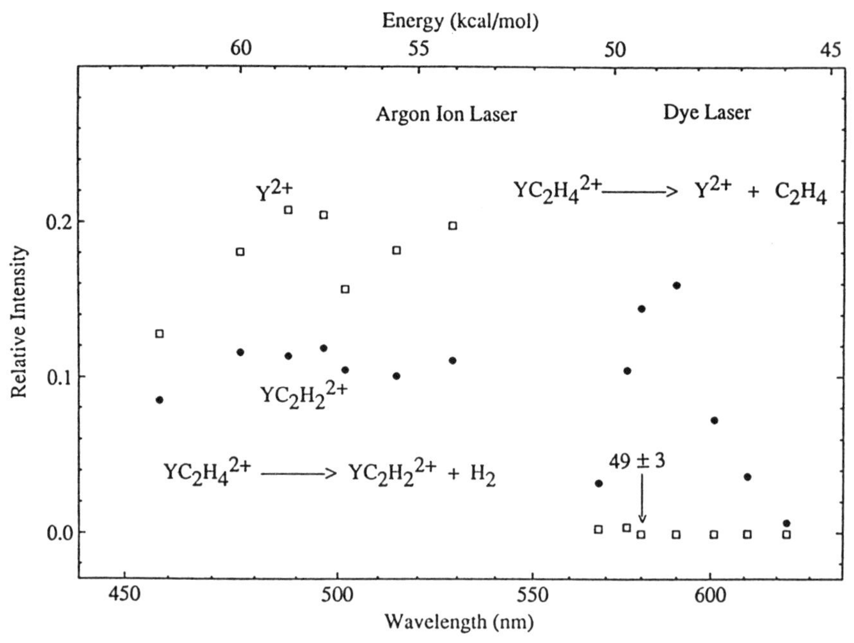

Fig. 4. Photoappearance spectra of Y^{2+} and $YC_2H_2^{2+}$ generated from $YC_2H_4^{2+}$ using the argon ion and dye lasers. The absolute intensities using the dye laser cannot be directly compared to those using the argon ion laser.

agreement with the bond energy calculated using $YC_3H_6^{2+}$. This lends strong support to the notion that thermodynamic thresholds and not spectroscopic thresholds are being observed. Furthermore, an average value of 42 ± 8 kcal/mol is in very good agreement with the theoretical bond energy $D(Y^{2+}-C_2H_2) = 47$ kcal/mol [56].

Table III shows bond energy comparisons with the corresponding lanthanum species. Doubly charged group 3 metal ions interact with C_2H_2, C_2H_4 and C_3H_6 primarily by electrostatic bonding, in contrast to the singly charged species, which insert. Hence, the ligand geometries are more like that of the free ligand for the doubly charged ions, as opposed to the singly charged ions, where a bond is broken. One could also argue that the bond strength for a given ligand will depend exclusively on the radial extent of M^{2+} and, hence, the bond strength for Y^{2+}-L > La^{2+}-L.

2.1.3. *Examples of spectroscopic photodissociation thresholds*

While, as stated above, most transition metal ion complexes have a high density of low lying excited states in the bond enthalpy region which allows the ions to readily absorb photons at this energy and dissociate, there is no guarantee that the presence of a transition metal center will lead to a thermodynamic threshold. Thus, it is necessary to obtain at least one and

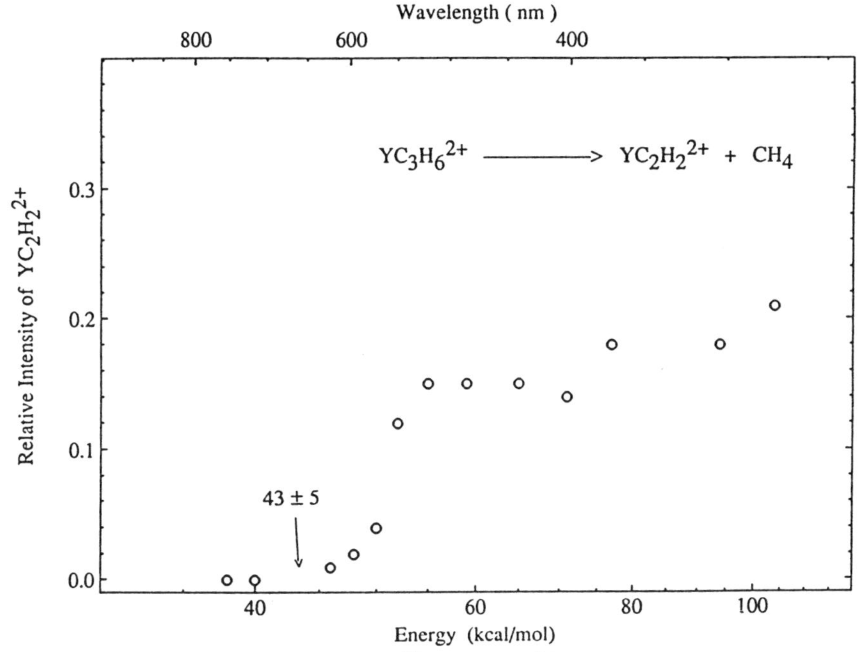

Fig. 5. Photoappearance spectrum of $YC_2H_2^{2+}$ from $YC_3H_6^{2+}$ as a function of wavelength. The threshold for the process is assigned as 43 ± 5 kcal/mol.

preferably two or more independent measures of the bond strength. A case in point is the heteronuclear dimer species $MgFe^+$ which photodissociates via reactions (13) and (14) [59]. The threshold for the

$$MgFe^+ \xrightarrow{h\nu} \quad >95\% \rightarrow Mg + Fe^+ \tag{13}$$

$$\quad <5\% \rightarrow Mg^+ + Fe \tag{14}$$

production of Fe^+ is at 49 kcal/mol (584 nm) indicating $D(Fe^+\text{-}Mg) \leqslant 49$ kcal/mol, which in turn yields $D(Mg^+\text{-}Fe) \leqslant 44$ kcal/mol. Bond energy information on $MgFe^+$ was also obtained using ion-molecule reaction bracketing techniques. In contrast to the photodissociation results, the observation that $MgFe^+$ reacts with ethylene by displacement of magnesium to form $FeC_2H_4^+$ implies $D(Fe^+\text{-}Mg) < D(Fe^+\text{-}C_2H_4) = 34.7 \pm 1.4$ kcal/mol [60] from which $D(Mg^+\text{-}Fe) < 30 \pm 1.4$ kcal/mol is derived. It is evident that the photodissociation threshold in this case is spectroscopic and yields only an upper limit. Furthermore, the predominant formation of the higher energy product Fe^+ (IE Mg = 7.647eV; IE Fe = 7.87 eV) [57] over the wavelength range studied (roughly 340–820 nm), clearly indicates that dissociation proceeds from an excited state surface and not via internal conversion to a vibrationally excited ground state. This is in contrast to collision-induced dissociation, where at lower energies Mg^+ dominates in accordance with Stevenson's rule and at

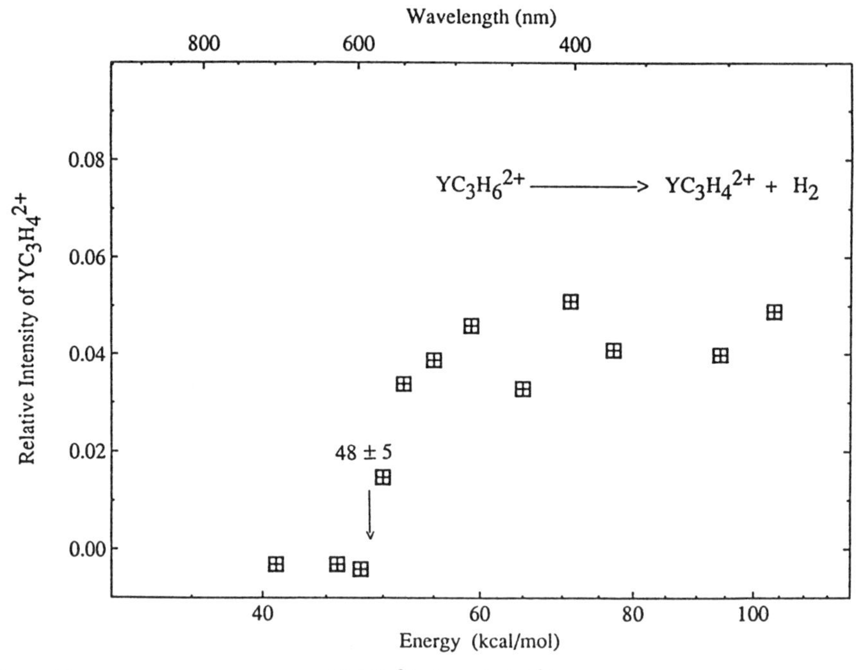

Fig. 6. Photoappearance spectrum of $YC_3H_4^{2+}$ from $YC_3H_6^{2+}$ as a function of wavelength. The threshold for the process is assigned as 48 ± 5 kcal/mol.

higher energies the intensities of Fe^+ and Mg^+ approach each other, as would be expected for dissociation from the ground state surface.

The photodissociation results were beautifully explained by a theoretical treatment on $MgFe^+$ by Bauschlicher and co-workers [59], which showed that the photon absorption at ~49 kcal/mol corresponds to an intense vertical transition from the ground $^6\Delta$ to the first excited $^6\Delta$ state, Figure 7. This excited state is repulsive and yields at the asymptote Fe^+ and Mg, in accordance with the experimental results. The binding energy was computed to be 29.4 kcal/mol, in good agreement with the experimental upper limit of 30 ± 1.4 kcal/mol deduced from the ion-molecule reactions. Thus, the absence of allowed excited states in the energy region corresponding to the thermodynamic thresholds for formation of both Mg^+ + Fe and Fe^+ + Mg leads to a spectroscopic photodissociation threshold on the order of 20 kcal/-mol higher than the bond dissociation energy.

A second interesting case is $Ag(benzene)^+$ [61]. Since Ag^+ is closed shell, one would expect a spectroscopic threshold considering that the first excited state of Ag^+ lies at 4.8 eV or 111 kcal/mol above the ground state [62], and most M^+-benzene bond energies reported lie around 50 ± 10 kcal/mol [63]. We became interested in this ion from the work of Duncan and coworkers [64] in which they observed a novel dissociative charge transfer pathway for

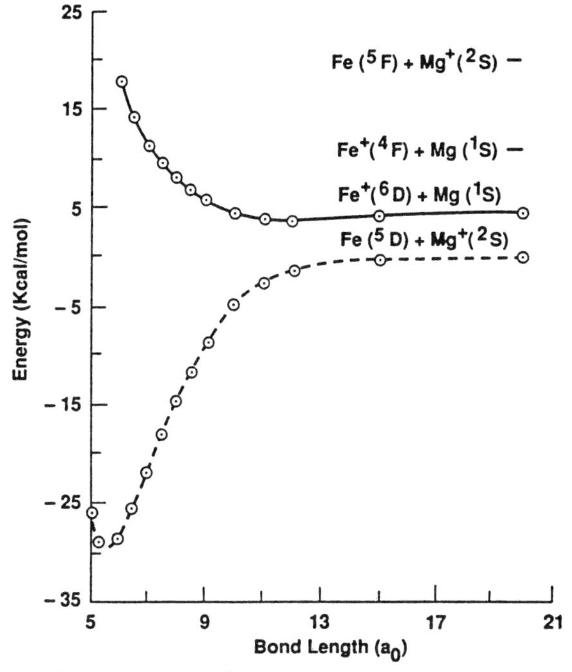

Fig. 7. $X^6\Delta$ and $(2)^6\Delta$ states of $MgFe^+$. The higher asymptotes that give rise to the quartet states derived from the Fe and Fe^+ $3d^7$ occupations are also noted. Figure copied from reference 59 with permission.

$Ag(benzene)^+$ and $Ag(toluene)^+$, reaction (15). Their results, obtained using a reflectron time-of-flight (TOF) mass spectrometer equipped with a super-

$$M(arene)^+ \xrightarrow{h\nu} arene^+ + M \quad (15)$$

$$ \longrightarrow arene + M^+ \quad (16)$$

sonic cluster source, were intriguing because such a charge transfer process had not been observed in our earlier studies on metal-benzene ion complexes. In addition, their reported value of $D(Ag^+\text{-benzene}) \leqslant 30$ kcal/mol was uncharacteristically low given that several metal-benzene bond energies are greater than 45 kcal/mol [63]. While the observation of a 100% yield of $C_6H_6^+$ via reaction (15) for $Ag(benzene)^+$ was reasonably explained since the charge transfer state is, presumably, the only allowed excited electronic state at low energy and involves a weak interaction of the delocalized benzene cation with Ag, we still expected that some internal conversion might be observed forming Ag^+. This prompted us to study this photochemistry under the longer time scales accessible by FTMS (msec to sec versus μsec for TOF).

Interestingly, our results on $Ag(benzene)^+$ and $Ag(toluene)^+$ proved to be significantly different than those obtained from the TOF experiments.

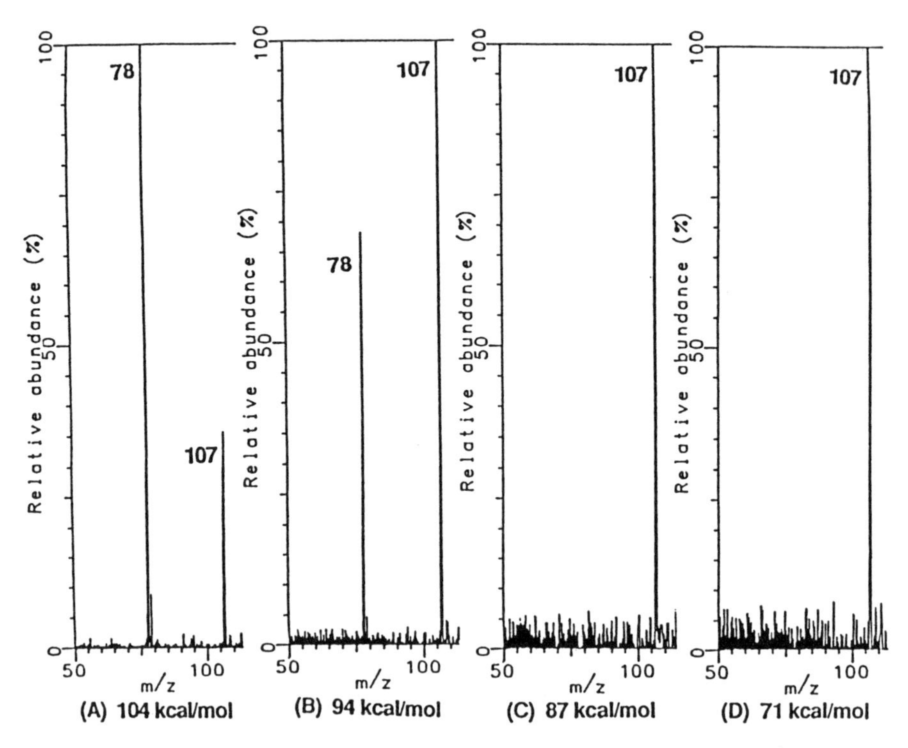

Fig. 8. Blank subtracted mass spectra of the photodissociation of $Ag(benzene)^+$ using light from the arc lamp and energy cutoff filters ranging from 71 to 104 kcal/mol. Figure copied from reference 61 with permission.

Figure 8, which shows the mass spectra of the photoproducts obtained when $Ag(benzene)^+$ is irradiated by broad band light from the arc lamp at different energy cut-offs, demonstrates two of these striking differences. First, both photoprocesses (15) and (16) are clearly observed in contrast to the TOF experiment in which only process (15) is detected. Secondly, the $C_6H_6^+$ intensity which dominates at higher energies is completely absent below 87 kcal/-mol, compared to the TOF experiment which pins the $C_6H_6^+$ threshold at 68 kcal/mol (418 nm). Similar differences were observed for $Ag(toluene)^+$.

With the use of several light sources, it was determined that the formation of $C_6H_6^+$ by process (15) and Ag^+ by process (16) are state selective and occur over different energy ranges [61]. Thus, while only $C_6H_6^+$ is observed as a product in the FTMS using a high energy excimer laser, and only Ag^+ is observed at the lower energies available with an argon ion laser, the broad band arc lamp produced both product ions with the high energy cut-off filters passing low and high energy light. Consequently, the reason no Ag^+ was observed in the TOF experiments was simply due to the irradiation wavelength and not the relative time scale of reactions (15) and (16).

Unfortunately, the discrepancy in the observed thresholds for $C_6H_6^+$ remains a mystery. While the observation by Duncan and coworkers of D(Ag-$C_6H_6^+$) < 68 kcal/mol, which in turn yields (Ag^+-C_6H_6) < 30 kcal/mol, seems too low considering recent calculations by Bauschlicher and coworkers yielding a value of 38 kcal/mol [65] and ion beam experiments by Armentrout and coworkers yielding 37 ± 1 kcal/mol [66], it is considerably closer than our value of (Ag^+-C_6H_6) < 55 kcal/mol obtained by monitoring reaction (15) directly and from the threshold of reaction (16) of D(Ag-$C_6H_6^+$) < 92 kcal/-mol. The possibilities of a two-photon process being involved in the TOF experiment, or poor sensitivity in the FTMS experiment were ruled out. In addition ions generated in the supersonic source should be colder than the direct method used in the FTMS and, hence, would be expected to give a higher appearance threshold, as opposed to the lower one observed. Therefore, it is difficult to fully rationalize the differences in the thresholds observed in the TOF and FTMS experiments at this time.

3. Infrared multiphoton dissociation (IRMPD)

One of the criteria mentioned earlier for observing photodissociation process (1) was that the photon have sufficient energy to break bonds and cause dissociation. In the case of ligand loss from a metal-ligand ion complex, for example, energies in the UV-VIS are required, limiting the usefulness of this spectroscopic probe. Freiser and Beauchamp demonstrated, however, that a trapped ion can in some instances absorb a second photon in a sequential fashion involving a long-lived vibrationally excited ground state intermediate [17]. Beauchamp and coworkers extended this method into the infrared, demonstrating that trapped ions can absorb on the order of 20 infrared photons in a sequential manner ultimately leading to dissociation via the lowest activation barrier process [21, 67, 68]. In the past 15 years, IRMPD has been employed to obtain fundamental information on a broad range of topics such as unimolecular and bimolecular reaction dynamics, vibrational relaxation, vibrationally induced electron detachment, fluence saturation, wavelength dependent spectra in the quasicontinuum, and isotopic and isomeric selectivity [12, 21, 25, 68].

While IRMPD has been applied successfully to a wide variety of organic ions in several laboratories, surprisingly few papers have appeared dealing with metal containing ions [69–71]. Beauchamp and coworkers demonstrated the utility of IRMPD for investigating organometallic ions in their benchmark study of various $Co(C_5H_{10})^+$ association complexes [69]. With the exception of $Co(cyclopentane)^+$, which did not absorb, each complex gave a single photodissociation product. $Co(1\text{–}pentene)^+$, for example, was observed to undergo IRMPD by loss of C_2H_4, while $Co(2\text{–}pentene)^+$ yielded CH_4 loss, exclusively. These results indicate that (1) the barrier to rearrangement between the 1 and 2–pentene complexes lies above that for dissociation, (2) since CH_4 loss is the lowest energy process for both ions, for 1–pentene loss of C_2H_4 has a lower activation barrier than CH_4, and (3) compared to most other activation methods, IRMPD yields fewer products and is a sensitive indicator of the lowest activation energy process.

Following this work, only a few papers have appeared on the IRMPD of metal containing ions leaving a real dearth of information in this area. Thus, we have initiated an intense effort in this wide open field. The systems we have studied to date include:

1. $M(acetone)_n^+$ M = Al, Fe, Co, Cu, ScO; n = 1 and 2
2. $MC_2H_2^+$ M = Fe, Co, Ni
3. $MC_nH_{2n}^+$ M = Fe, Co, Ni; n = 2–5
4. $MC_4H_6^+$ M = Fe, Co, Ni
5. $M(C_2H_4)(C_2H_2)^+$ M = Fe, Co, Ni
6. $M(c\text{-}C_5H_6)^+$ M = Fe, Co

$CoC_5H_6^+$ is observed to undergo multiphoton infrared dissociation to generate two products, reactions (17) and (18) [20]. Assuming that the

$$Co(c\text{-}C_5H_6)^+ \xrightarrow{nh\nu} Co^+ + c\text{-}C_5H_6 \quad (17)$$

$$Co(c\text{-}C_5H_6)^+ \xrightarrow{nh\nu} Co^+\text{-}C_5H_5 + H \quad (18)$$

reverse activation barriers for these two processes are negligible, observation of both reactions (17) and (18) during irradiation suggests that $D(Co^+\text{-}C_5H_6)$ and $D(CoC_5H_5{}^+\text{-}H)$ are within ~3 kcal/mol of each other. Interestingly, while the branching ratio for reactions 17 and 18 does not vary significantly with laser power (15–25 W), a dramatic deuterium isotope effect is observed. Irradiation of various $CoC_5H_nD_{6-n}{}^+$ (n = 1–6) species yields 100% H loss by process (18), and an increase in the whole ligand loss process (17) relative to H loss with increasing deuterium substitution. Large isotope effects have been observed previously in organic ion systems [21] and arise due to differences in activation and bond energies on the order of 1 kcal/mol, demonstrating the sensitivity of this technique to subtle energy differences.

The ability to probe subtle energy differences should find real use in obtaining relative and absolute bond energies via the method of competitive fragmentation. In analogy to competitive CID introduced by Cooks and coworkers [72, 73], IRMPD of the doubly ligated metal ion, $L\text{-}M^+\text{-}L'$, will result in competitive dissociation, reactions (19) and (20), with the most intense ion corresponding to the most tightly bound ligand. This method has

$$L\text{-}M^+\text{-}L' \xrightarrow{nh\nu} ML^+ + L' \quad (19)$$

$$L\text{-}M^+\text{-}L' \xrightarrow{nh\nu} ML'^+ + L \quad (20)$$

the advantage over CID and SORI in that it does not require consideration of angular momentum [74]. Compared to threshold UV-VIS photodissociation methods, where the dissociation signals may be small near the onsets, large dissociation signals can be obtained without compromising the selectivity. A recent example from our laboratory was a study of the IRMPD of $M(C_2H_2)(C_2H_4)^+$ for M = Fe, Co, Ni. The purpose of studying this ion was twofold: to use competitive ligand loss as a means of determining absolute and relative metal-ligand bond strengths, as stated above, and to complement our study on M(1,3–butadiene)$^+$ as an isomer of $MC_4H_6^+$ [75].

IRMPD of $Co(C_2H_4)(C_3H_6)^+$ yields exclusive loss of C_2H_4 in accordance with C_3H_6 being more polarizable and, therefore, more strongly bound than C_2H_4 to the metal center. The difference in the bond energies cannot exceed about 3 kcal/mol, however, as evidenced by the successful displacement of propene by ethene from $MC_3H_6^+$(M = Fe,Co,Ni). These results can be compared to values of $D(Co^+\text{-}C_2H_4) = 42.9 \pm 1.6$ kcal/mol and $D(Co^+\text{-}C_3H_6) = 43.1 \pm 1.6$ kcal/mol, obtained recently by Armentrout and coworkers [76]. While the values for $D(M^+\text{-}C_2H_2)$ are available for M = early transition metals, there is only one experimental study reported in the literature for the late transition metals. Fisher and Armentrout report $D(M^+\text{-}C_2H_2)$ of $> 33.9 \pm 4$, $> 30.0 \pm 2$, and $> 30.4 \pm 3$ kcal/mol for M = Co, Ni and Cu, respectively [77]. These values can be compared to theoretical predictions of 37, 39, and 36 kcal/mol for Co, Ni and Cu, respectively, by Bauschlicher and coworkers [78]. This same group also predicted the bond strengths of $M^+\text{-}C_2H_4$ to be 40, 41, and 40 kcal/mol for M = Co, Ni and Cu, respectively [79]. They attributed the greater bond strength of C_2H_4 compared to C_2H_2 as being due to the larger polarizability of C_2H_4.

The acetylene-ethylene complexes of cobalt and nickel were readily generated via reactions (21) and (22). Unfortunately, difficulty was encountered

$$M^+ + \text{n-butane} \longrightarrow M(C_2H_4)_2^+ + H_2 \quad (21)$$

$$M(C_2H_4)_2^+ + C_2H_2 \longrightarrow M(C_2H_2)(C_2H_4)^+ + C_2H_4 \quad (22)$$

$$(M = Co, Ni)$$

in generating sufficient intensities of the iron analog to study. IRMPD of $Co(C_2H_2)(C_2H_4)^+$ synthesized in this fashion yields equal loss of C_2H_2 and C_2H_4, within experimental error. $CoC_2H_4^+$ goes on to form Co^+ by IRMPD, while $CoC_2H_2^+$ is photoinactive. SORI [80] of $Co(C_2H_2)(C_2H_4)^+$, which is a collisional activation method, produces $CoC_2H_2^+$ and $CoC_2H_4^+$ in a roughly 1.7:1 ratio, while CID forms these ions in about a 2:1 ratio, as well as Co^+ at higher energies.

Surprisingly, $Ni(C_2H_2)(C_2H_4)^+$, prepared from reactions (21) and (22), did not undergo IRMPD during up to 3 seconds of irradiation time at 27 W laser power. In contrast $Ni(C_2H_4)_2^+$ is observed to photodissociate. The SORI and CID of $Ni(C_2H_2)(C_2H_4)^+$ is similar to that of cobalt with the amount of $NiC_2H_2^+$ produced greater than that of $NiC_2H_4^+$. Taken in total, these results suggest $D(M^+\text{-}C_2H_2) \geqslant D(M^+\text{-}C_2H_4)$ for M = Co and Ni, but within 2–3 kcal/mol of each other. Thus, we assign $D(Co^+\text{-}C_2H_2) = 43 \pm 3$ kcal/mol and

$D(Ni^+\text{-}C_2H_2) \geqslant 33 \pm 6$ kcal/mol based on the ion beam values of $D(Co^+\text{-}C_2H_4) = 42.9 \pm 1.6$ kcal/mol and $D(Ni^+\text{-}C_2H_4) \geqslant 33.0 \pm 4.5$ kcal/mol [76]. The IRMPD results in particular indicate that C_2H_2 and C_2H_4 have nearly equivalent bond energies. The small discrepancy with the collisional results, where $MC_2H_2^+$ is in greater abundance, may arise due to angular momentum effects favoring C_2H_4 loss. The SORI technique, however, is clearly important in probing ions like $Ni(C_2H_2)(C_2H_4)^+$ which are photoinactive. While the experimental results differ somewhat from the theoretical calculations in indicating C_2H_2, if anything, is bound somewhat more strongly than C_2H_4, both experiment and theory indicate that they are within a few kcal/mol of each other.

4. Photoinduced ion-molecule reactions

Thus far, the study of gas-phase ion photochemistry under the low pressure conditions used in FTMS has been largely limited to unimolecular photodissociation and, for negative ions, photodetachment processes. In contrast to the extensive literature on condensed-phase photochemistry, reports of bimolecular photochemistry in the gas phase are rare. This is due in part to the long times between collisions at the low pressures used in FTMS which permit other processes such as unimolecular dissociation, electron detachment, and even radiative relaxation to dominate. In addition, even if the activated ion is long-lived, collisional relaxation is more likely to occur than a reaction. Recently, however, several examples from our group and Buckner's group have demonstrated that these photoinduced reactions may be considerably more prevalent for metal-containing ions than previously thought.

In the most general definition, a photoinduced reaction is one in which irradiation speeds up or slows down a reaction. The most interesting case is when a new reaction channel is observed. In order to observe a photoinduced reaction, the lifetime of the photoactivated reagent ion must be at least on the order of milliseconds (considering typical operating pressures in FTMS of 10^{-9}–10^{-5} torr) and, of course, the activated ion must exhibit a different reactivity than the thermalized ion.

Long-lived excited intermediates are accessible by three routes: (1) intersystem crossing to a state spin-forbidden to the ground state; (2) infrared absorption or internal conversion to a vibrationally excited ground state; and (3) isomerization. Long-lived electronically and vibrationally excited ions, created directly in the ionization process (electron impact, photoionization, laser desorption), have been observed to exhibit unique chemistry [81–84]. In fact, efforts are usually taken to allow thermalization of "hot" ions by introducing a collisional cooling period. Thus, it was not a question of whether a photoinduced reaction would be observed, but simply of finding the right photochemical systems.

Using photon energies below the energy required for dissociation, and for negative ions below the energy for detachment, eliminates these competing channels. The observation of a wide variety of sequential multiphoton dissociation [12, 21] (vide supra) and detachment [25] processes would seem to provide ideal systems for observing photoinduced reactivity, since these processes occur through long-lived excited intermediates. However, in the systems tested this proves not to be the case since, as stated above, collisional deactivation generally occurs. In fact, Dunbar and co-workers have studied these processes extensively and have obtained detailed fundamental information on radiative and collisional relaxation processes [12, 85, 86]. This in part explains why only one example of a photoinduced reaction was reported prior to 1989.

In that early study, Bomse and Beauchamp reported that the reverse of

$$(CH_3OH)H(H_2O) + CH_3OH \underset{k_r}{\overset{k_f}{\rightleftharpoons}} (CH_3OH)_2H^+ + H_2O \qquad (23)$$

reaction (23) was increased by on the order of an impressive 10^3 under low-intensity CW infrared laser irradiation [19]. More recently, our studies on the chemistry and photochemistry of metal-containing ions indicated that these species might exhibit photoinduced reactivity for several reasons: first, the ions absorb broadly in the UV-VIS and ir; second, metal centers catalyze ligand rearrangements and other chemistry; and third, several cases have been reported in which the products from exothermic reactions exhibit a different reactivity after they have been permitted to undergo thermalizing collisions [87, 88]. This indeed has proved to be a fruitful approach with now several examples reported falling into five general categories: (1) deceleration of exothermic reactions; (2) photoinduced secondary fragmentation, whereby an internally excited product ion is generated which can undergo further fragmentation; (3) photoactivated ligand switching; (4) photoinduced isomerization; and (5) photoenhanced hydrogen/deuterium exchange. Examples of each are presented below.

4.1. Deceleration of exothermic reactions

In the presence of propane, $Co(c\text{-}C_5H_6)^+$ undergoes reaction (24) to generate, presumably, cyclopentadienyl-Co^+-allyl ion [20]. Upon irradiation

$$Co^{+}\text{—}(c\text{-}C_5H_6) + C_3H_8 \longrightarrow (\eta^3\text{-}C_3H_5)\text{—}Co^{+}\text{—}(C_5H_5) + 2H_2 \qquad (24)$$

with the CW CO_2 infrared laser at 10.6 μ, this reaction is dramatically slowed. The cyclopentadienyl-allyl complex, in a separate experiment, was found to be unaffected by the laser light and, in particular, does not photodissociate back to the reactant ion. There are two possible explanations for slowing of reaction (24) upon irradiation. First, reaction (24) is exothermic and will be slower upon "heating" one or both of the reactants and second, irradiation has caused rearrangement of the ion to a less reactive form. Similarly, these two possibilities have been used to explain the interesting observation that Si_n^+ (n = 39, 45) become less reactive upon absorption of light [89]. Unfortunately, these two possibilities are not easily distinguishable.

4.2. Photoinduced secondary fragmentation

Two examples of this process have been reported in which a portion of the excess energy of the reactant ion is carried over to the product ion resulting in additional fragmentation. The first example involved irradiation of Rh(c-$C_5H_6)^+$ using visible light in the presence of cyclopentane [15]. In the absence of light, Rh(c-$C_5H_6)^+$ undergoes the double dehydrogenation reaction (25). Upon irradiation with an argon ion laser at 488 nm, $RhC_{10}H_{10}^+$ (rhodacenium) is also observed coming from two channels, reactions (26) and (27). In order

$$Rh(c\text{-}C_5H_6)^+ + c\text{-}C_5H_{10} \longrightarrow RhC_{10}H_{12}^+ + 2H_2 \qquad (25)$$

to verify reaction (27), process (26) was eliminated by continuously ejecting $Rh(C_{10}H_{12})^+$. While $Rh(C_{10}H_{10})^+$ product is noticeably reduced, an easily measurable signal arising from reaction (27) can be seen, Figure 9.

The second example involved irradiation of Fe(isobutene)$^+$ in the infrared [15]. In the absence of light, Fe (isobutene)$^+$ undergoes reaction (28) with chlorobenzene. Upon irradiation a new product, Fe(toluene)$^+$, is observed coming from, again, two processes, reactions (29) and (30). Elimination of

$$RhC_{10}H_{12}^+ + h\nu \longrightarrow RhC_{10}H_{10}^+ + H_2 \qquad (26)$$

$$[Rh(c\text{-}C_5H_6)^+]^* + c\text{-}C_5H_{10} \longrightarrow RhC_{10}H_{10}^+ + 3H_2 \qquad (27)$$

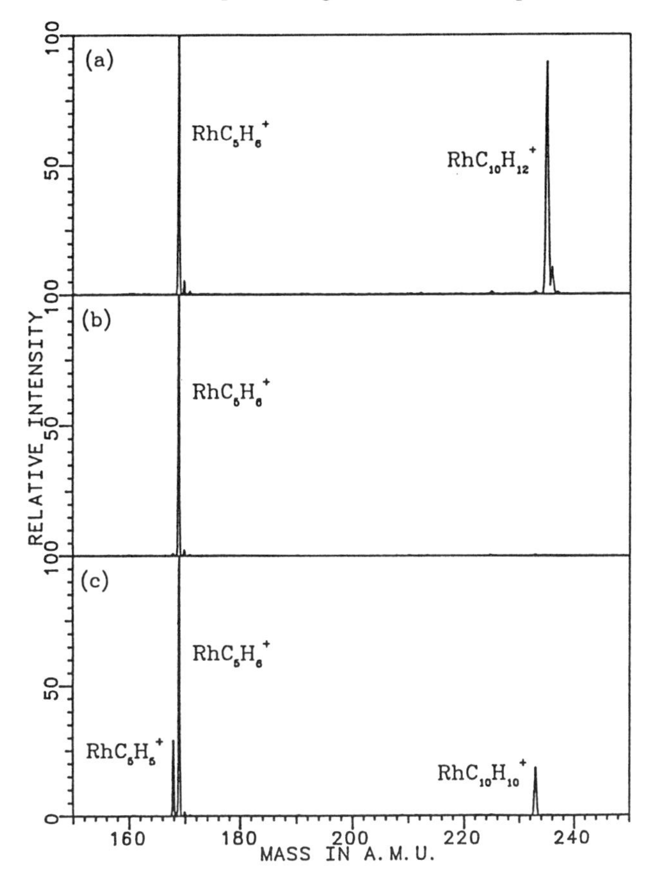

Fig. 9. (a) Reaction of $RhC_5H_6^+$ and cyclopentane (~10^{-7} torr) for 0.75 s with no irradiation, (b) with no irradiation and continuous ejection of $RhC_{10}H_{12}^+$, and (c) with 3 W 514 nm irradiation and continuous ejection of $RhC_{10}H_{12}^+$. Figure copied from reference 15 with permission.

Fe^+–(isobutene) + C_6H_5Cl ⟶ Fe^+–(2-methyl-1-propenyl)benzene + HCl (28)

Fe^+–(2-methyl-1-propenyl)benzene + nhν ⟶ Fe^+–toluene + C_3H_4 (29)

Fe^+–(isobutene) + nhν ⟶ [Fe^+–(isobutene)]* ⟶ (C_6H_5Cl) ⟶ Fe^+–toluene + HCl + C_3H_4 (30)

reaction (29) by continuous ejection of the Fe(7–methyl-1–phenyl propene)$^+$ resulted in a reduced, but still measurable Fe(toluene)$^+$ signal arising from process (30).

4.3. Photoinduced ligand switching

Considering the wealth of information on this process in solution, one could predict that it would be an important class of reactions in the gas phase as well. Pope and Buckner recently reported the first example in which irradiation of $V(C_6H_6)^+$ at 604 nm in the presence of CH_3CN produces $V(CH_3CN)^+$,

$$[V(C_6H_6)^+]^* + CH_3CN \rightarrow V(CH_3CN)^+ + C_6H_6 \qquad (31)$$

reaction (31) [90]. This reaction is normally endothermic and would not be observed with thermalized reactants. They explained this reaction as proceeding through an excited complex $[V(C_6H_6)(CH_3CN)^+]^*$ in which the internal energy is equal to the sum of the photo-excitation energy and the coordination energy of CH_3CN to the $V(C_6H_6)^+$ complex. The excited complex then fragments by competitive loss of CH_3CN and C_6H_6, the latter producing the ligand-switched product. In the absence of irradiation, only the coordination energy is available in the complex and fragmentation may only proceed back to the $V(C_6H_6)^+$ reactant ion. This mechanism is analogous to the one proposed by Bomse and Beauchamp to explain reaction (23). We observed two additional examples using infrared also involving aryl complexes, reactions (32) and (33) [15]. Both of these processes are normally endothermic and proceed readily in the opposite direction.

$$(\text{toluene})Co^+(\text{toluene}) + nh\nu \longrightarrow [(\text{toluene})Co^+(\text{toluene})]^* \xrightarrow{CH_3CN} (\text{toluene})Co^+{-}CH_3CN + \text{toluene} \qquad (32)$$

$$Fe^+(\text{toluene}) + nh\nu \longrightarrow [Fe^+(\text{toluene})]^* \xrightarrow{\text{benzene}} Fe^+(\text{benzene}) + \text{toluene} \qquad (33)$$

4.4. Photoinduced rearrangement

This is among the most interesting and potentially useful processes. In particular, observation of this process indicates that the barrier to rearrangement lies near or below that of dissociation and provides a means of measuring that barrier height, although this has yet to be done. At the same time, it may also be difficult to unambiguously prove that the process involves rearrangement and not simply internally excited ions.

Two examples have been reported. The first involves irradiation of Co_2NO^+ in the visible [16]. Klaassen and Jacobson first reported that this ion, generated by reacting Co^+ with $CoNO(CO)_3$ followed by collision-induced dissociation on the subsequent binuclear reaction product to cleave off the remaining CO ligands, exhibits biexponential kinetics in its reaction with O_2 [91]. In particular, they observed that about 55% of the ions reacted with O_2 to form $Co_2O_2^+$ and 45% remained unreactive. This ratio was not significantly altered by varying the background pressure of the argon present as a collision gas. Thus, they concluded that the reactive ion was a stable isomer, as opposed to being simply internally hot, and proposed structure I containing a molecularly bound NO. The unreactive ion they postulated as being a dissociated nitride-oxide structure II. Interestingly, irradiation

$$\underset{\text{I}}{Co^+\text{—}NO} \qquad \underset{\text{II}}{N\text{—}Co^+\text{—}O}$$

at 514.5 nm using the argon ion laser resulted in a quantitative conversion of the unreactive ions into the reactive form. At shorter wavelengths (higher energies), photodissociation by exclusive loss of NO was observed for the entire population of ions. Thus, the barrier to rearrangement for structures I and II lies below the dissociation threshold. In accordance with our observations, Jacobson and co-workers observed that collisional activation of the unreactive species also resulted in it becoming reactive. This system, together with other metal cluster-ligand systems, should provide useful models for studying photocatalytic transformations on surfaces.

Another interesting example is $Fe(c\text{-}C_5H_6)^+$ generated from the reaction of Fe^+ with cyclopentene by dehydrogenation. Previous H/D exchange and reactivity studies suggest that this species has a hydrido-cyclopentadienyl structure III which is close in energy to its isomer Fe^+-cyclopentadiene,

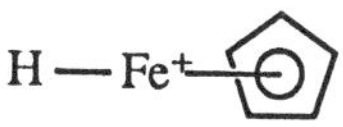

H—Fe⁺–(C₅H₅)   Fe⁺–(C₅H₆)

III    IV

structure IV [92, 93]. An experiment was performed to monitor the effect of infrared irradiation on the reaction of $FeC_5H_6^+$ with benzene. In the absence of light, thermalized $FeC_5H_6^+$ reacts readily with benzene to displace one H atom, exclusively, reaction (34). This result is certainly consistent

$$Fe(c\text{-}C_5H_6)^+ + C_6H_6 \longrightarrow (C_6H_6)Fe^+(C_5H_5) + H^\cdot \qquad (34)$$

with the hydrido structure III. Upon irradiation with the CO_2 laser at 10.6 μ, an entirely different product, $FeC_6H_6^+$, is generated apparently from photoactivated ligand switching reaction (35). Representative spectra from these experiments are shown in Figure 10. That the change in reactivity was not

$$Fe(c\text{-}C_5H_6)^+ + nh\nu \longrightarrow [Fe(c\text{-}C_5H_6)^+]^* \xrightarrow{C_6H_6} Fe^+\text{-}C_6H_6 + c\text{-}C_5H_6 \quad (35)$$

simply due to the amount of internal energy in the collision complex was tested by performing CID on $Fe(C_5H_6)(C_6H_6)^+$, which was generated by condensing cyclopentadiene on to $Fe(C_6H_6)^+$ and then permitted to undergo thermalizing collisions prior to CID. At the lowest CID energies, this complex was observed to lose H and at higher energies an additional benzene was lost yielding $FeC_5H_5^+$. Thus, observation of $FeC_6H_6^+$ upon irradiation is reasonably explained as arising from rearrangement of structure III to structure IV.

4.5. Photoenhanced H/D exchange

In the presence of D_2, $Co(c\text{-}C_5H_6)^+$ is observed to undergo one slow H/D exchange (~1/20 Langevin rate) followed by five considerably slower exchanges [92]. The first exchange is with the hydrogen on the metal in the hydrido-M^+-cyclopentadienyl structure III. As shown in Scheme 1, structure IV must be accessible where 1,5–sigmatropic shifts are required to exchange the remaining hydrogens. The presence of an energy barrier is evident, since the first exchange is slow but faster than the subsequent exchanges. This barrier might be associated with the isomerization step between structures III and IV, the sigmatropic shifts, or both. Infrared irradiation of $CoC_5H_5D^+$ in the presence of D_2 dramatically increases the rate of additional H/D exchanges, as might be expected from ions which possess internal energy above the barrier for isomerization.

4.6. Wavelength dependence studies in the visible

One of the promises of observing photoinduced reactions is that absorption information should be obtainable at wavelengths longer than that required for photodissociation. This possibility is clearly evident in the above examples involving absorption of infrared photons. While we are limited at this time to a single wavelength CO_2 infrared laser, we can explore the dependence of the photoinduced reactions observed in the visible using the argon-ion

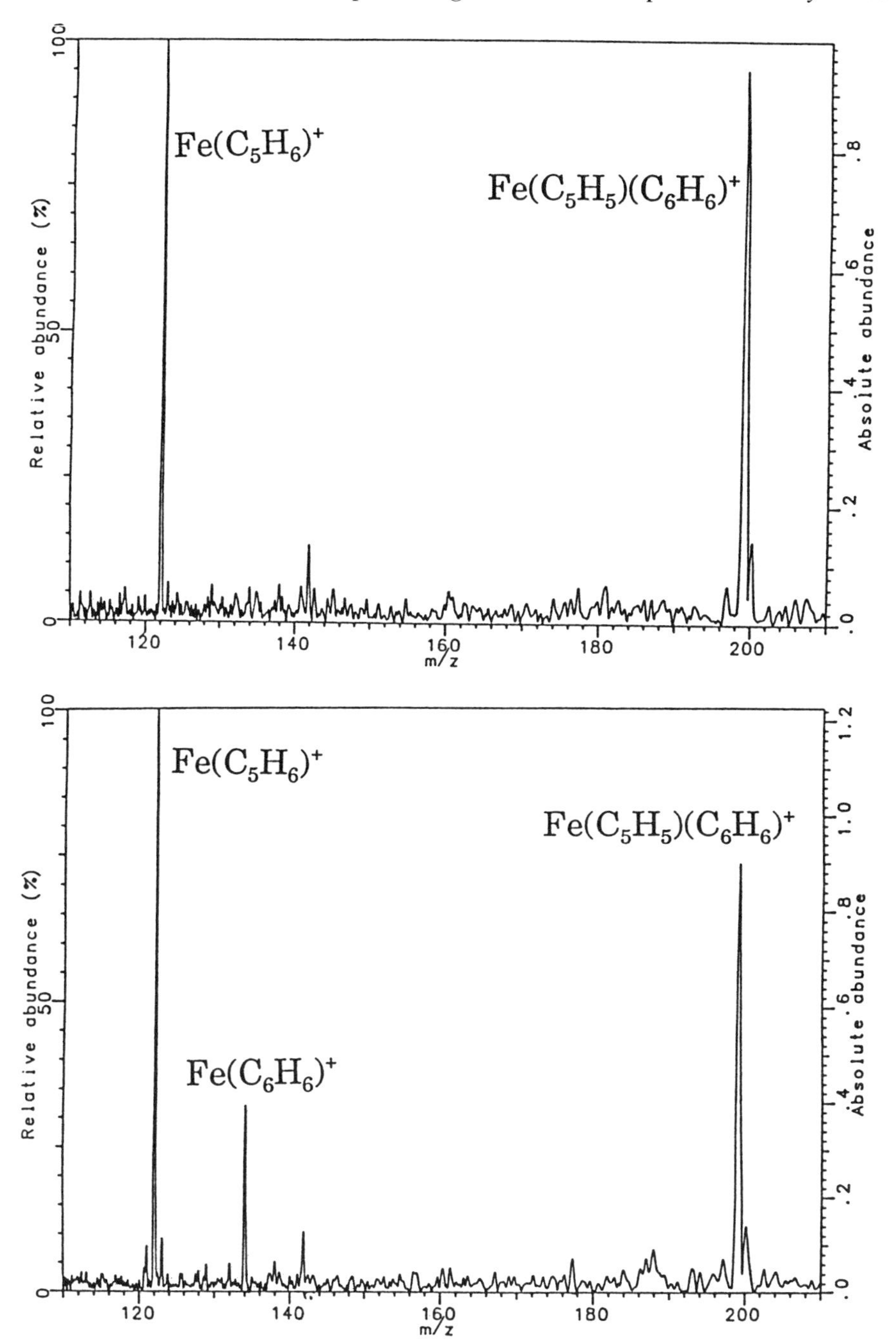

Fig. 10. (a) Reaction of $FeC_5H_6^+$ with benzene, 1s at 4×10^{-8} torr. (b) Same as (a) except with 15 W infrared radiation. Figure copied with permission from reference 20.

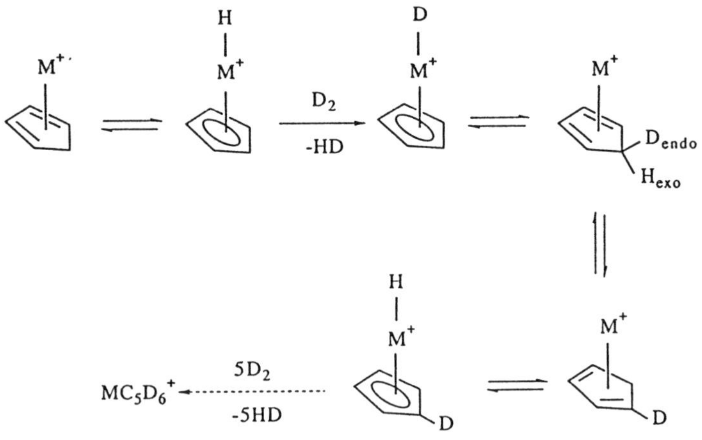

Scheme 1.

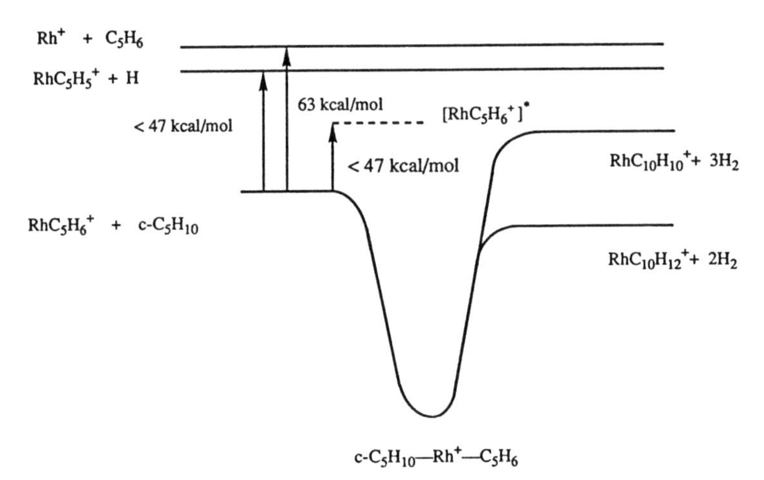

Fig. 11. Simple energy diagram for the photoinduced reaction of $RhC_5H_6^+$ with cyclopentane.

dye laser system. Our first attempts at this are presented here for the $RhC_6H_6^+$ system.

The observed photoinduced reactivity of $RhC_5H_6^+$ is attributed to the absorption of a visible photon to generate a long-lived vibrationally excited ground state ion, by internal conversion, which can collide with cyclopentane to form $RhC_{10}H_{10}^+$, reaction (27). A simplified energy diagram for this reaction is shown in Figure 11. By increasing the wavelength (decreasing the energy), thresholds could in theory be obtained for both the photodissociation and photoinduced pathways. In addition, as depicted in Figure 11, the photoinduced reactivity should still continue even after the dissociation processes have terminated. Unfortunately, this was not observed at the available wavelengths nor could a trend be observed for the ratio of $RhC_{10}H_{10}^+$: $RhC_5H_5^+$. We suspect that the intensity of $RhC_{10}H_{10}^+$ was lower than the actual yield because during ejection of $RhC_{10}H_{12}^+$ (65.7055 kHz) to prevent reaction (26), inadvertently some rhodocenium $RhC_{10}H_{10}^+$ (66.2435 kHz) was ejected. However, this is unavoidable and the reduced intensity precludes the determination of an accurate ratio.

While H loss is observed to occur out to 610 nm (47 kcal/mol), loss of the entire ligand has a threshold near 458 nm, reactions (25) and (26). This

		Threshold (kcal/mol)	
$RhC_5H_6^+ \xrightarrow{h\nu}$	$RhC_5H_5^+ + H$	< 47	(25)
	$Rh^+ + C_5H_6$	63±3	(26)

yields $D^o(Rh^+\text{-}C_5H_6) = 63 \pm 3$ kcal/mol. In a separate study, competitive CID and photodissociation gave 59 ± 3 kcal/mol [94]. Exothermic dehydrogenation of cyclopentane implies $D^o(Rh^+\text{-}C_5H_6) > 50$ kcal/mol [57]. In addition, the thresholds for reactions (25) and (26) yield a lower limit $D^o(Rh^+\text{-}C_5H_5) > 95$ kcal/mol. Exothermic dehydrogenation of cyclopentene yields $D^o(Rh^+\text{-}C_5H_5) > 101$ kcal/mol, in agreement with the results from this study [94].

While this first attempt at obtaining wavelength information by monitoring a photoinduced reaction proved somewhat disappointing, this general approach will undoubtedly yield useful information on other systems. Thus, monitoring photoinduced reactions provides a complementary method to multiphoton photodissociation for obtaining absorption information on ions at longer wavelengths than would be available by single photon photodissociation methods. In addition the photoinduced reactions are interesting and worthy of study in their own right.

Acknowledgements

Acknowledgment is made to the Division of Chemical Sciences in the Office of Basic Energy Sciences in the United States Department of Energy for supporting this research and to the National Science Foundation for providing funds for the advancement of FTICRMS methodology.

References

1. M. V. Buchanan (Ed.), *Fourier Transform Mass Spectrometry; Evolution, Innovation, and Applications*, Vol. 359 (American Chemical Society, Washington, D.C., 1987), p. 205.
2. B. S. Freiser, in *Techniques for the Study of Gas-Phase Ion-Molecule Reactions: Techniques of Chemistry*, Vol. 20, edited by J. M. Farrar and W. H. Saunders (Wiley-Interscience, New York, 1988), p. 61.

3. A. G. Marshall, *Adv. Mass. Spectrom.* **11A**, 651 (1989).
4. D. H. Russell, *Mass. Spec. Rev.* **5**, 167 (1986).
5. D. A. Laude, Jr., C. L. Johlman, R. S. Brown, D. A. Weil, and C. L. Wilkins, *Mass Spectrom. Rev.* **5**, 107 (1986).
6. B. Asamoto and R. C. Dunbar, *Analytical Applications of Fourier Transform Ion Cyclotron Resonance Mass Spectrometry* (VCH, New York, 1991), p. 306.
7. C. B. Jacoby, C. L. Holliman, and M. L. Gross, *NATO ASI Ser., Ser. C* **353**, 93 (1992).
8. R. B. Cody, R. C. Burnier, W. D. Reents, Jr., T. J. Carlin, D. A. McCrery, R. K. Lengel, and B. S. Freiser, *Int. J. Mass Spectrom. Ion Phys.* **33**, 37 (1980).
9. D. A. McCrery, E. B. Ledford, Jr., and M. L. Gross, *Anal. Chem.* **54**, 1435 (1982).
10. B. S. Freiser, *Talanta* **32**, 697 (1985).
11. B. S. Freiser, *Chemtracts* **1**, 65 (1989).
12. R. C. Dunbar, in *Gas Phase Ion Chemistry*, Vol. 3, edited by M. T. Bowers (Academic Press, New York, 1984), p. 130.
13. R. C. Dunbar, in *Gas Phase Inorganic Chemistry*, edited by D. H. Russell (Plenum, New York, 1989), p. 323.
14. R. L. Hettich and B. S. Freiser, in *Fourier Transform Mass Spectrometry: Evolution, Innovation, and Applications*, edited by M. V. Buchanan (American Chemical Society, Washington, D. C. 1987), Ch. 10.
15. J. R. Gord, S. W. Buckner, and B. S. Freiser, *J. Am. Chem. Soc.* **11**, 3753 (1989).
16. J. R. Gord and B. S. Freiser, *J. Am. Chem. Soc.* **111**, 3754 (1989).
17. B. S. Freiser and J. L. Beauchamp, *Chem. Phys. Lett.*, **35**, 35 (1975).
18. J. D. Faulk and R. C. Dunbar, *J. Am. Chem. Soc.* **114**, 8596 (1992).
19. D. S. Bomse and J. L. Beauchamp, *J. Am. Chem. Soc.*, **102**, 3967 (1980).
20. Y. Huang and B. S. Freiser, *J. Am. Chem. Soc.* **115**, 737 (1993).
21. L. R. Thorne and J. L. Beauchamp, in *Gas Phase Ion Chemistry*, Vol. 3, edited by M. T. Bowers (Academic Press, New York, 1984), p. 42.
22. C. H. Watson, J. A. Zimmerman, J. E. Bruce, and J. R. Eyler, *J. Phys. Chem.* **95**, 6081 (1991).
23. B. S. Freiser and J. L. Beauchamp, *J. Am. Chem. Soc.* **98**, 3136 (1976).
24. B. S. Freiser and J. L. Beauchamp, *J. Am. Chem. Soc.* **99**, 3214 (1977).
25. P. S. Drzaic, J. Marks, and J. I. Brauman, in *Gas Phase Ion Chemistry*, Vol. 3, edited by M. T. Bowers (Academic Press, New York, 1984), p. 168.
26. J. A. Castoro, L. M. Nuwaysir, C. F. Ijames, and C. L. Wilkins, *Anal. Chem.* **64**, 2238 (1992).
27. J. D. Faulk and R. C. Dunbar, *J. Am. Soc. Mass Spectrom.* **2**, 97 (1991).
28. D. F. Hunt, J. Shabanowitz, J. R. Yates, N.-Z. Zhu, D. H. Russell, and M. E. Castro, *Proc. Natl. Acad. Sci. USA* **84**, 620 (1987).
29. C. J. Cassady, S. Afzaal, and B. S. Freiser, *Org. Mass Spectrom.* **29**, 30 (1994).
30. C. J. Cassady and S. W. McElvaney, *Org. Mass Spectrom.*, **28**, 1650 (1993).
31. L. M. Nuwaysir and C. L. Wilkins, *Anal. Chem.* **61**, 689 (1989).
32. R. E. Tecklenburg, Jr., M. N. Miller, and D. H. Russell, *J. Am. Chem. Soc.* **111**, 1161 (1989).
33. P. H. Hemberger, N. S. Nogar, J. D. Williams, R. G. Cooks, and J. E. P. Syka, *Chem. Phys. Lett.* **191**, 405 (1992).
34. R. J. Hughes, R. E. March, and A. B. Young, *Int. J. Mass Spectrom. Ion Phys.* **47**, 85 (1983).
35. T. M. Miller, *Adv. Electron. Electron Phys.* **55**, 119 (1981).
36. M. L. Vestal and J. H. Futrell, *Chem. Phys. Lett.* **28**, 559 (1974).
37. J. S. Pilgrim and M. A. Duncan, *J. Am. Chem. Soc.* **115**, 4395 (1993).
38. D. E. Lessen, R. L. Asher, and P. J. Brucat, *J. Chem. Phys.* **95**, 1414 (1991).
39. P. J. Brucat, L.-S. Zheng, C. L. Pettiette, S. Yang, and R. E. Smalley, *J. Chem. Phys.* **84**, 3078 (1986).

40. F. M. Harris and J. H. Beynon, in *Gas Phase Ion Chemistry*, Vol. 3, edited by M. T. Bowers (Academic Press, New York, 1984), p. 100.
41. R. E. Tecklenburg, Jr., M. E. Castro, and D. H. Russell, *Anal. Chem.* **61**, 153 (1989).
42. J. R. Gord and B. S. Freiser, *Anal. Chim. Acta* **225**, 11 (1989).
43. R. B. Cody, J. A. Kinsinger, S. Ghaderi, I. J. Amster, F. W. McLafferty, and C. E. Brown, *Anal. Chim. Acta* **178**, 43 (1985).
44. S. Maruyama, L. R. Anderson, and R. E. Smalley, *Rev. Sci. Instrum.* **61**, 3686 (1990).
45. T. J. Carlin and B. S. Freiser, *Anal. Chem.* **55**, 571 (1983).
46. Y. Huang, Y. D. Hill, and B. S. Freiser, *J. Am. Chem. Soc.* **113**, 840 (1991).
47. Y. Ranasinghe, T. J. MacMahon, and B. S. Freiser, *J. Am. Chem. Soc.* **114**, 9112 (1992).
48. C. W. Bauschlicher, Jr. and S. R. Langhoff, *J. Phys. Chem.* **95**, 2278 (1991).
49. L. S. Sunderlin and P. B. Armentrout, *J. Am. Chem. Soc.* **111**, 3845 (1989).
50. M. B. Comisarow, V. Grassi, and G. Parisod, *Chem. Phys. Lett.* **57**, 413 (1978).
51. Y. A. Ranasinghe and B. S. Freiser, *Chem. Phys. Lett.* **200**, 135 (1992).
52. L. S. Sunderlin, N. Aristov, and P. B. Armentrout, *J. Am. Chem. Soc.* **109**, 78 (1987).
53. L. M. Roth and B. S. Freiser, *Mass Spectrom. Rev.* **10**, 303 (1991).
54. Y. D. S. Hill, Ph. D. Thesis, Purdue University (1992).
55. Y. A. Ranasinghe, Ph. D. Thesis, Purdue University (1992).
56. M. Rosi and C. W. Bauschlicher, Jr., *Chem. Phys. Lett.* **166**, 189 (1990).
57. S. G. Lias, J. E. Bartmess, J. F. Liebman, J. L. Holmes, R. D. Levin, and W. G. Mallard, *J. Phys. and Chem. Data* **17**, Suppl. 1, (1988).
58. H. Clauberg, D. W. Minsek, and P. Chen, *J. Am. Chem. Soc.* **114**, 99 (1992).
59. L. M. Roth, B. S. Freiser, C. W. Bauschlicher, Jr., H. Partridge, and S. R. Langhoff, *J. Am. Chem. Soc.* **113**, 3274 (1991).
60. R. H. Schultz and P. B. Armentrout (in preparation).
61. S. Afzaal and B. S. Freiser, *Chem. Phys. Lett.* **218**, 254 (1994).
62. C. E. Moore, *Atomic Energy Levels* (U. S. National Bureau of Standards, Washington, D.C., 1952), Circ. 467.
63. B. S. Freiser, in *ACS Symposium Series*: *Bonding Energetics in Organometallic Compounds*, Vol. 428, edited by T. J. Marks (Am. Chem. Soc., Washington, D.C., 1990), p. 55.
64. K. F. Willey, C. S. Yeh, D. L. Robbins, and M. A. Duncan, *J. Phys. Chem.* **96**, 9106 (1992).
65. C. W. Bauschlicher, Jr., H. Partridge, and S. R. Langhoff, *J. Phys. Chem.* **96**, 3273 (1992).
66. P. B. Armentrout (private communication).
67. D. S. Bomse, R. L. Woodin, and J. L. Beauchamp, *J. Am. Chem. Soc.* **101**, 5503 (1979).
68. D. S. Bomse, R. L. Woodin, and J. L. Beauchamp, in *Advances in Laser Chemistry*, edited by A. H. Zewail (Springer-Verlag, New York, 1978), p. 362.
69. M. A. Hanratty, C. M. Paulsen, and J. L. Beauchamp, *J. Am. Chem. Soc.* **107**, 5074 (1985).
70. S. K. Shin and J. L. Beauchamp, *J. Am. Chem. Soc.* **112**, 2066 (1990).
71. I. B. Surjasasmita and B. S. Freiser, *J. Am. Soc. Mass Spectrom.* **4**, 135 (1993).
72. S. A. McLuckey, A. E. Shoen, and R. G. Cooks, *J. Am. Chem. Soc.* **104**, 848 (1982).
73. S. A. McLuckey, D. Cameron, and R. G. Cooks, *J. Am. Chem. Soc.* **103**, 1313 (1981).
74. K. Ervin and P. B. Armentrout, *J. Chem. Phys.* **84**, 6750 (1986).
75. I. B. Surjasasmita and B. S. Freiser (paper in preparation).
76. C. L. Haynes and P. B. Armentrout, *Organometallics* (submitted).
77. E. R. Fisher and P. B. Armentrout, *J. Phys. Chem.* **94**, 1674 (1990).
78. M. Sodupe and C. W. Bauschlicher, *J. Phys. Chem.* **95**, 8640 (1991).
79. M. Sodupe, C. W. Bauschlicher, S. R. Langhoff, and H. Partridge, *J. Phys. Chem.* **96**, 2118 (1992).
80. J. W. Gautheir, T. R. Trautman, and D. B. Jacobson, *Anal. Chim. Acta* **246**, 211 (1991).
81. P. A. M. van Koppen, P. R. Kemper, and M. T. Bowers, *J. Am. Chem. Soc.* **114**, 1083 (1992).

82. T. Wyttenbach and M. T. Bowers, *J. Phys. Chem.* **96**, 8920 (1992).
83. S. D. Hanton, R. J. Noll, and J. C. Weisshaar, *J. Chem. Phys.* **96**, 5176 (1992).
84. W. D. Reents, Jr., F. Strobel, R. B. Freas, III, J. Wronka, and D. P. Ridge, *J. Phys. Chem.* **89**, 5666 (1985).
85. R. C. Dunbar, *J. Chem. Phys.* **90**, 2192 (1989).
86. R. C. Dunbar, *Mass Spectrom. Rev.* **11**, 309 (1992).
87. Y. Huang and B. S. Freiser, *J. Am. Chem. Soc.* **112**, 5085 (1990).
88. C. J. Cassady and B. S. Freiser, *J. Am. Chem. Soc.* **107**, 1573 (1985).
89. S. Maruyama, L. R. Anderson, and R. E. Smalley, *J. Chem. Phys.* **93**, 5349 (1990).
90. R. M. Pope and S. W. Buckner, *Organometallics* **11**, 1959 (1992).
91. J. J. Klaassen and D. B. Jacobson, *J. Am. Chem. Soc.* **110**, 974 (1988).
92. D. B. Jacobson and B. S. Freiser, *J. Am. Chem. Soc.* **105**, 7484 (1983).
93. D. B. Jacobson and B. S. Freiser, *J. Am. Chem. Soc.* **107**, 72 (1985).
94. J. H. Ng, Ph. D. Thesis, Purdue University (1992).

8. Applications of gas-phase electron-transfer equilibria in organometallic redox thermochemistry

DAVID E. RICHARDSON
Department of Chemistry, University of Florida, P.O. Box 117200, Gainesville, Florida 32611–7200, U.S.A.

1. Introduction

Ionization energies and electron affinities are key quantities in characterizing the thermodynamics of oxidation-reduction reactions for molecules. Much is known about these intrinsic properties for atoms and organic molecules in the gas phase [1, 2]. Far fewer organometallic compounds have had their ionization energies determined and even fewer electron affinities are known for metal-containing molecules.

In this chapter, the results from a gas-phase technique for determining ionization and electron attachment energetics for organometallic compounds and coordination complexes will be described. The *electron-transfer equilibrium* method (or ETE) has been used to obtain gas-phase redox energetics for a variety of neutral complexes of the transition elements. This ETE data complement and extend the vast amount of electron binding energy data determined by photoelectron spectroscopy. However, as will be shown, the ETE method yields thermochemical quantities that can readily be incorporated into thermodynamic cycles for extracting fundamental solvation energetics and bond disruption energies. Furthermore, the ability to extract electron attachment energetics via the ETE method has led to a sharp increase in our knowledge of gas-phase electron affinities of metal complexes. In turn, we have seen previously unrevealed trends in the intrinsic ligand substituent effects on redox thermochemistry.

2. Vertical vs. adiabatic energies

When considering the thermochemistry of electron removal (i.e. ionization, equation (1)) and electron attachment (equation (2)) for a molecule M, it is important to distinguish between *vertical* energies and *adiabatic* energies.

$$M(g) \rightleftharpoons M^{+}(g) + e^{-} \tag{1}$$

Ben S. Freiser (ed.), Organometallic Ion Chemistry, 259–282.

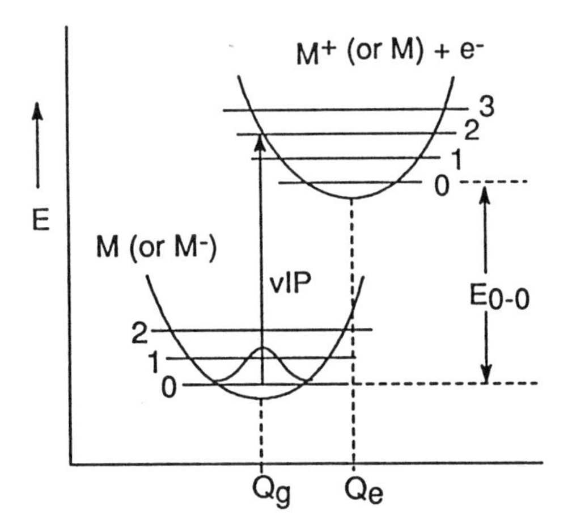

Fig. 1. Potential energy diagrams illustrating the difference between vertical ionization energies (vIP) and adiabatic ionization energies (aIP or E_{0-0}) for the removal of an electron from a neutral or anionic molecule M. The vIP is often higher than the aIP because the higher lying state is shifted in geometry along the coordinate Q compared to the ground state (e.g. an increase in Q could represent an increase in metal-ligand bond distances). If $Q_e \cong Q_g$ since little bond distortion occurs upon electron removal, then the observed vIP=aIP. This diagram is really only qualitatively complete for a diatomic molecule, which has one internal degree of freedom (the bond length). For a polyatomic molecule, a multidimensional diagram would be required for the many vibrational degrees of freedom. In metal complexes, it is also extremely common to have many "nested" electronic states that lead to overlapping bands in the photoelectron spectrum.

$$M(g) + e^- \rightleftharpoons M^-(g) \tag{2}$$

Vertical energies are normally derived from photoelectron spectroscopic measurements [3, 4] in which the excited state corresponds to the removal of an electron from either a neutral molecule or an anion by interaction with a fixed-frequency light source (X-rays, He lamp, synchrotron radiation, etc.). The kinetic energies of the ejected electrons are determined, and the electron binding energies are derived from the simple relationship $h\nu = E_{kin} + E_{binding}$. Because of the vertical nature of the excitation (Figure 1), a consequence of the Franck-Condon principle, the most probable excited states can have molecular geometries that are significantly different from the most stable structure for the ground state (indicated by Q_g). Therefore, the maximum intensity in the photoelectron spectrum can be at an energy quite different from the "0–0" transition energy in which the excited state is formed in its ground vibrational state(s). If the excited state geometry is fortuitously close to that of the ground electronic state, then the most intense transition can be the 0–0 transition, but this is only rarely observed in PES of organometallics.

In an adiabatic process, the initial electronic state is converted to the final state with both states in their in thcir lowest energy vibrational and rotational states. The energy change for the process of removing an electron from a neutral molecule is the adiabatic ionization energy. This quantity is given by the energy of the 0–0 transition in the photoelectron spectrum if it can be identified. Unfortunately, the broad and usually featureless peaks observed for PES of organometallics makes identification of the 0–0 energy impossible except in a few cases. The adiabatic ionization energy of an anion can also be determined from the anion's photoelectron spectrum if the 0–0 transition is observed, but PES of anions is much more technically challenging [5] and subject to the same lack of vibrational structure for larger organometallic species.

Chemists usually consider the energetics of ionization and electron attachment for a collection of molecules at some experimental temperature. In most thermochemical cycles of interest to chemists, a specific temperature is defined (usually 298 K) for all of the molecules and ions in the cycle. Of course, at temperatures above 0 K the occupancy of higher lying vibrational, rotational, and possibly electronic states must be considered, and a Boltzmann distribution over these states is observed. For an ionization or electron attachment reaction at temperature T, it is then necessary to give the thermochemistry using quantities such as the Gibbs energy change ($\Delta G_T{}^\circ$), the enthalpy change ($\Delta H_T{}^\circ$), and the entropy change ($\Delta S_T{}^\circ$). For example, one would use the "thermal ionization free energy change" $\Delta G_{i,298}{}^\circ$ for the one-electron oxidation reaction of equation (1) in a thermochemical cycle involving free energies at 298 K.

The 0–0 energies obtained in a photoelectron experiment are equal to the enthalpy of electron removal at $T = 0$ K [6]. Thus, the adiabatic ionization energy and the electron affinity of a molecule are equal to $\Delta H_{i,\,T=0}{}^\circ$ (equation (1)) and $-\Delta H_{a,\,T=0}{}^\circ$ (equation (2)), respectively. In order to obtain the thermochemical quantities for $T > 0$ K from the photoelectron data, it is necessary to use statistical mechanics coupled with a knowledge all of the degrees of freedom for both the molecule and the ion including vibrational frequencies and their degeneracies, rotational moments, electronic states, etc. Note that a standard state convention must also be applied consistently for the electron in equations (1) and (2) [7]. It is relatively rare that all of the required information is available for larger molecules and ions such as organometallics. This problem can be addressed by moving from spectroscopic methods to thermal equilibrium techniques as described in the next section.

3. Gas-phase electron-transfer equilibria

If one is primarily interested in deriving thermochemical quantities for application in cycles near room temperature, the electron-transfer equilibrium method is an attractive alternative to photoelectron spectroscopy coupled to statistical mechanics. If the experiment is done appropriately, both neutrals and ions in equations (1) and (2) are at or near thermal equilibrium with the surroundings, so thermodynamic quantities derived can be given for experimental temperatures.

The ETE method uses mass spectrometry to follow the equilibration of reactions of the type given in equations (3) and (4).

$$M(g) + R^{+}(g) \rightleftharpoons M^{+}(g) + R(g) \tag{3}$$

$$M(g) + R^{-}(g) \rightleftharpoons M^{-}(g) + R(g) \tag{4}$$

Here M is the molecule of interest and R is the reference compound. The reactions are allowed to equilibrate at known pressures P_{M} and P_{R}, and the ion populations (I) of the source or ion trap are determined by obtaining the mass spectrum. The standard free energy change for the reaction at temperature T as written can be obtained from the equilibrium constant as in equation (5).

$$\Delta G_{et,T}^{\circ} = -RT \ln K_{eq} = -\mathrm{RT} \ln [(I(M^{\pm})P_{R})/(I(R^{\pm})P_{M})] \tag{5}$$

Temperature dependence of the K_{eq} value can be used to obtain enthalpy and entropy changes in the vicinity of the experimental temperatures by using van't Hoff plots.

The ETE reaction can be followed in an ion trap such as those used in Fourier transform mass spectrometry (FTMS) [8–11] where the pressures of neutrals are in the 10^{-8} to 10^{-5} torr range. After several seconds of trapping and reaction, the ions have undergone hundreds of collisions (for higher pressures in the range) and are assumed to be near thermalized. This thermalization is quite important since the ions can be formed with high internal energy due to the exoergicity of reactions used to produce them (by chemical ionization or electron impact, for example). Figure 2 shows an example of FTMS ETE data for a typical experiment involving a ruthenocene derivative.

Alternatively, electron-transfer equilibrium constants have been determined by using pulsed high pressure mass spectrometry (PHPMS) [12]. In this technique, the reaction comes to equilibrium in a high pressure source ($\sim$ 1 torr total pressure including a background gas) following ion production by a pulsed high energy electron beam. The ions exiting the source region are mass analyzed, and a time plot of ion currents can be used to extract the K_{eq} value at long reaction times. This technique has the advantage of a much higher ion/molecule collision frequency than FTMS, so thermalization of the ions can be assured. Temperature dependence studies are relatively easy, and most published work includes determination of ΔS_{T}° and ΔH_{T}° values.

For whatever method is used, the maximum experimental value of K_{eq} is determined by practical limitations on measurable relative pressures of neutrals. Typically, the accessible range is $0.001 < K_{eq} < 1000$, although the PHPMS method has a greater range under many circumstances.

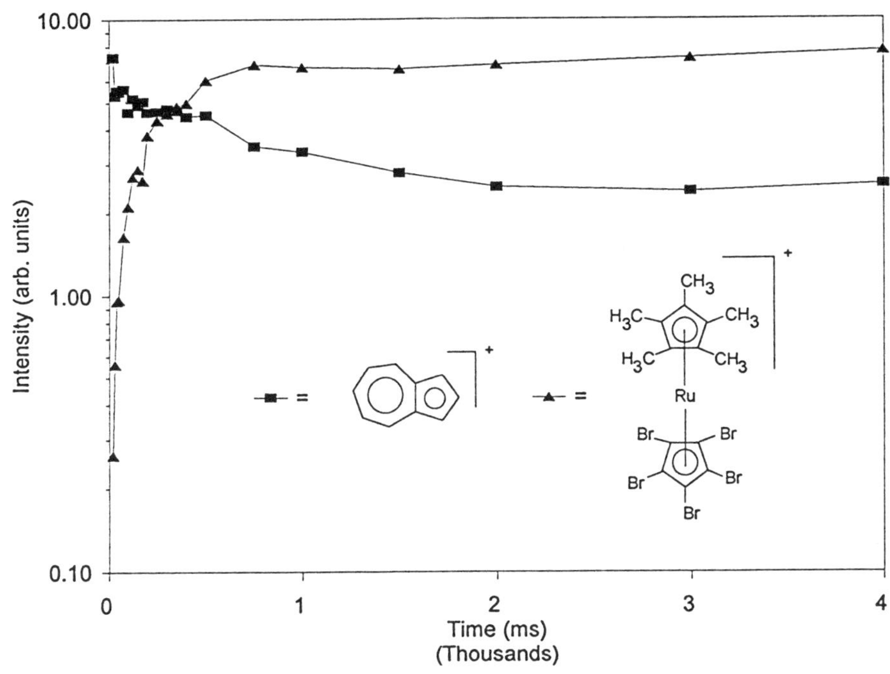

Fig. 2. Typical ion intensity vs. time data for an electron transfer equilibrium experiment using a Fourier transform mass spectrometer in the author's laboratory. After a period of thermalization of the ions via ion/molecule collisions with background gas, one ion is ejected from the ion trap (here the ruthenium metallocenium ion) and the ion intensities are recorded after various time delays (up to 4 s here). Electron transfer from the neutral metallocene to the azulene cation regenerates the metallocenium ion. After approximately 2 s, the ion intensities for the two ions are stable indicating that equilibrium has been attained. Equation (5) can then be used to obtain the equilibrium constant for the reaction under the experimental conditions.

4. Free energies of ionization and electron attachment

To derive the free energy changes for equations (1) and (2) from ETE results, one must know the relevant quantities for the reference compounds R. If the free energy of ionization for the reference compound, $\Delta G_{i,T}°(R)$, is known, then the corresponding energy for the new compound can be obtained from equation (6).

$$\Delta G_{i,T}°(M) = \Delta G_{i,T}°(R) + \Delta G_{et,T}°(\text{eq } 3) \tag{6}$$

In the same way, the free energy of electron attachment for a new compound, $\Delta G_{a,T}°(M)$ for equation (2), can be obtained from the free energy change for the reaction in equation (4) and application of equation (7).

$$\Delta G_{a,T}°(M) = \Delta G_{a,T}°(R) + \Delta G_{et,T}°(\text{eq } 4) \tag{7}$$

The thermodynamics for ionization and electron attachment reactions of anchor reference compounds are generally deduced from high resolution photoelectron spectroscopy coupled to statistical mechanics as described above. Once compounds are brought to equilibrium with this primary reference compound, an equilibrium ladder can be built with reactions involving secondary reference compounds. For example, the $SO_2^{0/-}$ couple is used as an anchor for electron attachment ($\Delta G_{a,423}°(SO_2) = -26.1$ kcal mol^{-1}) [13]. In a similar manner, we have recently introduced $(benzene)_2Cr^{0/+}$ as a reference compound in the low ionization energy range (5–6 eV) [14]. The latter compound has very little structural change accompanying ionization, so the 0–0 transition energy can be obtained with high accuracy by photoelectron spectroscopy; furthermore, the negligible changes in rotational moments and vibrational frequencies upon ionization mean that only the electronic degeneracy needs to be considered in the corrections for the entropy of ionization and heat capacity changes to obtain $\Delta G_{i,T}°$.

In some cases it is difficult or impossible to establish an accurately known pressure for the neutral M in the ion trap or source, but quite often one can produce the ion M^+ (or M^-) by some means (for example, by using an external source such as FAB, electrospray or laser desorption). In this case, one can still estimate adiabatic ionization energies or electron affinities by the *electron-transfer bracketing* (ETB) method. In this approach, various reference compounds are introduced and the appearance of R^+ (or R^-) due to the reverse reactions in equations (1) and (2) is taken as an indication that R has a lower ionization energy (or a higher electron affinity than M). On the other hand, if no electron transfer is observed with R, its ionization energy is presumed to be higher than that of M (or a lower electron affinity in the negative ion experiment). In this way it is often possible to estimate $\Delta G_{i,T}°$ and $\Delta G_{a,T}°$ values within ca. ± 5 kcal mol^{-1} or better depending on the number of reference compounds available in that part of the ladder.

Both the ETE and ETB methods can be inaccurate when the required electron-transfer reactions are too slow compared to loss of ions from the ion trap or source. In the ETE experiment, reactions in equation (1) or (2) may not reach equilibrium before the ion intensity is lost thus preventing the determination of K_{eq}. In the ETB experiment, the absence of a reaction with R may not mean that the reaction is endoergic, it may be instead exoergic but too slow to observe. In our work with organometallics and coordination compounds, it is sometimes found that electron-transfer reactions involving certain compounds become quite inefficient as the driving force for the reaction approaches zero. This inefficiency hampers determination of the adiabatic energetics in these cases. The explanation for the low efficiency of the reactions can be elusive, but they are usually associated with large geometrical rearrangements attending ionization or electron attachment [15]. In some cases, poor electronic interaction between reactant molecules and ions may also contribute [16].

5. Results for organometallic compounds

Tables I and II compile free energies of ionization and electron attachment for a variety of organometallics and coordination compounds. Table I con-

Table I. Ionization energetics for gas-phase metal complexes $M \rightleftharpoons M^+ + e^-$

Compound[a]	ΔG_i° [b]	Ref.	vIP[c]	Ref.
(TTFMH)Cp*Ru	192	20		
(TTFOSi)Cp*Ru	171.7	20		
(C_5F_5)Cp*Ru	170.8	22	179.2	25
azulene[d]	167.8	19		
(C_5Cl_5)Cp*Ru	165.4	20		
Cp_2Ru	164.6	19	171.8	26
N,N-dimethylaniline[d]	160.1	19		
(NO_2Cp)Cp*Ru	161.9	20		
Cp_2Os	160.6	19	164.9	27
$(TMSCp)_2Ru$	158.4	20		
N,N-diethyltoluidine[d]	156.3	19		
Cp_2V	154.5	19	156.3	28
Cp_2Fe	153.1	19	158.7	28
CpCp*Ru	152.3	20		
(TMSCp)Cp*Ru	151.3	20		
$(Ind)_2Ru$	151.0	20		
($CpCH_2NMe_2$)CpFe	150.2	19		
(EtCp)CpFe	150.2	19		
(Ind)Cp*Ru	149.4	20		
(n-BuCp)CpFe	149.1	19		
(t-BuCp)CpFe	149.1	19		
$Ru(acac)_3$	148.7	23	161	29
$(MeCp)_2Fe$	147.7	19	155.0	28
$Ru(Et,H,Et)_3$	146.6	23		
$(TMSCp)_2Fe$	146.5	23		
$Ru(i\text{-}Pr,H,i\text{-}Pr)_3$	145.3	23		
$Ru(n\text{-}Pr,H,n\text{-}Pr)_3$	145.2	23		
$Ru(n\text{-}Bu,H,n\text{-}Bu)_3$	145.1	23		
$Ru(t\text{-}Bu,H,t\text{-}Bu)_3$	145.2	23	156	29
Cp_2Ni	143.8	21,19	150.1	28
(Flu)Cp*Ru	143.1	20		
Cp_2Mn	142.5	19	144.4	30
(EtCp)CpNi	141.5	21		
$(MeCp)_2Ni$	139.6	21	146.7	28
$(EtCp)_2Ni$	138.2	21		
Cp^*_2Ru	137.9	14	150.8	14, 26
$(t\text{-}BuCp)_2Ni$	136.4	21		
Cp^*_2Os	136.4	14	145.5	14, 26
Cp_2Cr	127.5	14	131.4	30
Cp^*_2Fe	126.7	14	135.6	30
Bz_2Cr[d]	125.6	14	126.2	30
Cp_2Co	123.5	14	128.0	30
Cp^*_2Mn	121.6	14	122.9	30
Cp^*_2Ni	121	14	134.2	30
Cp^*_2Cr	(104)	14	113.7	30
Cp^*_2Co	(100)	14	108.5	30

[a] Ligand abreviations: TTFMH, η^5-1,2,3,4–triflurومethylcyclopentadienyl; TTFOSi, η^5-1,2,3,4–trifluormethyl-5–triethylsiloxycyclopentadienyl; Cp*, η^5-pentamethylcyclopentadienyl; NO_2Cp, η^5-nitrocyclopentadienyl; Cp, η^5-cyclopentadienyl; TMSCp, η^5-trimethylsilylcyclopentadienyl; Ind, indenyl; acac, acetylacetonate or 2,4–pentanedionate; (Et,H, Et-dk), 3,5–n-heptanedionate; (i-Pr,H,i-Pr-dk), 2,5–dimethyl-3,5–heptanedionate; (n-Pr,H,n-Pr-dk), 4,6–nonanedionate; (n-Bu,H,n-Bu-dk), 5,7–undecanedionate; (t-Bu,H,t-Bu-dk), 2,2,6,6–tetramethyl-3,5–heptanedionate; Flu, fluorenyl; EtCp, η^5-ethylcyclopentadienyl; MeCp, η^5-methylcyclopentadienyl; t-BuCp, η^5-tert-butylcyclopentadienyl; Bz, benzene. [b] Free energy of ionization at 350 K. Units are kcal mol^{-1}. References in column to right. [c] Vertical ionization energy from photoelectron spectrum. Units are kcal mol^{-1}. References in next column. [d] Reference compounds.

Table II. Electron attachment energies for $M + e^- \rightleftharpoons M^-$

Compound[a]	$-\Delta G_{a,350K}^{\circ}$ or $-\Delta H_{a,0K}^{\circ b}$	Method[c]	Ref.
$Mn(hfac)_3$	(109)	est	32
$Co(hfac)_3$	(97)	est	32
$Fe(hfac)_3$	(93)	est	32
$FeTPPF_{20}\beta Cl_8Cl$	77.3	ETE	38
$FeTPPF_{20}\beta Cl_8$	74.0	ETE	38
$V(hfac)_3$	73	ETB	32
$FeTPPF_{20}$ Cl	72.4	ETE	38
$Ti(hfac)_3$	69	ETB	32
$FeTPPCl_{20}Cl$	67.6	ETE	38
$Cr(hfac)_3$	67	ETB	32
$Mn(CO)_x(PF3)_y$ (x + y = 5, x < 4)	65.7	est	34
$FeTPPCl_8\beta Cl_8Cl$	65.0	ETE	38
$Sc(hfac)_3$	64	ETB	32
$Co(PF3)_4$	61.3	est	34
$Ga(hfac)_3$	60.4	ETE	32
$FeTPPCl_{20}\beta Cl_8$	59.7	ETE	38
$Mn(acac)_3$	59	ETB	32
$Mn(CO)_4PF3$	56.7	est	34
$Co(CO)_4$	55	PES	33
$Fe(CO)_4$	55	PES	33
$Mn(CO)_5$	51	PES	33
$FeTPPF_{20}$	49.6	ETE	38
FeTPPCl	49.6	ETE	38
$FeTPPoCl_8Cl$	48.4	ETE	38
FeTPP-piv	47.7	ETE	38
$Co(acac)_3$	47	ETB	32
FeTPP-val	45.4	ETE	38
$Ru(t\text{-}Bu,H,t\text{-}Bu)_3$	45.4	ETE	23
$Ru(n\text{-}Bu,H,n\text{-}Bu)_3$	45.3	ETE	23
$Ru(n\text{-}Pr,H,n\text{-}Pr)_3$	45.2	ETE	23
$Ru(i\text{-}Pr,H,i\text{-}Pr)_3$	44.2	ETE	23
FeTPP	43.1	ETE	38
$Fe(acac)_3$	43.0	ETE	32
$W(CO)_3$	42.9	PES	33
$FeTPPoCl_8$	42.8	ETE	38
$Ru(Et,H,Et)_3$	42.4	ETE	23
$Fe(CO)_3$	42	PES	33
$Ru(Me,H,Me)_3$	40.7	ETE	23
$Ru(acac)_3$	40.5	ETE	23, 32
Ni(TPPCHO)	38.0	ETE	39
NiTPP	34.8	ETE	39
$Fe(CO)_3COT$	33.0	ETE	36
$Ni(SALEN\text{-}H_2)$	31.6	ETE	39
$Cr(CO)_3$	31.1	PES	33
$Mo(CO)_3$	30.8	PES	33
Fe(CO)	29.1	PES	33
$Fe(CO)_2$	28.1	PES	33
$Fe(CO)_3CHT$	26.9	ETE	36

Table II. Continued

Compound[a]	$-\Delta G_{a,350K}{}^{\circ}$ or $-\Delta H_{a,0K}{}^{ob}$	Method[c]	Ref.
$V(acac)_3$	24.9	ETE	32
$Ni(CO)_3$	24.8	PES	33
$Fe(CO)_3$1,3–Butadiene	24.6	ETE	37
NiSALEN	24.2	ETE	39
$Cr(CO)_3$CHT	23.3	ETE	36
$Fe(CO)_3$CHD	22.0	ETE	36
$(t\text{-}BuCp)_2Ni$	21.2	ETE	21
$(EtCp)_2Ni$	20.3	ETE	21
(EtCp)CpNi	20.2	ETE	21
$Cr(acac)_3$	20	ETB	32
Cp_2Ni	19.7	ETE	21
$(MeCp)_2Ni$	19.0	ETE	21
Ni(CO)	18.5	PES	33
$Fe(CO)_3$TMM	17.0	ETE	36
$Cp^*{}_2Ni$	(16)	est	21
$Ni(CO)_2$	14.8	PES	33
$Ti(acac)_3$	($>$ 0)	est	32

[a] See Table I for many of the ligand abbreviations. Others: hfac, hexafluoroacetylacetonate; TPP, tetraphenylporphyrin; COT, cyclooctatetraene; SALEN-H2; CHT, 1,3,5–cycloheptatriene; CHD, cyclohexadiene; TMM, trimethylenemethane; see ref. 38 for abbreviations for: various TPP derivatives, piv, val. [b] Free energy of electron attachment at ~ 350 K or, for PES data, estimated electron affinity (enthalpy of electron attachment at 0 K). Units are kcal mol^{-1}.
[c] Methods abbreviations: ETE, electron transfer equilibrium; ETB, electron transfer bracketing; est, estimated, see reference for method; PES, anion photoelectron spectroscopy (PES vertical IP reported, equivalent to $-\Delta H_{a,T}{}^{\circ}$).

tains only the $\Delta G_{i,T}{}^{\circ}$ values obtained by ETE methods [14, 17–24]. Where available, vertical ionization energies are given for comparison [25–30]. Table II is a summary of known electron affinities for metal complexes and contains data obtained by negative ion photoelectron spectroscopy as well as ETE experiments [31–39]. Therefore, some of the compounds listed in Table II are not known or are not stable in condensed phases (e.g. Fe(CO), $Fe(CO)_2$), but all of the neutral compounds in Table I are well characterized in the solid state and/or solution. In both tables, several reference compounds used in the studies are included, but the original literature should be consulted for a full accounting of equilibria used in constructing the ladders.

The range of ionization energies in Table I covers ~ 100 kcal mol^{-1} (4eV), and comparison to tabulations of ETE energetics for organic compounds shows that many of the organometallics have exceptionally low ionization energies. The electron attachment energies in Table II also cover a wide range up to electron affinities in excess of 100 kcal mol^{-1}. In general, high electron affinities are obtained in metal compounds with extensive halogenation and highly conjugated ligands such as porphyrins or diketon-

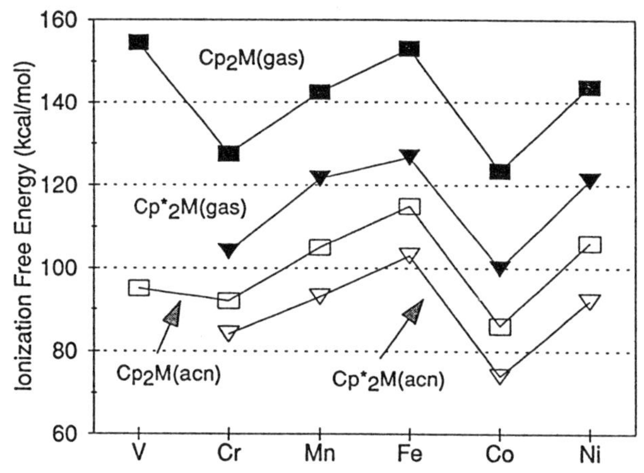

Fig. 3. Ionization free energies of metallocenes and decamethylmetallocenes as a function of metal for the first transition series. Gas-phase values (350 K) are shown with filled markers and the solution free energies derived from electrochemical potentials are shown with the open markers. Solvent is acetonitrile except for M = V, where the solvent is THF. The differential solvation free energy is the energy difference between the gas and solution points for a given metal and ligand combination.

ates. The electronic configurations of the anions are not clear for these compounds since the ligands themselves have high electron affinities.

6. Trends as a function of metal

The most complete ionization energy data sets as a function of metal are available for the metallocenes based on the parent ligand η^5-cyclopentadienyl (Cp) and its permethylated derivative η^5-pentamethylcyclopentadienyl (Cp*).

CH_3 H_3C CH_3 H_3C CH_3

Cp Cp*

Figure 3 shows the gas-phase ionization free energy trends for the Cp and Cp* metallocenes. From the comparisons in Table I, it is clear that the trends are also followed in the vertical IP data from PES. The PES data have been extensively analyzed in the literature primarily through a ligand field theory analysis [28], and the models will not be repeated here. The permethylation of the Cp ligand reduces the $\Delta G_{i,T}^{o}$ values by an average of 24 kcal mol^{-1}, which is only somewhat more than the average reduction in vIP values (~ 21 kcal mol^{-1}).

Also shown in Figure 3 are the estimated ionization free energies for the metallocenes in acetonitrile solution (except for Cp_2V, which is in THF). These values are obtained from the electrochemical $E_{1/2}$ values by assuming an absolute value for the reference electrode with the electron in the same standard state as the gas-phase (i.e. the ion convention). The recommended value for the NHE (4.44 V) is used along with well-established values of E° for the experimental reference electrodes to make the correction and allow direct comparisons to gas-phase data. It is obvious from the shift in the points in Figure 3 that solvation reduces the free energy of ionization more for the Cp complexes (~ 38 kcal mol^{-1}) than for the Cp* complexes (~ 24 kcal mol^{-1}). The lower differential solvation energy, $-\Delta\Delta G_{\mathrm{solv}}°$, for the Cp* complexes is predicted from the simple Born charging model, where the increased sizes of the molecule and ion diminish the solvation energy (as a function of 1/r, where r is the effective radius of the molecule). Quantitative estimates of $-\Delta\Delta G_{\mathrm{solv}}°$ based on the Born equation [40] and structural data (r ≈ 3.8 Å for the Cp complexes and r ≈ 5 Å for the Cp* complexes, Figure 4) lead to somewhat larger magnitudes than deduced from the experimental data (by 10–30%), but the general conclusion is that the metallocene (0/+) couples adhere closely in behavior to the spherical Born model.

Some electron attachment data are plotted as a function of metal in Figure 5. We have provided a thorough analysis of the trend in the gas-phase $M(acac)_3$ (0/−) $\Delta G_{\mathrm{a},350}°$ values based on the ligand field model (acac = acetylacetonate) [32].

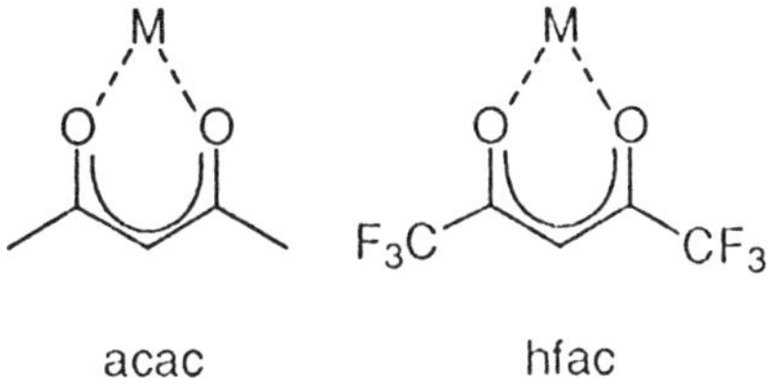

acac hfac

The detailed model generally accounts for the observed trends, but even a simple ligand field stabilization energy (LFSE) approach is adequate. Note that fluorination of the acac methyl groups (to yield hfac complexes) leads to an increase in the electron affinity of the complex by ~ 50 kcal mol^{-1} (the values for M = Mn, Fe, and Co are estimated from the increase observed for M = V and Cr, but the estimates may be too high based on electrochemical data for the Co case [41]). It is also noteworthy that many of the transition metal hfac complexes have electron affinities close to that of $Ga(hfac)_3$, which has a $3d^{10}$ configuration for Ga(III). In the latter case, the added electron must formally reside in a ligand-based orbital; therefore, there must be significant delocalization of the unpaired electron density in the complexes with M = Sc, Ti, V, and Cr. For comparison, the electron affinity of $Ga(acac)_3$ is either very low or negative.

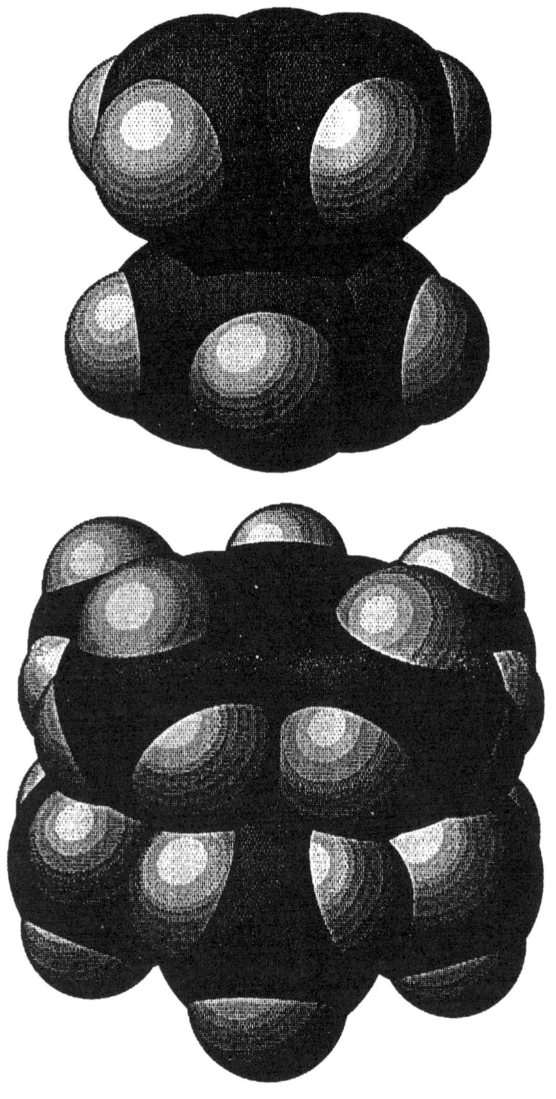

Fig. 4. Spacefill models for Cp metallocenes (top) and the permethylated Cp* derivatives (bottom).

Nickelocene is the only simple metallocene that forms a stable negative ion in the FTMS experiment, and we were able to determine its $\Delta G_{a,350}°$ value by ETE methods (indicated in Figure 5). Since the $Cp_2Ni^{0/-}$ couple is quasi-reversible in cyclic voltammetry, we were able to show that the $-\Delta\Delta G_{solv}°$ value for the (0/−) couple is ~ 38 kcal mol^{-1} and therefore comparable in magnitude to those of the (0/+) couples discussed earlier.

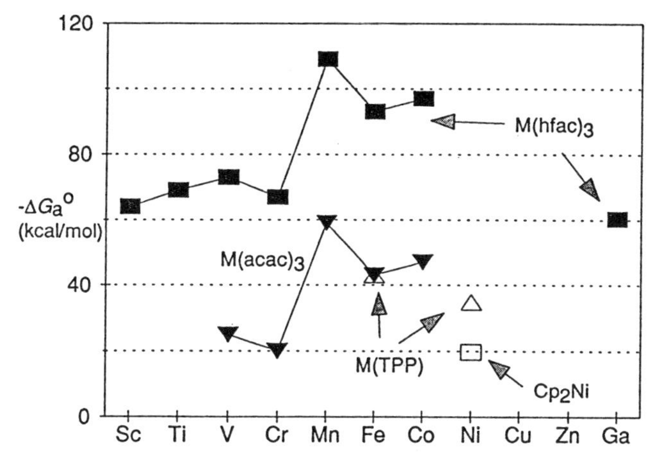

Fig. 5. Gas-phase electron attachment free energies of various metal complexes as a function of metal for the 3d metals. The values for the $M(hfac)_3$ complexes with M = Mn, Fe, and Co are estimated from the observed increase for the M = V and Cr in going from L = acac to L = hfac. Ligand abbreviations in Tables I and II.

This again confirms the general applicability of the Born model to metallocene couples since the Born energy is independent of the sign of the charge.

The emerging ETE data for metalloporphyrins and other macrocylic metal complexes are notable given the widespread applications of these complexes in catalysis as well as their biological relevance [42]. As seen in Figure 5 and the data in Table II, Fe^{II}(TPP) has an electron affinity comparable to that of $Fe^{III}(acac)_3$ and Ni^{II}(TPP) has a greater electron affinity than nickelocene.

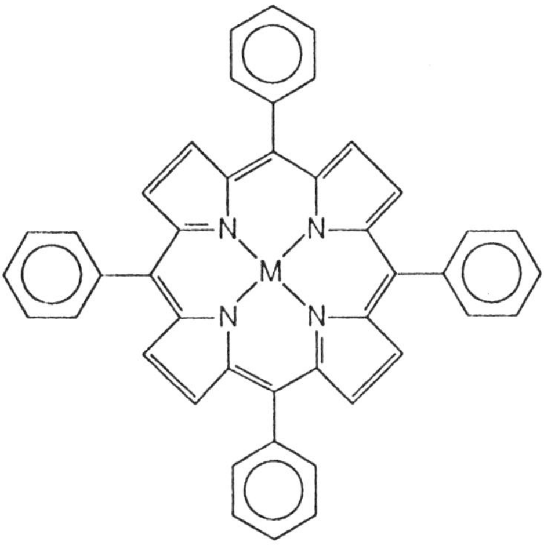

M(TPP)

Increased halogenation of the porphyrin increases the electron affinity markedly. The highly chlorinated and/or fluorinated iron porphyrins have electron affinities comparable to the reference compound tetracyanoethylene, which has the highest electron affinity of common organic acceptors. It is not known how the additional electron in the anionic complexes is distributed, and either the metal or the ligand can be considered the formal site of electron attachment. Ridge and coworkers have also demonstrated good correlations of gas-phase $\Delta G_a°$ values of the porphyrin complexes with solution redox potentials. The thermochemical radii based on the Born equation are in the range of 3.5±0.5 Å (surprisingly comparable to the metallocenes).

7. Ligand substituent effects

Electron attachment and ionization energetics provide a useful probe of the electronic effects of various substituents on ligands in metal complexes. If the oxidation or reduction process does not lead to significant structural rearrangement, then the thermochemical effect on redox energies in the gas phase can be modeled as purely electronic perturbations. Some other methods for assessing substituent effects, such as reactivity, may be influenced by steric interactions between substituents and reactants. In addition, substituent effects on redox thermochemistry in solution will not necessarily parallel intrinsic effects in the gas phase due to differences in differential solvation energies, especially when substituents being compared are much different in size or electronic effect. We have investigated gas-phase substituent effects in both ionization and electron attachment electron-transfer equilibria of organometallics and coordination compounds.

Table I has many entries for ferrocene, nickelocene, and ruthenocene derivatives, and these ionization free energies have been used to derive a single parameter (termed γ) that describes the effect of a given modified Cp *ligand* on the ionization energetics for the d^6 (and d^8) metallocenes. This parameter is not assigned to a given substituent but rather the entire ligand since we do not want to presuppose that multi-substituted Cp ligands will be accurately described by the sum of individual substituent parameters (i.e. substituent additivity is not assumed). Furthermore, there is no reasonable method for assigning a substituent constant for a phenyl ring fused to the Cp, as in indenyl and fluorenyl. However, we do assume ligand additivity, so the effect of two different Cp ligands is assumed to be described by the sum of the individual γ parameters.

From the data in Table I, γ parameters are derived by using equation (8), where

$$\Delta G_i°(\mathrm{LL'M}) = a(\gamma_L + \gamma_{L'}) + \Delta G_i°(\mathrm{Cp_2M}) \quad (7)$$

L and L′ are Cp derivatives. To anchor the parameters, $\gamma_{Cp} = 0$ and $\gamma_{Cp^*} = -1$ are set by definition. Therefore, a negative valuc for γ_L implies that ligand L stabilizes the ionized form of the metallocene relative to the neutral,

Table III. Cyclopentadienyl derivative γ parameters

Ligand[a]	γ
TTFMH	3.1
TTFMOSi	1.6
C_5F_5	1.5
C_5Cl_5	1.1
C_5Br_5	1.03
NO_2Cp	0.79
Cp	0[b]
MeCp	−0.20
EtCp	−0.22
$CpCH_2NMe_2$	−0.22
TMSCp	−0.24
n-BuCp	−0.30
t-BuCp	−0.32
Ind	−0.41
Flu	−0.65
Cp*	−1[b]

[b] By definition.

lowering the ionization energy. The coefficient *a* in equation (8) is a "sensitivity constant" that depends on the metal M and is conceptually similar to the Taft reaction constant ρ [43, 44]. In some cases both mixed ligand and homoleptic complexes are available for a given ligand, so the value of γ is chosen to give the best fit to equation (8).

Ligand parameters are compiled in Table III. Values for some of the Cp derivatives have been deduced from alkylnickelocene data, which is strongly correlated with alkylferrocenes. The values of *a* in equation (8) for M = Ru, Fe, and Ni are 13 kcal mol^{-1}, 13 kcal mol^{-1}, and 11 kcal mol^{-1}, respectively. (For Fe and Ni, *a* is determined by half the shift in ionization energy from Cp_2M to Cp^*_2M.) For methyl substituents, the net effect of five methyl substituents (in Cp*) is five times the effect of a single methyl group within the error limits on γ_{MeCp}. The similarity of n-Bu and t-Bu effects are explained by the probable "wrapping" of the n-Bu chain around the metallocene to maximize the ion-induced dipolar stabilization of the ion.

It is initially surprising to see that the η^5-C_5X_5 (X = F, Cl, Br) ligands all have comparable γ values, with the pentafluoro ligand being far less "electron withdrawing" than the η^5-$(CF_3)_4C_5H$ ligand. However, it is commonly found that π resonance effects must be considered in rationalizing halo substituent effects on aromatic rings. In the typical multiparameter classical model, a fluoro group is assumed to be a powerful σ withdrawing group due to its electronegativity, but this is compensated in aromatic compounds by π donation from the fluoro group into the ring. For the CF_3 group, the π effect is negligible and a strong σ effect dominates. The net result is that all halo substituents lead to similar effects on metallocene ionization. It is likely that halo group effects are not additive, and studies to assess γ parameters for C_5F_x ligands with x = 1–4 would be informative.

Generally, the γ parameters in Table III are in accord with other experimental trends in Cp substituent effects in organometallic chemistry. However, one significant difference is noted for L = Flu.

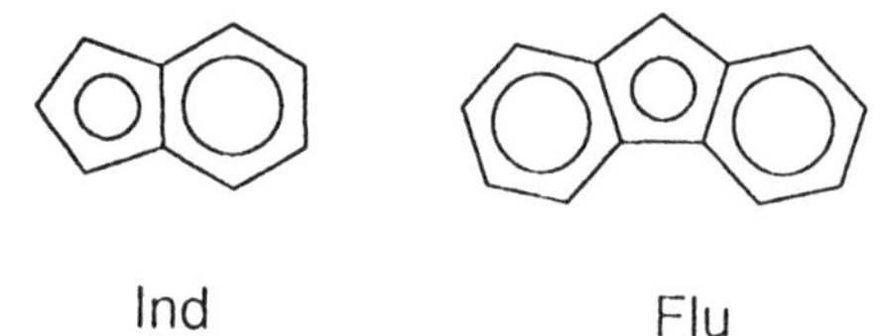

Evidence from electrochemistry and XPS core Ru 3d studies [45] suggested that Flu is more electron donating than Cp* (i.e. Flu would have an expected parameter $\gamma_{Flu} < -1$). However, the effect of Flu (γ_{Flu} = − 0.65) on the ionization energy of ruthenocene is only ~ 2/3 of that of Cp* ($\gamma_{Cp^*} = -1$), and, not surprisingly, somewhat less than twice the effect of Ind, which has a single fused ring on the Cp. The comparison to the electrochemical result must be considered in light of the possible differences in the solvation energies for complexes of the two ligands and the irreversibility of the Cp* $FluRu^{0/+}$ couple (which makes accurate determination of the $E_{1/2}$ difficult). On the other hand, the XPS and ETE results are generally well correlated except for the Flu vs. Cp* comparison. The origin of the discrepancy is not known at this time.

Several cautions must be introduced here concerning possible applications of γ parameters to correlations with organometallic reactivity. The parameters relate to the tendency of a given ligand to stabilize metallocene cations relative to the neutrals. These tendencies can be generalized to the oxidation thermochemistry of other metallocenes only to the extent the behavior for M = Ru and Fe (and perhaps Ni) is representative of the other metals. The parameters will probably be useful in correlating reactivity for processes in which the metal center loses electron density in the rate determining step and steric effects do not predominate. Thus the trend for such a reaction would be increasing rate for lower values of γ. The trends for the opposite case, where electron density increases at the metal, will generally be the opposite, but some exceptions are likely for alkyls (see the next section). The parameters are intrinsic to the ligand since they are obtained in gas-phase experiments, so the effects of solvation are not included. The quality of correlations in solution will therefore depend on the properties of the solvent (e.g. polar vs. nonpolar), and better correlations might be expected for solvents with low dielectric constants (i.e. less strongly solvating) assuming that specific metal-solvent interactions are not present.

8. Alkyl substituent effects

As is clear from the data in Tables I and III, alkyl groups tend to stabilize cationic metallocenes relative to the neutrals and thus have negative γ values.

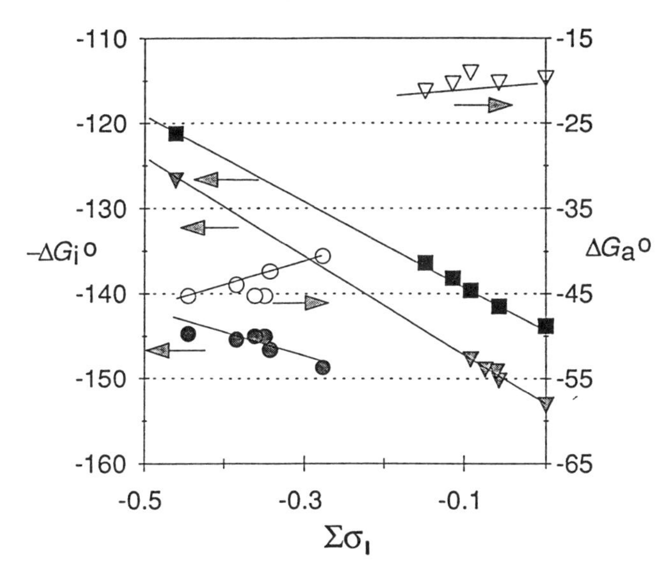

Fig. 6. Dependence of ionization free energies and electron attachment free energies of various metal complexes with alkyl substituents. The energies are plotted against the sum of the Taft σ_I parameters for the complex (these parameters range from σ_I H = 0 to σ_I t-Bu = − 0.075 for a single substituent). Thus larger alkyl groups are found to the left of a given series. Legend: ■, alkylferrocenes (left scale), ▼, alkylnickelocenes (left scale); ▽, alkylnickelocenes (right scale); ○, tris(β-diketonate)ruthenium (right scale), ●, tris(β-diketonate)ruthenium (left scale).

As the alkyl group increases in size and substitution at the α carbon, γ gets increasingly negative. This is in accord with the usual attribution of "electron donating" character to alkyl groups in metal complex redox chemistry (as seen in solution electrochemical potential measurements, for example). One often sees rationales along the lines of "increased electron donation by the alkyl substituted ligand to the metal decreases the ionization energy of the primarily metal-based HOMO". The most well studied example of this is found in permethylation of Cp complexes mentioned earlier. The decamethylmetallocenes have ionization energies ~ 24 kcal mol^{-1} lower than the parent metallocenes.

Close examination of the electron attachment data in Table II reveals a notable lack of consistency with the idea that alkyl groups are "electron donating". In the case of the alkylnickelocenes, variation of the alkyl substitution makes a large difference in the ionization energies, but $\Delta G_a{}^\circ$ is almost invariant with RCp (Figure 6). The t-BuCp substituted complex has a higher electron affinity than the parent nickelocene, suggesting that the alkyl group effect is somewhat "electron withdrawing". The data for the Ru β-diketonate complexes (below) are more compelling; as seen in Figure 6, a given alkyl

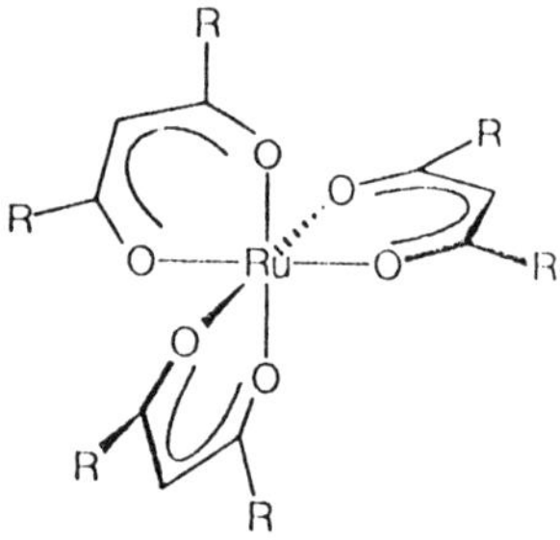

Ru β-diketonate complexes

group R increases the electron affinity compared to R = Me by the same extent the ionization free energy is lowered. The direction of an alkyl group substituent effect clearly depends on the nature of the reaction.

The unusual nature of alkyl substituent effects was recognized in earlier gas-phase/solution comparisons of acidities and proton affinities of organic acids and bases [46]. Solution studies of acid-base strength indicated that alkyl groups were uniformly electron donating (i.e. stabilizing the formation of cations and destabilizing the formation of anions). It soon became clear that a single parameter for describing alkyl effects was inadequate since some trends reversed from the gas-phase to solution. A useful electrostatic model for alkyl effects must include more than inductive effects, and it was obvious that the polarizability of the alkyl groups might be dominant in some cases. Thus, an alkyl group stabilizes formation of nearby cationic and anionic charges through ion-induced dipole interactions with equal efficacy since the stabilization is independent of the sign of the charge (all other things being equal).

It appears in the Ru β-diketonate complexes that a polarizability model is almost entirely adequate for rationalizing the data in Figure 6 (circles). Figure 7 illustrates the polarization model for R = t-butyl. Since the effect of a given alkyl group relative to R = Me is equal in magnitude for electron attachment and ionization, it appears that any "through bond" inductive effect is minimal (since the latter would supposedly stabilize ionization but destabilize electron attachment). In the case of the alkylnickelocenes, it has been surmised that the inductive effect and the polarizability effects are roughly equal in magnitude since the electron affinity varies little with the nature of R.

Finally, it is noteworthy that in polar solvents such as DMF and acetonitrile the alkyl group effects for the 0/− couples of the Ru diketonate complexes follow the trend found for the +/0 couples; i.e. the alkyl group now appears to be "electron donating" for both couples. Of course, the reversal from the gas-phase trend for the 0/− couples can be traced to the differential solvation energies (Figure 8). The larger alkyl groups restrict access of the polar solvent to the vicinity of the charge, in essence coating the ion with a larger layer

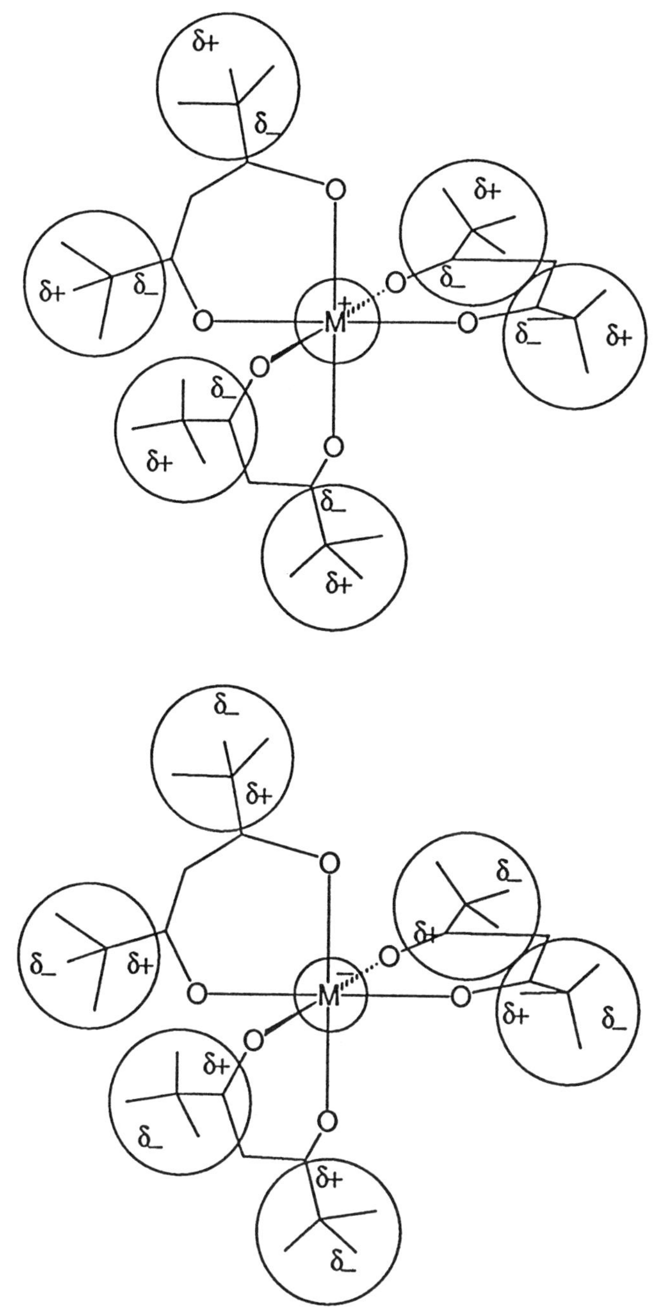

Fig. 7. Illustration of the stabilization of positive charge and negative charge at a central atom by polarizable t-butyl groups on the ligands. The induced dipoles of the alkyl substituents lower both the ionization and electron attachment energies of the neutral.

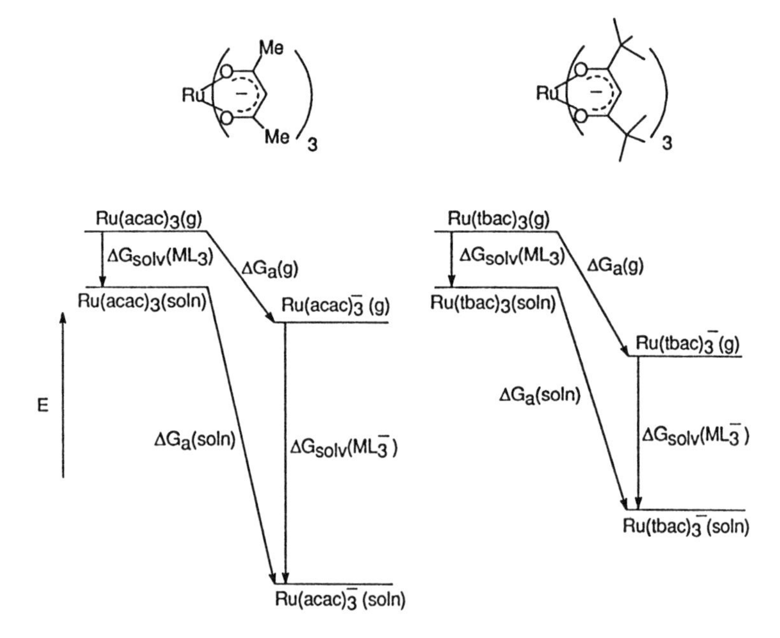

Fig. 8. Qualitative free energy cycles illustrating the origin of the reversal of trends in electron attachment to tris(β-diketonate)ruthenium derivatives in going from the gas phase to solution. The electron attachment free energy is more negative for R = t-Bu than for R = Me (L = acac) in the gas phase, but the $Ru(acac)_3$ complex anion is much better solvated in polar solvents. The result is a higher electron affinity for acac complex in solution (i.e. the acac complex is easier to reduce).

of nonpolar and poorly "solvating" hydrocarbon. At the extreme of very large and bulky alkyl groups, the gas-phase ion would essentially be "dissolved" in a molecular droplet of hydrocarbon. Since smaller alkyl substituents allow the solvent to get closer to the charge center (Figure 9), the formation of the ion is encouraged to a greater degree by solvation and the R = Me complex has the higher electron affinity in solution. Completely analogous observations and explanations are encountered in the gas-phase vs. solution acid-base chemistry mentioned above.

It should be noted that the Born solvation free energy model breaks down when applied to the diketonate complexes. Essentially, the solvation of the 0/− couples is far more exoergic than predicted while the 0/+ ionization solvation has a much smaller $-\Delta\Delta G_{solv}^{\circ}$ value than expected. The former observation was rationalized to be a consequence of solvent interpenetration between the chelating ligands making the effective Born thermochemical radius smaller. The same interpenetration should occur for the 0/+ couples as well, but the effective Born thermochemical radius is larger than the distance from the metal center to the periphery of the molecule. The differ-

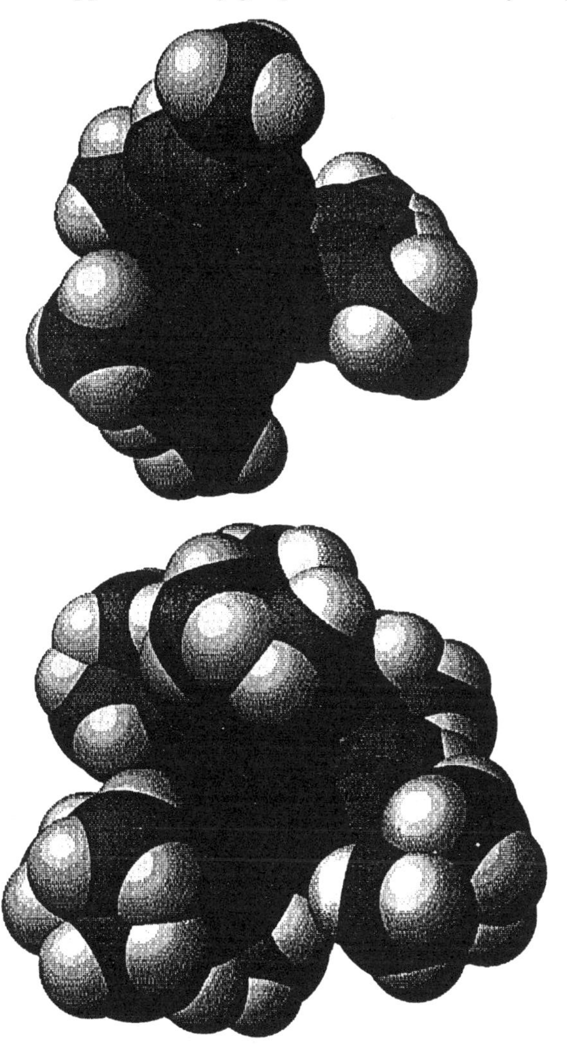

Fig. 9. Spacefill models of ruthenium tris(β-diketonate) complexes with R = Me (top) and R = t-butyl (bottom). Access of solvent to the vicinity of the metal center is greatly reduced by the larger substituent.

ences in solvation energies probably arise from significant differences in electronic charge distribution in the anions, neutrals and cations or from specific solute-solvent interactions. Classically, the nonuniform distribution of charge is handled by adding multipolar terms to the monopole solvation energy expression of the Born equation, which assumes spheres of uniform charge.

9. Future directions

The techniques described here are generally applicable for metal-containing compounds that are sufficiently volatile to establish a neutral pressure in the trap so ionization or electron attachment can be used to form ions. A large percentage of neutral organometallics and coordination compounds can be studied within this limitation. As more data become available, the nature of solvation and metal-ligand interactions will certainly be revealed in greater detail. It would be desirable to determine intrinsic redox ligand substituent effects in organometallics where the corresponding solution electrochemistry is reversible so detailed models for solvation thermochemistry could be compared for a large series of substituted ligands. Furthermore, it would be helpful to assess such effects for a large range of substituents as considered in the ruthenocene studies. Now that reference compounds are available in the range of ionization energies < 6 eV, it is reasonable to consider widely available ferrocene derivatives for these studies.

Many redox couples of interest in inorganic chemistry involve species with charges greater than $+1$ or less than -1. Although the electrochemical potentials are available for a vast array of such couples, essentially nothing is known about the gas-phase energetics of ionization or electron attachment for such ions. We and others have recently begun to exploit the external generation of multi-charged ions for introduction into the ion trap of an FTMS, and external sources such as electrospray ionization (ESI) can produce such ions easily from solutions of salts. The limitation to volatile compounds disappears, and even very large metal complex ions (including metalloproteins!) can be isolated for reactivity and spectroscopy studies following injection into the ion trap. Although equilibrium methods are no longer applicable, ETB studies can still be used to determine energetics of the gas-phase species (with due consideration of the electrostatics when both products are charged). Thus, it appears that gas-phase methods are on the verge of providing glimpses of the reactivity and thermodynamics of multi-charged metal complex ions in the absence of solvent. Since solvation energies increase roughly as the square of ion charge, the dramatic influence of solvation on redox thermochemistry should be even more obvious in these studies than those reviewed here.

Acknowledgements

Research in physical inorganic chemistry in the author's laboratory has been supported by National Science Foundation grants, which are gratefully acknowledged. The many contributions of coworkers at the University of Florida and elsewhere are noted in the references. Particular thanks are due to Dr. Matthew Ryan and Dr. Paul Sharpe, who carried out most of the

ETE studies, and Professor John Eyler (University of Florida), who has contributed enormously to our work over the years.

References

1. Reasonably complete listings for ionization energies and electron affinities for atoms and ions can be found in compilations such as the *Handbook of Chemistry and Physics*, D. R. Lide (Ed.), published by the CRC Press, Boca Raton, Florida.
2. Some data for molecules can be found in compilations such as that cited in reference 1. More complete listings are available in the following publications: (a) J. L Franklin, J. G. Dillard, H. M. Rosenstock, J. T. Herron, K. Draxl, and F. H. Field, *Ionization Potentials, Appearance Potentials, and Heats of Formation of Gaseous Positive Ions Nat. Stand. Ref. Data Ser., Nat. Bur. Stand. (U.S.)* **26**, 1 (1969), (b) H. M. Rosenstock, K. Draxl, B. W. Steiner, and J. T. Herron, *J. Phys, Chem. Ref. Data*, **6**, Suppl. 1, 1 (1977), (c) R. D. Levin and S. G. Lias, *Ionization Potential and Appearance Potential Measurements*, 1971–1981, *Nat. Stand. Ref. Data Ser., Nat. Bur. Stand. (U.S.)* **71**, 1 (1982).
3. J. W. Rabalais, *Principles of Ultraviolet Photoelectron Spectroscopy* (Wiley-Interscience, New York, 1977).
4. D. L. Lichtenberger and G. E. Kellogg, *Accts. Chem. Res.* **20**, 379 (1987).
5. H. Hotop and W. C. Lineberger, *J. Phys. Chem. Ref. Data* **14**, 731 (1985).
6. S. G. Lias, J. A. Bartmess, J. F. Liebman, J. L. Holmes, R. D. Levin, and W. G. Mallard (Eds.), *Gas-Phase Ion and Neutral Thermochemistry* (American Institute of Physics, New York, 1988).
7. P. Sharpe and D. E. Richardson, *Thermochimica Acta* **202**, 173 (1992).
8. A. G. Marshall, *Acc. Chem. Res.* **18**, 316 (1985).
9. M. B. Comisarow, *Analytica Chim. Acta* **178**, 1 (1985).
10. J. R. Eyler and G. Baykut, *Trends Anal. Chem.* **6**, 44 (1986).
11. M. L. Gross and D. L Rempel, *Science* **226**, 261 (1984).
12. S. Chowdhury and P. Kebarle, *Chem. Rev.* **87**, 513 (1987).
13. S. Chowdhury, T. Heinis, E. P. Grimsrud, and P. Kebarle, *J. Phys. Chem.* **90**, 2747 (1986).
14. M. F. Ryan, D. E. Richardson, D. L. Lichtenberger, and N. E. Gruhn, *Organometallics* **13**, 1190 (1994).
15. D. E. Richardson, *J. Phys. Chem.* **90**, 3697 (1986).
16. D. E. Richardson and J. R. Eyler, *Chem. Phys.* **176**, 457 (1993).
17. D. E. Richardson, C. S. Christ, Jr., P. Sharpe, M. F. Ryan, and J. R. Eyler, in *ACS Symposium Series*, Vol. 428, *Bond Energetics in Organometallic Compounds*, edited by T. Marks (American Chemical Society, Washington, D. C., 1990), pp. 70–83.
18. D. E. Richardson, in *Energetics of Organometallic Species*, Vol. 367, edited by J. Martinho Simões (NATO-Advanced Study Institute), pp. 233–251.
19. M. Ryan, J. R. Eyler, and D. E. Richardson, *J. Am. Chem. Soc.* **114**, 8611 (1992).
20. M. F. Ryan, A. Siedle, M. Burk, and D. E. Richardson, *Organometallics* **11**, 4231 (1992).
21. D. E. Richardson, M. F. Ryan, K. Maxwell, and M. N. Khan, *J. Am. Chem. Soc.* **114**, 10482 (1992).
22. D. E. Richardson, M. F. Ryan, W. E. Geiger, T. T. Chin, R. P. Hughes, and O. J. Curnow, *Organometallics* **12**, 613 (1993).
23. P. Sharpe, N. G. Alameddin, and D. E. Richardson (submitted).
24. M. Meot-Ner (Mauntner), *J. Am. Chem. Soc.* **111**, 2830 (1989).
25. D. L. Lichtenberger (private communication).
26. D. O'Hare, J. C. Green, T. P. Chadwick, and J. S. Miller, *Organometallics* **7**, 1335 (1988).
27. D. L. Lichtenberger and A. S. Copenhaver, *J. Chem. Phys.* **91**, 663 (1989).
28. J. C. Green, *Struct. Bonding (Berlin)* **43**, 37 (1986).

29. S. Aynetchi, P. B. Hitchcock, E. A. Seddon, K. R. Seddon, Y. Z. Yousif, J. A. Zora, and K. Stuckey, *Inorg. Chim. Acta* **113**, L7 (1986).
30. C. Cauletti, J. C. Green, R. Kelly, P. Powell, J. van Tilborg, J. Robbins, and J. Smart, *Electron. Spectrosc. Relat. Phenom.* **19**, 327 (1980).
31. P. Sharpe and D. E. Richardson, *J. Am. Chem. Soc.* **113**, 8339 (1991).
32. P. Sharpe, J. R. Eyler, and D. E. Richardson, *Inorg. Chem.* **29**, 2779 (1990).
33. P.C. Engelking and W. C. Lineberger, *J. Am. Chem. Soc.* **101**, 5569 (1979).
34. A. E. Stevens-Miller and T. M. Miller, in reference 18, p. 253.
35. P. S. Drzaic, J. Marks, and J. I. Brauman, in *Gas Phase Ion Chemistry*, Vol. 3, edited by M. T. Bowers (Academic Press, Orlando, 1984), p. 167.
36. P. Sharpe and P. Kebarle, *J. Am. Chem. Soc.* **115**, 782 (1993).
37. G. W. Dillow and P. Kebarle, *J. Am. Chem. Soc.* **114**, 5742 (1992).
38. H. L. Chen, P. E. Ellis, T. Wijesekera,T. E. Hagan, S. E. Groh, J. E. Lyons, and D. P. Ridge, *J. Am. Chem. Soc.* **116**, 1086 (1994).
39. H. L. Chen, Y. H. Pan, S. Groh, T. E. Hagan, and D. P. Ridge, *J. Am. Chem. Soc.* **113**, 2766 (1991).
40. M. Z. Born, *Z. Phys.* **1**, 45 (1920).
41. C. Tsiamis, S. Cambanis, and C. Hadjikostas, *Inorg. Chem.* **26**, 26 (1987).
42. D. Dolphin (Ed.), *The Porphyrins*, Vols. 1–5 (Academic Press, New York, 1978).
43. L. S. Levitt and H. F. Widing, *Prog. Phys. Org. Chem.* **12**, 119 (1976).
44. C. Hansch, A. Leo, and R. W. Taft, *Chem. Rev.* **91**, 165 (1991).
45. P. G. Gassman and C. H. Winter, *J. Am. Chem. Soc.* **110**, 6130 (1988).
46. For leading references see J. March, *Advanced Organic Chemistry*, 4th Edition (Wiley-Interscience, New York, 1992), Chapter 8.

TABLE OF BOND ENERGIES

Ben S. Freiser (ed.), Organometallic Ion Chemistry, 283–332.

Metal-Ligand Bond Dissociation Energies

Bond	kcal/mol	kJ/mol	Method	T	References
Ag^+-CH_3	15.9 ± 1.2	67 ± 5	Ion Beam	0 K	110
Ag^+-CH_3	21.9	91.7	Theory	0 K	121
$Ag(CH_3)^+-CH_3$	28.1	118	Theory	0 K	122
Ag^+-CO	21.2 ± 1.2	89 ± 5	Ion Beam	0 K	96
Ag^+-CO	19.2	80.4	Theory	0 K	131
$Ag(CO)^+-CO$	26.1 ± 0.9	109 ± 4	Ion Beam	0 K	96
$Ag(CO)^+-CO$	19.5	81.6	Theory	0 K	131
$Ag(CO)_2^+-CO$	13.1 ± 1.8	55 ± 8	Ion Beam	0 K	96
$Ag(CO)_3^+-CO$	10.8 ± 1.8	45 ± 8	Ion Beam	0 K	96
$Ag^+-C_2H_2$	28	117	Theory	0 K	124
$Ag^+-C_2H_4$	33.7	141	van't Hoff Plot	298 K	209
$Ag(C_2H_4)^+-C_2H_4$	32.4	136	van't Hoff Plot	298 K	209
$Ag^+-C_2H_5$	15.7 ± 1.8	66 ± 8	Ion Beam	0 K	110
$Ag^+-CH_3COCH_3$	≤31	≤ 130	Photodissociation	298 K	227
$Ag(C_5H_5N)_2^+-C_5H_5N$	16.7 ± 0.2	69.9 ± 0.8	van't Hoff Plot	298 K	213
$Ag(C_5H_5N)_3^+-C_5H_5N$	17.9 ± 0.2	74.9 ± 0.8	van't Hoff Plot	298 K	213
$Ag^+-C_6H_6$	37.4 ± 1.7	156 ± 7	Ion Beam	0 K	112
$Ag^+-C_6H_6$	36.5	153	Theory	0 K	126
$Ag^+-C_6H_5CH_3$	≤45	≤ 188	Photodissociation	298 K	227
Ag^+-H	9.5 ± 1.4	40 ± 6	Ion Beam	0 K	98
Ag^+-H	10.6	44.4	Theory	0 K	120
Ag^+-H	2.6	11	Theory	0 K	172
Ag^+-H	3.0	13	Theory	298 K	172
Ag^+-H_2O	33.3 ± 0.2	139 ± 1	van't Hoff Plot	298 K	213
$Ag(H_2O)^+-H_2O$	25.4 ± 0.3	106 ± 1	van't Hoff Plot	298 K	213
$Ag(H_2O)_2^+-H_2O$	15.0	62.8	Equilibrium	298 K	213
$Ag(H_2O)_2^+-H_2O$	15.0 ± 0.1	62.8 ± 0.4	van't Hoff Plot	298 K	213
$Ag(H_2O)_3^+-H_2O$	16.7	69.9	Equilibrium	298 K	213
$Ag(H_2O)_3^+-H_2O$	14.9	62.4	van't Hoff Plot	298 K	213
$Ag(H_2O)_4^+-H_2O$	13.7 ± 0.2	57.4 ± 0.8	van't Hoff Plot	298 K	213
$Ag(H_2O)_5^+-H_2O$	13.3 ± 0.6	55.7 ± 2.5	van't Hoff Plot	298 K	213
$Ag(NH_3)^+-NH_3$	36.9 ± 0.8	154 ± 3	van't Hoff Plot	298 K	213
$Ag(NH_3)_2^+-NH_3$	14.6 ± 0.1	61.1 ± 0.4	van't Hoff Plot	298 K	213
$Ag(NH_3)_3^+-NH_3$	13.0	54.4	van't Hoff Plot	298 K	213
$Ag(NH_3)_4^+-NH_3$	12.8 ± 0.2	53.6 ± 0.8	van't Hoff Plot	298 K	213
Ag^+-O	28.4 ± 1.2	119 ± 5	Ion Beam	0 K	106
$Ag-CH_3$	32.1 ± 1.6	134 ± 7	Ion Beam	0 K	110
$Ag-H$	47.5 ± 2.3	199 ± 10	Ion Beam	0 K	110
$Ag_2^+-C_6H_6$	≤42	≤ 176	Photodissociation	298 K	223, 226, 227
Ag_2-Ar	0.79	3.3	R2PI/TOF	298 K	225, 228
Ag_2-Kr	1.13	4.73	R2PI/TOF	298 K	225

[1] A temperature of 298° implies ambient temperature.

[2] Although this Bond Energy Table has been extensively proofread, the reader is urged to consult the original references for the *exact* value and description of the method.

Bond	kcal/mol	kJ/mol	Method	T	References
Ag_2-Xe	3.53	14.8	R2PI/TOF	298 K	219
Al^+-CH_2O	27.2	114	Theory	0 K	160, 161
$Al(CH_2O)^+$-CH_2O	18.6	77.9	Theory	0 K	161
Al^+-CH_3OH	34.4	144	Theory	0 K	160
Al^+-CN	29.2	122	Theory	0 K	160
Al^+-CO	6.7	28	Theory	0 K	160
Al^+-C_2H_2	15	63	Theory	0 K	302
Al^+-CH_3CHO	35.7	149	Theory	0 K	160
Al^+-CH_3CH_2OH	35.9	150	Theory	0 K	160
Al^+-CH_3COCH_3	41.7	174	Theory	0 K	161
$Al(CH_3COCH_3)^+$-CH_3CO	23	96	Theory	0 K	148
Al^+-C_6H_6	39	163	Theory	0 K	303
$Al(OH)^+$-H	74	310	Theory	0 K	301
Al^+-H_2O	24.9 ± 3.6	104 ± 15	Ion Beam	0 K	111
Al^+-H_2O	27.0	113	Theory	0 K	160
$Al(H_2O)^+$-H_2O	16.1 ± 1.2	67 ± 5	Ion Beam	0 K	111
$Al(H_2O)^+$-H_2O	20.1	84.2	Theory	0 K	160
$Al(H_2O)_2^+$-H_2O	15.2 ± 1.8	64 ± 8	Ion Beam	0 K	111
$Al(H_2O)_2^+$-H_2O	19.3	80.8	Theory	0 K	152
$Al(H_2O)_3^+$-H_2O	12.5 ± 1.4	52 ± 6	Ion Beam	0 K	111
$Al(H_2O)_3^+$-H_2O	18.0	75.4	Theory	0 K	152
Al^+-NH_3	34.9	146	Theory	0 K	160
Al^+-O	34.6 ± 2.8	145 ± 12	Ion Beam	0 K	66, 10
Al^+-O	41.03 ± 0.92	172 ± 4	Ion Beam	0 K	184
Al^+-O	39.42 ± 3.69	165 ± 15	Thermochemical	298 K	165
Al^+-OH	104	435	Theory	0 K	301
Al^+-OH_2	33	138	Theory	0 K	301
Al-O	123.1 ± 1	515 ± 4	Ion Beam	0 K	66, 10
Al-OH	36	151	Theory	0 K	301
Au^+-CH_2	94 ± 3	393 ± 13	Theory	0 K	259
Au^+-CO	50 ± 5	209 ± 21	Theory	0 K	299
Au^+-C_2H_4	73 ± 5	305 ± 21	Theory	0 K	299
Au^+-C_3H_6	>75	>314	Ion Molecule Reactions	298 K	299
Au^+-C_4H_6	>75	>314	Ion Molecule Reactions	298 K	299
Au^+-C_6H_6	~70	~293	Ion Molecule Reactions	298 K	299
Au^+-H	33.4	140	Theory	0 K	273
Au^+-H_2O	39 ± 5	163 ± 21	Theory	0 K	299
Au^+-I	>59	>247	Ion Molecule Reactions	0 K	299
Au^+-NH_3	69 ± 5	289 ± 21	Theory	0 K	299
Au-F	73	305	Theory	0 K	312
Au-F	76 ± 8	318 ± 33	Ion Molecule Reactions	298 K	311
Ba^+-D	58 ± 2	243 ± 8	Ion Beam	0 K	181
Ba^+-H	50.9	213	Theory	0 K	273

BOND	KCAL/MOL	KJ/MOL	METHOD	T	REFERENCES
Be^+-F	144.1	603.3	Theory	0 K	147
Be^+-O	93.2	390	Theory	0 K	147
Be^+-OH	138.4	579.4	Theory	0 K	147
Bi^+-C_6H_6	≤35.5	≤ 149	Photodissociation	298 K	223, 227
Ca^+-F	130.8	547.6	Theory	0 K	147
Ca^+-H	45.4 ± 2.8	190 ± 12	Ion Beam	0 K	113
Ca^+-H	44.7	187	Theory	0 K	177
Ca^+-H	38.1	159	Theory	298 K	177
Ca^+-H_2O	27.9	117	Theory	0 K	159
$Ca(H_2O)^+$-H_2O	24.2	101	Theory	0 K	159
Ca^+-O	82.3 ± 1.2	344 ± 5	Ion Beam	0 K	38, 10
Ca^+-O	75.9	318	Theory	0 K	147
Ca^+-OH	109.1	456.8	Theory	0 K	147
Ca^{2+}-H_2O	55.0	230	Theory	0 K	159
$Ca(H_2O)^{2+}$-H_2O	48.9	205	Theory	0 K	159
Cd^+-CH_3	48.5	203	Theory	298 K	173
Cd^+-CH_3	40.4	169	Theory	298 K	173
Cd^+-H	42.0	176	Theory	0 K	172
Cd^+-H	42.9	179	Theory	298 K	172
Co^+-Ar	11.8 ± 0.014	49±.06	Photodissociation	0 K	207
Co^+-Ar	9.89	41.4	Theory	0 K	143
$Co(Ar)^+$-Ar	10.03	41.99	Theory	0 K	142
Co^+-Br	>69	289	Ion Molecule Reactions	298 K	2
Co^+-C	82.9 ± 6.9	347 ± 29	Ion Beam	0 K	100
Co^+-C	90 ± 7	377 ± 21	Photodissociation	298 K	240
Co^+-CD	100.9 ± 8.8	422 ± 37	Ion Beam	0 K	100
Co^+–CH	120.3 ± 7	503 ± 29	Theory	0 K	269
Co^+-CH	100 ± 7	418 ± 21	Photodissociation	298 K	240
Co^+-CH_2	75.9 ± 1.2	318 ± 5	Ion Beam	0 K	46, 48, 10
Co^+-CH_2	79 ± 4	331 ± 17	Theory	0 K	123
Co^+-CH_2	84 ± 5	351 ± 21	Photodissociation	298 K	240
Co^+-CH_3	48.5 ± 0.9	203 ± 4	Ion Beam	0 K	45, 52, 34, 10
Co^+-CH_3	53.3 ± 2	223 ± 8	Kinetic Measurements	0 K	190
Co^+-CH_3	48.3	202	Theory	0 K	121
Co^+-CH_3	<69	<289	Ion Molecule Reactions	298 K	2
Co^+-CH_3	57 ± 7	238 ± 21	Photodissociation	298 K	242
$Co(CH_3)^+$-CH_3	39.1	164	Theory	0 K	122
Co^+-CH_3OD	35.3 ± 1.8	148 ± 8	Ion Beam	0 K	108
Co^+-CH_4	22.9 ± 0.7	96 ± 3	Equilibrium	0 K	188
Co^+-CH_4	21.4 ± 1.4	90 ± 6	Ion Beam	0 K	54, 10
Co^+-CH_4	49 ± 11	205 ± 46	Ion Molecule Reactions	298 K	230
$Co(CH_4)^+$-CH_4	24.8 ± 0.8	104 ± 3	Equilibrium	0 K	188
$Co(CH_4)^+$-CH_4	23.1 ± 1.2	97 ± 5	Ion Beam	0 K	54, 10

BOND	KCAL/MOL	KJ/MOL	METHOD	T	REFERENCES
$Co(CH_4)_2^+$-CH_4	~11	~46	Equilibrium	0 K	188
$Co(CH_4)_2^+$-CH_4	9.7 ± 1.2	41 ± 5	Ion Beam	0 K	54, 10
$Co(CH_4)_2^+$-CH_4	9.5 ± 1.2	40 ± 5	Ion Beam	0 K	103
$Co(CH_4)_3^+$-CH_4	16.1 ± 1.4	67 ± 6	Ion Beam	0 K	54, 10
$Co(CH_4)_3^+$-CH_4	15.5 ± 1.4	65 ± 6	Ion Beam	0 K	103
$Co(C_2H_6)^+$-CH_4	24.0 ± 1.2	100 ± 5	Equilibrium	0 K	188
$Co(H_2)^+$-CH_4	22.6 ± 1.2	95 ± 5	Equilibrium	0 K	188
$Co(H_2O)^+$-CH_4	25.9 ± 0.8	108 ± 3	Equilibrium	0 K	188
Co^+-CO	42.8 ± 2	179 ± 8	Ion Beam	0 K	69
Co^+-CO	41.5 ± 1.6	174 ± 7	Ion Beam	0 K	97
Co^+-CO	39 ± 5	163 ± 21	KERD	0 K	191
Co^+-CO	37.3	156	Theory	0 K	131
$Co(CO)^+$-CO	40.9 ± 2	171 ± 8	Ion Beam	0 K	69
$Co(CO)^+$-CO	36.4 ± 2.1	152 ± 9	Ion Beam	0 K	97
$Co(CO)^+$-CO	32.3	135	Theory	0 K	131
$Co(CO)_2^+$-CO	19.6 ± 2.8	82 ± 12	Ion Beam	0 K	97
$Co(CO)_3^+$-CO	18.0 ± 1.4	75 ± 6	Ion Beam	0 K	97
$Co(CO)_4^+$-CO	18.0 ± 1.2	75 ± 5	Ion Beam	0 K	97
Co^+-C_2H_2	>6 ± 3	>25 ± 13	Ion Beam	0 K	46, 10
Co^+-C_2H_2	37	155	Theory	0 K	124
Co^+-C_2H_3	48.4 ± 1.8	203 ± 8	Ion Beam	0 K	48, 10
Co^+-C_2H_4	42.9 ± 1.6	179 ± 7	Ion Beam	0 K	48, 10
Co^+-C_2H_4	42 ± 5	176 ± 21	KERD	0 K	198
Co^+-C_2H_4	40	167	Theory	0 K	125
Co^+-C_2H_4	37 ± 2	155 ± 8	Ion Molecule Reactions	298 K	229
Co^+-C_2H_5	46.1 ± 2.6	193 ± 11	Ion Beam	0 K	45, 10
Co^+-C_2H_6	28.0 ± 1.6	117 ± 7	Equilibrium	0 K	188
Co^+-C_2H_6	24.0 ± 1.2	100 ± 5	Ion Beam	0 K	54, 10
Co^+-$(CH_3)_2$	84.1	352	Ion Molecule Reactions	0 K	274
Co^+-$(CH_3)_2$	110 ± 3	460 ± 13	KERD	298 K	191
$Co(CH_4)^+$-C_2H_6	28.4 ± 1.3	119 ± 5	Equilibrium	0 K	188
$Co(C_2H_6)^+$-C_2H_6	26.8 ± 1	112 ± 4	Equilibrium	0 K	188
$Co(C_2H_6)_2^+$-C_2H_6	<12	<50	Equilibrium	0 K	188
Co^+-C_3H_3	89 ± 17	372 ± 71	Ion Molecule Reactions	298 K	237
Co^+-C_3H_4	>18.7 ± 2.1	>78 ± 9	Ion Beam	0 K	48, 10
Co^+-C_3H_5	>79	>331	Ion Molecule Reactions	298 K	3
Co^+-C_3H_6	43.1 ± 1.6	180 ± 7	Ion Beam	0 K	48, 10
Co^+-$CH_2CH_2CH_2$ (to Δ)	29.9 ± 1.7	125 ± 7	Ion Beam	0 K	48, 10
Co^+-C_3H_6 (metallocyclic)	81 ± 5	339 ± 21	KERD	0 K	192, 195
Co^+-C_3H_6	44 ± 5	184 ± 21	KERD	0 K	191
Co^+-C_3H_6	48 ± 3	201 ± 13	KERD	298 K	191
Co^+-C_3H_8	30.9 ± 1.4	129 ± 6	Ion Beam	0 K	54, 10

BOND	KCAL/MOL	KJ/MOL	METHOD	T	REFERENCES
Co^+-C_4H_6	>85.9	>359	Ion Molecule Reactions	298 K	4
Co^+-C_4H_6	52 ± 4	218 ± 17	Theory	298 K	231
Co^+-$N(C_2H_5)_2$	>59	>247	Ion Molecule Reactions	298 K	6
Co^+-C_5H_8	85 ± 5	356 ± 21	Ion Molecule Reactions	298 K	237
$Co(C_5H_8)^+$-C_5H_8	118 ± 10	494 ± 42	Ion Molecule Reactions	298 K	238
Co^+-C_6H_6	61.1 ± 2.5	256 ± 11	Ion Beam	0 K	114
Co^+-C_6H_6	62.6	262	Theory	0 K	126
Co^+-C_6H_6	68 ± 5	285 ± 21	Photodissociation	298 K	242
$Co(C_6H_6)^+$-C_6H_6	39.9 ± 3.2	167 ± 14	Ion Beam	0 K	114
Co^+-C_{60}	58 ± 16	243±67	Ion Molecule Reactions	298 K	256
Co^+-Cl	67.8 ± 2.8	284 ± 12	Ion Beam	0 K	52, 10
Co^+-Cl	>168	>703	Ion Molecule Reactions	298 K	2
Co^+-Co	63.76	267.0	Ion Beam	0 K	82
Co^+-H	45.7 ± 1.4	191 ± 6	Ion Beam	0 K	42
Co^+-H	44.5	186	Theory	0 K	120
Co^+-H	55.5 ± 2.3	232 ± 10	Theory	0 K	269
Co^+-H	52 ± 5	218 ± 21	Ion Molecule Reactions	298 K	235
Co^+-H	<71	<297	Ion Molecule Reactions	298 K	2
Co^+-H	24.1	101	Theory	298 K	177
$Co(C_5H_5)(CO)_2^+$-H	73 ± 5	305 ± 21	Thermochemical	298 K	166
$Co(C_5H_8)(CO)_2^+$-H	55.5 ± 2.3	232 ± 10	Theory	0 K	269
$Co(O)^+$-H	107 ± 4	448 ± 17	Thermochemical	298 K	171
Co^+-H_2	18.2 ± 1	76 ± 4	Equilibrium	0 K	188, 189
Co^+-H_2	17.5 ± 2.3	73 ± 10	Ion Beam	0 K	115
Co^+-H_2	16.9	70.8	Theory	0 K	135
$Co(H_2)^+$-H_2	17.0 ± 0.7	71 ± 3	Equilibrium	0 K	188, 189
$Co(H_2)^+$-H_2	15.1	63.2	Theory	0 K	135
$Co(H_2)_2^+$-H_2	9.6 ± 0.5	40 ± 2	Equilibrium	0 K	189
$Co(H_2)_2^+$-H_2	10.7	44.8	Theory	0 K	134
$Co(H_2)_3^+$-H_2	9.6 ± 0.6	40 ± 3	Equilibrium	0 K	189
$Co(H_2)_4^+$-H_2	4.3 ± 0.7	18 ± 3	Equilibrium	0 K	189
$Co(H_2)_5^+$-H_2	4.0 ± 0.7	17 ± 3	Equilibrium	0 K	189
$Co(H_2)_6^+$-H_2	0.8 ± 0.5	3 ± 2	Equilibrium	0 K	189
$Co(H_2O)^+$-H_2	19.6 ± 0.5	82 ± 2	Equilibrium	0 K	189
$Co(H_2O)^+$-H_2	18.2	76.2	Theory	0 K	135
Co^+-H_2O	38.5 ± 1.4	161 ± 6	Ion Beam	0 K	67
Co^+-H_2O	38.2	160	Theory	0 K	128
Co^+-H_2O	40.1	168	CID	298 K	315
Co^+-H_2O	37.1	155	CID	298 K	294
$Co(H_2O)^+$-H_2O	38.7 ± 1.6	162 ± 7	Ion Beam	0 K	67
$Co(H_2O)^+$-H_2O	39.3	164	Theory	0 K	128
$Co(H_2O)^+$-H_2O	41.9	175	CID	298 K	315

BOND	KCAL/MOL	KJ/MOL	METHOD	T	REFERENCES
$Co(H_2O)_2^+$-H_2O	15.5 ± 1.2	65 ± 5	Ion Beam	0 K	67
$Co(H_2O)_3^+$-H_2O	13.8 ± 1.4	58 ± 6	Ion Beam	0 K	67
Co^+-He	3.02 ± 0.1	13 ± 0	Equilibrium	0 K	185
Co^+-He	3.99	16.7	Theory	0 K	143
$Co(He)^+$-He	3.41 ± 0.1	14.3 ± 0.4	Equilibrium	0 K	185
$Co(He)^+$-He	4.27	17.9	Theory	0 K	143
$Co(He)_2^+$-He	1.22 ± 0.1	5.1 ± 0.4	Equilibrium	0 K	185
Co^+-I	50.6 ± 2	212 ± 8	Kinetic Measurements	0 K	190
Co^+-I	>71	>297	Ion Molecule Reactions	298 K	2
Co^+-Kr	15.4	64.5	Photodissociation	298 K	205
Co^+-$NHCH_3$	>42.6	>178	Ion Molecule Reactions	298 K	6
Co^+-NH_2	58.9 ± 3.2	246 ± 13	Ion Beam	0 K	58, 10
Co^+-NH_3	50.3	210	Theory	0 K	130
Co^+-NH_3	58.8	246	CID	298 K	315
$Co(NH_3)^+$-NH_3	52.5	220	Theory	0 K	130
$Co(NH_3)^+$-NH_3	61.1	256	CID	298 K	315
Co^+-NO	<71	<297	Ion Molecule Reactions	298 K	5
Co^+-N_2	23.1	96.7	Theory	0 K	134
Co^+-Ne	2.18 ± 0.1	9.1 ± 0.4	Equilibrium	0 K	185
Co^+-Ne	2.63	11.0	Theory	0 K	143
$Co(Ne)^+$-Ne	1.95 ± 0.1	8.2 ± 0.4	Equilibrium	0 K	185
$Co(Ne)^+$-Ne	2.67	11.2	Theory	0 K	143
Co^+-O	75.0 ± 1	314 ± 4	Ion Beam	0 K	38, 46, 10
Co^+-O	63.63 ± 2.31	266 ± 10	Ion Beam	0 K	184
Co^+-O	60.63 ± 13.83	254 ± 58	Thermochemical	298 K	184
Co^+-OH	71.7 ± 0.9	300 ± 4	Ion Beam	0 K	34, 10, 72
Co^+-OH	72 ± 2	301 ± 8	CID	298 K	293
Co^+-OH	72.2	302	CID	298 K	294
Co^+-OH	<91	<381	Ion Molecule Reactions	298 K	2
Co^+-OH	71 ± 3	297 ± 13	Photodissociation	298 K	232
Co^+-S	62 ± 5	259 ± 21	Photodissociation	298 K	241
Co^+-Si	74.9 ± 1.6	314 ± 7	Ion Beam	0 K	116
Co^+-$SiCH_3$	>60	>251	Ion Beam	0 K	170
Co^+-$Si(CH_3)_2$	>69	>289	Ion Beam	0 K	170
Co^+-SiH	69.9 ± 3.7	292 ± 15	Ion Beam	0 K	116
Co^+-SiH_2	67 ± 6	280 ± 25	Ion Beam	0 K	170
Co^+-SiH_2	51.9 ± 1.8	217 ± 8	Ion Beam	0 K	116
Co^+-SiH_3	45.2 ± 3	189 ± 13	Ion Beam	0 K	116
Co-Ar	11.7	49.0	Photodissociation	298 K	205
Co-CH_2	84 ± 5	351 ± 21	Theory	0 K	269
Co-CH_2	79 ± 9	331 ± 38	Proton Transfer	298 K	249
Co-CH_3	42.5 ± 2	178 ± 8	Ion Beam	0 K	45, 46, 10
Co-Co	≤30.4	≤ 127	Ion Beam	0 K	90
Co-H	43.0 ± 1.2	180 ± 5	Ion Beam	0 K	45, 46, 15, 10

BOND	KCAL/MOL	KJ/MOL	METHOD	T	REFERENCES
Co-H	42.2 ± 3	177 ± 13	Ion Molecule Reactions	298 K	236
Co^{2+}-OH	44 ± 12	184 ± 50	Charge Stripping FAB	298 K	288
Co_2^+-Co	48.13	201.5	Ion Beam	0 K	81, 72
Co_2-Co	≥33.4	≥140	Ion Beam	0 K	92, 83
Co_3^+-Co	48.9	205	Ion Beam	0 K	77, 72
Co_3-Co	55.6	233	Ion Beam	0 K	93, 83
Co_4^+-Co	65.6	274	Ion Beam	0 K	77, 72
Co_4-Co	65.5	274	Ion Beam	0 K	93, 83
Co_5^+-Co	76.3	319	Ion Beam	0 K	77, 72
Co_5-Co	76.3	319	Ion Beam	0 K	93, 83
Co_6^+-Co	67.6	283	Ion Beam	0 K	77, 72
Co_6-Co	61.1	256	Ion Beam	0 K	85, 83
Co_7^+-Co	72.4	303	Ion Beam	0 K	77, 72
Co_7-Co	67.6	283	Ion Beam	0 K	85, 83
Co_8^+-Co	67.6	283	Ion Beam	0 K	77, 72
Co_8-Co	66.6	279	Ion Beam	0 K	85, 83
Co_9^+-Co	68.0	285	Ion Beam	0 K	77, 72
Co_9-Co	70.3	294	Ion Beam	0 K	85, 83
Co_{10}^+-Co	74.9	313	Ion Beam	0 K	77, 72
Co_{10}-Co	71.9	301	Ion Beam	0 K	85, 83
Co_{11}^+-Co	78.6	329	Ion Beam	0 K	77, 72
Co_{11}-Co	78.6	329	Ion Beam	0 K	85, 83
Co_{12}^+-Co	83.9	351	Ion Beam	0 K	77, 72
Co_{12}-Co	84.9	355	Ion Beam	0 K	85, 83
Co_{13}^+-Co	72.4	303	Ion Beam	0 K	77, 72
Co_{13}-Co	71.9	301	Ion Beam	0 K	85, 83
Co_{14}^+-Co	90.9	380	Ion Beam	0 K	77, 72
Co_{14}-Co	88.6	371	Ion Beam	0 K	85, 83
Co_{15}^+-Co	84.9	355	Ion Beam	0 K	77, 72
Co_{15}-Co	88.1	369	Ion Beam	0 K	85, 83
Co_{16}^+-Co	82.8	346	Ion Beam	0 K	77, 72
Co_{16}-Co	79.3	332	Ion Beam	0 K	85, 83
Co_{17}^+-Co	88.8	372	Ion Beam	0 K	77, 72
Co_{17}-Co	88.6	371	Ion Beam	0 K	85, 83
Cr^+-Ar	6.55 ± 0.4	27 ± 2	Equilibrium	0 K	185
Cr^+-Ar	6.69 ± 0.9	28 ± 4	Theory	0 K	200
Cr^+-Ar	5.49	23.0	Theory	0 K	143
$Cr(Ar)^+$-Ar	6.48	27.1	Theory	0 K	143
Cr^+-C	127	531	Theory	0 K	280
Cr^+-CH	70 ± 7	293 ± 29	Ion Beam	0 K	24, 10
Cr^+-CH	150	628	Theory	0 K	276
Cr^+-CH_2	51.7 ± 0.9	216 ± 4	Ion Beam	0 K	24, 26, 10
Cr^+-CH_2	103	431	Theory	0 K	276

BOND	KCAL/MOL	KJ/MOL	METHOD	T	REFERENCES
Cr^+-CH_2	38.7	162	Theory	0 K	271
Cr^+-CH_2	57 ± 4	238 ± 17	Theory	0 K	123
Cr^+-CH_3	26.2 ± 0.8	110 ± 3	Ion Beam	0 K	24, 27, 10
Cr^+-CH_3	25.5	107	Theory	0 K	121
Cr^+-CH_3	52	218	Theory	0 K	276
Cr^+-CH_3	24.1	101	Theory	298 K	173
Cr^+-CH_3	9.2	38	Theory	298 K	173
$Cr(CH_3)^+$-CH_3	24.2	101	Theory	0 K	122
Cr^+-CO	21.4 ± 0.9	90 ± 4	Ion Beam	0 K	68
Cr^+-CO	21.4	89.6	Theory	0 K	131
Cr^+-CO	21.5	90.0	Theory	0 K	284
$Cr(CO)^+$-CO	22.6 ± 0.7	95 ± 3	Ion Beam	0 K	68
$Cr(CO)^+$-CO	22.7	95.0	Theory	0 K	131
$Cr(CO)_2{}^+$-CO	12.9 ± 1.4	54 ± 6	Ion Beam	0 K	68
$Cr(CO)_3{}^+$-CO	12.2 ± 1.8	51 ± 8	Ion Beam	0 K	68
$Cr(CO)_4{}^+$-CO	14.8 ± 0.7	62 ± 3	Ion Beam	0 K	68
$Cr(CO)_5{}^+$-CO	31.1 ± 1.8	130 ± 8	Ion Beam	0 K	68
Cr^+-C_2H_2	44 ± 5	184 ± 21	Ion Beam	0 K	24, 10
Cr^+-C_2H_2	25	105	Theory	0 K	124
Cr^+-C_2H_3	54.1 ± 1.4	226 ± 6	Ion Beam	0 K	27, 10
Cr^+-C_2H_4	≥30 ± 5	≥126 ± 21	Ion Beam	0 K	24, 10
Cr^+-$CHCH_3$	43.0 ± 1.8	180 ± 8	Ion Beam	0 K	27, 10
Cr^+-C_2H_4	26	109	Theory	0 K	125
Cr^+-C_2H_5	30.6 ± 1.2	128 ± 5	Ion Beam	0 K	27, 10
Cr^+-C_3H_4	33.6 ± 1.6	141 ± 7	Ion Beam	0 K	27, 10
Cr^+-$C(CH_3)_2$	29.7 ± 2.1	124 ± 9	Ion Beam	0 K	27, 10
Cr^+-$CHCH_2CH_3$	27.1 ± 2.5	113 ± 10	Ion Beam	0 K	27, 10
Cr^+-n- C_3H_7	27.6 ± 1.4	115 ± 6	Ion Beam	0 K	27, 10
Cr^+-i- C_3H_7	24.2 ± 1.2	101 ± 5	Ion Beam	0 K	27, 10
Cr^+-C_6H_6	40.6 ± 2.3	170 ± 10	Ion Beam	0 K	114
Cr^+-C_6H_6	37.4	156	Theory	0 K	126
$Cr(C_6H_6)^+$-C_6H_6	55.3 ± 4.4	232 ± 18	Ion Beam	0 K	114
Cr^+-Cl	47.3	198	Theory	0 K	277
Cr^+-Cl	>50.5	>211	Ion Molecule Reactions	298 K	298
Cr^+-Cr	30.0	126	Ion Beam	0 K	75, 72
Cr^+-Cr	49.1	205	Photodissociation	298 K	201
Cr^+-F	73	305	Theory	0 K	279
Cr^+-H	31.6 ± 2.1	132 ± 9	Ion Beam	0 K	14
Cr^+-H	27.7	116	Theory	0 K	120
Cr^+-H	60.7 ± 2	254 ± 8	Theory	0 K	269
Cr^+-H	25.3	106	Theory	298 K	173
$Cr(CO)_6{}^+$-H	60.7 ± 2	254 ± 8	Theory	0 K	269
$Cr(CO)_6{}^+$-H	58 ± 3	243 ± 13	Thermochemical	298 K	166

BOND	KCAL/MOL	KJ/MOL	METHOD	T	REFERENCES
$Cr(C_5H_5)(CH_3)(CO)_3^+$-H	61 ± 4	255 ± 17	Thermochemical	298 K	166
$Cr(C_5H_5)(CO)_2(NO)^+$-H	52 ± 3	218 ± 13	Thermochemical	298 K	166
$Cr(C_6H_6)(CO)_3^+$-H	56 ± 4	234 ± 17	Thermochemical	298 K	166
$Cr(O)^+$-H	89 ± 5	372 ± 21	Ion Beam	0 K	171
Cr^+-H_2O	30.9 ± 2.1	129 ± 9	Ion Beam	0 K	67
Cr^+-H_2O	30.1	126	Theory	0 K	128
Cr^+-H_2O	21.9	91.7	CID	298 K	315
Cr^+-H_2O	29.0	121	CID	298 K	294
$Cr(H_2O)^+$-H_2O	33.9 ± 1.4	142 ± 6	Ion Beam	0 K	67
$Cr(H_2O)^+$-H_2O	32.5	136	Theory	0 K	128
$Cr(H_2O)^+$-H_2O	29.5	123	CID	298 K	315
$Cr(H_2O)_2^+$-H_2O	12.0 ± 1.2	50 ± 5	Ion Beam	0 K	67
$Cr(H_2O)_3^+$-H_2O	12.2 ± 1.6	51 ± 7	Ion Beam	0 K	67
Cr^+-He	0.98 ± 0.1	4.1 ± 0.4	Equilibrium	0 K	185
Cr^+-He	1.01	4.23	Theory	0 K	143
$Cr(He)^+$-He	1.15	4.81	Theory	0 K	143
Cr^+-N	49	205	Theory	0 K	279
Cr^+-NH_2	65.1 ± 2.3	272 ± 10	Ion Beam	0 K	34, 10
Cr^+-NH_2	40.1	168	Theory	0 K	281
Cr^+-NH_3	38.9	163	Theory	0 K	130
Cr^+-NH_3	37	155	CID	298 K	315
$Cr(NH_3)^+$-NH_3	40.7	170	Theory	0 K	130
$Cr(NH_3)^+$-NH_3	40.8	171	CID	298 K	315
Cr^+-N_2	14.1 ± 0.9	59 ± 4	Photodissociation	0 K	200
Cr^+-N_2	12.2	51.1	Theory	0 K	134
Cr^+-Ne	1.38 ± 0.1	5.8 ± 0.4	Equilibrium	0 K	185
Cr^+-Ne	1.20	5.02	Theory	0 K	143
$Cr(Ne)^+$-Ne	0.90 ± 0.2	4 ± 1	Equilibrium	0 K	185
$Cr(Ne)^+$-Ne	1.38	5.78	Theory	0 K	143
Cr^+-O	85.8 ± 2.8	359 ± 12	Ion Beam	0 K	38, 10
Cr^+-O	85 ± 1.3	356 ± 5	Ion Beam	0 K	170
Cr^+-O	57	238	Theory	0 K	279
Cr^+-O	51	213	Theory	0 K	267
Cr^+-O	63.17 ± 13.83	264 ± 58	Thermochemical	298 K	184
$Cr(O)^+$-O	83 ± 2	347 ± 8	Ion Beam	0 K	184
$Cr(O)^+$-O	82.99 ± 23.1	347 ± 97	Thermochemical	298 K	184
Cr^+-OH	71.3 ± 3.2	298 ± 13	Ion Beam	0 K	39, 10
Cr^+-OH	73 ± 5	305 ± 21	Ion Beam	0 K	171
Cr^+-OH	73 ± 5	305 ± 21	CID	298 K	293
Cr^+-OH	74.3	311	CID	298 K	294
$Cr(O)^+$-OH	73 ± 5	305 ± 21	Ion Beam	0 K	171
Cr^+-Si	48.4 ± 3.7	203 ± 15	Ion Beam	0 K	94
Cr^+-SiH	23.5 ± 6.7	98 ± 28	Ion Beam	0 K	94

Bond	kcal/mol	kJ/mol	Method	T	References
Cr^+-SiH_2	22.8 ± 3.2	95 ± 13	Ion Beam	0 K	94
Cr^+-SiH_3	18.0 ± 4.1	75 ± 17	Ion Beam	0 K	94
Cr^+-Xe	16.3 ± 2.4	68 ± 10	Ion Beam	0 K	68
Cr-CH_3	33.4 ± 1.6	140 ± 7	Ion Beam	0 K	27, 10
Cr-CO	343	1435	Theory	298 K	264
$Cr(CO)_6$-C_2H_4	14.334	60.01	CID	298 K	291
Cr-Cr	40.8	171	Photodissociation	298 K	201
Cr-H	44.6 ± 1.6	187 ± 7	Ion Beam	0 K	15, 17, 10
Cr-H	41.2 ± 3	172 ± 13	Ion Molecule Reactions	298 K	236
Cr-O	103.5 ± 1.2	433 ± 5	Ion Beam	0 K	24, 10
$Cr(O)_5$-O_2	30.66 ± 3.46	128 ± 14	CID	298 K	290
$Cr(CO)_3(O)^-$-CO	38.04 ± 3.46	159 ± 14	CID	298 K	290
$Cr(CO)_5^-$-CO	44.72 ± 3.46	187 ± 14	CID	298 K	290
$Cr(CO)_3(O)_2^-$-$(CO)_2$	28.12 ± 2.31	118 ± 10	CID	298 K	290
Cr_2^+-Cr	46.4	194	Ion Beam	0 K	75, 72
Cr_3^+-Cr	24.0	100	Ion Beam	0 K	75, 72
Cr_4^+-Cr	51.4	215	Ion Beam	0 K	75, 72
Cr_5^+-Cr	40.8	171	Ion Beam	0 K	75, 72
Cr_6^+-Cr	58.8	246	Ion Beam	0 K	75, 72
Cr_7^+-Cr	51.9	217	Ion Beam	0 K	75, 72
Cr_8^+-Cr	59.5	249	Ion Beam	0 K	75, 72
Cr_9^+-Cr	55.3	231	Ion Beam	0 K	75, 72
Cr_{10}^+-Cr	59.3	248	Ion Beam	0 K	75, 72
Cr_{11}^+-Cr	61.6	258	Ion Beam	0 K	75, 72
Cr_{12}^+-Cr	69.2	290	Ion Beam	0 K	75, 72
Cr_{13}^+-Cr	70.3	294	Ion Beam	0 K	75, 72
Cr_{14}^+-Cr	63.9	267	Ion Beam	0 K	75, 72
Cr_{15}^+-Cr	67.3	282	Ion Beam	0 K	75, 72
Cr_{16}^+-Cr	65.3	273	Ion Beam	0 K	75, 72
Cr_{17}^+-Cr	49.8	208	Ion Beam	0 K	75, 72
Cr_{18}^+-Cr	55.6	233	Ion Beam	0 K	75, 72
Cr_{19}^+-Cr	69.2	290	Ion Beam	0 K	75, 72
Cr_{20}^+-Cr	60.4	253	Ion Beam	0 K	75, 72
Cu^+-Ar	9.34	39.1	Theory	0 K	143
Cu^+-CH_2	61.2 ± 1.2	256 ± 5	Ion Beam	0 K	46, 10
Cu^+-CH_2	61 ± 4	255 ± 17	Theory	0 K	123
Cu^+-CH_2O	34.7 ± 2.3	145 ± 10	Ion Beam	0 K	46, 10
Cu^+-CH_3	26.4 ± 1.6	110 ± 7	Ion Beam	0 K	45, 10
Cu^+-CH_3	28.8	120	Theory	0 K	121
$Cu(CH_3)^+$-CH_3	40.5	169	Theory	0 K	122
Cu^+-CH_4	21.4	89.6	Theory	0 K	135
Cu^+-CO	35.5 ± 1.6	149 ± 7	Ion Beam	0 K	96
Cu^+-CO	33.4	140	Theory	0 K	131

Bond	kcal/mol	kJ/mol	Method	T	References
$Cu(CO)^+$-CO	41.0 ± 0.7	172 ± 3	Ion Beam	0 K	96
$Cu(CO)^+$-CO	35.5	149	Theory	0 K	131
$Cu(CO)_2^+$-CO	18.0 ± 0.9	75 ± 4	Ion Beam	0 K	96
$Cu(CO)_3^+$-CO	12.7 ± 0.7	53 ± 3	Ion Beam	0 K	96
Cu^+-C_2H_2	> 2 ± 2	>8 ± 8	Ion Beam	0 K	46, 10
Cu^+-C_2H_2	37	155	Theory	0 K	124
Cu^+-C_2H_4	≥22.7 ± 2.5	≥95 ± 10	Ion Beam	0 K	46, 10
Cu^+-$CHCH_3$	19 ± 6	79 ± 25	Ion Beam	0 K	45, 10
Cu^+-C_2H_4	40	167	Theory	0 K	125
Cu^+-C_2H_6	15.5	64.9	Theory	0 K	146
Cu^+-C_3H_6	39.9	167	Theory	0 K	146
Cu^+-C_3H_8	21.4	89.6	Theory	0 K	146
Cu^+-C_6H_6	52.1 ± 2.3	218 ± 10	Ion Beam	0 K	114
Cu^+-C_6H_6	51.1	214	Theory	0 K	126
Cu^+-C_6H_6	≤45.7	≤ 191	Photodissociation	298 K	223, 227
$Cu(C_6H_6)^+$-C_6H_6	37.1 ± 2.8	155 ± 12	Ion Beam	0 K	114
Cu^+-H	21.2 ± 3	89 ± 13	Ion Beam	0 K	42
Cu^+-H	18.5	77.5	Theory	0 K	120
Cu^+-H	21.5	90.0	Theory	0 K	177
Cu^+-H	20.9	87.5	Theory	0 K	177
Cu^+-H	23.5	98.4	Theory	298 K	260
$Cu(CH_4)^+$-H_2	15.6	65.3	Theory	0 K	135
Cu^+-H_2	14.2	59.4	Theory	0 K	135
$Cu(H_2)^+$-H_2	14.8	62.0	Theory	0 K	135
$Cu(H_2O)^+$-H_2	16.5	69.1	Theory	0 K	135
Cu^+-H_2O	37.6 ± 1.8	157 ± 8	Ion Beam	0 K	67
Cu^+-H_2O	37.3	156	Theory	0 K	128
Cu^+-H_2O	35.0	146	CID	298 K	294
$Cu(H_2O)^+$-H_2O	40.6 ± 1.6	170 ± 7	Ion Beam	0 K	67
$Cu(H_2O)^+$-H_2O	37.8	158	Theory	0 K	128
$Cu(H_2O)_2^+$-H_2O	13.6 ± 1.8	57 ± 8	Ion Beam	0 K	67
$Cu(H_2O)_2^+$-H_2O	15.4	64.5	Theory	0 K	129
$Cu(H_2O)_2^+$-H_2O	16.4	68.7	CID	298 K	315
$Cu(H_2O)_2^+$-H_2O	16.4 ± 0.2	69 ± 1	van't Hoff Plot	298 K	213
$Cu(H_2O)_3^+$-H_2O	13.0 ± 1	54 ± 4	Ion Beam	0 K	67
$Cu(H_2O)_3^+$-H_2O	13.3	55.7	Theory	0 K	129
$Cu(H_2O)_3^+$-H_2O	16.7	69.9	CID	298 K	315
$Cu(H_2O)_3^+$-H_2O	16.7	69.9	van't Hoff Plot	298 K	213
$Cu(H_2O)_4^+$-H_2O	14.0 ± 0.1	58.6 ± 0.4	van't Hoff Plot	298 K	213
Cu^+-He	2.54	10.6	Theory	0 K	143
$Cu(He)^+$-He	2.21	9.25	Theory	0 K	142
Cu^+-Kr	13.40	56.10	Theory	0 K	142
$Cu(Kr)^+$-Kr	14.55	60.92	Theory	0 K	142

BOND	KCAL/MOL	KJ/MOL	METHOD	T	REFERENCES
Cu^+-NH_2	45.9 ± 3	192 ± 13	Ion Beam	0 K	58, 10
Cu^+-NH_3	51.8	217	Theory	0 K	130
$Cu(NH_3)^+$-NH_3	51.6	216	Theory	0 K	130
$Cu(NH_3)_2^+$-NH_3	17.2	72.0	Theory	0 K	129
$Cu(NH_3)_2^+$-NH_3	14.0 ± 0.2	59 ± 1	van't Hoff Plot	298 K	213
$Cu(NH_3)_3^+$-NH_3	12.2	51.1	Theory	0 K	129
$Cu(NH_3)_3^+$-NH_3	12.8 ± 0.2	54 ± 1	van't Hoff Plot	298 K	213
$Cu(NH_3)_4^+$-NH_3	12.8 ± 0.1	53.6 ± 0.4	van't Hoff Plot	298 K	213
Cu^+-Ne	2.21	9.25	Theory	0 K	143
Cu^+-O	37.4 ± 3.5	156 ± 15	Ion Beam	0 K	38, 10
Cu^+-S	61.1 ± 1.8	256 ± 8	Ion Beam	0 K	118
Cu^+-SiH	58.8 ± 6.5	246 ± 27	Ion Beam	0 K	118
Cu^+-SiH_2	≥55.1 ± 1.6	≥ 231 ± 7	Ion Beam	0 K	118
Cu^+-SiH_3	23.1 ± 6	97 ± 25	Ion Beam	0 K	118
Cu-CH_3	53.3 ± 1.2	223 ± 5	Ion Beam	0 K	45, 46, 10
Cu-C_6H_{12}	10 ± 1.4	49±.06	Kinetic Measurements	0 K	275
CuC_6H_{12}-C_6H_{12}	18 ± 2.4	75±10	Kinetic Measurements	0 K	275
Cu-H	60.0 ± 1.4	251 ± 6	Ion Beam	0 K	15, 45, 46, 10
Cu-H	60.6 ± 1.4	254 ± 6	Ion Beam	0 K	118
Fe^+-Ar	4.04	16.9	Theory	0 K	143
$Fe(Ar)^+$-Ar	5.10	21.4	Theory	0 K	142
Fe^+-Br	72 ± 5	301 ± 21	Theory	0 K	163
Fe^+-Br	>69	> 289	Ion Molecule Reactions	298 K	2
Fe^+-C	94 ± 7	393 ± 21	Photodissociation	298 K	240
Fe^+-CH	106.8 ± 7	447 ± 29	Theory	0 K	269
Fe^+-CH	101 ± 7	423 ± 21	Photodissociation	298 K	240
Fe^+-CH_2	81.5 ± 0.9	341 ± 4	Ion Beam	0 K	47, 10
Fe^+-CH_2	74 ± 5	310 ± 21	Theory	0 K	123
Fe^+-CH_2	82 ± 7	343 ± 21	Photodissociation	298 K	240
Fe^+-CH_3	54.6 ± 1.1	228 ± 5	Ion Beam	0 K	50, 51, 10
Fe^+-CH_3	53.1	222	Theory	0 K	121
Fe^+-CH_3	69 ± 5	289 ± 21	Ion Molecule Reactions	298 K	233
Fe^+-CH_3	<69	< 289	Ion Molecule Reactions	298 K	2
Fe^+-CH_3	65 ± 5	272 ± 21	Photodissociation	298 K	242
$Fe(CH_3)^+$-CH_3	43.1 ± 2.5	180 ± 10	Ion Beam	0 K	50, 51, 56, 10
$Fe(CH_3)^+$-CH_3	33.0	138	Theory	0 K	122
Fe^+-CH_4	13.7 ± 0.8	57 ± 3	Ion Beam	0 K	53
Fe^+-CH_4	13.6	56.9	Theory	0 K	162
$Fe(CH_4)^+$-CH_4	23.3 ± 1	97 ± 4	Ion Beam	0 K	53
$Fe(CH_4)_2^+$-CH_4	23.6 ± 1.4	99 ± 6	Ion Beam	0 K	53
$Fe(CH_4)_3^+$-CH_4	17.6 ± 1.4	74 ± 6	Ion Beam	0 K	53
Fe^+-CNH	39 ± 2	163 ± 8	Ion Molecule Reactions	298 K	308
Fe^+-CO	31.3 ± 1.8	131 ± 8	Ion Beam	0 K	70

BOND	KCAL/MOL	KJ/MOL	METHOD	T	REFERENCES
Fe^+-CO	32 ± 5	134 ± 21	KERD	0 K	194
Fe^+-CO	30.3	127	Theory	0 K	131
Fe^+-CO	37 ± 6	155 ± 25	Ion Molecule Reactions	298 K	229
$Fe(CO)^+$-CO	36.1 ± 1.2	151 ± 5	Ion Beam	0 K	70
$Fe(CO)^+$-CO	34.5	144	Theory	0 K	131
$Fe(CO)_2{}^+$-CO	15.9 ± 1.2	67 ± 5	Ion Beam	0 K	70
$Fe(CO)_3{}^+$-CO	24.7 ± 1.4	103 ± 6	Ion Beam	0 K	70
$Fe(CO)_4{}^+$-CO	26.8 ± 0.9	112 ± 4	Ion Beam	0 K	70
Fe^+-CO_2	8 ± 2	33±8	Ion Molecule Reactions	298	300
Fe^+-CO_2	9.5 ± 2	40 ± 8	Ion Molecule Reactions	298 K	304
Fe^+-C_2H_2	28	117	Theory	0 K	124
Fe^+-C_2H_2	32 ± 6	134 ± 25	Photodissociation	298 K	245
Fe^+-C_2H_3	56.8 ± 2.5	238 ± 10	Ion Beam	0 K	47, 10
Fe^+-C_2H_4	34.7 ± 1.4	145 ± 6	Ion Beam	0 K	55, 10
Fe^+-C_2H_4	35 ± 5	146 ± 21	KERD	0 K	192
Fe^+-C_2H_4	30	126	Theory	0 K	125
Fe^+-C_2H_4	34 ± 2	142 ± 8	Ion Molecule Reactions	298 K	229
Fe^+-C_2H_5	55.7 ± 2.1	233 ± 9	Ion Beam	0 K	50, 10
Fe^+-C_2H_6	15.2 ± 1.4	64 ± 6	Ion Beam	0 K	56, 10
Fe^+-C_2H_6	18 ± 5	75 ±21	KERD/Theory	0 K	191
Fe^+-C_3H_3	92 ± 15	385 ± 63	Ion Molecule Reactions	298 K	237
Fe^+-C_3H_6	34.7 ± 1.7	145 ± 7	Ion Beam	0 K	47, 10
Fe^+-$CH_2CH_2CH_2$ (to Δ)	31.8 ± 0.9	133 ± 4	Ion Beam	0 K	47, 10
Fe^+-C_3H_6 (metallocyclic)	85 ± 5	356 ± 21	KERD	0 K	192, 195
Fe^+-C_3H_6	37 ± 5	155 ± 21	KERD	0 K	192
Fe^+-C_3H_6	37 ± 2	155 ± 8	Ion Molecule Reactions	298 K	229
Fe^+-C_3H_8	17.9 ± 1	75 ± 4	Ion Beam	0 K	55, 57, 10
Fe^+-C_4H_6	48 ± 5	201 ± 21	Photodissociation	298 K	242
Fe^+-C_5H_5	77 ± 10	322 ± 42	Theory	0 K	138
Fe^+-C_5H_5	88 ± 7	368 ± 29	Photodissociation	298 K	251
Fe^+-C_5H_6	50 ± 5	209± 21	KERD	0 K	193
Fe^+-C_5H_6	55 ± 5	230± 21	Photodissociation	298 K	251
Fe^+-C_{60}	44 ± 7	184±29	Ion Molecule Reactions	298 K	257
Fe^+-C_6H_4	73 ± 10	305 ± 42	Theory	0 K	137
Fe^+-C_6H_4	>66	> 276	Ion Molecule Reactions	298 K	296
Fe^+-C_6H_4	76 ± 10	318 ± 42	Photodissociation	298 K	250
Fe^+-C_6H_5	77.6	325	Theory	0 K	164
Fe^+-C_6H_5	>64	> 268	Ion Molecule Reactions	298 K	296
Fe^+-C_6H_5	71 ± 5	297±21	Photodissociation	298 K	164
Fe^+-C_6H_6	49.6 ± 2.3	207 ± 10	Ion Beam	0 K	114
Fe^+-C_6H_6	51.1	214	Theory	0 K	126
Fe^+-C_6H_6	58 ± 5	243 ± 21	Ion Molecule Reactions	298 K	233

BOND	KCAL/MOL	KJ/MOL	METHOD	T	REFERENCES
Fe^+-C_6H_6	55 ± 5	230 ± 21	Photodissociation	298 K	242
Fe^+-C_6H_6	≤62.1	≤ 260	Photodissociation	298 K	223, 227
$Fe(C_6H_6)^+$-C_6H_6	44.7 ± 3.9	187 ± 16	Ion Beam	0 K	114
Fe^+-C_6H_8	70 ± 5	293 ± 21	KERD	0 K	193
Fe^+-C_7H_7 (tropylium)	86.8	363	Theory	0 K	164
Fe^+-C_7H_7 (tolyl)	84.4	353	Theory	0 K	164
Fe^+-C_7H_7 (benzyl)	69.4	290	Theory	0 K	164
Fe^+-Cl	80 ± 5	335 ± 21	Theory	0 K	163
Fe^+-F	103 ± 5	431 ± 21	Theory	0 K	163
Fe^+-F	101	423	Theory	0 K	314
Fe^+-F	97 ± 8	406 ± 33	Knudsen Cell MS	298 K	309
Fe^+-Fe	64.0	268	Ion Beam	0 K	10, 76, 72
Fe^+-Fe	56.03	234.6	Photodissociation	298 K	204
Fe^+-H	48.9 ± 1.4	205 ± 6	Ion Beam	0 K	41
Fe^+-H	59 ± 5	247 ± 21	Ion Beam	0 K	167
Fe^+-H	52.3	219	Theory	0 K	120
Fe^+-H	57.0 ± 4	238 ± 17	Theory	0 K	269
Fe^+-H	58 ± 2	243 ± 8	Ion Molecule Reactions	298 K	235
Fe^+-H	<71	< 297	Ion Molecule Reactions	298 K	2
Fe^+-H	31.2	131	Theory	298 K	177
$Fe(CO)_5^+$-H	57.0 ± 4	238 ± 17	Theory	0 K	269
$Fe(CO)_5^+$-H	74 ± 5	310 ± 21	Thermochemical	298 K	166
$Fe(C_5H_5)(CH_3)(CO)_2^+$-H	53 ± 3	222 ± 13	Thermochemical	298 K	166
$Fe(C_5H_5)_2^+$-H	54 ± 5	226 ± 21	Thermochemical	298 K	166
$Fe(C_5H_8)^+$-H	46 ± 5	192 ± 21	Photodissociation	298 K	251
$Fe(C_5H_8)(CH_3)(CO)_2^+$-H	57.0 ± 4	238 ± 17	Theory	0 K	269
$Fe(C_5H_8)_2^+$-H	57.0 ± 4	238 ± 17	Theory	0 K	269
$Fe(O)^+$-H	106 ± 4	444 ± 17	Thermochemical	298 K	171
$Fe(OH)^+$-H	25	105	Theory	0 K	307
Fe^+-H_2O	30.6 ± 1.2	128 ± 5	Ion Beam	0 K	67
Fe^+-H_2O	33.7	141	Theory	0 K	128
Fe^+-H_2O	32.8	137	CID	298 K	315
Fe^+-H_2O	28.8	120	CID	298 K	294
$Fe(H_2O)^+$-H_2O	39.3 ± 1	164 ± 4	Ion Beam	0 K	67
$Fe(H_2O)^+$-H_2O	37.0	155	Theory	0 K	128
$Fe(H_2O)^+$-H_2O	40.8	171	CID	298 K	315
$Fe(H_2O)_2^+$-H_2O	18.2 ± 0.9	76 ± 4	Ion Beam	0 K	67
$Fe(H_2O)_3^+$-H_2O	12.0 ± 1.6	50 ± 7	Ion Beam	0 K	67
Fe^+-He	0.30	1.3	Theory	0 K	143
Fe^+-I	>71	> 297	Ion Molecule Reactions	298 K	2
Fe^+-NCH	36 ± 2	151 ± 8	Ion Molecule Reactions	298 K	308
Fe^+-NH	54 ± 14	226 ± 59	Ion Molecule Reactions	298 K	246
Fe^+-NH	61 ± 5	255 ± 21	Photodissociation	298 K	246

Bond	kcal/mol	kJ/mol	Method	T	References
Fe^+-NH_2	73.9 ± 2.3	309 ± 10	Ion Beam	0 K	39, 10
Fe^+-NH_3	42.8	179	Theory	0 K	130
$Fe(NH_3)^+$-NH_3	50.3	210	Theory	0 K	130
Fe^+-N_2	10.5 ± 2	44 ± 8	Ion Molecule Reactions	298 K	304
Fe^+-Ne	0.65	2.7	Theory	0 K	143
Fe^+-O	80.1 ± 1.2	335 ± 5	Ion Beam	0 K	59, 10
Fe^+-O	69.39 ± 2.31	290 ± 10	Ion Beam	0 K	184
Fe^+-O	72.62 ± 5.53	304 ± 23	Thermochemical	298 K	184
$Fe(O)^+$-O	90 ± 14	377 ± 59	Ion Beam	0 K	184
$Fe(O)^+$-O	66	276	Theory	0 K	305
$Fe(O)^+$-O	89.91 ± 13.83	376 ± 58	Thermochemical	298 K	184
Fe^+-OH	87.5 ± 2.9	366 ± 12	Ion Beam	0 K	61, 10
Fe^+-OH	85.3	357	CID	298 K	294
Fe^+-OH	>91	> 381	Ion Molecule Reactions	298 K	2
Fe^+-OH	73 ± 3	305 ± 13	Photodissociation	298 K	232
Fe^+-OH	77 ± 6	322 ± 25	Proton Transfer	298 K	232
Fe^+-O_2	24	100	Theory	0 K	305
Fe^+-S	65 ± 5	272 ± 21	Photodissociation	298 K	241
Fe^+-S	61 ± 6	255 ± 25	Photodissociation	298 K	245
Fe^+-S_2	48 ± 5	201 ± 21	Photodissociation	298 K	245
$Fe(S)^+$-S_2	49 ± 5	205 ± 21	Photodissociation	298 K	245
$Fe(S)_2^+$-S_2	49 ± 5	205 ± 21	Photodissociation	298 K	245
$Fe(S)_3^+$-S_2	43 ± 5	180 ± 21	Photodissociation	298 K	245
$Fe(S)_4^+$-S_2	38 ± 5	159 ± 21	Photodissociation	298 K	245
Fe^+-Sc	79 ± 5	331 ± 21	Photodissociation	298 K	243
Fe^+-Si	66.2 ± 2.1	277 ± 9	Ion Beam	0 K	116
Fe^+-$Si(CH_3)_2$	>69	> 289	Ion Beam	0 K	170
Fe^+-SiH	60.6 ± 3	254 ± 13	Ion Beam	0 K	116
Fe^+-SiH_2	<72	< 301	Ion Beam	0 K	170
Fe^+-SiH_2	43.4 ± 2.1	181 ± 9	Ion Beam	0 K	116
Fe^+-SiH_3	43.8 ± 2.1	183 ± 9	Ion Beam	0 K	116
Fe^+-V	101 ± 5	423 ± 21	Photodissociation	298 K	239
Fe^+-Xe	9.1 ± 1.4	38 ± 6	Ion Beam	0 K	53, 71
Fe-CH_2	106.6 ± 5	446 ± 21	Theory	0 K	269
Fe-CH_2	87 ± 7	364 ± 29	Proton Transfer	298 K	249
Fe-CH_3	32 ± 7	134 ± 29	Ion Beam	0 K	28, 10
Fe-CO	10.5	44	PE-Spectroscopy	298 K	313
Fe-Fe	27.2	114	Ion Beam	0 K	89, 10, 83
Fe-Fe	30.44	127.4	Photodissociation	298 K	204
Fe-H	34.5 ± 0.8	144 ± 3	Ion Beam	0 K	15, 44, 10
Fe-H	29.6 ± 3	124 ± 13	Ion Molecule Reactions	298 K	236
Fe-O	101.4 ± 3.7	424 ± 15	Ion Beam	0 K	60, 10
Fe^{2+}-OH	58 ± 12	243 ± 50	Charge Stripping FAB	298 K	288
Fe_2^+-Fe	40.4	169	Ion Beam	0 K	10, 76, 10

Bond	kcal/mol	kJ/mol	Method	T	References
Fe_2^+-Fe	29.3	123	Photodissociation	298 K	204
Fe_2^+-H	52 ± 16	218 ± 67	CID	298 K	244
Fe_2-Fe	44.0	184	Ion Beam	0 K	89, 10, 83
Fe_3^+-Fe	51.6	216	Ion Beam	0 K	10, 76, 72
Fe_3-Fe	50.4	211	Ion Beam	0 K	89, 10, 83
Fe_4^+-Fe	62.1	260	Ion Beam	0 K	10, 76, 72
Fe_4-Fe	51.9	217	Ion Beam	0 K	89, 10, 83
Fe_5^+-Fe	75.3	315	Ion Beam	0 K	76, 10, 72
Fe_5^+-Fe	26.98	113	Photodissociation	298 K	204
Fe_5-Fe	75.6	316	Ion Beam	0 K	85, 10, 83
Fe_6^+-Fe	76.5	320	Ion Beam	0 K	76, 10, 72
Fe_6-Fe	71.9	301	Ion Beam	0 K	85, 10, 83
Fe_7^+-Fe	60.2	252	Ion Beam	0 K	76, 10, 72
Fe_7-Fe	53.8	225	Ion Beam	0 K	85, 10, 83
Fe_8^+-Fe	67.2	281	Ion Beam	0 K	76, 10, 72
Fe_8-Fe	67.4	282	Ion Beam	0 K	85, 10, 83
Fe_9^+-Fe	67.9	284	Ion Beam	0 K	10, 76, 72
Fe_9-Fe	66.0	276	Ion Beam	0 K	85, 10, 83
Fe_{10}^+-Fe	73.6	308	Ion Beam	0 K	10, 76, 72
Fe_{10}-Fe	74.3	311	Ion Beam	0 K	85, 10, 83
Fe_{11}^+-Fe	79.8	334	Ion Beam	0 K	10, 76, 72
Fe_{11}-Fe	81.3	340	Ion Beam	0 K	85, 10, 83
Fe_{12}^+-Fe	97.5	408	Ion Beam	0 K	10, 76, 72
Fe_{12}-Fe	99.7	417	Ion Beam	0 K	85, 10, 83
Fe_{13}^+-Fe	67.2	281	Ion Beam	0 K	10, 76, 72
Fe_{13}-Fe	69.3	290	Ion Beam	0 K	85, 10, 83
Fe_{14}^+-Fe	90.1	377	Ion Beam	0 K	10, 76, 72
Fe_{14}-Fe	86.8	363	Ion Beam	0 K	85, 10, 83
Fe_{15}^+-Fe	76.2	319	Ion Beam	0 K	10, 76, 72
Fe_{15}-Fe	77.9	326	Ion Beam	0 K	85, 10, 83
Fe_{16}^+-Fe	77.2	323	Ion Beam	0 K	10, 76, 72
Fe_{16}-Fe	74.1	310	Ion Beam	0 K	85, 10, 83
Fe_{17}^+-Fe	74.3	311	Ion Beam	0 K	10, 76, 72
Fe_{17}-Fe	72.2	302	Ion Beam	0 K	85, 10, 83
Fe_{18}^+-Fe	89.1	373	Ion Beam	0 K	10, 76, 72
Fe_{18}-Fe	84.1	352	Ion Beam	0 K	85, 10, 83
Gd^+-C_6H_6	>60	> 251	Ion Molecule Reactions	298 K	191
Gd^+-H_2	120 ± 5	502 ± 21	Ion Molecule Reactions	298 K	191
Gd^+-O	>179	> 749	Ion Molecule Reactions	298 K	191
Hf^+-H	54.9	230	Theory	0 K	273
Hf-CH_2	104 ± 5	435 ± 21	Theory	0 K	259
Hg^+-H	48.6	203	Theory	0 K	273
Ir^+-CH_2	123 ± 5	515 ± 21	Theory	0 K	259

BOND	KCAL/MOL	KJ/MOL	METHOD	T	REFERENCES
Ir^+-$(CH_3)_2$	126.1	528	Theory	0 K	274
Ir^+-H	65.8	275	Theory	0 K	273
K^+-Ar	1.41	5.9	Theory	0 K	141
K^+-H_2O	15.2	63.6	CID	298 K	315
K^+-H_2O	17.9	74.9	Equilibrium	298 K	287
$K(H_2O)^+$-H_2O	16.1	67.4	Equilibrium	298 K	287
$K(H_2O)_2^+$-H_2O	13.2	55.3	Equilibrium	298 K	287
$K(H_2O)_3^+$-H_2O	11.8	49.4	Equilibrium	298 K	287
K^+-NH_3	16.9	70.8	CID	298 K	315
K^+-NH_3	20.1	84.2	Equilibrium	298 K	215
$K(NH_3)^+$-NH_3	16.3	68.2	Equilibrium	298 K	215
$K(NH_3)_2^+$-NH_3	13.5	56.5	Equilibrium	298 K	215
$K(NH_3)_3^+$-NH_3	11.6	48.6	Equilibrium	298 K	215
La^+-C	102 ± 8	427 ± 33	Photodissociation	298 K	243
La^+-CH	125 ± 8	523 ± 33	Photodissociation	298 K	243
La^+-CH_2	95.8 ± 1.6	401 ± 7	Ion Beam	0 K	25, 10
La^+-CH_2	98 ± 1.5	410 ± 6	Theory	0 K	259
La^+-CH_2	106 ± 5	444 ± 21	Photodissociation	298 K	243
La^+-CH_3	51.8 ± 3.5	217 ± 15	Ion Beam	0 K	25, 10
La^+-C_2H_2	63 ± 7	264 ± 29	Ion Beam	0 K	25, 10
La^+-C_2H_2	66	276	Theory	0 K	140
La^+-C_2H_2	52 ± 3	218 ± 13	Photodissociation	298 K	254
La^+-C_2H_4	≥ 22	≥ 92	Ion Beam	0 K	25
La^+-C_2H_4	46	192	Theory	0 K	140
La^+-C_3H_6	51	213	Theory	0 K	140
La^+-H	57.2 ± 2.1	239 ± 9	Ion Beam	0 K	11
La^+-H	60.4	253	Theory	0 K	273
$La(H)^+$-H	60.6 ± 2.6	254 ± 11	Ion Beam	0 K	11, 25, 10
La^+-Si	66.2 ± 2.3	277 ± 10	Ion Beam	0 K	119
La^+-SiH	63.6 ± 5.8	266 ± 24	Ion Beam	0 K	119
La^+-SiH_2	$\geq 55.1 \pm 1.6$	$\geq 231 \pm 7$	Ion Beam	0 K	119
La^+-SiH_3	46.1 ± 6.5	193 ± 27	Ion Beam	0 K	119
La^{2+}-C_2H_2	44	184	Theory	0 K	140
La^{2+}-C_2H_4	41	172	Theory	0 K	140
La^{2+}-C_3H_6	48	201	Theory	0 K	140
Li^+-Ar	5.84	24.4	Theory	0 K	141
$Li(Ar)^+$-Ar	5.93	24.8	Theory	0 K	142
Li^+-C_3H_6	≥ 17.3	≥ 72.4	Ion Molecule Reactions	298 K	8
Li^+-C_4H_6	>31	> 130	Ion Molecule Reactions	298 K	8
Li^+-C_4H_7Br	>14.7	> 61.5	Ion Molecule Reactions	298 K	8
Li^+-C_4H_7Cl	>37.3	> 156	Ion Molecule Reactions	298 K	8
Li^+-C_6H_6	37	155	Ion Molecule Reactions	298 K	165
Li^+-C_6H_{12}	24	100	Ion Molecule Reactions	298 K	165

Bond	kcal/mol	kJ/mol	Method	T	References
Li^+-Kr	7.49	31.4	Theory	0 K	141
Li^+-PH_3	28 ± 1	117 ± 4	Ligand displacement	298 K	8
Lu^+-CH_2	≥55.0 ± 1.4	≥ 230 ± 6	Ion Beam	0 K	25, 10
Lu^+-CH_3	42.1 ± 4.8	176 ± 20	Ion Beam	0 K	25, 10
Lu^+-H	48.7 ± 3.7	204 ± 15	Ion Beam	0 K	11
$Lu(H)^+$-H	45.6 ± 5.7	191 ± 24	Ion Beam	0 K	11, 25, 10
Lu^+-Si	25.6 ± 3.2	107 ± 13	Ion Beam	0 K	119
Lu^+-SiH_2	22.6 ± 2.3	95 ± 10	Ion Beam	0 K	119
Mg^+-Ar	3.25	13.6	Theory	0 K	143
Mg^+-Ar	3.67	15.4	Photodissociation	298 K	220, 221
$Mg(Ar)^+$-Ar	3.14	13.2	Theory	0 K	142
Mg^+-CH_3	44.4	186	Theory	0 K	150
Mg^+-CH_4	7.6	32	Theory	0 K	156
Mg^+-CH_3OH	41 ± 5	172 ± 21	Theory	0 K	149
$Mg(CH_3OH)^+$-CH_3OH	31.2	131	Theory	0 K	152
Mg^+-CN	64.0	268	Theory	0 K	150
Mg^+-CO	9.1	38.1	Theory	0 K	150
Mg^+-CO_2	16.8	70.3	Theory	0 K	154
Mg^+-CO_2	14.7	61.6	Photodissociation	298 K	222, 218
$Mg(CO_2)^+$-CO_2	11.0	46.0	Theory	0 K	154
Mg^+-C_2H_2	18.8	78.7	Theory	0 K	157
Mg^+-C_2H_4	18.6	77.9	Theory	0 K	157
Mg^+-CH_3CHO	43 ± 5	180 ± 21	Theory	0 K	149
Mg^+-CH_3CHO	37.9	159	Theory	0 K	153
Mg^+-C_2H_6	9.7	41	Theory	0 K	153
Mg^+-C_2H_5OH	38.3	160	Theory	0 K	153
Mg^+-$(CH_3)_2O$	38.3	160	Theory	0 K	153
Mg^+-C_2H_5CHO	39.2	164	Theory	0 K	153
Mg^+-$(CH_3)_2CO$	43.1	180	Theory	0 K	153
Mg^+-i-C_3H_7OH	40.3	169	Theory	0 K	153
Mg^+-n-C_3H_7OH	39.7	166	Theory	0 K	153
Mg^+-THF	44.0	184	Theory	0 K	153
Mg^+-n-C_3H_7CHO	40.0	167	Theory	0 K	153
Mg^+-$(C_2H_5)_2O$	42.5	178	Theory	0 K	153
Mg^+-n-C_4H_9OH	40.5	169	Theory	0 K	153
Mg^+-C_6H_6	30.4	127	Theory	0 K	150
Mg^+-C_6H_6	≤ 26.9	≤ 113	Photodissociation	298 K	223
$Mg(C_6H_6)^+$-C_6H_6	12.4	51.9	Theory	0 K	152
Mg^+-F	102.8	430.4	Theory	0 K	147
Mg^+-Fe	29.4 ± 5	123 ± 21	Theory	0 K	139
Mg^+-H_2	2.2	9.2	Theory	0 K	151
$Mg(H_2)^+$-H_2	1.9	8.0	Theory	0 K	134
Mg^+-H_2CO	35 ± 5	146 ± 21	Theory	0 K	149

BOND	KCAL/MOL	KJ/MOL	METHOD	T	REFERENCES
Mg^+-H_2O	28.4 ± 3	119 ± 13	Ion Beam	0 K	111
Mg^+-H_2O	36 ± 5	151 ± 21	Theory	0 K	149
Mg^+-H_2O	24.3	102	Photodissociation	298 K	217, 224
$Mg(H_2O)^+$-H_2O	22.4 ± 1.6	94 ± 7	Ion Beam	0 K	111
$Mg(H_2O)^+$-H_2O	29 ± 5	121 ± 21	Theory	0 K	149
$Mg(H_2O)_2{}^+$-H_2O	17.3 ± 2.1	72 ± 9	Ion Beam	0 K	111
$Mg(H_2O)_2{}^+$-H_2O	22	92	Theory	0 K	152
$Mg(H_2O)_3{}^+$-H_2O	11.5 ± 2.1	48 ± 9	Ion Beam	0 K	111
$Mg(H_2O)_3{}^+$-H_2O	15.3	64.1	Theory	0 K	149
Mg^+-He	0.21	.88	Theory	0 K	143
Mg^+-Kr	5.50	23.0	Photodissociation	298 K	221
Mg^+-NH_3	37.7	158	Theory	0 K	150
$Mg(NH_3)^+$-NH_3	29.9	125	Theory	0 K	159
$Mg(NH_3)_2{}^+$-NH_3	24.2	101	Theory	0 K	159
Mg^+-N_2	6.9	29	Theory	0 K	134
Mg^+-Ne	0.48	2.0	Theory	0 K	143
Mg^+-O	57.7 ± 2.3	241 ± 10	Ion Beam	0 K	105
Mg^+-O	53.3	223	Theory	0 K	147
Mg^+-OH	83.5	349	Theory	0 K	147
Mg^+-O_2	23.3	97.6	Theory	0 K	155
Mg^+-Xe	11.96	50.07	Photodissociation	298 K	221
Mg^{2+}-H_2O	81.9	343	Theory	0 K	159
$Mg(H_2O)^{2+}$-H_2O	73.8	309	Theory	0 K	159
$Mg(H_2O)_2{}^{2+}$-H_2O	59.8	250	Theory	0 K	159
$Mg_2{}^+$-CO_2	8.1	34	Theory	0 K	154
Mn^+-Ar	3.02	12.6	Theory	0 K	143
Mn^+-CH_2	68.3 ± 2.1	286 ± 9	Ion Beam	0 K	43, 10
Mn^+-CH_2	61 ± 5	255 ± 21	Theory	0 K	123
$Mn(CO)_5{}^+$-CH_2	101.2	423	Theory	0 K	269
Mn^+-CH_3	49.1 ± 0.9	205 ± 4	Ion Beam	0 K	43, 49, 10
Mn^+-CH_3	43.7	183	Theory	0 K	121
Mn^+-CH_3	40.3	169	Theory	298 K	173
Mn^+-CH_3	29.8	125	Theory	298 K	173
$Mn(CH_3)^+$-CH_3	26.4	110	Theory	0 K	122
Mn^+-CH_3OH	32 ± 7	134 ± 29	Ion Molecule Reactions	298 K	297
Mn^+-CO	6.0 ± 2.4	25 ± 10	Ion Beam	0 K	69
Mn^+-CO	11.6	48.6	Theory	0 K	131
Mn^+-CO	>7	> 29	KERD	298 K	174
$Mn(CO)^+$-CO	15.0 ± 2.4	63 ± 10	Ion Beam	0 K	69
$Mn(CO)^+$-CO	12.0	50.2	Theory	0 K	131
$Mn(CO)^+$-CO	<25	< 105	KERD	298 K	174
$Mn(CO)_2{}^+$-CO	17.6 ± 2.4	74 ± 10	Ion Beam	0 K	69
$Mn(CO)_2{}^+$-CO	31 ± 6	130 ± 25	KERD	298 K	174

Bond	kcal/mol	kJ/mol	Method	T	References
$Mn(CO)_3^+$-CO	15.6 ± 2.4	65 ± 10	Ion Beam	0 K	69
$Mn(CO)_3^+$-CO	20 ± 3	84 ± 13	KERD	298 K	174
$Mn(CO)_4^+$-CO	29.0 ± 2.4	121 ± 10	Ion Beam	0 K	69
$Mn(CO)_4^+$-CO	16 ± 3	67 ± 13	KERD	298 K	174
$Mn(CO)_5^+$-CO	33.9 ± 2.5	142 ± 10	Ion Beam	0 K	69
$Mn(CO)_5^+$-CO	32 ± 5	134 ± 21	KERD	298 K	174
$Mn(C_6H_6)^+$-CO	30	126	Thermochemical	298 K	179
$Mn(C_6H_6)(CO)^+$-CO	25	105	Thermochemical	298 K	179
$Mn(C_6H_6)(CO)_2^+$-CO	12	50	Thermochemical	298 K	179
$Mn(C_6H_6)(CO)_3^+$-CO	21	88	Thermochemical	298 K	179
$Mn(C_6H_6)(CO)_4^+$-CO	10	42	Thermochemical	298 K	179
Mn^+-C_2H_2	19	79	Theory	0 K	124
Mn^+-C_2H_4	20	84	Theory	0 K	125
Mn^+-C_2H_5OH	34 ± 7	142 ± 29	Ion Molecule Reactions	298 K	297
Mn^+-$(CH_3)_2CO$	39 ± 7	163 ± 29	Ion Molecule Reactions	298 K	297
Mn^+-n-C_3H_7OH	35 ± 7	146 ± 29	Ion Molecule Reactions	298 K	297
Mn^+-$(CH_3)_3CO$	39 ± 7	163 ± 29	Ion Molecule Reactions	298 K	297
Mn^+-$(CH_3)_3COH$	38 ± 7	159 ± 29	Ion Molecule Reactions	298 K	297
Mn^+-C_6H_6	31.8 ± 2.1	133 ± 9	Ion Beam	0 K	114
Mn^+-C_6H_6	35.1	147	Theory	0 K	126
$Mn(CO)_5^+$-C_6H_6	32 ± 5	134 ± 21	Thermochemical	298 K	179
$Mn(C_6H_6)^+$-C_6H_6	48.4 ± 3.9	203 ± 16	Ion Beam	0 K	114
Mn^+-Cl	>50.5	> 211	Ion Molecule Reactions	298 K	298
Mn^+-H	47.5 ± 3.5	199 ± 15	Ion Beam	0 K	40
Mn^+-H	43.7	183	Theory	0 K	120
Mn^+-H	51.5 ± 3.4	215 ± 14	Theory	0 K	269
Mn^+-H	53 ± 5	222 ± 21	Ion Molecule Reactions	298 K	180
Mn^+-H	25.9	108	Theory	298 K	177
Mn^+-H	38.7	162	Theory	298 K	173
Mn^+-H	28.0	117	Theory	298 K	173
$Mn(CH_3)(CO)_5^+$-H	59.5 ± 3.4	249 ± 14	Theory	0 K	269
$Mn(CH_3)(CO)_5^+$-H	76 ± 3	318 ± 13	Thermochemical	298 K	166
$Mn(CH_3C_5H_4)(CO)_3^+$-H	71 ± 3	297 ± 13	Thermochemical	298 K	166
$Mn(CH_3C_5H_4)(CO)_3^+$-H	59.5 ± 3.4	249 ± 14	Theory	0 K	269
$Mn(H)(CO)_5^+$-H	87 ± 3	364 ± 13	Thermochemical	298 K	166
Mn^+-H_2O	28.4 ± 1.4	119 ± 6	Ion Beam	0 K	67
Mn^+-H_2O	28.5	119	Theory	0 K	128
Mn^+-H_2O	26.5	111	CID	298 K	315
Mn^+-H_2O	32.5	136	CID	298 K	294
$Mn(H_2O)^+$-H_2O	21.4 ± 1.2	90 ± 5	Ion Beam	0 K	67
$Mn(H_2O)^+$-H_2O	21.5	90.0	Theory	0 K	128
$Mn(H_2O)^+$-H_2O	17.8	74.5	CID	298 K	315

BOND	KCAL/MOL	KJ/MOL	METHOD	T	REFERENCES
$Mn(H_2O)_2^+$-H_2O	25.8 ± 1.4	108 ± 6	Ion Beam	0 K	67
$Mn(H_2O)_3^+$-H_2O	12.0 ± 1.2	50 ± 5	Ion Beam	0 K	67
Mn^+-He	0.23	.96	Theory	0 K	124, 143
Mn^+-I	>50.5	> 211	Ion Molecule Reactions	298 K	298
Mn^+-Mn	30.0	126	Theory	0 K	132
Mn^+-NH_2	61 ± 5	255 ± 21	Ion Beam	0 K	39
Mn^+-NH_2	42.6	178	Theory	0 K	281
Mn^+-NH_3	35.5	149	Theory	0 K	130
Mn^+-NH_3	36.9	154	CID	298 K	315
$Mn(NH_3)^+$-NH_3	28.1	118	Theory	0 K	130
$Mn(NH_3)^+$-NH_3	40.8	171	CID	298 K	315
$Mn(NH_3)^+$-NH_3	34.1	143	CID	298 K	315
$Mn(NH_3)_2^+$-NH_3	11.8	49	CID	298 K	315
Mn^+-Ne	0.53	2.2	Theory	0 K	143
Mn^+-O	68.0 ± 3	285 ± 13	Ion Beam	0 K	38
Mn^+-O	57.17 ± 2.31	239 ± 10	Ion Beam	0 K	184
Mn^+-O	58.32 ± 13.82	244 ± 58	Thermochemical	298 K	184
Mn^+-OH	79 ± 6	331 ± 25	Ion Beam	0 K	39
Mn^+-OH	74	310	Theory	0 K	306
Mn^+-OH	82.0	343	CID	298 K	294
Mn^+-OH	81 ± 4	339 ± 17	Ion Molecule Reactions	298 K	306
Mn-CH_2	101.6 ± 3	425 ± 13	Theory	0 K	269
Mn-CH_3	>8.3 ± 2.8	> 35 ± 12	Ion Beam	0 K	43, 10
Mn-Co	10.99 ± 1.91	46 ± 8	CID	298 K	292
Mn-H	38.9 ± 1.5	163 ± 6	Ion Beam	0 K	15, 43
Mn-O	82.8 ± 6	346 ± 25	Ion Beam	0 K	40
$Mn_2(CO)_{10}(H)^+$-CO	19.11 ± 1.67	80 ± 7	CID	298 K	291
Mo^+-CH_2	71 ± 4	297 ± 17	Theory	0 K	123
Mo^+-CH_2	71	297	Theory	0 K	267
Mo^+-CH_3	31.9	133	Theory	0 K	121
Mo^+-CH_3	30.2	126	Theory	298 K	173
Mo^+-CH_3	18.3	76.6	Theory	298 K	173
$Mo(CH_3)^+$-CH_3	37.4	156	Theory	0 K	122
Mo^+-CO	18.8	78.7	Theory	0 K	131
$Mo(CO)^+$-CO	21.1	88.3	Theory	0 K	131
Mo^+-C_2H_2	24	100	Theory	0 K	124
Mo^+-H	39.7 ± 1.4	166 ± 6	Ion Beam	0 K	99
Mo^+-H	35.3	148	Theory	0 K	120
Mo^+-H	67.8 ± 3	284 ± 13	Theory	0 K	269
Mo^+-H	19.6	82.1	Theory	298 K	172
$Mo(CO)_6^+$-H	67.8 ± 3	284 ± 13	Theory	0 K	269
$Mo(CO)_6^+$-H	65 ± 3	272 ± 13	Thermochemical	298 K	166
$Mo(H)^+$-H	35.1	147	Theory	298 K	176
Mo^+-O	71	297	Theory	0 K	267

BOND	KCAL/MOL	KJ/MOL	METHOD	T	REFERENCES
Mo^+-O	79	331	Theory	0 K	267
Mo-H	46.0 ± 3	192 ± 13	Ion Molecule Reactions	298 K	236
$Mo(CO)_5^-$-CO	41.49 ± 2.31	174 ± 10	CID	298 K	290
$Mo(CO)_3(O)_2^-$-$(CO)_2$	9.68 ± 1.15	41 ± 5	CID	298 K	290
$Mo(O)_3^-$-O	>73	> 305	Ion Molecule Reactions	298 K	208
$Mo(O)_4^-$-O	<50	< 209	Ion Molecule Reactions	298 K	208
$Mo(CO)_4^-$-O_2	13.83 ± 1.15	57.9 ± 4.8	CID	298 K	290
$Mo(O)_5^-$-O_2	35.04 ± 4.6	147 ± 19	CID	298 K	290
Na^+-Ar	3.30	13.8	Theory	0 K	143
$Na(Ar)^+$-Ar	3.16	13.2	Theory	0 K	142
Na^+-H_2	3.1	13	Theory	0 K	134
$Na(H_2)^+$-H_2	2.6	11	Theory	0 K	134
$Na(H_2)_2^+$-H_2	2.2	9.2	Theory	0 K	134
Na^+-H_2O	22.6 ± 1.8	95 ± 8	Ion Beam	0 K	111
Na^+-H_2O	23.3	97.6	Theory	0 K	127
Na^+-H_2O	21	88	CID	298 K	315
Na^+-H_2O	24.0	100	Equilibrium	298 K	286
$Na(H_2O)^+$-H_2O	19.6 ± 1.4	82 ± 6	Ion Beam	0 K	111
$Na(H_2O)^+$-H_2O	20.6	86.2	Theory	0 K	127
$Na(H_2O)^+$-H_2O	14.3	59.9	CID	298 K	315
$Na(H_2O)^+$-H_2O	19.8	82.9	Equilibrium	298 K	286
$Na(H_2O)_2^+$-H_2O	16.8 ± 1.4	70 ± 6	Ion Beam	0 K	111
$Na(H_2O)_2^+$-H_2O	16.6	69.5	Theory	0 K	127
$Na(H_2O)_2^+$-H_2O	15.8	66.2	Equilibrium	298 K	286
$Na(H_2O)_3^+$-H_2O	13.1 ± 1.4	55 ± 6	Ion Beam	0 K	111
$Na(H_2O)_3^+$-H_2O	13.9	58.2	Theory	0 K	127
$Na(H_2O)_3^+$-H_2O	13.0	54.4	Equilibrium	298 K	286
Na^+-He	0.76	3.2	Theory	0 K	143
Na^+-NH_3	28.4	119	CID	298 K	315
Na^+-NH_3	29.1	122	Equilibrium	298 K	214
$Na(NH_3)^+$-NH_3	22.9	95.9	Equilibrium	298 K	214
$Na(NH_3)_2^+$-NH_3	17.1	71.6	Equilibrium	298 K	214
$Na(NH_3)_3^+$-NH_3	14.7	61.6	Equilibrium	298 K	214
Na^+-N_2	7.2	30	Theory	0 K	134
Na^+-Ne	1.45	6.07	Theory	0 K	143
Nb^+-C	134.8 ± 31	564 ± 130	Photodissociation	298 K	243
Nb^+-CH	145 ± 8	607 ± 33	Photodissociation	298 K	243
Nb^+-CH_2	89 ± 3	372 ± 13	Theory	0 K	123
Nb^+-CH_2	109 ± 7	456 ± 29	Photodissociation	298 K	243
Nb^+-CH_3	49.5	207	Theory	0 K	121
$Nb(CH_3)^+$-CH_3	53.5	224	Theory	0 K	122
Nb^+-CH_3NH_2	32.1	134	van't Hoff Plot	298 K	210

BOND	KCAL/MOL	KJ/MOL	METHOD	T	REFERENCES
Nb^+-CO	28.6	120	Theory	0 K	131
$Nb(CO)^+$-CO	20.7	86.7	Theory	0 K	131
Nb^+-C_2H_2	59	247	Theory	0 K	124
Nb^+-C_2H_2	57 ± 3	238±13	Photodissociation	298 K	255
Nb^{2+}-C_2H_4	62.2	260	Theory	0 K	136
Nb^+-C_6H_4	≥79	≥ 331	Ion Molecule Reactions	298 K	248
Nb^+-C_6H_6	52.1	218	Theory	0 K	126
Nb^+-C_6H_6	64 ± 3	276 ± 29	Ion Molecule Reactions	298 K	247
Nb^+-C_7H_8	61 ± 12	255 ± 50	Ion Molecule Reactions	298 K	247
Nb^+-H	52.6 ± 1.6	220 ± 7	Ion Beam	0 K	99
Nb^+-H	52.5	220	Theory	0 K	120
Nb^+-H	71.3 ± 3	298 ± 13	Theory	0 K	269
Nb^+-H	39.3	164	Theory	298 K	172
Nb^+-H	49.6	208	Theory	298 K	172
$Nb(CH)^+$-H	64 ± 5	268 ± 21	Photodissociation	298 K	243
Nb^+-Nb	136.0	569.4	Ion Beam	0 K	10, 79
Nb-Nb	121.7	509.5	Ion Beam	0 K	87
Nb^{2+}-CH_2	94.6	396	Theory	0 K	136
Nb^{2+}-C_2H_2	60.2	252	Theory	0 K	136
Nb_2^+-CH_2	197 ± 10	824 ± 42	Charge Transfer	298 K	252
Nb_2^+-Nb	117.1	490.3	Ion Beam	0 K	10, 79
Nb_2-Nb	107.1	448.4	Ion Beam	0 K	87
Nb_3^+-Nb	141.0	590.3	Ion Beam	0 K	10, 79
Nb_3-Nb	136.5	571.5	Ion Beam	0 K	87
Nb_4^+-Nb	133.1	557.3	Ion Beam	0 K	10, 79
Nb_4-Nb	128.8	539.3	Ion Beam	0 K	87
Nb_5^+-Nb	135.8	568.6	Ion Beam	0 K	10, 79
Nb_5-Nb	134.1	561.4	Ion Beam	0 K	87
Nb_6^+-Nb	155.4	650.6	Ion Beam	0 K	10, 79
Nb_6-Nb	154.6	647.3	Ion Beam	0 K	87
Nb_7^+-N_2	<51.4	< 215	Ion Molecule Reactions	298 K	202
Nb_7^+-Nb	141.5	592.4	Ion Beam	0 K	10, 78
Nb_7-Nb	145.8	610.4	Ion Beam	0 K	87
Nb_8^+-N_2	<41.97	< 176	Ion Molecule Reactions	298 K	202
Nb_8^+-Nb	138.1	578	Ion Beam	0 K	10, 78
Nb_8-Nb	130.5	546.4	Ion Beam	0 K	87
Nb_9^+-Nb	144.4	604.6	Ion Beam	0 K	10, 78
Nb_9-Nb	150.8	631.4	Ion Beam	0 K	87
Nb_{10}^+-Nb	143.4	600.4	Ion Beam	0 K	10, 78
Nb_{10}-Nb	126.2	528.4	Ion Beam	0 K	87
Ni^+-Ar	11.35	47.52	Theory	0 K	143
Ni^+-Ar	12.68	53.09	Photodissociation	298 K	199
$Ni(Ar)^+$-Ar	11.83	49.53	Theory	0 K	143

BOND	KCAL/MOL	KJ/MOL	METHOD	T	REFERENCES
Ni^{+}-Br	>69	> 289	Ion Molecule Reactions	298 K	2
Ni^{+}-CF_2	47 ± 7	197 ± 29	Ion Beam	0 K	178
Ni^{+}-CH_2	73.1 ± 1	306 ± 4	Ion Beam	0 K	46
Ni^{+}-CH_2	65	272	Theory	0 K	265
Ni^{+}-CH_2	77 ± 4	322 ± 17	Theory	0 K	123
Ni^{+}-CH_3	44.7 ± 1.4	187 ± 6	Ion Beam	0 K	45
Ni^{+}-CH_3	39.8	167	Theory	0 K	121
Ni^{+}-CH_3	60	251	Theory	0 K	265
Ni^{+}-CH_3	<56	< 234	Ion Molecule Reactions	298 K	2
$Ni(CH_3)^{+}$-CH_3	45.8	192	Theory	0 K	122
Ni^{+}-CO	42.5 ± 2.2	178 ± 9	Ion Beam	0 K	71
Ni^{+}-CO	41.7 ± 2.5	174 ± 11	Ion Beam	0 K	71
Ni^{+}-CO	38 ± 5	159 ± 21	KERD/Theory	0 K	191
Ni^{+}-CO	36.7	154	Theory	0 K	131
$Ni(CO)^{+}$-CO	40.9 ± 1	164 ± 7	Ion Beam	0 K	71
$Ni(CO)^{+}$-CO	40.1 ± 2.5	168 ± 11	Ion Beam	0 K	71
$Ni(CO)^{+}$-CO	34.7	145	Theory	0 K	131
$Ni(CO)_2^{+}$-CO	22.5 ± 1	92 ± 8	Ion Beam	0 K	71
$Ni(CO)_2^{+}$-CO	21.9 ± 1.4	92 ± 6	Ion Beam	0 K	71
$Ni(CO)_3^{+}$-CO	17.8 ± 1.2	74 ± 5	Ion Beam	0 K	71
$Ni(CO)_3^{+}$-CO	17.3 ± 0.7	72 ± 3	Ion Beam	0 K	71
Ni^{+}-CO_2	24.9 ± 0.2	104±1	Photodissociation	0 K	206
Ni^{+}-C_2H_2	>2 ± 4	> 8	Ion Beam	0 K	46
Ni^{+}-C_2H_2	39	163	Theory	0 K	124
Ni^{+}-C_2H_2	56	234	Theory	0 K	265
Ni^{+}-C_2H_2	57	238	Theory	0 K	265
Ni^{+}-C_2H_4	≥33.0 ± 4.5	≥ 138 ± 19	Ion Beam	0 K	46
Ni^{+}-$CHCH_3$	53 ± 6	222 ± 25	Ion Beam	0 K	45
Ni^{+}-C_2H_4	42	176	Theory	0 K	125
Ni^{+}-C_2H_5	53.1 ± 5.8	222 ± 24	Ion Beam	0 K	45
Ni^{+}-C_2H_6	28 ± 5	117 ±21	KERD/Theory	0 K	191
Ni^{+}-C_5H_5	>41	> 172	Ion Molecule Reactions	298 K	1
Ni^{+}-C_6H_6	58.1 ± 2.5	243 ± 10	Ion Beam	0 K	114
Ni^{+}-C_6H_6	59.3	248	Theory	0 K	126
$Ni(C_6H_6)^{+}$-C_6H_6	35.1 ± 2.8	147 ± 12	Ion Beam	0 K	114
Ni^{+}-Cl	45.0 ± 0.9	188 ± 4	Ion Beam	0 K	52
Ni^{+}-F	≥109	≥ 456	Ion Molecule Reactions	298 K	2
Ni^{+}-H	38.7 ± 1.8	162 ± 8	Ion Beam	0 K	42
Ni^{+}-H	41.0	172	Theory	0 K	120
Ni^{+}-H	38.5 ± 1.4	161 ± 6	Theory	0 K	269
Ni^{+}-H	<71	< 297	Ion Molecule Reactions	298 K	2
Ni^{+}-H	12.1	50.7	Theory	298 K	177
$Ni(CO)_4^{+}$-H	38.5 ± 1.4	161 ± 6	Theory	0 K	269

Bond	kcal/mol	kJ/mol	Method	T	References
$Ni(CO)_4^+$-H	54 ± 3	226 ± 13	Thermochemical	298 K	166
$Ni(C_5H_5)(NO)^+$-H	78 ± 3	326 ± 13	Thermochemical	298 K	166
Ni^+-H_2O	43.1 ± 0.7	180 ± 3	Ion Beam	0 K	67
Ni^+-H_2O	41.1	172	Theory	0 K	128
Ni^+-H_2O	39.7	166	CID	298 K	315
Ni^+-H_2O	36.5	153	CID	298 K	294
$Ni(H_2O)^+$-H_2O	40.1 ± 1.8	168 ± 8	Ion Beam	0 K	67
$Ni(H_2O)^+$-H_2O	38.2	160	Theory	0 K	128
$Ni(H_2O)^+$-H_2O	40.6	170	CID	298 K	315
$Ni(H_2O)_2^+$-H_2O	16.1 ± 1.4	67 ± 5	Ion Beam	0 K	67
$Ni(H_2O)_3^+$-H_2O	12.5 ± 1.4	52 ± 6	Ion Beam	0 K	67
Ni^+-He	3.12 ± 0.1	13 ± 0	Equilibrium	0 K	185
Ni^+-He	4.45	18.6	Theory	0 K	143
$Ni(He)^+$-He	3.42 ± 0.1	14.3 ± 0.4	Equilibrium	0 K	185
$Ni(He)^+$-He	4.64	19.4	Theory	0 K	143
$Ni(He)_2^+$-He	1.34 ± 0.1	5.6 ± 0.4	Equilibrium	0 K	185
Ni^+-I	>71	> 297	Ion Molecule Reactions	298 K	2
Ni^+-NH_2	53.2 ± 1.8	223 ± 8	Ion Beam	0 K	58
Ni^+-NH_3	54.6	228	Theory	0 K	130
Ni^+-NH_3	51.2	214	CID	298 K	315
$Ni(NH_3)^+$-NH_3	50.0	209	Theory	0 K	130
$Ni(NH_3)^+$-NH_3	55.1	231	CID	298 K	315
$Ni(NH_3)_2^+$-NH_3	17.8	74.5	CID	298 K	315
Ni^+-NO	54.4 ± 1.8	228 ± 8	Ion Beam	0 K	71
Ni^+-NO	43 ± 5	180 ± 21	Ion Molecule Reactions	298 K	235
$Ni(NO)^+$-NO	29.3 ± 1.6	123 ± 7	Ion Beam	0 K	71
$Ni(NO)_2^+$-NO	27.4 ± 1.2	115 ± 5	Ion Beam	0 K	71
Ni^+-N_2	26.5 ± 2.5	111 ± 11	Ion Beam	0 K	71
$Ni(N_2)^+$-N_2	26.5 ± 2.5	111 ± 11	Ion Beam	0 K	71
$Ni(N_2)_2^+$-N_2	13.4 ± 0.9	56 ± 4	Ion Beam	0 K	71
$Ni(N_2)_3^+$-N_2	10.1 ± 2.3	42 ± 10	Ion Beam	0 K	71
$(Ni^+)^*$-Ne	0.73 ± 0.2	3 ± 1	Equilibrium	0 K	185, 196
Ni^+-Ne	2.37 ± 0.1	9.92 ± 0.4	Equilibrium	0 K	185
Ni^+-Ne	3.37	14.1	Theory	0 K	143
$Ni(Ne)^+$-Ne	2.16 ± 0.1	9.0 ± 0.4	Equilibrium	0 K	185
$Ni(Ne)^+$-Ne	3.22	13.5	Theory	0 K	143
Ni^+-Ni	48.8	204	Ion Beam	0 K	10, 78
Ni^+-Ni	53.5 ± 0.46	224±2	Photodissociation	0 K	207
Ni^+-Ni	52	218	Theory	0 K	133
Ni^+-Ni	69.2	290	Ion Molecule Reactions	<65	199
Ni^+-O	63.3 ± 1.2	265 ± 5	Ion Beam	0 K	38, 46
Ni^+-O	44.95 ± 2.31	188 ± 10	Ion Beam	0 K	184
Ni^+-OH	56.3 ± 4.6	236 ± 19	Ion Beam	0 K	39

Bond	kcal/mol	kJ/mol	Method	T	References
Ni^+-OH	42.2	177	CID	298 K	294
Ni^+-OH	<91	< 381	Ion Molecule Reactions	298 K	2
Ni^+-S	60 ± 5	251 ± 21	Photodissociation	298 K	241
Ni^+-Si	77.0 ± 1.6	322 ± 7	Ion Beam	0 K	116
Ni^+-$Si(CH_3)_2$	>69	> 289	Ion Beam	0 K	170
Ni^+-$SiHCH_3$	>60	> 251	Ion Beam	0 K	170
Ni^+-SiH	77.9 ± 3.5	326 ± 15	Ion Beam	0 K	116
Ni^+-SiH_2	67 ± 6	280 ± 25	Ion Beam	0 K	170
Ni^+-SiH_2	≥55.1 ± 1.6	≥ 231 ± 7	Ion Beam	0 K	116
Ni^+-SiH_3	44.0 ± 2.8	184 ± 12	Ion Beam	0K	116
Ni-CH_2	117 ± 6	490 ± 25	Theory	0 K	269
Ni-CH_2	65	272	Theory	298 K	261
Ni-CH_3	49.7 ± 1.8	208 ± 8	Ion Beam	0 K	45, 46
Ni-CH_3	60	251	Theory	298 K	261
Ni-CO	27	113	Theory	0 K	265
Ni_2-CO	34	142	Theory	0 K	265
Ni-CO	40.5 ± 5.8	169 ± 24	CID	298 K	316
Ni-C_2H_4	14.2	59.4	Theory	298 K	262
Ni-H	56.5 ± 2	236 ± 8	Ion Beam	0 K	15, 45, 46
Ni-H	64	268	Theory	0 K	265
Ni-Ni	47.8	200	Ion Beam	0 K	91
Ni-O	91	381	Theory	0 K	265
Ni-OH	50	209	Theory	0 K	265
Ni-S	76	318	Theory	0 K	265
Ni-C_2H_4	35.2 ± 4.2	147 ± 18	Kinetic Measurements	0 K	295
Ni_2^+-Ni	56.2	235	Ion Beam	0 K	10, 78
Ni_2-C	91	381	Theory	0 K	265
Ni_2-CH_2	122	510	Theory	0 K	265
Ni_2-CO	34	142	Theory	0 K	265
Ni_2-Ni	20.8	87.1	Ion Beam	0 K	86
Ni_2-O	101	423	Theory	0 K	265
Ni_2-S	124	519	Theory	0 K	265
Ni_3^+-Ni	47.6	199	Ion Beam	0 K	10, 78
Ni_3-Ni	37.8	158	Ion Beam	0 K	86
Ni_4^+-Ni	53.3	223	Ion Beam	0 K	10, 78
Ni_4-Ni	65.0	272	Ion Beam	0 K	86
Ni_5^+-Ni	65.7	275	Ion Beam	0 K	10, 78
Ni_5-Ni	79.8	334	Ion Beam	0 K	93
Ni_6^+-Ni	70.7	296	Ion Beam	0 K	10, 78
Ni_6-Ni	54.3	227	Ion Beam	0 K	86
Ni_7^+-Ni	63.3	265	Ion Beam	0 K	10, 78
Ni_7-Ni	64.8	271	Ion Beam	0 K	86
Ni_8^+-Ni	67.6	283	Ion Beam	0 K	10, 78
Ni_8-Ni	65.5	274	Ion Beam	0 K	86

BOND	KCAL/MOL	KJ/MOL	METHOD	T	REFERENCES
Ni_9^+-Ni	67.6	283	Ion Beam	0 K	10, 78
Ni_9-Ni	65.7	275	Ion Beam	0 K	86
Ni_{10}^+-Ni	70.0	293	Ion Beam	0 K	10, 78
Ni_{10}-Ni	67.6	283	Ion Beam	0 K	86
Ni_{11}^+-Ni	80.1	335	Ion Beam	0 K	10, 78
Ni_{11}-Ni	79.8	334	Ion Beam	0 K	86
Ni_{12}^+-Ni	84.1	352	Ion Beam	0 K	10, 78
Ni_{12}-Ni	84.4	353	Ion Beam	0 K	86
Ni_{13}^+-Ni	71.5	299	Ion Beam	0 K	10, 78
Ni_{13}-Ni	71.2	298	Ion Beam	0 K	86
Ni_{14}^+-Ni	79.8	334	Ion Beam	0 K	10, 78
Ni_{14}-CO	29.7	124	Theory	298 K	263
Ni_{14}-Ni	79.6	333	Ion Beam	0 K	86
Ni_{15}^+-Ni	80.8	338	Ion Beam	0 K	10, 78
Ni_{15}-Ni	81.3	340	Ion Beam	0 K	86
Ni_{16}^+-Ni	84.6	354	Ion Beam	0 K	10, 78
Ni_{16}-Ni	84.8	355	Ion Beam	0 K	86
Ni_{17}^+-Ni	81.5	341	Ion Beam	0 K	10, 78
Ni_{17}-Ni	81.0	339	Ion Beam	0 K	86
Os^+-CH_2	113 ± 3	473 ± 13	Theory	0 K	259
Os^+-H	56.2	235	Theory	0 K	273
Pb^+-CH_3NH_2	35.4	148	van't Hoff Plot	298 K	210
Pb^+-CH_3OH	23.3	97.6	van't Hoff Plot	298 K	210
Pb^+-C_6H_6	26.2 ± 0.4	110 ± 2	van't Hoff Plot	298 K	211
$Pb(H_2O)^+$-H_2O	16.9	70.8	van't Hoff Plot	298 K	216
$Pb(H_2O)_2^+$-H_2O	12.2	51.1	van't Hoff Plot	298 K	216
$Pb(H_2O)_3^+$-H_2O	10.8	45.2	van't Hoff Plot	298 K	216
$Pb(H_2O)_4^+$-H_2O	10.0	41.9	van't Hoff Plot	298 K	216
$Pb(H_2O)_5^+$-H_2O	9.6	40	van't Hoff Plot	298 K	216
$Pb(H_2O)_6^+$-H_2O	9.4	39	van't Hoff Plot	298 K	216
$Pb(H_2O)_7^+$-H_2O	9.2	38	van't Hoff Plot	298 K	216
Pb^+-NH_3	28.3 ± 0.2	118 ± 1	van't Hoff Plot	298 K	212
$Pb(NH_3)^+$-NH_3	19.2 ± 0.1	80.4 ± 0.4	van't Hoff Plot	298 K	212
$Pb(NH_3)_2^+$-NH_3	13.0 ± 0.1	54.4 ± 0.4	van't Hoff Plot	298 K	212
$Pb(NH_3)_3^+$-NH_3	10.7 ± 0.2	44.8 ± 0.8	van't Hoff Plot	298 K	212
Pd^+-H	50.1	210	Theory	0 K	272
Pd^+-CH_2	70 ± 3	293 ± 13	Theory	0 K	123
Pd^+-CH_3	59 ± 5	247 ± 21	Ion Beam	0 K	168
Pd^+-CH_3	47.2	197	Theory	0 K	121
Pd^+-CH_3	35.8	150	Theory	0 K	272
Pd^+-CH_3	47.1	197	Theory	298 K	173
Pd^+-CH_3	35.0	146	Theory	298 K	173
$Pd(CH_3)^+$-CH_3	36.1	151	Theory	0 K	122

Bond	kcal/mol	kJ/mol	Method	T	References
$Pd(PH_3)_2^+$-CH_3	22.3	93.4	Theory	0 K	272
Pd^+-CO	31.5	132	Theory	0 K	131
$Pd(CO)^+$-CO	27.6	115	Theory	0 K	131
Pd^+-C_2H_2	36	151	Theory	0 K	124
Pd^+-H	47.7 ± 0.9	200 ± 4	Ion Beam	0 K	98
Pd^+-H	49.0	205	Theory	0 K	120
Pd^+-H	45 ± 3	188 ± 13	Theory	0 K	269
Pd^+-O	33.7 ± 2.5	141 ± 11	Ion Beam	0 K	106
Pr^+-O	>179	> 749	Ion Molecule Reactions	298 K	191
Pt^+-CH_2	123 ± 5	515 ± 21	Theory	0 K	259
Pt^+-CH_2	115 ± 4	481 ± 17	Ion Molecule Reactions	298 K	310
Pt^+-CH_3	53.0	222	Theory	0 K	272
$Pt(Cl)_2(PH_3)_2^+$-CH_3	31.7	133	Theory	0 K	273
$Pt(PH_3)_2^+$-CH_3	39.6	166	Theory	0 K	272
Pt^+-H	62.9	263	Theory	0 K	273
$Pt(PH_3)_2^+$-H	58.3	244	Theory	0 K	272
Re^+-CH_2	97 ± 5	406 ± 21	Theory	0 K	259
Re^+-H	44.5	186	Theory	0 K	273
$Re(CH_3)(CO)_5^+$-H	67.1	281	Theory	0 K	269
$Re(CH_3)(CO)_5^+$-H	73 ± 3	305 ± 13	Thermochemical	298 K	166
Rh^+-C	98.0 ± 4.2	410 ± 17	Ion Beam	0 K	109
Rh^+-C	>120	> 502	Photodissociation	298 K	243
Rh^+-CH	106.1 ± 2.8	444 ± 12	Ion Beam	0 K	95
Rh^+-CH	102 ± 7	427 ± 29	Photodissociation	298 K	243
Rh^+-CH_2	85.1 ± 1.8	356 ± 8	Ion Beam	0 K	109
Rh^+-CH_2	84 ± 4	351 ± 17	Theory	0 K	123
Rh^+-CH_2	91 ± 5	393 ± 21	Ion Molecule Reactions	298 K	234
Rh^+-CH_2	89 ± 5	372 ± 21	Photodissociation	298 K	243
Rh^+-CH_3	47 ± 5	197 ± 21	Ion Beam	0 K	168
Rh^+-CH_3	33.9 ± 1.4	142 ± 6	Ion Beam	0 K	109
Rh^+-CH_3	37.1	155	Theory	0 K	121
$Rh(CH_3)^+$-CH_3	43.9	184	Theory	0 K	122
Rh^+-CO	30.8	129	Theory	0 K	131
$Rh(CO)^+$-CO	28.1	118	Theory	0 K	131
Rh^+-C_2H_2	38.5 ± 0.7	161 ± 3	Ion Beam	0 K	109
Rh^+-C_2H_2	33	138	Theory	0 K	124
Rh^+-C_2H_4	≥ 30.9	≥ 129	Ion Beam	0 K	109
Rh^+-C_2H_5	41.5 ± 4.2	174 ± 17	Ion Beam	0 K	109
Rh^+-$(CH_3)_2$	70.0	293	Theory	0 K	274
Rh^+-C_3H_4	53.0 ± 4.6	222 ± 19	Ion Beam	0 K	109
Rh^+-C_3H_4	52 ± 5	218 ± 20	Ion Beam	0 K	109
Rh^+-C_3H_6	28.1	118	Ion Beam	0 K	109
Rh^+-H	42 ± 3	176 ± 13	Ion Beam	0 K	168

BOND	KCAL/MOL	KJ/MOL	METHOD	T	REFERENCES
Rh^+-H	38.5 ± 0.9	161 ± 4	Ion Beam	0 K	98
Rh^+-H	41.5	174	Theory	0 K	120
Rh^+-H	50.1 ± 3	210 ± 13	Theory	0 K	269
Rh^+-H	22.5	94.2	Theory	298 K	172
Rh^+-H	35.7	149	Theory	298 K	172
$Rh(CH)^+$-H	89 ± 5	372 ± 21	Photodissociation	298 K	243
$Rh(C_5H_5)(CO)_2^+$-H	80 ± 5	335 ± 21	Thermochemical	298 K	166
$Rh(C_5H_8)(CO)_2^+$-H	50.1 ± 3	210 ± 13	Theory	0 K	269
Rh^+-NO	40 ± 5	167 ± 21	Ion Molecule Reactions	298 K	235
Rh^+-O	69.6 ± 1.4	291 ± 6	Ion Beam	0 K	106
Rh-CH_2	94 ± 5	393 ± 21	Theory	0 K	269
Rh-H	55.8 ± 1.4	233 ± 6	Ion Beam	0 K	109
Ru^+-CH_2	80 ± 4	335 ± 17	Theory	0 K	123
Ru^+-CH_2	68	285	Theory	0 K	266
$Ru(H)(Cl)^+$-CH_2	84.7	354	Theory	0 K	268
$Ru(H)(Cl)^+$-CH_2	88.2	369	Theory	0 K	269
Ru^+-CH_3	54 ± 5	226 ± 21	Ion Beam	0 K	168
Ru^+-CH_3	39.6	166	Theory	0 K	121
$Ru(CH_3)^+$-CH_3	45.7	191	Theory	0 K	122
$Ru(Cl)^+$-CH_3	54.3	227	Theory	0 K	268
Ru^+-CO	30.9	129	Theory	0 K	131
$Ru(CO)^+$-CO	29.0	121	Theory	0 K	131
Ru^+-C_2H_2	32	134	Theory	0 K	124
Ru^+-C_6H_6	48.7	204	Theory	0 K	126
Ru^+-H	41 ± 3	172 ± 13	Ion Beam	0 K	168
Ru^+-H	37.4 ± 1.2	156 ± 5	Ion Beam	0 K	98
Ru^+-H	37.8	158	Theory	0 K	120
Ru^+-H	39.5	165	Theory	0 K	268
Ru^+-H	55.6 ± 3	233 ± 13	Theory	0 K	269
Ru^+-H	18.6	77.9	Theory	298 K	172
Ru^+-H	47.2	197	Theory	298 K	172
$Ru(CH_2)(Cl)^+$-H	55.6 ± 3	233 ± 13	Theory	0 K	269
$Ru(C_5H_5)_2^+$-H	79 ± 5	331 ± 21	Thermochemical	298 K	166
$Ru(C_5H_8)_2^+$-H	55.6 ± 3	233 ± 13	Theory	0 K	269
$Ru(Cl)^+$-H	54.1	226	Theory	0 K	268
Ru^+-O	87.9 ± 1.2	368 ± 5	Ion Beam	0 K	106
Ru^+-O	67.1	281	Theory	0 K	270
Ru-CH_2	88.2	369	Theory	0 K	269
Sc^+-Ar	3.44	14.4	Theory	0 K	143
Sc^+-C	77.0 ± 1.4	322 ± 6	Ion Beam	0 K	19
Sc^+-CH	96.0	402	Theory	0 K	278
Sc^+-CH	100	418	Theory	0 K	276
Sc^+-CH_2	96.0 ± 5.3	402 ± 22	Ion Beam	0 K	25
Sc^+-CH_2	85 ± 3	356 ± 13	Theory	0 K	123

Bond	kcal/mol	kJ/mol	Method	T	References
Sc^+-CH_2	68.0	285	Theory	0 K	278
Sc^+-CH_2	72	301	Theory	0 K	276
Sc^+-CH_2	95 ± 6	397 ± 25	Theory	0 K	269
Sc^+-CH_3	55.8 ± 2.3	246 ± 10	Ion Beam	0 K	18
Sc^+-CH_3	65 ± 5	272 ± 21	Ion Beam	0 K	169
Sc^+-CH_3	52.4	219	Theory	0 K	121
Sc^+-CH_3	40.1	168	Theory	0 K	278
Sc^+-CH_3	49	205	Theory	0 K	276
Sc^+-CH_3	54.5	228	Theory	298 K	173
Sc^+-CH_3	45.5	190	Theory	298 K	173
$Sc(CH_3)^+$-CH_3	55.0 ± 2.6	230 ± 11	Ion Beam	0 K	18, 30
$Sc(CH_3)^+$-CH_3	56.5	236	Theory	0 K	122
$Sc(H)^+$-CH_3	58.2 ± 2.4	244 ± 10	Ion Beam	0 K	30
Sc^+-CO	15.2	63.6	Theory	0 K	131
Sc^+-CO	13.5	56.5	Theory	0 K	284
$Sc(CO)^+$-CO	17.8	74.5	Theory	0 K	131
Sc^+-C_2H_2	57.3 ± 4.6	240 ± 19	Ion Beam	0 K	18
Sc^+-C_2H_2	51	213	Theory	0 K	124
Sc^+-C_2H_2	52 ± 3	218 ± 13	Photodissociation	298 K	254
Sc^+-C_2H_4	≥31.3	≥ 131	Ion Beam	0 K	18
Sc^+-C_2H_4	32	134	Theory	0 K	125
Sc^+-C_3H_4	61 ± 5	255 ± 21	Ion Beam	0 K	30
Sc^+-C_3H_6	41	172	Theory	0 K	140
Sc^+-C_6H_4	94 ± 10	393 ± 42	Theory	0 K	145
Sc^+-C_6H_4	88 ± 5	368 ± 21	Photodissociation	298 K	253
Sc^+-C_6H_6	44.1	185	Theory	0 K	126
Sc^+-Fe	48 ± 5	201 ± 21	Photodissociation	298 K	243
Sc^+-H	56.3 ± 2.1	236 ± 9	Ion Beam	0 K	11
Sc^+-H	54 ± 4	226 ± 17	Ion Beam	0 K	169
Sc^+-H	56.0	234	Theory	0 K	120
Sc^+-H	50.7	212	Theory	0 K	278
Sc^+-H	59.0 ± 2	247 ± 8	Theory	0 K	269
Sc^+-H	56.6	237	Theory	298 K	173
Sc^+-H	47.5	199	Theory	298 K	173
Sc^+-H_2	5.5 ± 0.3	23 ± 1	Equilibrium	0 K	186, 197
$Sc(H_2)^+$-H_2	6.9 ± 0.3	29 ± 1	Equilibrium	0 K	186
$Sc(H_2)_2{}^+$-H_2	5.4 ± 0.3	23 ± 1	Equilibrium	0 K	186
$Sc(H_2)_3{}^+$-H_2	5.0 ± 0.4	21 ± 2	Equilibrium	0 K	186
Sc^+-H_2O	34.5	144	Theory	0 K	128
Sc^+-H_2O	31.4	131	CID	298 K	294
$Sc(H_2O)^+$-H_2O	30.2	126	Theory	0 K	128
Sc^+-He	0.14	0.59	Theory	0 K	143
Sc^+-N	89	372	Theory	0 K	280

BOND	KCAL/MOL	KJ/MOL	METHOD	T	REFERENCES
Sc^+-NH	115.3 ± 2.3	482 ± 10	Ion Beam	0 K	32
Sc^+-NH_2	82.9 ± 2.3	347 ± 10	Ion Beam	0 K	32
Sc^+-NH_2	79	331	Theory	0 K	282
Sc^+-NH_3	39.8	167	Theory	0 K	130
$Sc(NH_3)^+$-NH_3	34.5	144	Theory	0 K	130
Sc^+-Ne	0.35	1.5	Theory	0 K	143
Sc^+-O	164.6 ± 1.3	689 ± 5	Ion Beam	0 K	19, 35, 36
Sc^+-OH	119.2 ± 2.1	499 ± 9	Ion Beam	0 K	35
Sc^+-OH	87.8	367	CID	298 K	294
Sc^+-Si	57.9 ± 2.5	242 ± 11	Ion Beam	0 K	119
Sc^+-SiH	53.7 ± 2.5	225 ± 11	Ion Beam	0 K	119
Sc^+-SiH_2	50.0 ± 1.8	209 ± 8	Ion Beam	0 K	119
Sc^+-SiH_3	40.6 ± 3.7	170 ± 15	Ion Beam	0 K	119
Sc-CH_3	28 ± 7	117 ± 29	Ion Beam	0 K	28
Sc-H	55.2	231	Theory	0 K	260
Sc-H	56.1	235	Theory	298 K	260
$Sc(H)^+$-H	55.4 ± 1.2	232 ± 5	Ion Beam	0 K	11, 18
Sc^{2+}-CH_4	27.6	115	Theory	0 K	146
Sc^{2+}-C_2H_2	50	209	Theory	0 K	140
Sc^{2+}-C_2H_4	51	213	Theory	0 K	140
Sc^{2+}-C_2H_6	39.2	164	Theory	0 K	146
Sc^{2+}-C_3H_6	64	268	Theory	0 K	140
Sc^{2+}-C_3H_8	49.3	206	Theory	0 K	146
Sc^{2+}-OH	108 ± 7	452 ± 29	Charge Stripping FAB	298 K	288
Si^+-H	90.0	377	Photodissociation	298 K	317
Si-Br	96.0	402	Photodissociation	298 K	317
Si-C	88.0	368	Photodissociation	298 K	317
Si-Cl	111.0	464.7	Photodissociation	298 K	317
Si-F	160	669	Photodissociation	298 K	317
Si-I	77	322	Photodissociation	298 K	317
Si-N	100.0	418.7	Photodissociation	298 K	317
Si-O	128.0	536.9	Photodissociation	298 K	317
Si-S	99.0	414	Photodissociation	298 K	317
Si-Si	74.0	310	Photodissociation	298 K	317
Sr^+-CO_2	10.1	42.3	Theory	0 K	154
$Sr(CO_2)^+$-CO_2	8.2	34	Theory	0 K	154
Sr^+-F	131.7	551.4	Theory	0 K	147
Sr^+-H	50.0 ± 1.3	209 ± 5	Ion Beam	0 K	65
Sr^+-H	44.1	185	Theory	0 K	172
Sr^+-H	37.1	155	Theory	298 K	172
Sr^+-H	45.0	188	Theory	298 K	172
Sr^+-H_2O	25 ± 3	105 ± 13	Theory	0 K	158
$Sr(H_2O)^+$-H_2O	23 ± 3	96 ± 13	Theory	0 K	158
$Sr(H_2O)_2^+$-H_2O	17.0	71.2	Theory	0 K	159

Bond	kcal/mol	kJ/mol	Method	T	References
Sr^+-NH_3	25.6	107	Theory	0 K	159
$Sr(NH_3)^+$-NH_3	22.6	94.6	Theory	0 K	159
$Sr(NH_3)_2^+$-NH_3	17.1	71.6	Theory	0 K	159
Sr^+-O	80.0 ± 1.4	335 ± 6	Ion Beam	0 K	105
Sr^+-O	75.6	316	Theory	0 K	147
Sr^+-OH	107.7	450.9	Theory	0 K	147
Sr^{2+}-H_2O	46.9	196	Theory	0 K	159
$Sr(H_2O)^{2+}$-H_2O	42.0	176	Theory	0 K	159
$Sr(H_2O)_2^{2+}$-H_2O	38.0	159	Theory	0 K	159
Ta^+-CH_2	115 ± 5	481 ± 21	Theory	0 K	259
Ta^+-H	54.0	226	Theory	0 K	273
Ta^+-Ta	159.2	666.5	Ion Beam	0 K	80
Ta_2^+-Ta	153.9	644.4	Ion Beam	0 K	80
Ta_3^+-Ta	177.8	744.4	Ion Beam	0 K	80
Tc^+-CH_2	83 ± 5	347 ± 21	Theory	0 K	123
Tc^+-CH_3	47.7	200	Theory	0 K	121
Tc^+-CH_3	43.0	180	Theory	298 K	173
Tc^+-CH_3	30.2	126	Theory	298 K	173
$Tc(CH_3)^+$-CH_3	44.4	186	Theory	0 K	122
Tc^+-CO	17.6	73.7	Theory	0 K	131
$Tc(CO)^+$-CO	26.9	113	Theory	0 K	131
Tc^+-C_2H_2	25	105	Theory	0 K	124
Tc^+-H	50.7	212	Theory	0 K	120
Tc^+-H	67.1	281	Theory	0 K	269
Tc^+-H	47.2	197	Theory	298 K	172
Ti^+-Ar	6.92	29.0	Theory	0 K	144
Ti^+-C	93.4 ± 5.5	391 ± 23	Ion Beam	0 K	19
Ti^+-CH	114.2 ± 1.3	478 ± 5	Ion Beam	0 K	21, 22
Ti^+-CH	133	556	Theory	0 K	276
Ti^+-CH_2	90.9 ± 2.1	380 ± 9	Ion Beam	0 K	21
Ti^+-CH_2	83 ± 3	347 ± 13	Theory	0 K	123
Ti^+-CH_2	82	343	Theory	0 K	276
Ti^+-CH_2	91.7 ± 6	384 ± 25	Theory	0 K	269
Ti^+-CH_3	51.1 ± 0.7	214 ± 3	Ion Beam	0 K	22
Ti^+-CH_3	51.6	216	Theory	0 K	121
Ti^+-CH_3	46	192	Theory	0 K	276
$Ti(CH_3)^+$-CH_3	62 ± 6	259 ± 25	Ion Beam	0 K	22
$Ti(CH_3)^+$-CH_3	45.0	188	Theory	0 K	122
Ti^+-CO	28.1 ± 1.4	118 ± 6	Ion Beam	0 K	102
Ti^+-CO	24.7	103	Theory	0 K	131
Ti^+-CO	8.9	37	Theory	0 K	284
$Ti(CO)^+$-CO	27.0 ± 0.9	113 ± 4	Ion Beam	0 K	102
$Ti(CO)^+$-CO	20.6	86.2	Theory	0 K	131

BOND	KCAL/MOL	KJ/MOL	METHOD	T	REFERENCES
$Ti(CO)_2^+$-CO	24.0 ± 0.9	100 ± 4	Ion Beam	0 K	102
$Ti(CO)_3^+$-CO	20.8 ± 1.2	87 ± 5	Ion Beam	0 K	102
$Ti(CO)_4^+$-CO	16.6 ± 0.9	69 ± 4	Ion Beam	0 K	102
$Ti(CO)_5^+$-CO	17.8 ± 0.7	74 ± 3	Ion Beam	0 K	102
$Ti(CO)_6^+$-CO	12.5 ± 1.6	52 ± 7	Ion Beam	0 K	102
Ti^+-C_2H_2	60.4 ± 4.6	253 ± 19	Ion Beam	0 K	22
Ti^+-C_2H_2	47	197	Theory	0 K	124
Ti^+-C_2H_3	79.9 ± 5.8	334 ± 24	Ion Beam	0 K	22
Ti^+-C_2H_4	≥28.5	≥ 119	Ion Beam	0 K	22
Ti^+-C_2H_4	31	130	Theory	0 K	125
Ti^+-C_2H_5	49.5 ± 1.6	207 ± 7	Ion Beam	0 K	22
Ti^+-C_3H_4	60 ± 3	251 ± 13	Ion Beam	0 K	22
Ti^+-C_4H_6	>28.1	> 118	Ion Molecule Reactions	298 K	7
Ti^+-C_6H_6	61.8 ± 2.1	259 ± 9	Ion Beam	0 K	114
Ti^+-C_6H_6	62.8	263	Theory	0 K	126
$Ti(C_6H_6)^+$-C_6H_6	60.4 ± 4.4	253 ± 18	Ion Beam	0K	114
Ti^+-F	≥109	≥ 456	Ion Molecule Reactions	298 K	9
Ti^+-H	53.3 ± 2.5	223 ± 10	Ion Beam	0 K	12
Ti^+-H	53.3	223	Theory	0 K	120
Ti^+-H	58.9	246	Theory	0 K	285
Ti^+-H	60.8 ± 2	254 ± 8	Theory	0 K	269
Ti^+-H	43.7	183	Theory	298 K	177
Ti^+-H_2O	36.9 ± 1.4	154 ± 6	Ion Beam	0 K	67
Ti^+-H_2O	37.5	157	Theory	0 K	128
Ti^+-H_2O	38.0	159	CID	298 K	294
$Ti(H_2O)^+$-H_2O	32.5 ± 1.2	136 ± 5	Ion Beam	0 K	67
$Ti(H_2O)^+$-H_2O	33.8	141	Theory	0 K	128
$Ti(H_2O)_2^+$-H_2O	15.9 ± 1.6	67 ± 7	Ion Beam	0 K	67
$Ti(H_2O)_3^+$-H_2O	20.1 ± 1.8	84 ± 8	Ion Beam	0 K	67
Ti^+-He	1.15	4.81	Theory	0 K	144
Ti^+-N	116.3 ± 2.8	487 ± 12	Ion Beam	0 K	32
Ti^+-N	123	515	Theory	0 K	280
Ti^+-NH	107.7 ± 2.8	451 ± 12	Ion Beam	0 K	32
Ti^+-NH_2	82.8 ± 3	346 ± 13	Ion Beam	0 K	32
Ti^+-NH_2	69.2	290	Theory	0 K	281
Ti^+-NH_3	44.1	185	Theory	0 K	130
$Ti(NH_3)^+$-NH_3	37.8	158	Theory	0 K	130
Ti^+-Ne	1.53	6.41	Theory	0 K	144
Ti^+-O	158.7 ± 1.6	664 ± 7	Ion Beam	0 K	19
Ti^+-O	140	586	Theory	0 K	267
Ti^+-OH	111.2 ± 2.8	465 ± 12	Ion Beam	0 K	35
Ti^+-OH	113.0	473	CID	298 K	294
Ti^+-Si	58.6 ± 3.9	245 ± 16	Ion Beam	0 K	94

Bond	kcal/mol	kJ/mol	Method	T	References
Ti^+-$Si(CH_3)_2$	>60	> 251	Ion Beam	0 K	170
Ti^+-$SiHCH_3$	>60	> 251	Ion Beam	0 K	170
Ti^+-SiH	53.0 ± 2.5	222 ± 10	Ion Beam	0 K	94
Ti^+-SiH_2	>58	> 243	Ion Beam	0 K	170
Ti^+-SiH_2	50.0 ± 1.6	209 ± 7	Ion Beam	0 K	94
Ti^+-SiH_3	39.0 ± 4.2	163 ± 17	Ion Beam	0 K	94
Ti^+-Ti	56.15	235.1	Ion Beam	0 K	81
Ti-CH_3	42 ± 7	176 ± 29	Ion Beam	0 K	28
Ti-H	45.2 ± 1.4	189 ± 6	Ion Beam	0 K	15, 16
Ti_2^+-Ti	55.0	230	Ion Beam	0 K	10, 73
Ti_3^+-Ti	80.8	338	Ion Beam	0 K	10, 73
Ti_4^+-Ti	81.3	340	Ion Beam	0 K	10, 73
Ti_5^+-Ti	84.6	354	Ion Beam	0 K	10, 73
Ti_6^+-Ti	95.6	400	Ion Beam	0 K	10, 73
Ti_7^+-Ti	66.4	278	Ion Beam	0 K	10, 73
Ti_8^+-Ti	82.7	346	Ion Beam	0 K	10, 73
Ti_9^+-Ti	80.1	335	Ion Beam	0 K	10, 73
Ti_{10}^+-Ti	82.2	344	Ion Beam	0 K	10, 73
Ti_{11}^+-Ti	97.0	406	Ion Beam	0 K	10, 73
Ti_{12}^+-Ti	112.3	470.2	Ion Beam	0 K	10, 73
Ti_{13}^+-Ti	76.7	321	Ion Beam	0 K	10, 73
Ti_{14}^+-Ti	95.6	400	Ion Beam	0 K	10, 73
Ti_{15}^+-Ti	86.0	360	Ion Beam	0 K	10, 73
Ti_{16}^+-Ti	82.2	344	Ion Beam	0 K	10, 73
Ti_{17}^+-Ti	76.0	318	Ion Beam	0 K	10, 73
Ti_{18}^+-Ti	107.6	450.5	Ion Beam	0 K	10, 73
Ti_{19}^+-Ti	97.8	409	Ion Beam	0 K	10, 73
Ti_{20}^+-Ti	94.4	395	Ion Beam	0 K	10, 73
Ti_{21}^+-Ti	98.0	410	Ion Beam	0 K	10, 73
U^+-Cl	>84 ± 1	> 351 ± 4	Ion Beam	0 K	175
U^+-D	66.85 ± 2.3	280 ± 10	Ion Beam	0 K	183
U^+-D	67 ± 2	280 ± 8	Ion Beam	0 K	183
U^+-D	177.51 ± 9.2	743 ± 38	CID	298 K	181
U^+-F	113 ± 12	473 ± 50	Ion Beam	0 K	175
U^+-N	108.3 ± 4.6	453 ± 19	Ion Beam	0 K	183
U^+-O	184 ± 7	770 ± 29	Ion Beam	0 K	182
U^+-O	184.42 ± 6.92	772 ± 29	CID	298 K	182
$U(O)^+$-O	178 ± 9	745 ± 38	Ion Beam	0 K	182
V^+-Ar	5 ± 5	21 ± 21	Ion Beam	0 K	37
V^+-Ar	7.22	30.2	Theory	0 K	143
V^+-Ar	8.76	36.7	Photodissociation	<65K	203
$V(Ar)^+$-Ar	6.85	28.7	Theory	0 K	142
V^+-C	88.4 ± 1.2	370 ± 5	Ion Beam	0 K	19, 20

Bond	kcal/mol	kJ/mol	Method	T	References
V^+-CH	112.4 ± 1.2	470 ± 5	Ion Beam	0 K	20, 23
V^+-CH	108.0	452.2	Theory	0 K	283
V^+-CH	146	611	Theory	0 K	276
V^+-CH_2	77.8 ± 1.4	326 ± 6	Ion Beam	0 K	23
V^+-CH_2	79 ± 3	331 ± 13	Theory	0 K	123
V^+-CH_2	103	431	Theory	0 K	276
V^+-CH_2	132.8 ± 2	556 ± 8	Theory	0 K	269
V^+-CH_3	46.0 ± 1.7	192 ± 7	Ion Beam	0 K	20
V^+-CH_3	43.0	180	Theory	0 K	121
V^+-CH_3	62	259	Theory	0 K	276
$V(CH_3)^+$-CH_3	47.5 ± 2.4	199 ± 10	Ion Beam	0 K	20, 29
$V(CH_3)^+$-CH_3	43.6	182	Theory	0 K	122
$V(H)^+$-CH_3	46 ± 4	192 ± 17	Ion Beam	0 K	29
V^+-CO	27.0 ± 1.4	120 ± 14	Ion Beam	0 K	29
V^+-CO	27.0 ± 0.7	113 ± 3	Ion Beam	0 K	101
V^+-CO	27.2	114	Theory	0 K	131
V^+-CO	11.4	47.7	Theory	0 K	284
$V(CO)^+$-CO	25.3 ± 2	106 ± 8	Ion Beam	0 K	29
$V(CO)^+$-CO	21.7 ± 0.9	91 ± 4	Ion Beam	0 K	101
$V(CO)^+$-CO	23.8	100	Theory	0 K	131
$V(CO)_2^+$-CO	14.6 ± 3	61 ± 13	Ion Beam	0 K	29
$V(CO)_2^+$-CO	16.6 ± 0.9	69 ± 4	Ion Beam	0 K	101
$V(CO)_3^+$-CO	22.7 ± 3.3	95 ± 14	Ion Beam	0 K	29
$V(CO)_3^+$-CO	20.5 ± 2.3	86 ± 10	Ion Beam	0 K	101
$V(CO)_4^+$-CO	21.3 ± 2	93 ± 8	Ion Beam	0 K	29
$V(CO)_4^+$-CO	21.7 ± 0.7	91 ± 3	Ion Beam	0 K	101
$V(CO)_5^+$-CO	24.9 ± 2	124 ± 8	Ion Beam	0 K	29
$V(CO)_5^+$-CO	23.8 ± 1.6	99 ± 7	Ion Beam	0 K	101
$V(CO)_6^+$-CO	12.0 ± 2.1	50 ± 9	Ion Beam	0 K	101
$V(O)^+$-CO	36.9	154	Photodissociation	37K	203
V^+-CO_2	17.3 ± 0.9	72 ± 4	Ion Beam	0 K	104
V^+-C_2	125.3 ± 3.6	524 ± 15	Ion Beam	0 K	20
V^+-C_2H	117.4 ± 1.8	491 ± 8	Ion Beam	0 K	20
V^+-C_2H_2	48.9 ± 4.6	205 ± 19	Ion Beam	0 K	20
V^+-C_2H_2	35	146	Theory	0 K	124
V^+-C_2H_3	88.1 ± 4.6	369 ± 19	Ion Beam	0 K	20
V^+-$CHCH_3$	67 ± 5	280 ± 21	Ion Beam	0 K	31
V^+-C_2H_4	≥23.2 ± 1.4	≥ 97 ± 6	Ion Beam	0 K	29
V^+-C_2H_4	28	117	Theory	0 K	125
V^+-C_2H_5	54 ± 4	226 ± 17	Ion Beam	0 K	31
V^+-C_6H_6	51.1	214	Theory	0 K	126
V^+-C_6H_6	55.8 ± 2.3	233 ± 10	Ion Beam	0K	114
V^+-C_6H_6	62 ± 5	259 ± 21	Photodissociation	298 K	242

Bond	kcal/mol	kJ/mol	Method	T	References
$V(C_6H_6)^+-C_6H_6$	58.8 ± 4.4	246 ± 18	Ion Beam	0K	114
V^+-Fe	75 ± 5	314 ± 21	Photodissociation	298 K	239
V^+-H	47.3 ± 1.4	198 ± 6	Ion Beam	0 K	13
V^+-H	48.6	203	Theory	0 K	120
V^+-H	63.4 ± 1.4	265 ± 6	Theory	0 K	269
V^+-H	33.8	141	Theory	298 K	177
$V(CO)_6{}^+-H$	56 ± 3	234 ± 13	Thermochemical	298 K	166
V^+-H_2	10.2 ± 0.5	43 ± 2	Equilibrium	0 K	187
V^+-H_2	10.2	42.7	Theory	0 K	135
$V(H_2)^+-H_2$	10.7 ± 0.5	45 ± 2	Equilibrium	0 K	187
$V(H_2)^+-H_2$	9.7	41	Theory	0 K	135
$V(H_2)_2{}^+-H_2$	8.8 ± 0.4	37 ± 2	Equilibrium	0 K	187
$V(H_2)_3{}^+-H_2$	9.0 ± 0.4	38 ± 2	Equilibrium	0 K	187
$V(H_2)_4{}^+-H_2$	4.2 ± 0.5	18 ± 2	Equilibrium	0 K	187
$V(H_2)_5{}^+-H_2$	9.6 ± 0.5	40 ± 2	Equilibrium	0 K	187
$V(H_2)_6{}^+-H_2$	<2.5	< 10	Equilibrium	0 K	187
$V(H_2O)^+-H_2$	11.0	46.0	Theory	0 K	135
V^+-H_2O	35.1 ± 1.2	147 ± 5	Ion Beam	0 K	67
V^+-H_2O	34.7	145	Theory	0 K	128
V^+-H_2O	35.1	147	CID	298 K	315
V^+-H_2O	36.2	151	CID	298 K	294
$V(H_2O)^+-H_2O$	36.0 ± 2.3	151 ± 10	Ion Beam	0 K	67
$V(H_2O)^+-H_2O$	37.9	159	Theory	0 K	128
$V(H_2O)^+-H_2O$	35.5	149	CID	298 K	315
$V(H_2O)_2{}^+-H_2O$	16.1 ± 1.2	67 ± 5	Ion Beam	0 K	67
$V(H_2O)_2{}^+-H_2O$	12.1	50.7	CID	298 K	315
$V(H_2O)_3{}^+-H_2O$	16.1 ± 1.8	67 ± 8	Ion Beam	0 K	67
V^+-He	3.04	12.7	Theory	0 K	143
V^+-Kr	9 ± 5	38 ± 21	Ion Beam	0 K	37
V^+-Kr	10.6	44.4	Photodissociation	<65K	203
V^+-N	103.4 ± 1.6	433 ± 7	Ion Beam	0 K	33
V^+-N	131	548	Theory	0 K	280
V^+-NH	95.1 ± 3.7	398 ± 15	Ion Beam	0 K	33
V^+-NH	101 ± 7	423 ± 29	Ion Molecule Reactions	298 K	246
V^+-NH_2	70.0 ± 1.4	293 ± 6	Ion Beam	0 K	33
V^+-NH_2	65.5	274	Theory	0 K	281
V^+-NH_3	43.6	182	Theory	0 K	130
V^+-NH_3	51.9	217	CID	298 K	315
$V(NH_3)^+-NH_3$	40.4	169	Theory	0 K	130
$V(NH_3)^+-NH_3$	45	188	CID	298 K	315
$V(NH_3)_2{}^+-NH_3$	22.6	94.6	CID	298 K	315
$V(NH_3)_3{}^+-NH_3$	18.7	78.3	CID	298 K	315
V^+-Ne	2.01	8.42	Theory	0 K	143

Bond	kcal/mol	kJ/mol	Method	T	References
V^+-O	134.8	564.4	Ion Beam	0 K	19, 37
V^+-OD	101.7 ± 4.4	426 ± 18	Ion Beam	0 K	107
V^+-OH	103.8 ± 3.5	434 ± 15	Ion Beam	0 K	35
V^+-OH	107.0	448.0	CID	298 K	294
V^+-Si	54.7 ± 3.5	228 ± 15	Ion Beam	0 K	94
V^+-$SiHCH_3$	>60	> 251	Ion Beam	0 K	170
V^+-$Si(CH_3)_2$	>60	> 251	Ion Beam	0 K	170
V^+-SiH	48.2 ± 2.8	202 ± 12	Ion Beam	0 K	94
V^+-SiH_2	46.6 ± 2.1	195 ± 9	Ion Beam	0 K	94
V^+-SiH_3	35.5 ± 3.7	149 ± 15	Ion Beam	0 K	94
V^+-V	72.41	303.2	Ion Beam	0 K	81
V^+-Xe	19 ± 4	79 ± 17	Ion Beam	0 K	37
V^+-Xe	19.1 ± 1.6	80 ± 7	Ion Beam	0 K	101
V-CH_2	101.4 ± 2	424 ± 8	Theory	0 K	269
V-CH_3	40 ± 4	167 ± 17	Ion Beam	0 K	29
$V(C_6H_6)$-C_6H_6	57 ± 5	238 ± 21	Photodissociation	298 K	242
V-H	49.0 ± 1.6	205 ± 7	Ion Beam	0 K	15, 17
V-H	37.9 ± 3	159 ± 13	Ion Molecule Reactions	298 K	236
V-V	63.49	265.8	Ion Beam	0 K	88
VO^+-CO	24.2 ± 2.3	101 ± 10	Ion Beam	0 K	104
VO^+-O	70.6 ± 9.2	295 ± 39	Ion Beam	0 K	104
V_2^+-V	52.3	219	Ion Beam	0 K	74
V_2-V	32.7	137	Ion Beam	0 K	84
V_3^+-V	81.4	341	Ion Beam	0 K	74
V_3-V	84.6	354	Ion Beam	0 K	84
V_4^+-V	74.7	313	Ion Beam	0 K	74
V_4-V	71.0	297	Ion Beam	0 K	84
V_5^+-V	95.2	398	Ion Beam	0 K	74
V_5-V	92.9	389	Ion Beam	0 K	84
V_6^+-V	89.0	372	Ion Beam	0 K	74
V_6-V	86.0	360	Ion Beam	0 K	84
V_7^+-V	92.0	385	Ion Beam	0 K	74
V_7-V	94.8	397	Ion Beam	0 K	84
V_8^+-V	84.6	354	Ion Beam	0 K	74
V_8-V	80.9	338	Ion Beam	0 K	84
V_9^+-V	91.3	382	Ion Beam	0 K	74
V_9-V	90.6	379	Ion Beam	0 K	84
V_{10}^+-V	91.3	382	Ion Beam	0 K	74
V_{10}-V	87.2	365	Ion Beam	0 K	84
V_{11}^+-V	95.2	398	Ion Beam	0 K	74
V_{11}-V	94.3	395	Ion Beam	0 K	84
V_{12}^+-V	107.2	448.8	Ion Beam	0 K	74
V_{12}-V	106.8	447.2	Ion Beam	0 K	84
V_{13}^+-V	92.7	388	Ion Beam	0 K	74

BOND	KCAL/MOL	KJ/MOL	METHOD	T	REFERENCES
V_{13}-V	93.4	391	Ion Beam	0 K	84
V_{14}^{+}-V	107.9	451	Ion Beam	0 K	74
V_{14}-V	<101.9	< 426.6	Ion Beam	0 K	84
V_{15}^{+}-V	89.9	376	Ion Beam	0 K	74
V_{15}-V	>96.6	> 404	Ion Beam	0 K	84
V_{16}^{+}-V	96.6	404	Ion Beam	0 K	74
V_{16}-V	96.9	405	Ion Beam	0 K	84
V_{17}^{+}-V	91.1	381	Ion Beam	0 K	74
V_{17}-V	89.0	372	Ion Beam	0 K	84
V_{18}^{+}-V	97.5	408	Ion Beam	0 K	74
V_{18}-V	<92.7	< 388	Ion Beam	0 K	84
V_{19}^{+}-V	105.8	443.0	Ion Beam	0 K	74
W^{+}-CH_2	111 ± 5	464 ± 21	Theory	0 K	259
W^{+}-H	49.9	209	Theory	0 K	273
$W(CO)_6^{+}$-H	64 ± 3	268 ± 13	Thermochemical	298 K	166
$W(CO)_3(O)^{-}$-CO	1.66 ± 0.1	7.0 ± 0.4	CID	298 K	290
$W(CO)_5^{-}$-CO	38.04 ± 2.31	159 ± 10	CID	298 K	290
$W(CO_3)(O)_2^{-}$-$(CO)_2$	≤2.31	≤ 9.67	CID	298 K	290
$W(O)_5^{-}$-O_2	40.80 ± 3.46	171 ± 14	CID	298 K	290
Y^{+}-CH_2	92.8 ± 3	388 ± 13	Ion Beam	0 K	25
Y^{+}-CH_2	90 ± 3	377 ± 13	Theory	0 K	123
Y^{+}-CH_3	56.4 ± 1.2	236 ± 5	Ion Beam	0 K	25
Y^{+}-CH_3	58.6	245	Theory	0 K	121
Y^{+}-CH_3	58.3	244	Theory	298 K	173
Y^{+}-CH_3	48.2	202	Theory	298 K	173
$Y(CH_3)^{+}$-CH_3	57.2	239	Theory	0 K	122
Y^{+}-CO	11.6	48.6	Theory	0 K	131
$Y(CO)^{+}$-CO	13.5	56.5	Theory	0 K	131
Y^{+}-C_2H_2	60 ± 7	251 ± 29	Ion Beam	0 K	25
Y^{+}-C_2H_2	56	234	Theory	0 K	124
Y^{+}-C_2H_2	52 ± 3	218 ± 13	Photodissociation	298 K	254
Y^{+}-C_2H_4	≥26	109	Ion Beam	0 K	25
Y^{+}-C_2H_4	33	138	Theory	0 K	140
Y^{+}-C_3H_6	43	180	Theory	0 K	140
Y^{+}-C_6H_6	40.8	171	Theory	0 K	126
Y^{+}-H	61.3 ± 1.4	256 ± 6	Ion Beam	0 K	11
Y^{+}-H	61.1 ± 1.8	256 ± 8	Ion Beam	0 K	99
Y^{+}-H	59.4	249	Theory	0 K	120
Y^{+}-H	66.7 ± 3	279 ± 13	Theory	0 K	269
Y^{+}-H	58.8	246	Theory	298 K	173
$Y(H)^{+}$-H	62.0 ± 2.1	259 ± 9	Ion Beam	0 K	11, 25
Y^{+}-Si	58.1 ± 3	243 ± 13	Ion Beam	0 K	119
Y^{+}-SiH	63.6 ± 3.7	266 ± 15	Ion Beam	0 K	119

BOND	KCAL/MOL	KJ/MOL	METHOD	T	REFERENCES
Y^+-SiH_2	≥55.1 ± 1.6	≥ 231 ± 7	Ion Beam	0 K	119
Y^+-SiH_3	49.1 ± 3.7	206 ± 15	Ion Beam	0 K	119
Y^{2+}-CH_4	22.0	92.1	Theory	0 K	146
Y^{2+}-C_2H_4	47	197	Theory	0 K	140
Y^{2+}-C_2H_6	31.2	131	Theory	0 K	146
Y^{2+}-C_3H_6	54	226	Theory	0 K	140
Y^{2+}-C_3H_8	41.6	174	Theory	0 K	146
Zn^+-CH_3	66.9 ± 1.6	280 ± 4	Ion Beam	0 K	63
Zn^+-CH_3	60.7	254	Theory	298 K	173
Zn^+-CH_3	52.2	218	Theory	298 K	173
Zn^+-H	54.4 ± 3	228 ± 13	Ion Beam	0 K	62
Zn^+-H	51.7 ± 3.7	216 ± 15	Ion Beam	0 K	118
Zn^+-H	52.4	219	Theory	0 K	260, 177
Zn^+-H	54.7	229	Theory	298 K	173
Zn^+-H_2O	32.8	137	Theory	0 K	128
Zn^+-H_2O	39.0	163	CID	298 K	294
$Zn(H_2O)^+$-H_2O	22.2	93.0	Theory	0 K	128
Zn^+-O	38.4 ± 1.1	161 ± 5	Ion Beam	0 K	38, 64
Zn^+-OH	30.4	127	CID	298 K	294
Zn^+-Si	65.5 ± 2.3	274 ± 10	Ion Beam	0 K	118
Zn^+-SiH	77.7 ± 2.5	325 ± 10	Ion Beam	0 K	118
Zn^+-SiH_2	38.0 ± 2.1	159 ± 9	Ion Beam	0 K	118
Zn^+-SiH_3	71.7 ± 3.5	300 ± 15	Ion Beam	0 K	118
Zn-CH_3	16.6 ± 2.3	69 ± 10	Ion Beam	0 K	63
Zn-O	37.1 ± 0.9	155 ± 4	Ion Beam	0 K	64
Zr^+-CH_2	101 ± 3	423 ± 13	Theory	0 K	123
Zr^+-CH_3	57.3	240	Theory	0 K	121
$Zr(CH_3)^+$-CH_3	64.2	269	Theory	0 K	122
Zr^+-CO	22.6	95	Theory	0 K	131
$Zr(CO)^+$-CO	19.5	81.6	Theory	0 K	131
Zr^+-C_2H_2	68	285	Theory	0 K	124
Zr^+-C_2H_2	59 ± 3	247±13	Photodissociation	298 K	255
Zr^+-H	52.1 ± 1.8	218 ± 8	Ion Beam	0 K	99
Zr^+-H	56.7	237	Theory	0 K	120
Zr^+-H	66.7 ± 3	279 ± 13	Theory	0 K	269
Zr^+-H	55.5	232	Theory	298 K	172

LITERATURE REFERENCES

REF #	FULL LITERATURE REFERENCE	YEAR
1.)	R. Stepnowski and J. Allison, *J. Am. Chem. Soc.*, **111**, 449 (1989).	1989
2.)	J. Allison and D. P. Ridge, *J. Am. Chem. Soc.*, **101**, 4998 (1979).	1979
3.)	J. Allison and M. Lonbarski, *Int. J. Mass Spectrom. Ion Proc.*, **49**, 281 (1983).	1983
4.)	A. Tsarbopoulos and J. Allison, *Organometallics*, **3**, 86 (1984).	1984
5.)	C. J. Cassady, B. S. Freiser, S. W. McElvany and J. Allison, *J. Am. Chem. Soc.*, **106**, 6125 (1984).	1984
6.)	J. Allison and B. Radecki, *J. Am. Chem. Soc.*, **106**, 946 (1984).	1984
7.)	J. Allison and D. P. Ridge, *J. Am. Chem. Soc.*, **99**, 35 (1976).	1976
8.)	G. H. Weddle, J. Allison and D. P. Ridge, *J. Am. Chem. Soc.*, **99**, 105 (1977).	1977
9.)	J. Allison and D. P. Ridge, *J. Am. Chem. Soc.*, **100**, 163 (1978).	1978
10.)	P. B. Armentrout and B. L. Kickel, in *Organometallic Ion Chemistry*, edited by B. S. Freiser, (Values have been adjusted to 0 K).	1994
11.)	J. L. Elkind, L. S. Sunderlin, and P. B. Armentrout, *J. Phys. Chem.*, **93**, 3151 (1989).	1989
12.)	J. L. Elkind and P. B. Armentrout, *Int. J. Mass Spectrom. Ion Proc.* , **83**, 259 (1988).	1988
13.)	J. L. Elkind and P. B. Armentrout, *J. Phys. Chem.*, **89**, 5626 (1985).	1985
14.)	J. L. Elkind and P. B. Armentrout, *J. Phys. Chem.*, **86**, 1868 (1987).	1987
15.)	P. B. Armentrout and L. S. Sunderlin, in *Transition Metal Hydrides* , edited by A. Dedieu, (VCH), New York, 1992, pp.1-64. (These values are critically reviewed).	1992
16.)	Y.-M. Chen, D. E. Clemmer and P. B. Armentrout, *J. Chem. Phys,*, **95**, 1228, (1991).	1991
17.)	Y.-M. Chen, D. E. Clemmer and P. B. Armentrout, *J. Chem. Phys,*, **98**, 4929, (1993).	1993
18.)	L. S. Sunderlin, N. Aristov, and P. B. Armentrout, *J. Am. Chem. Soc.*, **109**, 78 (1987).	1987
19.)	D. E. Clemmer, J. L. Elkind, N. Aristov, and P. B. Armentrout, *J. Chem. Phys,*, **95**, 3387 (1991).	1991
20.)	N. Aristov and P. B. Armentrout, *J. Am. Chem. Soc.*, **108**, 1806 (1986).	1986
21.)	L. S. Sunderlin and P. B. Armentrout, *J. Phys. Chem.*, **92**, 1209 (1988).	1988
22.)	L. S. Sunderlin and P. B. Armentrout, *Int. J. Mass Spectrom .Ion Proc.* , **94**, 149 (1989).	1989
23.)	N. Aristov and P. B. Armentrout, *J. Phys. Chem.*, **91**, 6178 (1987).	1987
24.)	R. Georgiadis and P. B. Armentrout, *Int. J. Mass Spectrom .Ion Proc.* , **89**, 227 (1989).	1989
25.)	L. S. Sunderlin and P. B. Armentrout, *J. Am. Chem. Soc.*, **111**, 3845 (1989).	1989
26.)	R. Georgiadis and P. B. Armentrout, *J. Phys. Chem.*, **92**, 7066 (1988).	1988
27.)	E. R. Fisher and P. B. Armentrout, *J. Am. Chem. Soc.*, **114**, 2039 (1992).	1992
28.)	P. B. Armentrout, *ACS Symp. Ser.* , **428**, 18 (1990).	1990
29.)	M. R. Sievers and P. B. Armentrout, work in progress.	1994
30.)	L. S. Sunderlin and P. B. Armentrout, *Organometallics* , **9**, 1248 (1990).	1990
31.)	N. Aristov, Ph.D. Thesis, University of California, Berkeley, (1986).	1986
32.)	D. E. Clemmer, L. S. Sunderlin and P. B. Armentrout, *J. Phys. Chem.*, **94**, 3008 (1990).	1990
33.)	D. E. Clemmer, L. S. Sunderlin and P. B. Armentrout, *J. Phys. Chem.*, **94**, 208 (1990).	1990

REF #	FULL LITERATURE REFERENCE	YEAR
34.)	Y. -M. Chen and P. B. Armentrout, work in progress.	1990
35.)	D. E. Clemmer, N. Aristov and P. B. Armentrout, *J. Phys. Chem.*, **97**, 544 (1993).	1993
36.)	D. E. Clemmer, M. Knowles, N. Aristov and P. B. Armentrout, work in progress as reported in reference 26.	1994
37.)	N. Aristov and P. B. Armentrout, *J. Phys. Chem.*, **90**, 5135 (1986).	1986
38.)	E. R. Fisher, J. L. Elkind, D. E. Clemmer, R. Georgiadis, S. K. Loh, N. Aristov, L. S. Sunderlin, and P. B. Armentrout, *J. Chem. Phys.*, **93**, 2676 (1990).	1990
39.)	D. E. Clemmer and P. B. Armentrout, work in progress.	1994
40.)	J. L. Elkind and P. B. Armentrout, *J. Chem. Phys.*, **84**, 4862 (1986).	1986
41.)	J. L. Elkind and P. B. Armentrout, *J. Am. Chem. Soc.*, **108**, 2765 (1986); *J. Phys. Chem.*, **90**, 5736 (1986).	1986
42.)	J. L. Elkind and P. B. Armentrout, *J. Phys. Chem.*, **90**, 6576 (1986).	1986
43.)	L. S. Sunderlin and P. B. Armentrout, *J. Phys. Chem.*, **94**, 3589 (1990).	1990
44.)	R. H. Schultz and P. B. Armentrout, *J. Chem. Phys.*, **94**, 2262 (1991).	1991
45.)	R. Georgiadis, E. R. Fisher and P. B. Armentrout, *J. Am. Chem. Soc.*, **111**, 4251 (1989).	1989
46.)	E. R. Fisher and P. B. Armentrout, *J. Phys. Chem.*, **94**, 1674 (1990).	1990
47.)	R. H. Schultz and P. B. Armentrout, *Organometallics*, **11**, 828 (1992).	1992
48.)	C. L. Haynes and P. B. Armentrout, *Organometallics*, **13**, 3480 (1994).	1994
49.)	R. Georgiadis and P. B. Armentrout, *Int. J. Mass. Spectrom. Ion Proc.*, **91**, 123 (1989).	1989
50.)	R. H. Schultz, J. L. Elkind and P. B. Armentrout, *J. Am. Chem. Soc.*, **110**, 411 (1988).	1988
51.)	E. R. Fisher, R. H. Schultz, and P. B. Armentrout, *J. Phys. Chem.*, **93**, 7382 (1989).	1989
52.)	E. R. Fisher, L. S. Sunderlin, and P. B. Armentrout, *J. Phys. Chem.*, **93**, 7375 (1989).	1989
53.)	R. H. Schultz, and P. B. Armentrout, *J. Phys. Chem.*, **97**, 596 (1993).	1993
54.)	C. L. Haynes, E. R. Fisher and P. B. Armentrout, work in progress.	1994
55.)	R. H. Schultz, and P. B. Armentrout, manuscript in preparation.	1994
56.)	R. H. Schultz, and P. B. Armentrout, *J. Phys. Chem.*, **96**, 1662 (1992).	1992
57.)	R. H. Schultz, and P. B. Armentrout, *J. Am. Chem. Soc.*, **113**, 729 (1991).	1991
58.)	D. E. Clemmer and P. B. Armentrout, *J. Phys. Chem.*, **95**, 3084 (1991).	1991
59.)	S. K. Loh, E. R. Fisher, L. Lian, R. H. Schultz and P. B. Armentrout, *J. Phys. Chem.*, **93**, 3159 (1989).	1989
60.)	S. K. Loh, L. Lian and P. B. Armentrout, *J. Chem. Phys.*, **91**, 6148 (1989).	1989
61.)	Y.-M. Chen, D.E. Clemmer and P.B. Armentrout, work in progress.	1994
62.)	R. Georgiadis and P. B. Armentrout, *J. Am. Chem. Soc.*, **108**, 2119 (1986).	1986
63.)	R. Georgiadis and P. B. Armentrout, *J. Phys. Chem.*, **92**, 7060 (1988).	1988
64.)	D.E. Clemmer, N.F. Dalleska and P.B. Armentrout, *J. Chem. Phys.*, **95**, 7263 (1991).	1991
65.)	N.F. Dalleska, K.C. Crellin and P.B. Armentrout, *J. Phys. Chem.*, **97**, 3123 (1993).	1993
66.)	D.E. Clemmer, M.E. Weber and P.B. Armentrout, *J. Phys. Chem.*, **96**, 10888 (1992).	1992
67.)	N.F. Dalleska, K. Honma, L.S. Sunderlin and P.B. Armentrout, *J. Am. Chem. Soc.*, **116**, 3519 (1994).	1994

BOND DISSOCIATION ENERGIES: LITERATURE REFERENCES

REF #	FULL LITERATURE REFERENCE	YEAR
68.)	F.A. Khan, D.E. Clemmer, R.H. Schultz and P.B. Armentrout, *J. Phys. Chem.*, **97**, 7978 (1993).	1993
69.)	F.A. Khan and P.B. Armentrout, work in progress.	1994
70.)	R.H. Schultz, K. Crellin and P.B. Armentrout, *J. Am. Chem. Soc.*, **113**, 8590 (1991).	1991
71.)	F.A. Khan, D.A. Steele and P.B. Armentrout, *J. Phys. Chem.*, **99**, 7819 (1995).	1994
72.)	Absolute uncertainties gradually increase from ~2 kcal/mol for small cluster ions to ~12 kcal/mol for larger cluster ions. Relative uncertainties are less than 4 kcal/mol for all clusters.	1994
73.)	L. Lian, C.-X. Su and P.B. Armentrout, *J. Chem. Phys.*, **97**, 4084 (1992).	1992
74.)	C.-X. Su, D.A. Hales and P.B. Armentrout, *J. Chem. Phys.*, **99**, 6613 (1993).	1993
75.)	C.-X. Su and P.B. Armentrout, *J. Chem. Phys.*, **99**, 6506 (1993).	1993
76.)	L. Lian, C.-X. Su and P.B. Armentrout, *J. Chem. Phys.*, **97**, 4072 (1992).	1992
77.)	D.A. Hales, C.-X. Su, L. Lian and P.B. Armentrout, *J. Chem. Phys.*, **100**, 1049 (1993).	1993
78.)	L. Lian, C.-X. Su and P.B. Armentrout, *J. Chem. Phys.*, **96**, 7542 (1992).	1992
79.)	D.A. Hales, L. Lian and P.B. Armentrout, *Int. J. Mass Spectrom. Ion Proc.* , **102**, 269 (1990).	1990
80.)	Preliminary value from D.A. Hales, Ph.D. Thesis, University of California, Berkeley, 1990.	1990
81.)	L.M. Russon, S.A. Heidecke, M.K. Birke, J. Conceicao, M.D. Morse, and P.B. Armentrout, *J. Chem. Phys.*, **100**, 4747 (1994).	1994
82.)	L.M. Russon, S.A. Heidecke, M.K. Birke, J. Conceicao, P.B. Armentrout and M.D. Morse, *Chem. Phys. Lett.*, **204**, 235 (1993).	1993
83.)	Values are calculated by using equation (12) in Chapter 1 with cationic cluster ion bond energies and ionization energies (I.E.'s) from the references given. Absolute uncertainties gradually increase from ~2 kcal/mol for small cluster ions to ~12 kcal/mol for larger cluster ions. Relative uncertainties are less than 4 kcal/mol for all clusters.	1994
84.)	I.E.'s from D.M. Cox, R.L. Whetten, M.R. Zakin, D.J. Trevor, K.C. Reichmann and A. Kaldor, in W.C. Stwalley and M. Lapp (Eds.), *AIP Conference Proceedings* , 146, *Optical Science and Engineering Ser.* , 6, Advances in Laser Science , Vol. I, AIP, New York, 9186, p. 527.	1994
85.)	Except where noted, I.E.'s are taken from S. Yang and M.B. Knickelbein, *J. Chem. Phys.*, **93**, 1533 (1990).	1990
86.)	Except where noted, I.E.'s are taken from M.B. Knickelbein, S. Yang and S.J. Riley, *J. Chem. Phys.*, **93**, 94 (1990).	1990
87.)	I.E.'s taken from M.B. Knickelbein and S. Yang, *J. Chem. Phys.*, **93**, 5760 (1990).	1990
88.)	E.M. Spain and M.D. Morse, *Int. J. Mass Spectrom. Ion Proc.* , **102** 183 (1990).	1990
89.)	I.E.'s taken from E.A. Rohlfing, D.M. Cox, A. Kaldor and K.H. Johnson, *J. Chem. Phys.*, **81**, 3846 (1984).	1984
90.)	I.E. (Co_2) ≤ 6.42 eV, M.D. Morse and Z.-W. Fu, personal communication.	1994
91.)	M.D. Morse, G.P. Hansen, P.R.R. Langridge-Smith, L.-S. Zheng, M.E. Geusic, D.L. Michalopoulos and R.E. Smalley, *J. Chem. Phys.*, **80**, 5400 (1984).	1984
92.)	The sum of D(Co_2) and D(Co_2-Co) equals D(Co_2^+) + D(Co_2^+-Co) + I.E.(Co_3) - I.E.(Co) = 67.3 ± 3 kcal/mol.	1994
93.)	I.E.'s taken from E.K. Parks, T.D. Klots and S.J. Riley, *J. Chem. Phys.*, **92**, 3813 (1990).	1990
94.)	B. L. Kickel and P. B. Armentrout, *J. Am. Chem. Soc.*, **116**, 10742 (1994).	1994
95.)	Y.-M. Chen and P. B. Armentrout, *J. Am. Chem. Soc.*, in press.	1995
96.)	F. Meyer, Y.-M. Chen and P.B. Armentrout, *J. Am. Chem. Soc.* , **117**, 4071 (1995).	1995

REF #	FULL LITERATURE REFERENCE	YEAR
97.)	S. Goebel, C.L. Haynes, F.A. Khan and P.B. Armentrout,*J. Am. Chem. Soc.*, **117**, 6994 (1995).	1995
98.)	Y.-M. Chen, J.L. Elkind, P.B. Armentrout, *J. Phys. Chem.*, **99**, 10438 (1995).	1995
99.)	M.R. Sievers, Y.-M. Chen, J.L. Elkind and P.B. Armentrout, *J. Phys. Chem.*, submitted for publication.	1995
100.)	C.L. Haynes, Y.-M. Chen and P.B. Armentrout, *J. Phys. Chem.*, **99**, 9110 (1995).	1995
101.)	M.R. Sievers and P.B. Armentrout, *J. Phys. Chem.*, **99**, 8135 (1995).	1995
102.)	F. Meyer and P.B. Armentrout, *Molec. Phys.*, submitted for publication.	1995
103.)	C.L. Haynes, P.B. Armentrout, J.K. Perry and W.A. Goddard, III, *J. Phys. Chem.*, **99**, 6340 (1995).	1995
104.)	M.R. Sievers and P.B. Armentrout, *J. Chem. Phys.*, **102**, 754 (1995).	1995
105.)	N.F. Dalleska and P.B. Armentrout, *Int. J. Mass Spectrom. Ion Processes*, **134**, 203 (1994).	1994
106.)	Y.-M. Chen, and P.B. Armentrout, *J. Chem. Phys.*, **103**, 618 (1995).	1995
107.)	D.E. Clemmer, Y.-M. Chen, N. Aristov and P.B. Armentrout, *J. Phys. Chem.*, **98**, 7538 (1994).	1994
108.)	Y.-M. Chen, D.E. Clemmer and P.B. Armentrout *J. Am. Chem. Soc.*, **116**, 7815 (1994).	1994
109.)	Y.-M. Chen, and P.B. Armentrout, *J. Am. Chem. Soc.*, **117**, 9291 (1995).	1995
110.)	Y.-M. Chen, and P.B. Armentrout, *J. Phys. Chem.*, **99**, 11424 (1995).	1995
111.)	N.F. Dalleska, B.L. Tjelta and P.B. Armentrout, *J. Phys. Chem.*, **98**, 4191 (1994).	1994
112.)	Y.-M. Chen and P.B. Armentrout, *Chem. Phys. Lett.*, **210**, 123 (1993).	1993
113.)	R. Georgiadis and P.B. Armentrout, *J. Phys. Chem.* **92**, 7060 (1988).	1988
114.)	F. Meyer, F.A. Khan and P.B. Armentrout, *J. Am. Chem. Soc.*, **117**, 9740 (1995).	1995
115.)	C. L. Haynes and P. B. Armentrout *Chem. Phys. Lett.*, submitted for publication.	1995
116.)	B. L. Kickel and P. B. Armentrout *J. Am. Chem. Soc.* **117**, 764 (1995).	1995
117.)	B. L. Kickel; P. B. Armentrout *J. Phys. Chem.* **1995**, *99*, 2024.	1995
118.)	B. L. Kickel and P. B. Armentrout *J. Phys. Chem.* **99**, 2024 (1995).	1995
119.)	B. L. Kickel and P. B. Armentrout *J. Am. Chem. Soc.* **117**, 4057 (1995).	1995
120.)	L. G. M. Petterson, C. W. Bauschlicher, S. R. Langhoff and H. Partridge, *J. Chem. Phys.*, **87**, 481 (1987).	1987
121.)	C. W. Bauschlicher, S. R. Langhoff, H. Partridge and L. A. Barnes, *J. Chem. Phys.*, **91**, 2399 (1989).	1987
122.)	M. Rosi, C. W. Bauschlicher, S. R. Langhoff and H. Partridge, *J. Chem. Phys.*,**94**, 8656 (1990).	1990
123.)	C. W. Bauschlicher, H. Partridge, J. A. Sheehy, S. R. Langhoff and M. Rosi, *J. Chem. Phys.*, **96**, 6969 (1992).	1992
124.)	M. Sodupe and C. W. Bauschlicher, *J. Chem. Phys.*,**95**, 8640 (1991).	1991
125.)	M. Sodupe, C. W. Bauschlicher, S. R. Langhoff and H. Partridge, *J. Chem. Phys.*,**96**, 2118 (1992).	1992
126.)	C. W. Bauschlicher, H. Partridge and S. R. Langhoff, *J. Chem. Phys.*, **96**, 3273 (1992).	1992
127.)	C. W. Bauschlicher, S. R. Langhoff, H. Partridge, J. E. Rice and A. Komornicki, *J. Chem. Phys.*, **95**, 5142 (1991).	1991
128.)	M. Rosi and C. W. Bauschlicher, *J. Chem. Phys.*, **92**, 1876 (1990).	1990
129.)	C. W. Bauschlicher, S. R. Langhoff and H. Partridge, *J. Chem. Phys.*, **94**, 2068 (1991).	1991

REF #	FULL LITERATURE REFERENCE	YEAR
130.)	S. R. Langhoff, C. W. Bauschlicher, H. Partridge and M. Sodupe, *J. Chem. Phys.*, **95**, 10677 (1991).	1991
131.)	L. A. Barnes, M. Rosi and C. W. Bauschlicher, *J. Chem. Phys.*, **93**, 609 (1990).	1990
132.)	C. W. Bauschlicher, *Chem. Phys. Lett.* , **156**, 95 (1989).	1989
133.)	C. W. Bauschlicher, H. Partridge and S. R. Langhoff, *Chem. Phys. Lett.* , **195**, 360 (1992).	1992
134.)	C. W. Bauschlicher, H. Partridge and S. R. Langhoff, *J. Phys. Chem.*, **96**, 2475 (1992).	1992
135.)	P. Maitre and C. W. Bauschlicher, *J. Phys. Chem.*, **97**, 11912 (1993).	1993
136.)	C. W. Bauschlicher, S. R. Langhoff and H. Partridge, *J. Phys. Chem.*, **95**, 6191 (1991).	1991
137.)	C. W. Bauschlicher, *J. Phys. Chem.*, **97**, 3709 (1993).	1993
138.)	M. Sodupe and C. W. Bauschlicher, *Chem. Phys. Lett.* **207**, 19 (1993).	1993
139.)	L. M. Roth, B. S. Freiser, C. W. Bauschlicher, H. Partridge and S. R. Langhoff, *J. Am. Chem. Soc.*, **113**, 3274 (1991).	1991
140.)	C. W. Bauschlicher and S. R. Langhoff, *J. Phys. Chem.*, **95**, 2278 (1991) and M. Rosi and C. W. Bauschlicher, *Chem. Phys. Lett.*, **166**, 189 (1990).	1991
141.)	C. W. Bauschlicher, H. Partridge and S. R. Langhoff, *J. Phys. Chem.*, **91**, 4733 (1989).	1989
142.)	C. W. Bauschlicher, H. Partridge and S. R. Langhoff, *Chem. Phys. Lett.*, **165**, 272 (1990).	1990
143.)	H. Partridge, C. W. Bauschlicher and S. R. Langhoff, *J. Phys. Chem.*, **96**, 5350 (1992).	1992
144.)	H. Partridge and C. W. Bauschlicher, *J. Phys. Chem.*, **98**, 2301 (1994).	1994
145.)	Y. Huang, Y. D. Hill, M. Sodupe, C. W. Bauschlicher and B. S. Freiser, *Inorganic Chem.*,**30**, 3822 (1991).	1992
146.)	Y. D. Hill, B. S. Freiser and C. W. Bauschlicher, *J. Am. Chem. Soc.*, **113**, 1507 (1991).	1991
147.)	H. Partridge, S. R. Langhoff and C. W. Bauschlicher, *J. Phys. Chem.*, **84**, 4489 (1986).	1986
148.)	C. W. Bauschlicher, L. A. Barnes and P. R. Taylor, *J. Phys. Chem.*, **93**, 2932 (1989).	1989
149.)	C. W. Bauschlicher and H. Partridge, *J. Phys. Chem.*, **95**, 3946 (1991).	1991
150.)	C. W. Bauschlicher and H. Partridge, *Chem. Phys. Lett.* , **181**, 129 (1991).	1991
151.)	C. W. Bauschlicher, *Chem. Phys. Lett.*, **201**, 11 (1993).	1993
152.)	C. W. Bauschlicher and H. Partridge, *J. Phys. Chem.*, **95**, 9694 (1991).	1991
153.)	H. Partridge and C. W. Bauschlicher, *J. Phys. Chem.*, **96**, 9694, (1991).	1991
154.)	M. Sodupe, C. W. Bauschlicher and H. Partridge, *Chem. Phys. Lett.*, **192**, 185 (1992).	1992
155.)	M. Sodupe and C. W. Bauschlicher, *Chem. Phys. Lett.*, **203**, 215 (1993).	1993
156.)	C. W. Bauschlicher and M. Sodupe, *Chem. Phys. Lett.*, **214**, 489 (1993).	1993
157.)	M. Sodupe and C. W. Bauschlicher, *Chem. Phys.*, in press.	1994
158.)	M. Sodupe, C. W. Bauschlicher and H. Partridge, *J. Chem. Phys.*, **95**, 9422 (1991).	1991
159.)	C. W. Bauschlicher, M. Sodupe and H. Partridge, *J. Chem. Phys.*, **96**, 4453 (1992).	1992
160.)	M. Sodupe and C. W. Bauschlicher,*Chem. Phys. Lett.* , **181**, 321 (1991).	1991
161.)	C. W. Bauschlicher, F. Bouchard, J. W. Hepburn, T. B. McMahon, P. Surjasamita, L. Roth, J. R. Gord and B. S. Freiser, *Int. J. Mass Spectrom. Ion Proc.* , **109**, 15 (1991).	1991
162.)	A. Ricca, C. W. Bauschlicher and M. Rosi, *J. Phys. Chem.*, submitted.	1994
163.)	C. W. Bauschlicher, Unpublished Results.	1994

REF #	FULL LITERATURE REFERENCE	YEAR
164.)	Y. Xu, E. Garcia, B. S. Freiser, and C. W. Bauschlicher, *J. Am. Chem. Soc.*, submitted.	1994
165.)	R.H. Staley and J.L. Beauchamp, *J. Am. Chem. Soc.*, **97**, 5920 (1975).	1975
166.)	A.E. Stevens and J.L. Beauchamp, *J. Am. Chem. Soc.*, **103**, 190 (1981).	1981
167.)	L.F. Halle, F.S. Klein and J.L. Beauchamp, *J. Am. Chem. Soc.*, **106**, 2543 (1984).	1984
168.)	M.L. Mandich, L.F. Halle and J.L. Beauchamp, *J. Am. Chem. Soc.*, **106**, 4403 (1984).	1984
169.)	M.A. Tolbert and J.L. Beauchamp, *J. Am. Chem. Soc.*, **106**, 8117 (1984).	1984
170.)	K. Hayashibara, G. H. Kruppa and J. L. Beauchamp, *J. Am. Chem. Soc.*, **108**, 5668 (1986).	1986
171.)	H. Kang and J.L. Beauchamp, *J. Am. Chem. Soc.*, **108**, 7502 (1986).	1986
172.)	J. B. Shilling, W. A. Goddard, III, and J. L. Beauchamp, *J. Am. Chem. Soc.*, **109**, 5565 (1987).	1987
173.)	J.B. Schilling, W.A. Goddard, III and J.L. Beauchamp, *J. Am. Chem. Soc.*, **109**, 5573 (1987).	1987
174.)	D.V. Dearden, K. Hayashibara, J.L. Beauchamp, N.J. Kirchner, P.A.M. van Koppen and M.T. Bowers, *J. Am. Chem. Soc.*, **111**, 2401 (1989).	1989
175.)	P.B. Armentrout and J.L. Beauchamp, *J. Phys. Chem.*, **85**, 4103 (1981).	1981
176.)	J.B. Schilling, W.A. Goddard, III and J.L. Beauchamp, *J. Phys. Chem.*, **91**, 4470 (1987).	1987
177.)	J. B. Shilling, W. A. Goddard, III, and J. L. Beauchamp, *J. Phys. Chem.*, **91**, 5616 (1987).	1987
178.)	L.F. Halle, P.B. Armentrout and J.L. Beauchamp, *Organometallics* , **2**, 1829 (1983).	1983
179.)	J.A.M. Simões, J.C. Schultz and J.L. Beauchamp, *Organometallics* , **4**, 1238 (1985).	1985
180.)	A.E. Stevens and J.L. Beauchamp, *Chem. Phys. Lett.*, **78**, 291 (1981).	1981
181.)	P.B. Armentrout and J.L. Beauchamp, *Chem. Phys.*, **48**, 315 (1980).	1980
182.)	P.B. Armentrout and J.L. Beauchamp, *Chem. Phys.*, **50**, 21 (1980).	1980
183.)	L.F. Halle and J.L. Beauchamp, *J. Chem. Phys.*, **66**, 4683 (1977).	1977
184.)	L.F. Halle and J.L. Beauchamp, *J. Chem. Phys.*, **76**, 2449 (1982).	1982
185.)	P. R. Kemper, M.-T. Hsu and M. T. Bowers, *J. Phys. Chem.*, **95**, 10600 (1991).	1991
186.)	J. E. Bushnell, P. R. Kemper and M. T. Bowers, *J. Am. Chem. Soc.*, in press.	1994
187.)	J. E. Bushnell, P. R. Kemper and M. T. Bowers, *J. Phys. Chem.*, **97**, 11628 (1993).	1993
188.)	P. R. Kemper, J. E. Bushnell, P. A. M. van Koppen, M. T. Bowers *J. Phys. Chem.*, **97**, 1810 (1993).	1993
189.)	P. R. Kemper, J. E. Bushnell, G. von Helden and M. T. Bowers, *J. Phys. Chem.*, **97**, 52 (1993).	1993
190.)	P. A. M. van Koppen, P. R. Kemper and M. T. Bowers, *J. Am. Chem. Soc.*, **115**, 5616 (1993).	1993
191.)	M. A. Hanratty, J. L. Beauchamp, A. J. Illies, P. A. M. van Koppen and M. T. Bowers, *J. Am. Chem. Soc.*, **110**, 1 (1988).	1988
192.)	P. A. M. van Koppen, D. B. Jacobson, A. J. Illies, M. T. Bowers, M. A. Hanratty and J. L. Beauchamp, *J. Am. Chem. Soc.*, **111**, 1991 (1989).	1989
193.)	D. V. Dearden, J. L. Beauchamp, P. A. M. van Koppen and M. T. Bowers, *J. Am. Chem. Soc.*, **112**, 9373 (1990).	1990
194.)	C. J. Carpenter, P. A. M. van Koppen and M. T. Bowers, work in progress.	1994
195.)	Note: Bond energy assuming the products formed are M^+ + trimethylene	1994
196.)	Note: $(Ni^+)^*$ refers to the excited state of Ni^+ which has a $4s3d^8$ electronic configuration.	1994

REF #	FULL LITERATURE REFERENCE	YEAR
197.)	Note: For Sc-H_2, Sc^+ inserts into the H-H bond with an average Sc^+-H bond energy of 54.2 kcal/mol.	1994
198.)	P. A. M. van Koppen, M. T. Bowers, and J. L. Beauchamp, *Organometallics*, **9**, 625 (1990).	1990
199.)	D.E. Lessen and P.J. Brucat, *Chem. Phys. Lett.*, **152**, 473 (1988)	1988
200.)	D.E. Lessen, R.L. Asher and P.J. Brucat, *Chem. Phys. Lett.*, **177**, 380 (1991)	1991
201.)	D.E. Lessen, R.L. Asher and P.J. Brucat, *Chem. Phys. Lett.*, **182**, 412 (1991)	1991
202.)	P.J. Brucat C.L. Pettiette, S. Yang, L.-S. Zheng, M.J. Craycraft and R.E. Smalley, *J. Chem. Phys.*, **85**, 4747 (1986)	1986
203.)	D.E. Lessen and P.J. Brucat, *J. Chem. Phys.*, **91**, 4522 (1989)	1989
204.)	P.J. Brucat, L.-S. Zheng, C.L. Pettiette, S. Yang, and R.E. Smalley, *J. Chem. Phys.*, **84**, 3078 (1986)	1986
205.)	D.E. Lessen and P.J. Brucat, *J. Chem. Phys.*, **90**, 6296 (1989)	1989
206.)	R. L. Asher, D. Bellert, T. Buthelezi, G. Weerasekera, and P. J. Brucat, J. *Chem. Phys. Lett.*, **224**, (1994).	1994
207.)	R. L. Asher, D. Bellert, T. Buthelezi, and P. J. Brucat, *Chem. Phys. Lett.*, **224** (1994).	1994
208.)	R.G. Keesee, B. Chen, A.C. Harms and A.W. Castleman, *Int. J. Mass Spectrom. Ion Proc.* , **123**, 225 (1993).	1993
209.)	B.C. Guo and A.W. Castleman, *Chem. Phys. Lett.* , **181**, 16 (1991).	1991
210.)	B.C. Guo and A.W. Castleman, *Int. J. Mass Spectrom. Ion Proc.* , **100**, 665 (1990).	1990
211.)	R.J. Stanley and A.W. Castleman, *J. Chem. Phys.*, **92**, 5770 (1990).	1990
212.)	K.L. Gleim, B.C. Guo, R.G. Keesee and A.W. Castleman, *J. Phys. Chem.*, **93**, 6805 (1989).	1989
213.)	P.M. Holland and A.W. Castleman, *J. Chem. Phys.*, **76**, 4195 (1982).	1982
214.)	A.W. Castleman, P.M. Holland, D.M. Lindsay and K.I. Peterson, *J. Am. Chem. Soc.*, **100**, 6039 (1978).	1978
215.)	A.W. Castleman, Jr.,*Chem. Phys. Lett.*, **53(3)**, 560-4 (1978).	1978
216.)	I.N. Tang and A.W. Castleman, *J. Chem. Phys.*, **57**, 3638 (1972).	1972
217.)	C. S. Yeh, K. F. Willey, D. L. Robbins, J. S. Pilgrim and M. A. Duncan, *Chem. Phys. Lett.*, **196**, 233 (1992).	1992
218.)	C. S. Yeh, K. F. Willey, D. L. Robbins, J. S. Pilgrim and M. A. Duncan, *J. Chem. Phys.*, **98**, 1867 (1993).	1993
219.)	D. L. Robbins, K. F. Willey, C. S. Yeh, and M. A. Duncan, *J. Phys. Chem.*, **96**, 4824 (1992).	1992
220.)	J. S. Pilgrim, C. S. Yeh and M. A. Duncan, *Chem. Phys. Lett.*, **210**, 322 (1993).	1993
221.)	J. S. Pilgrim, C. S. Yeh, K. R. Berry and M. A. Duncan, *J. Chem. Phys.*, **100**, 7945 (1994).	1994
222.)	K. F. Willey, C. S. Yeh, D. L. Robbins and M. A. Duncan, *Chem. Phys. Lett.*, **192**, 179 (1992).	1992
223.)	K. F. Willey, C. S. Yeh, D. L. Robbins, and M. A. Duncan, *J. Phys. Chem.*, **96**, 9106, (1992).	1992
224.)	K. F. Willey, C. S. Yeh, D. L. Robbins, J. S. Pilgrim and M. A. Duncan, *J. Chem. Phys.*, **97**, 8886 (1992).	1992
225.)	K. F. Willey, P. Y. Cheng, C. S. Yeh, D. L. Robbins, and M. A. Duncan, *J. Chem. Phys.*, **95**, 6249 (1991).	1991
226.)	K. F. Willey, P. Y. Cheng, K. D. Pearce, and M. A. Duncan, *J. Phys. Chem.*, **94**, 4769, (1990).	1990
227.)	K. F. Willey, P. Y. Cheng, M. B. Bishop, and M. A. Duncan, *J. Am. Chem. Soc.*, **113**, 4721 (1991).	1991

REF #	FULL LITERATURE REFERENCE	YEAR
228.)	P. Y. Cheng, K. F. Willey, and M. A. Duncan, *Chem. Phys. Lett.*, **163**, 469 (1989).	1989
229.)	D.B. Jacobson and B.S. Freiser, *J. Am. Chem. Soc.*, **105**, 7484 (1983).	1983
230.)	D.B. Jacobson and B.S. Freiser, *J. Am. Chem. Soc.*, **106**, 3891 (1984).	1984
231.)	D.B. Jacobson and B.S. Freiser, *J. Am. Chem. Soc.*, **106**, 3900 (1984).	1984
232.)	C. J. Cassady, and B. S. Freiser, *J. Am. Chem. Soc.*, **106**, 6176 (1984).	1984
233.)	D.B. Jacobson and B.S. Freiser, *J. Am. Chem. Soc.*, **106**, 4623 (1984).	1984
234.)	D.B. Jacobson and B.S. Freiser, *J. Am. Chem. Soc.*, **107**, 5870 (1985).	1985
235.)	C. J. Cassady and B. S. Freiser, *J. Am. Chem. Soc.*, **107**, 1566 (1985).	1985
236.)	L. Sallans, K.R. Lane, R.R. Squires and B.S. Freiser, *J. Am. Chem. Soc.*, **107**, 4379 (1985).	1985
237.)	D.B. Jacobson and B.S. Freiser, *J. Am. Chem. Soc.*, **107**, 5876 (1985).	1985
238.)	D.B. Jacobson and B.S. Freiser, *J. Am. Chem. Soc.*, **107**, 7399 (1985).	1985
239.)	R.L. Hettich and B.S. Freiser, *J. Am. Chem. Soc.*, **107**, 6222 (1985).	1985
240.)	R.L. Hettich and B.S. Freiser, *J. Am. Chem. Soc.*, **106**, 2537 (1986).	1986
241.)	T.C. Jackson and B.S. Freiser, *Int. J. Mass. Spectrom. Ion Proc.* , **72**, 169 (1986).	1986
242.)	R.L. Hettich, T.C. Jackson, E.M. Stanko and B.S. Freiser, *J. Am. Chem. Soc.*, **108**, 5086 (1986).	1986
243.)	R.L. Hettich, T.C. Jackson, E.M. Stanko and B.S. Freiser, *J. Am. Chem. Soc.*, **109**, 3543 (1987).	1987
244.)	R.A. Forbes, L.M. Lech and B.S. Freiser, *Int. J. Mass Spectrom. Ion Proc.* , **77**, 107 (1987).	1987
245.)	T.J. MacMahon, T.C. Jackson and B.S. Freiser, *J. Am. Chem. Soc.*, **111**, 421 (1989).	1989
246.)	S.W. Buckner, J.R. Gord and B.S. Freiser, *J. Am. Chem. Soc.*, **110**, 6606 (1988).	1988
247.)	S.W. Buckner and B.S. Freiser, *J. Phys. Chem.* , **93**, 3667 (1989).	1989
248.)	S.W. Buckner, J.R. Gord and B.S. Freiser, *J. Chem. Phys.*, **94**, 4282 (1991).	1991
249.)	D.B. Jacobson, J.R. Gord and B.S. Freiser, *Organometallics* , **8**, 2957 (1989).	1989
250.)	Y. Huang and B.S. Freiser, *J. Am. Chem. Soc.*, **111**, 2387 (1989).	1989
251.)	Y. Huang and B.S. Freiser, *J. Am. Chem. Soc.*, **112**, 5085 (1990).	1990
252.)	Y. Huang, and B.S. Freiser, *Inorg. Chem.*, **29**, 1102 (1990).	1990
253.)	Y. Huang, Y. D. Hill, M. Sodupe, C. W. Bauschlicher, Jr., and B. S. Freiser, *Inorg. Chem.* , **20**, 3822 (1991).	1991
254.)	Y.A. Ranasinghe and B.S. Freiser, *Chem. Phys. Letters* , **200**, 135 (1992).	1992
255.)	D. R. A. Ranatunga, and B. S. Freiser, *Chem. Phys. Lett.*, **233**, 319 (1995).	1994
256.)	S. Z. Kan, Y. G. Byun, and B. S. Freiser, *J. Am. Chem. Soc.*, **116**, 8815 (1994).	1994
257.)	S. Z. Kan, Y. G. Byun, S. A. Lee, and B. S. Freiser, B. S. *J. Mass Spectrom*, **30**, 194 (1995).	1994
258.)	D. R. A. Ranatunga, and B. S. Freiser, *Chem. Phys. Lett.*, submitted.	1994
259.)	Irikura and W.A. Goddard, III, *J. Am. Chem. Soc.*, submitted.	1994
260.)	J. B. Shilling and W. A. Goddard, III, *J. Am. Chem. Soc.*, **108**, 582-584 (1986)	1986
261.)	A. K. Rappe' and W. A. Goddard, III, *J. Am. Chem. Soc.*, **97**, 3966-3968 (1977)	1977
262.)	T. H. Upton and W. A. Goddard, III, *J. Am. Chem. Soc.*, **100**, 321-323 (1978)	1978

REF #	FULL LITERATURE REFERENCE	YEAR
263.)	J. N. Allison and W. A. Goddard, III, *Sur. Science*, **110**, 615-618 (1981)	1981
264.)	A. K. Rappe' and W. A. Goddard, III, *Nature*, **285**, 311-312 (1980)	1980
265.)	W. A. Goddard, III, S. P. Walsh, A. K. Rappe', T. H. Upton and C. F. Melius, *J. Vac. Sci. Technol.*., **14**, 416-418 (1977)	1977
266.)	E. A. Carter and W. A. Goddard, III, *J. Am. Chem. Soc.*, **108**, 2180 (1986)	1986
267.)	A. K. Rappe' and W. A. Goddard, III, *J. Am. Chem. Soc.*, **104**, 448 (1982)	1982
268.)	E. A. Carter and W. A. Goddard, III, *Organometallics.*, **7**, 675 (1988)	1988
269.)	E. A. Carter and W. A. Goddard, III, *J. Phys. Chem.*, **92**, 5679 (1988)	1988
270.)	E. A. Carter and W. A. Goddard, III, *J. Phys. Chem.*, **92**, 2109(1988)	1988
271.)	E. A. Carter and W. A. Goddard, III, *J. Am. Chem. Soc.*, **108**, 4746(1986)	1986
272.)	J. J. Low and W. A. Goddard, III, *J. Am. Chem. Soc.*, **108**, 6115(1986)	1986
273.)	G. Ohanessian, M. J. Brusich and W. A. Goddard, III, *J. Am. Chem. Soc.*, **112**, 7179(1990)	1990
274.)	J. K. Perry, W. A. Goddard, III and G. Ohanessian, *J. Am. Chem. Soc.*, **97**, 7560(1992)	1992
275.)	M. A. Blitz, S. A. Mitchell, and P. A. Hackett, *J. Phys. Chem.*, **97**, 5305 (1993).	1993
276.)	A. E. Alvarado-Swaisgood and J. F. Harrison, *J. Mol. Str. (Theochem.)*, **169**, 155 (1988).	1988
277.)	A. E. Alvarado-Swaisgood and J. F. Harrison, *J. Phys. Chem.*, **92**, 5896 (1988).	1988
278.)	A. E. Alvarado-Swaisgood and J. F. Harrison, *J. Phys. Chem.*, **92**, 2757 (1988).	1988
279.)	J. F. Harrison, *J. Phys. Chem.*, **90**, 3313 (1986).	1986
280.)	K. L. Kunze, and J. F. Harrison, *J. Phys. Chem.*, **93**, 2983 (1989).	1989
281.)	S. Kapellos, A. Mavridis and J. F. Harrison, *J. Phys. Chem.*, **95**, 6860 (1991).	1991
282.)	A. Mavridis, F. L. Herera, and J. F. Harrison, *J. Phys. Chem.*, **95**, 6854 (1991).	1991
283.)	A. Mavridis, A. E. Alvarado-Swaisgood, and J. F. Harrison, *J. Phys. Chem.*, **90**, 2584 (1986).	1986
284.)	A. Mavridis, J. F. Harrison and J. Allison, *J. Am. Chem. Soc.*, **111**, 2482 (1989).	1989
285.)	A. Mavridis, and J. F. Harrison, *J. Chem.* Soc., *Faraday Trans.* 2, **85**, 1391 (1989).	1989
286.)	P. Kebarle, *J. Phys. Chem.*, **74**, 1466 (1970).	1970
287.)	P. Kebarle, *Can. J. Chem.*, **47**, 2619 (1969).	1969
288.)	S. McCullough-Catalano, C. B. Lebrilla *J. Am. Chem. Soc.* **115**, 1441 (1993).	1993
289.)	F. Bouchard, J. W. Hepburn, T. B. McMahon, *J. Am. Chem. Soc.*, **111**, 8934-8935 (1989)	1989
290.)	C. E. C. A. Hop and T. B. McMahon, *J. Am. Chem. Soc.*, **114**, 1237-43 (1992)	1992
291.)	C. E. C. A. Hop and T. B. McMahon, *Inorg. Chem.*, **30**, 2828-30 (1991)	1991
292.)	C. E. C. A. Hop and T. B. McMahon, *J. Am. Chem. Soc.*, **113(1)**, 355-7 (1991)	1991
293.)	T. F. Magnera, D. E. David, D. Stulik, R. G. Orth, H. T. Jonkman, and J. Michl, *J. Am. Chem. Soc.*, **111**, 5036 (1989).	1989
294.)	T. F. Magnera, D. E. David, and J. Michl, *J. Am. Chem. Soc.*, **111**, 4100, (1989).	1989
295.)	S. A. Mitchell, *Int. J. Chem. Kin.*, **26**, 97 (1994).	1994
296.)	T. G. Dietz, D. S. Chatellier, and D. P. Ridge, *J. Am. Chem. Soc.*, **100**, 4905, (1978).	1978
297.)	B. S. Larsen, R. B. Freas, and D. P. Ridge, *J. Phys. Chem.*, **88**, 6014 (1984).	1984

REF #	FULL LITERATURE REFERENCE	YEAR
298.)	F. Strobel, and D. P. Ridge, *J. Phys. Chem.*, , **93**, 3635 (1989).	1989
299.)	D. Schröder, J. Hrusák, R. H. Hertwig, W. Koch, P. Schwerdtfeger, and H. Schwarz, *Organometallics*, submitted.	1994
300.)	J. Schwartz, and H. Schwarz, *Organometallics*, **13**, 1518 (1994).	1994
301.)	J. Hrusák, D. Stöckigt, and H. Schwarz, *Chem. Phys. Lett.*, **221**, 518 (1994)	1994
302.)	J. Hrusák and D. Stöckigt, *J. Phys. Chem.*, **98**, 3675 (1994)	1994
303.)	D. Stöckigt and H. Schwarz, *Int. J. Mass Spectrom. Ion Proc.*, in press.	1995
304.)	J. Schwarz, C. Heinemann, and H. Schwarz, *J. Phys. Chem.*, submitted.	1995
305.)	D. Schröder, A. Fiedler, J. Schwarz, and H. Schwarz, *Inorg. Chem.*, **33**, 5094 (1994).	1994
306.)	M. F. Ryan, A. Fiedler, D. Schröder, and H. Schwarz, *J. Am. Chem. Soc.*, **117**, 2033 (1995).	1995
307.)	A. Fiedler, D. Schröder, S. Shaik, and H. Schwarz, *J. Am. Chem. Soc.*, **117**, 10734 (1994).	1994
308.)	D. Stöckigt and H. Schwarz, *Chem. Ber.*, **127**, 791 (1994).	1992
309.)	L. N. Gorokhov, M. J. Ryzhov, J. S. Khodeev, *J. Phys. Chem.*, **59**, 1761 (1985).	1985
310.)	R. Wesendrup, D. Schröder, and H. Schwarz, *Angewandt Chem. Int. Ed. Engl.*, **33**, 1174 (1994).	1994
311.)	D. Schröder, J. Hrusák, I. C. Tornieporth-Ötting, T. M. Klapötke, and H. Schwarz, *Angewandt Chem. Int. Ed. Engl.*, **33**, 212 (1994).	1994
312.)	P. Schwerdtfeger, J. S. McFeaters, R. L. Stephens, M. J. Lidell, M. Dolg, and B. A. Hess, *Chem. Phys. Lett.*, **218**, 362 (1994).	1994
313.)	P. W. Villalta, D. G. Leopold, *J. Chem. Phys.*, **98**, 7730 (1993).	1993
314.)	D. Schröder, J. Hrusák, and H. Schwarz, *Helv. Chim. Acta*, **75**, 2215 (1992).	1992
315.)	R. R. Squires, *J. Am. Chem. Soc.*, **111**, 4101 (1989).	1989
316.)	L. S. Sunderlin, W. Dingneng, and R. R. Squires, *J. Am. Chem. Soc.*, **114**, 2788 (1992).	1992
317.)	R. Walsh, *Acc. Chem. Res.*, **14**, 246 (1981).	1981

Index

Understanding Chemical Reactivity

1. Z. Slanina: *Contemporary Theory of Chemical Isomerism.* 1986 ISBN 90-277-1707-9
2. G. Náray-Szabó, P.R. Surján, J.G. Angyán: *Applied Quantum Chemistry.* 1987 ISBN 90-277-1901-2
3. V.I. Minkin, L.P. Olekhnovich and Yu. A. Zhdanov: *Molecular Design of Tautomeric Compounds.* 1988 ISBN 90-277-2478-4
4. E.S. Kryachko and E.V. Ludeña: *Energy Density Functional Theory of Many-Electron Systems.* 1990 ISBN 0-7923-0641-4
5. P.G. Mezey (ed.): *New Developments in Molecular Chirality.* 1991 ISBN 0-7923-1021-7
6. F. Ruette (ed.): *Quantum Chemistry Approaches to Chemisorption and Heterogeneous Catalysis.* 1992 ISBN 0-7923-1543-X
7. J.D. Simon (ed.): *Ultrafast Dynamics of Chemical Systems.* 1994 ISBN 0-7923-2489-7
8. R. Tycko (ed.): *Nuclear Magnetic Resonance Probes of Molecular Dynamics.* 1994 ISBN 0-7923-2795-0
9. D. Bonchev and O. Mekenyan (eds.): *Graph Theoretical Approaches to Chemical Reactivity.* 1994 ISBN 0-7923-2837-X
10. R. Kapral and K. Showalter (eds.): *Chemical Waves and Patterns.* 1995 ISBN 0-7923-2899-X
11. P. Talkner and P. Hänggi (eds.): *New Trends in Kramers' Reaction Rate Theory.* 1995 ISBN 0-7923-2940-6
12. D. Ellis (ed.): *Density Functional Theory of Molecules, Clusters, and Solids.* 1995 ISBN 0-7923-3083-8
13. S.R. Langhoff (ed.): *Quantum Mechanical Electronic Structure Calculations with Chemical Accuracy.* 1995 ISBN 0-7923-3264-4
14. R. Carbó (ed.): *Molecular Similarity and Reactivity: From Quantum Chemical to Phenomenological Approaches.* 1995 ISBN 0-7923-3309-8
15. B.S. Freiser (ed.): *Organometallic Ion Chemistry.* 1996 ISBN 0-7923-3478-7
16. D. Heidrich (ed.): *The Reaction Path in Chemistry: Current Approaches and Perspectives.* 1995 ISBN 0-7923-3589-9
17. O. Tapia and J. Bertrán (eds.): *Solvent Effects and Chemical Reactivity.* 1996 ISBN 0-7923-3995-9

Kluwer Academic Publishers – Dordrecht / Boston / London